HADRON PHYSICS

Related Titles from the AIP Conference Proceedings Subseries on High Energy and Particle Physics

496 Workshop on Instabilities of High Intensity Hadron Beams in Rings
Edited by T. Roser and S. Y. Zhang, December 1999, 1-56396-910-6

494 New Directions in Quantum Chromodynamics
Edited by Chueng-Ryong Ji and Dong-Pil Min, November 1999, 1-56396-908-4

490 Particles and Fields: Eighth Mexican School
Edited by Juan Carlos D'Olivo, Gabriel López Castro, and Myriam Mondragón, November 1999, 1-56396-895-9

488 High Energy Physics at the Millennium: MRST'99
Edited by Pat Kalyniak, Stephen Godfrey, and B. Kamal, October 1999, 1-56396-902-5

482 RHIC Physics and Beyond: Kay Kay Gee Day
Edited by Berndt Müller and Robert Pisarski, July 1999, 1-56396-878-9

459 Heavy Quarks at Fixed Target
Edited by Harry W.K. Cheung and Joel N. Butler, January 1999, 1-56396-864-9

453 Particles, Fields, and Gravitation
Edited by Jakub Rembieliński, December 1998, 1-56396-837-1

452 Toward the Theory of Everything: MRST'98
Edited by James M. Cline, Marcia E. Knutt, Gregory D. Mahlon, and Guy D. Moore, November 1998, 1-56396-845-2

448 Workshop on Space Charge Physics in High Intensity Hadron Rings
Edited by A. U. Luccio and W. T. Weng, October 1998, 1-56396-824-X

432 Hadron Spectroscopy: Seventh International Conference
Edited by Suh-Urk Chung and Hans J. Willutzki, June 1998, 1-56396-765-0

To learn more about these titles, or the AIP Conference Proceedings Series, please visit the webpage
http://www.aip.org/catalog/aboutconf.html

Picture on Title Page

Title page illustration (opposite): Eighteenth century instrument used for illustrating the equilibrium of forces. Courtesy of the Physics Museum of the University of Coimbra.

HADRON PHYSICS

Effective Theories of Low Energy QCD

Coimbra, Portugal   September 1999

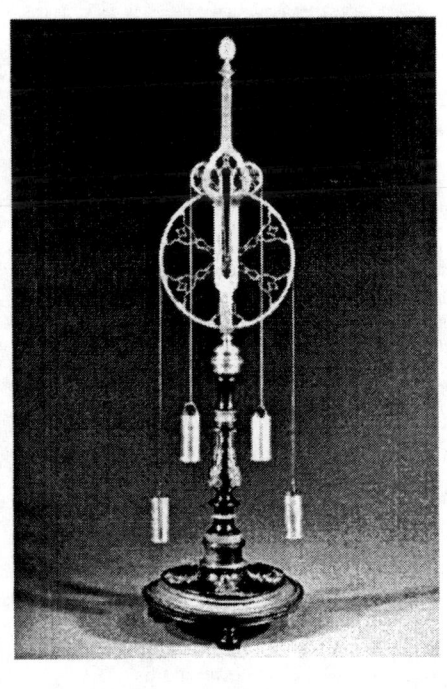

EDITORS
A. H. Blin
B. Hiller
M. C. Ruivo
C. A. Sousa
E. van Beveren
Universidade de Coimbra, Portugal

Melville, New York
AIP CONFERENCE PROCEEDINGS ■ 508

Editors:

A. H. Blin
B. Hiller
M. C. Ruivo
C. A. Sousa
E. van Beveren

Centro de Física Teórica
Universidade de Coimbra
P-3004-516 Coimbra
PORTUGAL

E-mail: alex@teor.fis.uc.pt
brigitte@teor.fis.uc.pt
maria@teor.fis.uc.pt
celia@teor.fis.uc.pt
eef@teor.fis.uc.pt

Authorization to photocopy items for internal or personal use, beyond the free copying permitted under the 1978 U.S. Copyright Law (see statement below), is granted by the American Institute of Physics for users registered with the Copyright Clearance Center (CCC) Transactional Reporting Service, provided that the base fee of $17.00 per copy is paid directly to CCC, 222 Rosewood Drive, Danvers, MA 01923. For those organizations that have been granted a photocopy license by CCC, a separate system of payment has been arranged. The fee code for users of the Transactional Reporting Service is: 1-56396-927-0/00/$17.00.

© 2000 American Institute of Physics

Individual readers of this volume and nonprofit libraries, acting for them, are permitted to make fair use of the material in it, such as copying an article for use in teaching or research. Permission is granted to quote from this volume in scientific work with the customary acknowledgment of the source. To reprint a figure, table, or other excerpt requires the consent of one of the original authors and notification to AIP. Republication or systematic or multiple reproduction of any material in this volume is permitted only under license from AIP. Address inquiries to Office of Rights and Permissions, Suite 1NO1, 2 Huntington Quadrangle, Melville, N.Y. 11747-4502; phone: 631-576-2268; fax: 631-576-2450; e-mail: rights@aip.org.

L.C. Catalog Card No. 00-100412
ISBN 1-56396-927-0
ISSN 0094-243X
Printed in the United States of America

CONTENTS

Preface ix
Committees xi

BARYONS

Non-Local Regularization of Chiral Quark Models in the Soliton Sector 3
    G. Ripka and B. Golli
Baryons as Solitons 13
    H. Walliser
Discriminating between Chiral Models of the Nucleon via Hard
Semi-Inclusive $\vec{e}N$ and $\vec{N}\vec{N}$ Scattering 23
    B. Dressler, K. Goeke, M. V. Polyakov, P. Schweitzer, M. Strikman, and C. Weiss
Stability of Multiquark Systems 34
    F. Stancu
Quantum Monte Carlo Studies of Relativistic Effects in ^{3}H and ^{4}He 43
    A. Arriaga
Computation of the Nucleon Properties in the Chiral Quark Model 53
    J. Baacke and H. Sprenger
Multibaryons in the Skyrme Model 63
    N. N. Scoccola
Form Factors of Baryons in a Confining and Covariant
Diquark-Quark Model 73
    M. Oettel, S. Ahlig, R. Alkofer, and C. Fischer
Nucleon-Nucleon Interaction in a Chiral Constituent Quark Model 83
    D. Bartz and F. Stancu
Chiral NN Interactions in Nuclear Matter 94
    B. Krippa
A Sketch of Two and Three Bodies 100
    H. W. Grießhammer

VARIOUS APPROACHES TO QCD

A Renormalisation-Group Treatment of Two-Body Scattering 113
    M. C. Birse, J. A. McGovern, and K. G. Richardson
QCD Dynamics at $\theta \sim \pi$ 122
    A.V. Smilga
Wilson's Area Law and String Effective Actions for Confining
Gauge Theories 132
    D. Antonov and D. Ebert
Magnetic Monopoles, Center Vortices, Confinement and Topology
of Gauge Fields 142
    H. Reinhardt, M. Engelhardt, K. Langfeld, M. Quandt, and A. Schäfke

How Could Quark Polarization be Measured 152
  A. V. Efremov
Heavy Flavor Contributions to QCD Sum Rules and the Running
Coupling Constant 162
  W. L. van Neerven
Color-Flavor Transformations and QCD Low-Energy Effective Action 172
  J. Budzcies and Y. Shnir

CHIRAL PERTURBATION THEORY

Chiral QCD Dynamics: Recent Results.................................... 185
  U.-G. Meißner
Up to Two-Loop Calculations in HBCPT.................................. 195
  J. A. McGovern, M. C. Birse, and K. B. V. Kumar
Nonequilibrium Chiral Perturbation Theory and Disoriented
Chiral Condensates 204
  A. Gómez Nicola

EFFECTS OF HOT AND DENSE MATTER

The Sigma Meson and Chiral Restoration in the Hot and/or
Dense Nuclear Matter 217
  T. Kunihiro
Matter-Induced Hadronic Processes 218
  W. Broniowski, W. Florkowski, and B. Hiller
The Deconfinement Phase Transition, Hadronization and the NJL Model 226
  S. Raha
Interplay Between Kaons and Kaon Like Excitations of the Medium 237
  M. C. Ruivo
ρ-Mass Modification in Dense Medium—Some Recent Studies
in Finite Nucleus... 246
  A. Bhattacharyya
Magnetic Oscillations in the Nambu-Jona-Lasinio Model 253
  D. Ebert and K. G. Klimenko

MESONS

On Valence Gluons in Heavy Vector Quarkonia: Implications
for Spectra and Decays 265
  S. B. Gerasimov
Brief Review of some Modified Versions of the Nambu-Jona-Lasinio
Model 273
  M. Jaminon

Invariant Regularization of One-Loop Determinant in Non-Renormalizable Models.. 283
    A. A. Osipov, B. Hiller, and A. H. Blin
Light Scalar Mesons.. 290
    D. Black, A. H. Fariborz, and J. Schechter
First Radial Excitations of Scalar Meson Nonet 300
    M. K. Volkov and V. L. Yudichev
Scalar Mesons as "Simple" $Q\bar{Q}$ States................................... 310
    E. van Beveren and G. Rupp
The Resonating Mean-Field Theoretical Approach to the Nambu-Jona-Lasinio Model - σ-π Sector- 318
    S. Nishiyama, J. da Providência, and O. Ohno
Extended Non-Chiral Quark Models Confronting QCD 328
    A. A. Andrianov and V. A. Andrianov
Strong Decays of the Scalar Glueball in a Confining Effective Lagrangian of QCD.. 338
    M. Jaminon, M. Mathot, and B. Van den Bossche
Models of Low Energy Effective Theory Applied to Kaon Non-Leptonic Decays and other Matrix Elements 348
    J. Bijnens
Anomalies for Nonlocal Dirac Operators 358
    E. Ruiz Arriola and L. L. Salcedo
Meson Properties in a Renormalizable Version of the NJL Model 368
    A. L. Mota, M. C. Nemes, B. Hiller, and H. Walliser
Mesons in Non-Local Chiral Quark Models................................. 380
    W. Broniowski

Workshop Program 391
List of Participants... 397
Author Index... 409

Preface

The *International Workshop on Hadron Physics of Low Energy QCD* was held in Coimbra from the 10th to the 15th of September, 1999, and was organized by the Centre of Theoretical Physics of the University of Coimbra. This conference emanated as a natural consequence from the research work, interests and major collaborations of members of the Centre of Theoretical Physics.

More than forty speakers addressed a variety of subjects, ranging from low-energy QCD dynamics, chiral perturbation theory and various approaches to meson and baryon phenomenology, to effects of hot and dense nuclear matter. Fruitful crosslinks between these topics had become clear in the recent past, which indeed led to a coherent overview of the state of low-energy hadron physics in the contributions to the conference and these proceedings.

We wish to express our gratitude to the members of the Program Advisory Committee for their valuable suggestions and advice in the organization of the Workshop Program. Thanks are due to all participants for the fruitful discussions and friendly atmosphere, and in particular to the speakers for their high quality presentations and for their effort to finish the manuscripts for the Proceedings in time.

The conference was organized with financial support of many organizations: Fundação para a Ciência e Tecnologia, PRAXIS grants PCEX/C/13/96, 2/2.1/FIS/451 and P/FIS/12247/98, Calouste Gulbenkian Foundation, Luso-American Foundation for Development, Grupo Teórico de Altas Energias, Orient Foundation, University of Coimbra, Faculty of Science and Technology of the University of Coimbra, Coimbra City Hall, and Tourist Office of Coimbra.

We wish to thank all those who assisted in organizing this conference and helped to make it a successful and memorable meeting.

We hope this Workshop to be the first in a series to be organized on a regular basis.

The Editors
December, 1999.

Organizing Committee

J. da Providência(Chair)
E. van Beveren
A. H. Blin*
L. P. Brito
B. Hiller*
J. P. da Providência
M. C. Ruivo*
C. A. de Sousa*

*Scientific Secretaries

Program Advisory Committee

W. Broniowski (Cracow)
M. Jaminon (Liège)
A. A. Osipov (Dubna)
G. Ripka (Saclay)
M. C. Ruivo (Coimbra)
H. Walliser (Siegen)

Secretary

Natália Cristina Silva

BARYONS

Non-local regularization of chiral quark models in the soliton sector

Georges Ripka* and Bojan Golli[†]

*Service de Physique Théorique, Centre d'Etudes de Saclay
F-91191 Gif-sur-Yvette Cedex, France[1]
[†]Faculty of Education, University of Ljubljana and J.Stefan Institute,
Jamova 39, P.O.Box 3000, 1001 Ljubljana, Slovenia[2]

Abstract. A chiral quark model is described which is regularized in terms of Lorentz invariant non-local interactions. The model is regularized to all loop orders and it ensures the proper quantization of the baryon number. It sustains bound hedgehog solitons which, after suitable centre of mass corrections, can adequately describe the nucleon.

I SOME SPECIFICITIES OF CHIRAL QUARK MODELS

This work was done in collaboration with Wojciech Broniowski from Krakow. We consider chiral quark models which encompass three sectors. The vacuum and soliton sectors, which are treated in the mean-field (leading order in N_c) approximation, and the meson sector, which describes the (next to leading order in N_c) vibrations of the vacuum sector. Not all models are applicable to the three sectors. For example, constituent quark models, in which quarks interact with confining forces, cannot describe the vacuum sector, that is, the Dirac sea. However, they can and do describe the excited states of baryons, a thing which the chiral quark models cannot do (except, possibly, the Δ) for lack of confinement.

Chiral quark models (nor any of the other low energy quark models) have not been derived from QCD. The only serious attempt to derive them from QCD is the instanton gas model [1,2]. In this approach, the chiral quark model is derived by calculating the propagation of quarks in a gas of instantons. A regularized effective theory results, as it should. It predicts both the value of the cut-off and the form of the regulator. The non-local regularization discussed here has the same form as the one derived from the instanton gas model. Unfortunately, the quark models derived from the instanton structure of the vacuum do not lead to quark or color

[1]) E-mail: ripka@spht.saclay.cea.fr
[2]) E-mail: Bojan.Golli@ijs.si

confinement. This serious limitation serves as a reminder that we have not really succeeded in deriving low energy effective theories from QCD.

Other so-called "derivations" of quark models from QCD involve more guesswork than derivation. Most telling is their inability to derive a regularized model. If infinities appear in an effective theory, one should seek the physical processes which prevent the infinities from occurring. Invoking the roughly 200 MeV QCD cut-off is not a serious argument. Nor does QCD imply in any sense that the quark-quark interaction at low energy should be a one-gluon exchange with a modified gluon propagator. The regularizations used so far in the Nambu Jona-Lasinio type models for example (proper-time regularization being the most commonly used one so far), are nothing but renormalization techniques in which a finite cut-off is maintained. Not only is this arbitrary but such regularizations are flawed with problems.

One might argue that the value of the cut-off should not matter. Indeed it would not if the effective theory consisted, for example, in eliminating some high energy degrees of freedom and using the remaining degrees of freedom to work out the dynamics of low energy phenomena. In such a case, one might expect the cut-off to be much larger than the inverse size of the composite particles and the results not to be sensitive to the cut-off. In chiral quark models, however, this is not the case. The cut-offs required to fit f_π are about 700 MeV, hardly larger than the ρ or the nucleon mass. This is a fact of life, whether we like it or not. One can of course simply discard such models, but better models do not seem to be forthcoming.

II THE SOLITON IN THE NON-LOCAL CHIRAL QUARK MODEL

The non-local chiral quark model is defined by the euclidean action:

$$I\left(q, q^\dagger\right) = \left\langle q \left| \partial_\tau + \frac{\vec{\alpha}.\vec{\nabla}}{i} + m \right| q \right\rangle - \frac{G^2}{2} \int d_4 x \left(\langle q |r| x \rangle \beta \Gamma_a \langle x |r| q \rangle \right)^2 . \quad (1)$$

In this expression, $\Gamma_a = (1, i\gamma_5 \tau_a)$, $q(x) \equiv \langle x | q \rangle$ is the quark field, and r is a regulator. The regulator is assumed to be diagonal in momentum space and it has a range which defines an effective euclidean cut-off Λ. For example, we could take $\langle k |r| k' \rangle = \delta_{k,k'} r(k^2)$ with $r(k^2) = e^{-\frac{k^2}{2\Lambda^2}}$, where k is a euclidean 4-vector $k_\mu = (\omega, \vec{k})$ with $k^2 = \omega^2 + \vec{k}^2$. The interaction term of the action (1) can be viewed as a contact 4-fermion interaction involving the *delocalized quark fields*:

$$\psi(x) = \langle x |r| q \rangle = \int d_4 y \, \langle x |r| y \rangle \, q(y) . \quad (2)$$

An action of the form (1) is derived from the instanton gas model of the QCD vacuum [1,2], which predicts a cut-off function of the form:

$$r\left(k^2\right) = f\left(k\rho/2\right), \qquad f(z) = -z\frac{d}{dz}\left(I_0(z) K_0(z) - I_1(z) K_1(z)\right) \qquad (3)$$

where ρ is the instanton size. The the cut-off is determined by the inverse instanton size ρ. The form (3) has $r(z=0) = 1$ and $r(z) \underset{z\to\infty}{\to} \frac{9}{2k^6\rho^6}$. However, at large euclidean momenta k, the form (3) is no longer valid and the cut-off function is dominated by one gluon exchange. It decreases as $\frac{1}{k^2}$ (with possible logarithmic corrections) and not as $\frac{1}{k^6}$. We find that the fall-off of the regulator at large euclidean k^2 does not affect the soliton properties very much. For this reason, we have felt free to use various simple forms of cut-off functions, such as a gaussian, which have an additional advantage in that they can be analytically (although arbitrarily) continued to negative values of k^2. We shall see below that the analytic continuation is required to include the valence orbit. Similar regularization has been used by the Manchester group [3] in the meson and vacuum sectors. Various regularization schemes are reviewed in chapter 6 of Ref. [4].

The euclidean action allows us to calculate the partition function $Z = \int D(a) D\left(a^\dagger\right) e^{-I\left(a,a^\dagger\right)}$ and the ground state energy $E = -\frac{\partial}{\partial \beta} \ln Z$. The partition function cannot be written in the form $Z = Tr\, e^{-\beta H}$ because the regulator in the action (1) prevents us from defining a hamiltonian H. We are also unable to quantize the quark fields but we shall see that the baryon number is nonetheless properly quantized.

We work with the equivalent bosonized form of the action:

$$I(\varphi) = -Tr \ln\left(\partial_\tau + \frac{\vec{\alpha}.\vec{\nabla}}{i} + \beta m + \beta r \varphi_a \Gamma_a r\right) + \frac{1}{2G^2} \int d_4 x\, \varphi_a^2(x) \qquad (4)$$

in which case the partition function is given by the path integral $Z = \int D(\varphi) e^{-I(\varphi)}$. We refer to $\varphi_a \Gamma_a = S + i\gamma_5 \tau_a P_a$, as the "chiral field" and we say that the chiral field is "on the chiral circle" if, for all x, we have $S^2(x) + P_a^2(x) = M_0^2$, where M_0 is an x-independent constant mass.

We have calculated a localized and time independent stationary point of the action (4), consisting of a chiral field with a hedgehog shape $S(r) + i\gamma_5 \hat{x}_a \hat{\tau}_a P(r)$ [5]. The shape of the fields and the soliton energy can be calculated in terms of the energies $e_\lambda(\omega)$ of the quark orbit. The "Dirac hamiltonian" is diagonal in the energy representation, although it remains energy dependent. The quark orbits $|\omega, \lambda_\omega\rangle$ satisfy the equations :

$$\partial_\tau |\omega, \lambda_\omega\rangle = i\omega |\omega, \lambda_\omega\rangle, \qquad \left(\frac{\vec{\alpha}.\vec{\nabla}}{i} + \beta m + \beta r \varphi_a \Gamma_a r\right) |\omega, \lambda_\omega\rangle = e_\lambda(\omega) |\omega, \lambda_\omega\rangle. \qquad (5)$$

The energy of the soliton is:

$$E_{sol} = N_c e_{val} + \frac{1}{2\pi} \int_{-\infty}^{\infty} \omega d\omega \sum_{\lambda_\omega} \frac{i + \frac{de_\lambda(\omega)}{d\omega}}{i\omega + e_\lambda(\omega)} + \frac{1}{2G^2} \int d_3 x\, \varphi_a^2(\vec{x}) - vac. \qquad (6)$$

where $-vac.$ means that we subtract the vacuum energy. In the vacuum, $P = 0$, $S = M_0$ and there is no valence orbit contribution e_{val}. The latter is discussed in the next section.

III THE QUANTIZATION OF THE BARYON NUMBER AND THE VALENCE ORBIT

We calculate the baryon number from the Noether current associated to the gauge transformation $q(x) \to e^{-i\alpha(x)} q(x)$. It turns out to be:

$$B = -\frac{1}{2\pi i N_c} \int_{-\infty}^{\infty} d\omega \sum_{\lambda_\omega} \frac{i + \frac{de_\lambda(\omega)}{d\omega}}{i\omega + e_\lambda(\omega)}. \qquad (7)$$

The extra term $\frac{de_\lambda(\omega)}{d\omega}$ in the numerator arises from the fact that the regulator r does not commute with $\alpha(x)$. Its effect is to make the residues of all the poles of the quark propagator $\frac{1}{i\omega + e_\lambda(\omega)}$ equal to unity. This effectively quantizes the baryon number in a manner which does not seem to be related to the topology of the hedgehog field.[3] This is most fortunate because, a priori, there is no reason to expect a theory, in which we cannot quantize the quark field, to yield a properly quantized baryon number.

The expression (7) suggests a way to include the valence orbit so as to ensure that the baryon number of the soliton, relative to the vacuum, is equal to unity. We calculate "on-shell" pole of the quark propagator in the hedgehog background field by searching for a solution of the equation $i\omega + e_\lambda(\omega)|_{\omega = ie_{val}} = 0$. Because of the regulator, the solutions are scattered all over the complex ω plane. However, it is well known that, in the local theory, where we set $r = 1$, and for a hedgehog field with winding number unity, a well separated bound orbit with grand spin and parity 0^+ occurs with energy e_{val} close to zero [6]. In the non-local theory, we find that a solution of the equation $\omega = ie_{val}(\omega)$ can always be found on the imaginary ω axis, close to the origin $\omega = 0$, and that no other pole occurs in the vicinity. We therefore ensure that the soliton has a baryon number $B = 1$ by deforming the integration path over ω in such a way as to include the contribution of this pole. This requires an analytic continuation of the regulator. Such a continuation is arbitrary but the analytic continuation does not extend as far from the origin as e_{val}. Indeed, since the soliton size is small, $\vec{k}^2 > 0$ is large and this, on the average, makes $k^2 = -e_{val}^2 + \vec{k}^2$ less negative. Unfortunately however, the form (3) of the regulator, predicted in the instanton model, does not allow any analytic continuation whatsoever, thereby, strictly, prohibiting its use in the soliton calculation.

---

[3] Nor is the soliton stabilized by the topology of the chiral field.

IV RESULTS OF SELF-CONSISTENT SOLITON CALCULATIONS

The model parameters are the coupling constant G appearing in the lagrangian, the cut-off Λ appearing in the regulator and the current quark mass m. The values of the three parameters are constrained by fitting the pion decay constant $f_\pi = 93$ MeV and the pion mass to $m_\pi = 139$ MeV. The expression used to calculate the pion decay constant f_π is:

$$f_\pi^2 = 2N_f N_c M_0^2 \int \frac{d_4 k}{(2\pi)^4} \frac{r_k^4 - k^2 r_k^2 \frac{dr_k^2}{dk^2} + k^4 \left(\frac{dr_k^2}{dk^2}\right)^2}{(k^2 + r_k^2 M_0^2)^2} \tag{8}$$

valid in the chiral limit $m \to 0$ and it is not identical to the Pagels-Stokar formula [7]. This leaves one undetermined parameter which we choose to be the constituent quark mass M_0 at zero 4-momentum. The pion decay constant f_π sets the scale. Grossly, soliton energies increase and soliton radii diminish as f_π increases (see table 2).

Figure 1 shows the soliton energy E_{sol} as a function of the free parameter M_0. A soliton is a bound state of $N_c = 3$ quarks which polarize the Dirac sea. With a gaussian regulator, it is formed if $M_0 \gtrsim 276$ MeV, that is, for a sufficiently strong coupling constant $G \gtrsim 4.7 \times 10^{-3}$ MeV^{-1}. The bound state occurs when the energy of the system is lower than the energy $N_c M_q$ of N_c free constituent quarks in the vacuum: $E_{sol} < N_c M_q$. The mass M_q is the on-shell constituent quark mass, obtained by searching for the pole of the quark propagator in the vacuum. It is the solution of the equation $k^2 + (r_k^2 M_0 + m)^2 \big|_{k^2 = -M_q^2} = 0$, which requires an analytic continuation of the regulator to negative values of k^2. Figure 1 also shows $N_c M_q$. At the critical value $M_0 \approx 276$ MeV, the two curves merge. The contribution $N_c e_{val}$ of the valence orbit is also shown. At the critical value of M_0, the energy e_{val} of the valence orbit, which is the on-shell mass of a quark propagating in the hedgehog field, becomes a well distinguished bound orbit.

At $M_0 \approx 309$ MeV, the curve displaying $N_c M_q$ on figure 1 abruptly stops. Indeed, for larger values of M_0, the poles of the quark propagator no longer occur for real values of k^2. This means that quarks can no longer materialize on-shell in the vacuum. This feature is discussed in chapter 6 of Ref. [4] and it has been considered by several authors as a sign of quark confinement [8-10]. In fact, when a pole of the quark propagator disappears from the real k^2 axis, it simply moves into the complex plane. Such poles indicate instability of the assumed vacuum state against the addition of a single quark.

However, our calculation shows that, in the background soliton field, the on-shell valence orbit continues to exist and so does the soliton. Unfortunately, the regulator also introduces extra unwanted poles in the propagators of colorless mesons, so that the model does not express color confinement. Similar unwanted poles occur

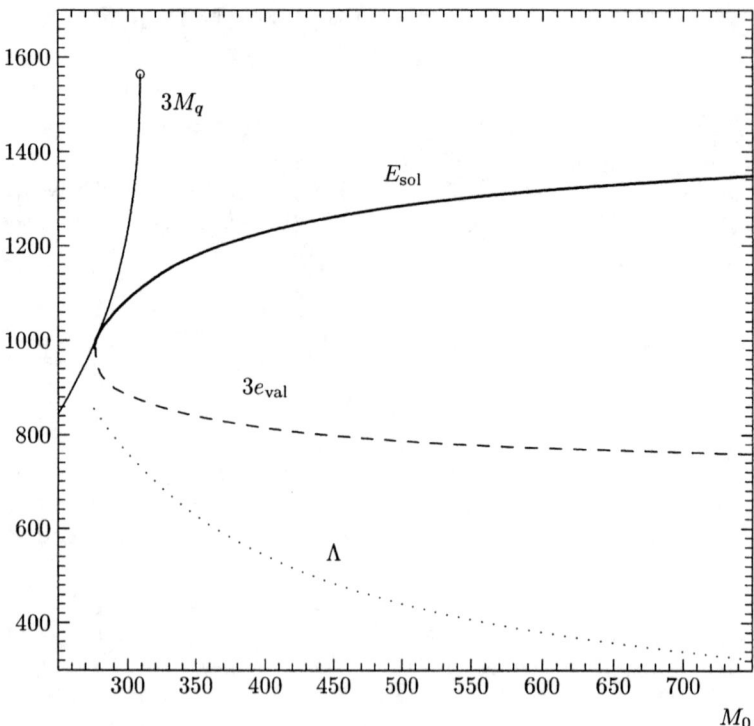

FIGURE 1. The energy of the soliton [in MeV] (bold solid line), N_c times the free-space quark mass (solid line) and the valence contribution to the soliton energy (dashed line) plotted as functions of the parameter M_0 [in MeV]. A Gaussian regulator is used; Λ (dots) is fitted to $f_\pi = 93$ MeV.

in proper-time regularization [11]. Our ignorance as how to continue propagators in the complex k^2 plane reflects our ignorance of the confining mechanism [12].

Apart from the solitons consisting of three valence quarks we find stable solitons consisting of a single valence quark in the background soliton field (see figure 2) as well as of two valence quarks. Similar solutions have been found in the linear sigma model with valence quarks [13].

Figure 3 shows the scalar and pseudoscalar fields $S(x)/M_0$ and $P(x)/M_0$ of the soliton obtained with several values of M_0, together with the soliton quark density $\rho(x)$. Note that, within the soliton, the fields *do not* lie on the chiral circle and $S^2(x) + P^2(x) < M_0^2$. Indeed, the pion component $P(x)$ never reaches the values $-M_0$. This is a new dynamical result. This is the only calculation, as far we know, in which one can check dynamically whether the chiral field remains or not on the chiral circle. It could not be checked in the renormalized linear sigma model, because close lying Landau poles occur which make the soliton unstable

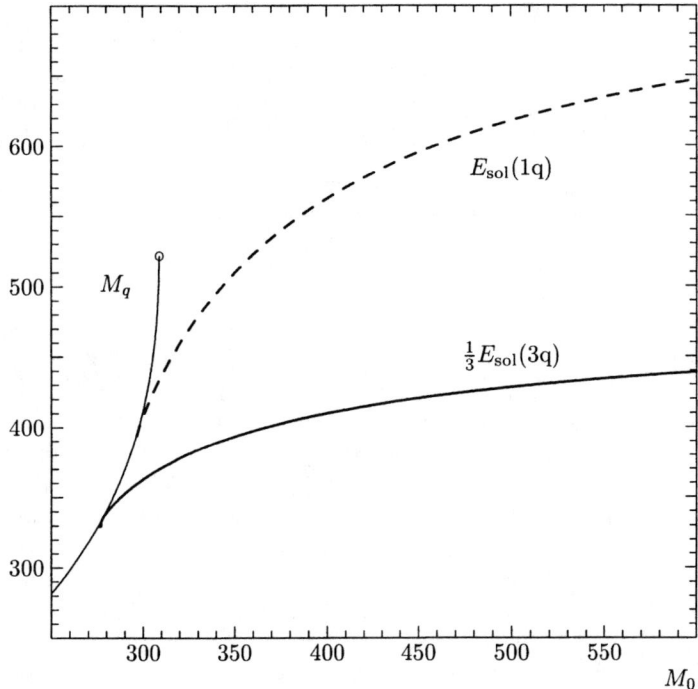

FIGURE 2. The energy per quark [in MeV] for the soliton with three valence quarks (bold line), the soliton with one valence quark (dashed line) and the free-space quark mass M_q plotted as functions of the parameter M_0 [in MeV].

against high gradients in the fields [14,15]. It could also not be checked in local theories which use proper-time regularization because, in such theories, the soliton is unstable unless the fields are constrained to remain on the chiral circle [16,17]. No such instability occurs with the non-local regularization.

The soliton we obtain with non-local regularization has a structure which lies midway between a Friedberg-Lee soliton [18,19] (in which the pion field has a vanishing classical value), and a Skyrmion [20,21] (in which the chiral field is constrained to remain on the chiral circle). This raises the problem of the collective rotational motion of the soliton. If the deformation in spin and isospin space is stable enough to sustain a rotation without significant distortion, then the Δ can be described as a rotation of the soliton and the $N - \Delta$ mass splitting can be estimated by cranking. If, however, the deformation is small, the Δ may be better described as a bound state of quarks with aligned spins and isospins. We have not tackled this problem yet.

Table 1 shows some properties of calculated solitons for various values of the mass parameter M_0. Rather good values of g_A are obtained. The soliton mass and

M_0 MeV	Λ MeV	m MeV	$\langle \bar{q}q \rangle^{1/3}$ MeV	$1/G$ MeV	e_{val} MeV	E_{Dirac} MeV	E_{sol} MeV	$\langle r^2 \rangle^{1/2}$ fm	g_A
300	760	7.62	−215	182	295	2360	1088	1.32	1.28
350	627	10.4	−200	140	280	1715	1180	1.04	1.16
400	543	13.2	−185	113	272	1433	1229	0.97	1.14
450	484	15.9	−173	94	266	1275	1261	0.96	1.12

TABLE 1. Properties of self-consistent soliton solutions obtained with a gaussian regulator.

energies need to be corrected for spurious centre of mass motion (see table 2).

The fields which describe the soliton break translational symmetry. The center of mass of the system is not at rest and it makes a spurious contribution both to the energy and to the mean square radius (more generally, to the form factor). This spurious contribution is not measured and it should be subtracted from the calculated values. The subtraction occurs at the next to leading order (in N_c) approximation. A rough estimate can be obtained from an oscillator model. If N_c particles of mass m move in a 1s state of a harmonic oscillator of frequency $\hbar\omega$, the centre of mass of the system is also in a 1s state and it contributes $\frac{3}{4}\hbar\omega = \langle P^2 \rangle / 2N_c m$ to the energy. We have therefore corrected the soliton energies by subtracting $\langle P^2 \rangle / 2E_{sol}$ from the calculated energy. Furthermore, in the oscillator model, the center of mass contributes a fraction $\frac{1}{N_c}$ of the mean square radius, so that we have corrected the mean square radius by multiplying the calculated value by a factor equal to $\left(1 - \frac{1}{N_c}\right)$.

Table 2 shows the result. The soliton energies and radii are then considerably closer to the experimental values observed in the nucleon.

M_0 MeV	E_{sol} MeV	$\langle r^2 \rangle_{sol}$ fm^2	E_{corr} MeV	$\langle r^2 \rangle_{corr}$ fm^2
300	1088	1.7	965	1.1
350	1180	1.08	990	0.72
400	1229	0.94	1000	0.62
450	1261	0.92	980	0.61
450*	1458	0.69	1200	0.43

TABLE 2. Elimination of spurious c.m. motion. Gaussian regulator, Λ fitted to $f_\pi = 93$ MeV; * Λ fitted to $f_\pi = 1.25 \times 93$ MeV.

V CONCLUSION: WHY TAKE THE TROUBLE?

The non-local regularization effectively cuts out of the quark propagators the 4-momenta which are larger than the cut-off. The non-local regularization makes the

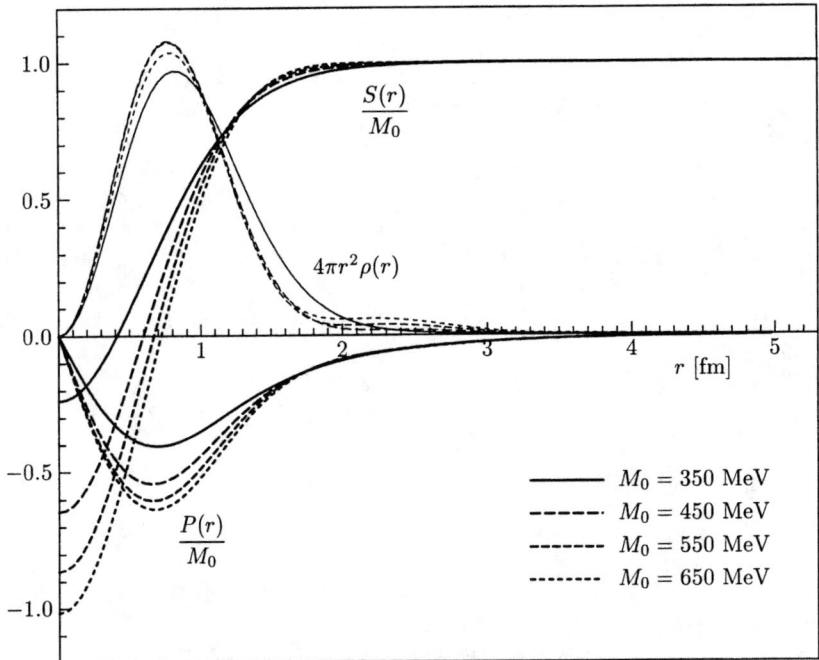

FIGURE 3. Self consistently determined fields and baryon densities ($4\pi r^2 \rho$) for various values of M_0; a gaussian regulator is used.

theory finite at all loop orders. The simpler proper-time and Pauli-Villars regularization schemes regularize the quark loop only and they require extra independent cut-offs when next to leading order meson loops are included. Both the real and the imaginary parts of the action are regularized, while the anomalous properties remain independent of the cut-off [22–24], and the baryon number remains properly quantized. In proper time and Pauli-Villars regularization schemes only the real part of the action is regularized and the imaginary part is left unregularized in order to enforce correct anomalous processes. Why not limit the 3-momenta of the quarks, thereby avoiding unwanted extra poles in the propagators? Breaking Lorentz covariance in the meson sector is annoying in that it requires to boost composite particles calculated in their rest frame.

REFERENCES

1. Diakonov, D. I. and Petrov, V. Y., *Nucl. Phys.* **B272**, 457 (1986).
2. Shuryak, E., *Nucl. Phys.* **B203**, 93,116,140 (1982).
3. Plant, R. S. and Birse, M. C., *Nucl. Phys.* **A628**, 607 (1998).

4. Ripka, Georges, *Quarks Bound by Chiral Fields*, Oxford University Press, Oxford 1997.
5. Broniowski, W., Golli, B. and Ripka, G., *Phys. Lett.* **B437**, 24 (1998).
6. Kahana, S. and Ripka, G., *Nucl. Phys.* **A429**, 462 (1984).
7. Pagels, H. and Stokar, S., *Phys. Rev.* **D20**, 2947 (1979).
8. Buballa, M. and Krewald, S., *Phys. Lett.* **B294**, 19 (1992).
9. Roberts, C. D. and Williams, A. G., *Prog.Part. And Nucl. Phys.* **33**, 475 (1994).
10. Bowler, R. D. and Birse, M. C., *Nucl. Phys.* **A582**, 655 (1995).
11. Nikolov, E. N., Broniowski, W., Ripka, G. and Goeke, K., *Zeit. Phys.* **A354**, 421 (1996).
12. Reusch, H., Stingl, M., Häbel, U., Könning, R. and Wizard, S., *Zeit. Phys.* **A336**, 423,435 (1990).
13. Golli, B. and Rosina, M., *Phys. Lett.* **B393**, 161 (1997).
14. Ripka, G. and Kahana, S., *Phys. Rev.* **D36**, 1233 (1987).
15. Perry, R., *Phys. Lett.* **B199**, 489 (1987).
16. Grümmer, F., Sieber, P., Meissner, Th. and Goeke, K., *Nucl. Phys.* **A547**, 459 (1992).
17. Wünsch, R., Sieber, P., Grümmer, F., Meissner, Th., Ripka, G. and Goeke, K., *Phys. Lett.* **B299**, 183 (1993).
18. Lee, T. D., *Particle Physics and Introduction to Field Theory*, Harwood Academic Press (New-York) 1981.
19. Wilets, L., *Nontopological solitons*, World Scientific, Singapore, 1989.
20. Skyrme, T. H. R., *Nucl. Phys.* **31**, 556 (1962).
21. Holzwarth, G., editor., *Baryons as Skyrme Solitons*, World Scientific, Singapore, 1993.
22. Cahill, R. T., Roberts, C. D. and Praschifka, J., *Ann. Phys. (NY)* **188**, 20 (1988).
23. Terning, J., Holdom, B., and Verbeek, K., *Phys. Lett.* **B232**, 351 (1989).
24. Ball, R. D. and Ripka, G., in Souza, C., Fiolhais, C., Fiolhais, M. and Urbano, J. N. editors, *Many Body Physics (Coimbra 1993)*, World Scientific, 1993.

Baryons as Solitons

Hans Walliser

Fachbereich Physik, Universität Siegen, D57068 Siegen, Germany

Abstract. Chiral lagrangians as effective field theories of QCD are sucessfully applied to meson physics in the framework of chiral perturbation theory. Because of their nonlinear structure these lagrangians allow for static soliton solutions interpreted as baryons. Their semiclassical quantization, which provides the leading order in an $1/N_C$ expansion with N_C the number of colors, turned out to be insufficient to obtain satisfactory agreement with empirical baryon observables. However with $N_C = 3$, large corrections are expected in the next-to-leading order carried by mesonic fluctuations around the soliton background, which require renormalization to 1-loop. In contrast to chiral perturbation theory, the low–energy lagrangian proves inapt and terms with an arbitrary number of gradients may in principle contribute. Assumptions about the a priori unknown higher chiral orders are tested by the scale–dependence of the results. For example, in the simple Sine–Gordon model with 1 scalar field in 1+1 dimensions, knowledge of the low–energy behaviour together with the mere existence of an underlying 1-loop renormalizable scale–independent solitonic theory is sufficient to regain the full solution. Baryonic observables calculated within that framework generally lead to better agreement with experiment except for the axial quantities. For these quantities the $1/N_C$ expansion does not converge sufficiently fast because the current algebra mixes different N_C orders.

INTRODUCTION

The need for effective field theories in hadron physics is brought about by the intractability of QCD at low energies. The energetically lowest lying degrees of freedom of such hadron effective field theories are pions (and kaons) which are considered to be the Goldstone modes of the spontaneously broken chiral symmetry in QCD. This concept leads to chiral perturbation theory (ChPT) pioneered by Weinberg [1] and consequently exploited by Gasser and Leutwyler [2]. Baryons may be introduced into this framework in different ways. One possiblity, baryon chiral perturbation theory (BChPT), is to treat them as external fields coupled in the most general chiral invariant way to pions [3]. Unfortunately the proliferation of terms allowed from chiral symmetry alone is enormous already at the lowest chiral orders. Therefore, the conjecture that baryons emerge as solitons from the meson fields themselves [4] (without explicit fermionic degrees of freedom) in the limit of a large number of colors N_C [5] is considered an attractive alternative built on the lagrangian established in ChPT in the meson sector. The link to ChPT is the consideration of loop corrections.

Loops in soliton models should not be confused with those in BChPT. Already the leading order N_C tree approximation sums up an infinite number of BChPT multi-loop graphs without closed meson loops [6]. After Moussallam has forwarded a first calculation of the soliton's Casimir energy [7], we used essentially the Gasser Leutwyler lagrangian in dimensional regularization in order to make close contact to ChPT. By coupling the corresponding external fields we calculated a variety of baryon observables in tree + 1-loop with the aim to promote the phenomenological soliton model towards a more quantitative model [8]. For all details I refer to this report which was published together with Frank Meier.

I would like to stress that in this approach there are no problems with confinement, the fundamental degrees of freedom are colorless. There are also no stability problems, the solitons are stabilized topologically.

FORMULATION

Starting point is the standard Gasser-Leutwyler chiral lagrangian [2]

$$\mathcal{L} = \mathcal{L}^{(2)} + \mathcal{L}^{(4)} + \mathcal{L}^{(higherChOs)}, \qquad (1)$$

$$\mathcal{L}^{(2)} = \frac{f^2}{4} tr\left[-(U^\dagger D_\mu U)^2 + \mathcal{M}(U + U^\dagger)\right], \qquad \mathcal{L}^{(4)} = \sum \ell_i(\mu)\mathcal{L}_i^{(4)}$$

where the chiral orders (ChOs) count the number of gradients (and mass matrices $\mathcal{M}$). The Goldstone bosons are parametrized by the matrix $U \in$ SU(2) or $U \in$ SU(3) and the covariant derivative D_μ contains couplings to external fields. A loop expansion of a theory like (1) necessarily brings about the need for renormalization which requires truncation at some finite chiral order. This distinguishes vacuum and soliton sector.

vacuum sector: $U = 1$
In the vacuum sector the external momenta can be made small by designing the experiment such that this requirement is fulfilled. The truncation of (1) is then justified with ChPT being the result. Calculating a process e.g. up to ChO4 we have to consider $\mathcal{L}^{(4)}$ in tree and $\mathcal{L}^{(2)}$ in tree + 1-loop approximation. The divergencies of the 1-loop graphs of $\mathcal{L}^{(2)}$ are absorbed into the tree graph coefficients of $\mathcal{L}^{(4)}$

$$\ell_i(\mu) = \ell_i(m_\rho) - \frac{\gamma_i}{16\pi^2}\ell n\frac{\mu}{m_\rho}, \qquad \ell_i(m_\rho) \sim N_C^1 \qquad (2)$$

which become dependent on the chiral scale μ (the γ_i are simple numerical factors).

soliton sector: $U = e^{i\boldsymbol{\tau}\hat{\boldsymbol{r}}F(r)}$ $\quad (B = 1$ hedgehog$)$
The lagrangian (1) possesses classical soliton solutions. For a first correction to this leading N_C tree approximation we consider 1-loop contributions, multi-loops are

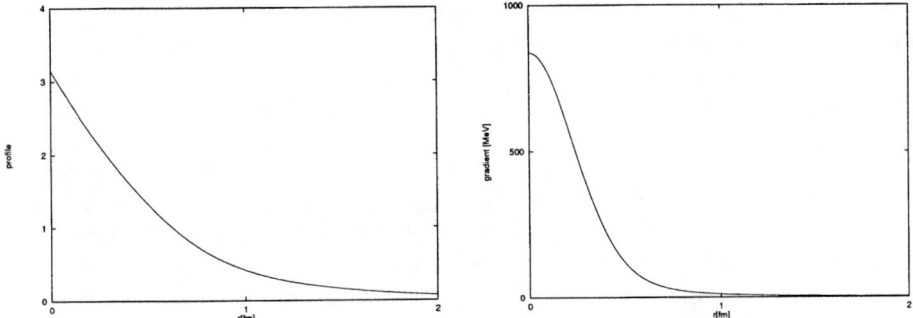

FIGURE 1. Typical soliton profile $F(r)$ and the corresponding gradient. In the center of the soliton the magnitude of the gradient becomes comparable to the chiral scale $\mu \simeq m_\rho$.

suppressed by further powers of $1/N_C$. The crucial difference to the vacuum sector is however that the soliton itself constitutes external fields which cannot be made weak by assumption. In fact, the gradients in a typical soliton come dangerously close to the chiral scale $\mu = m_\rho$ (Fig. 1) and we must expect *all* higher ChOs to contribute. However, it is noticed that large gradients appear only in a very limited region close to the center of the soliton. It turns out that the main effect of the essentially unknown higher ChOs is to supply sufficient repulsion to adjust the correct soliton size. There are various ways to simulate the higher ChOs. The simplest choice is to add an additional Skyrme term which is somewhat unfortunate because it destroys the chiral lagrangian at ChO4. Nevertheless, I will present

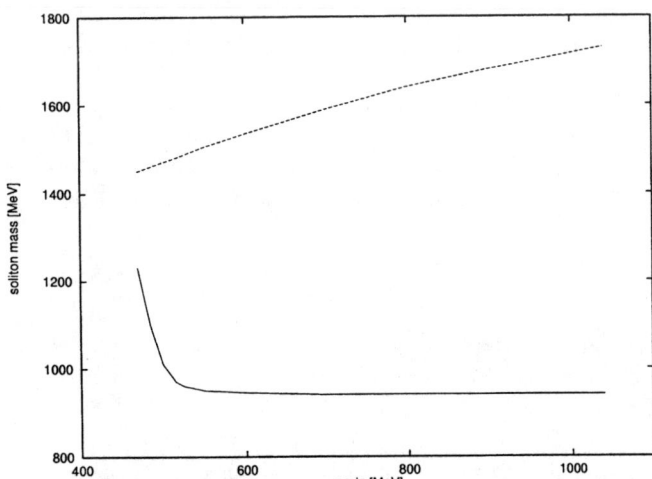

FIGURE 2. Scale dependence of the soliton mass in tree (dashed line) and tree + 1-loop (solid line) for the standard lagrangian with an additional Skyrme term ($e = 4.25$).

results for this case because it reasonably simulates the higher ChOs introduced by vector meson (which leave the chiral lagrangian to ChO4 untouched). Alternatives, such as ChO6 terms are discussed in [8]. Eventually one has to include vector and scalar mesons explicitely, however I do not expect that this will influence the results significantly.

The soliton mass in tree and tree + 1-loop is shown in Fig. 2 for the model with effective Skyrme parameter $e = 4.25$. It is exactly for this particular strength that in tree + 1-loop scale–independence is reached over a wide region of scales (until the soliton becomes unstable at very small scales). It will become immediately clear why approximate scale–independence is such an important requirement for soliton calculations.

Some typical baryon observables are listed in Table 1. Large 1-loop contributions are observed which generally improve the calculated values towards the experimental data, with one exception: the axial coupling constant g_A.

TABLE 1. Tree and 1-loop contributions for some typical adiabatic quantities. Model with effective Skyrme parameter $e = 4.25$.

		tree N_C^1	1-loop N_C^0	tree + 1-loop	experiment
nucleon mass	$M[MeV]$	1629	-683	946	939
sigma term	$\sigma\,[MeV]$	54	-21	33	45 ± 7
axial coupling	g_A	0.91	-0.25	0.66	1.26
mag. moment	$\mu^V[n.m.]$	1.62	$+0.62$	2.24	2.35
electric pol.	$\alpha\,[10^{-4}fm^3]$	17.8	-8.0	9.8	9.5 ± 5.0

SCALE-INVARIANCE IN SOLITON MODELS

Here we want to study to what extend the requirement of (almost) scale-invariance in the soliton sector is capable to fix the properties of the theory beyond low energy. For that purpose and also in order to illustrate the present method we choose the simple Sine-Gordon (SG) model with 1 scalar field in 1+1 dimensions

$$\mathcal{L} = \frac{1}{2g^2}\partial_\mu\phi\partial^\mu\phi - \mathcal{V}[\phi^2], \qquad \mathcal{V}[\phi^2] = \frac{m^2}{g^2}(\cos\phi + 1), \qquad (3)$$

whose properties in the vacuum as well as in the soliton sector are well-known [9]. Apart from the vacuum expectation values $\phi_{vac}(x) = v = \pm\pi, \ldots$ the classical e.o.m. possesses static soliton solutions $\phi_{vac}(x) = 4\arctan e^{mx} - \pi$ with classical soliton mass $M_0 = 8m/g^2$. This expression is clearly non–purturbative in the weak coupling expansion in powers of $g^2/4\pi$, which would correspond to the $1/N_C$ expansion in our chiral models. In order to calculate the next to leading order one considers fluctuations off the soliton background and solves the corresponding scattering equations for the phase–shifts

$$\delta(p) = 2\arctan\frac{m}{p} \xrightarrow{p\to\infty} \frac{a_1}{p}, \qquad a_1 = 2m, \tag{4}$$

which in case of the SG model are known analytically. With the phase–shift at hand the 1-loop correction may be evaluated via the formula for the Casimir energy

$$\begin{aligned}E_{cas} &= -\frac{1}{2\pi}\left[\int_0^\infty \frac{pdp}{\omega}(\delta(p)-\frac{a_1}{p}) - m\delta(0)\right] + \frac{a_1}{2\pi}\ell n\frac{\mu}{m}\\ &= -\frac{m}{\pi} + \frac{m}{\pi}\ell n\frac{\mu}{m},\end{aligned} \tag{5}$$

where the UV divergent part was subtracted and added again using dimensional regularization which renders the mass parameter

$$m^2(\mu) = m^2(1 - \frac{g^2}{4\pi}\ell n\frac{\mu}{m}) \tag{6}$$

dependent on a scale μ. The well-known exact result for the soliton mass in tree + 1-loop

$$E = M_0(\mu) + E_{cas}(\mu) = \frac{8m(\mu)}{g^2} - \frac{m}{\pi}(1 - \ell n\frac{\mu}{m}) \tag{7}$$

becomes then scale-independent to that order: the SG model is a renormalizable theory.

Now let us forget about the full theory and pretend that only its low–energy behaviour (in the meson sector) be known

$$\mathcal{V}[\phi^2] = \sum_{n=1}^{4} c_n(\phi^2 - v^2)^n. \tag{8}$$

This corresponds to the chiral expansion in our hadronic models. As the sum is truncated at some finite order, say at n=4, exact renormalizability (and also periodicity) is lost but the model may still be renormalized order by order as in ChPT. The 4 coefficients c_n are fixed such that the effective low–energy theory agrees with the SG model, i.e. the curvature and quartic terms at the vacuum expectation value $\phi_{vac} = v$, which is the only free parameter, are the same (cf. Fig. 3). As a consequence the mass renormalization eq.(6) is untouched. From Fig. 3 it is obvious that each of the effective theories possesses a soliton solution whose size is characterized by the parameter v (just put the Fig. on top and consider the motion of a classical point particle in the effective potentials). The soliton profile and its classical mass as well as the scattering phase–shifts are calculated numerically. Applying formula (5) for the Casimir energy it is implicitly assumed that the renormalized couplings of the "higher ChO terms", (here $c_n, n > 4$) remain zero and scale–independent. Through the mass parameter $m(\mu)$ the result for the soliton mass in tree + 1-loop becomes scale-dependent. This scale-dependence is shown in Fig. 4 for the various

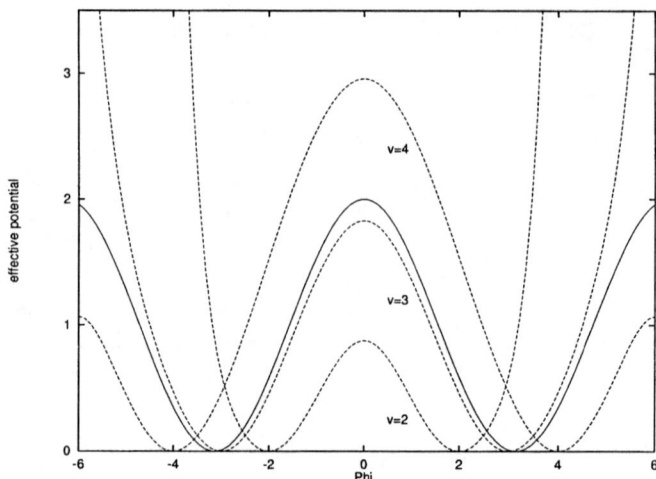

FIGURE 3. Exact Sine-Gordon potential in units of m^2/g^2 (solid line), and the corresponding effective potentials for various values of v (dashed lines).

effective models. It is noticed that approximate scale–independence requires a value $v \sim 3$ which provides an excellent approximation to the exact SG result. Thus it is demonstrated that the low-energy properties together with the scale-invariance of the results are sufficient to determine the SG soliton properly. Indeed, for this simple model with 1 scalar field it is easy to set up an algorithm in order to regain the exact solution, i.e. the (periodic) SG potential (3). Of course for chiral

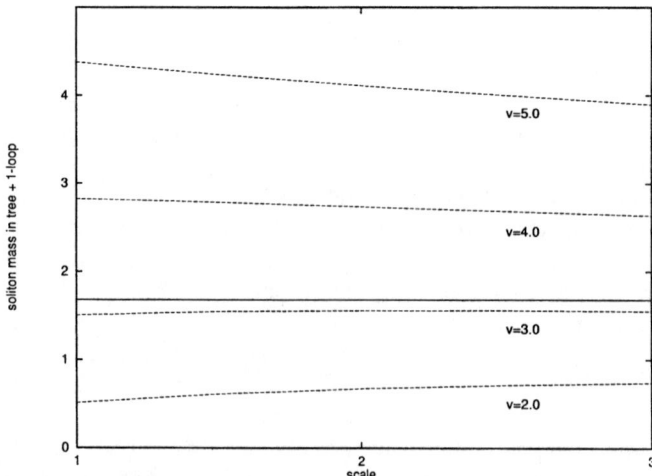

FIGURE 4. Scale-dependence of the soliton mass in tree + 1-loop for the various effective potentials (dashed lines). Exact SG result for $g = 2$, $m = 1$ (solid line).

lagrangians with 3 pion fields (and occasionally other meson fields) the situation is less restrictive, still the requirement of (almost) scale–independence sets severe constraints on the essentially unknown higher ChOs.

FURTHER APPLICATIONS

The present program has many applications which lead beyond the calculation of baryon observables in next to leading order in N_C as reported in [8]. Some of these issues will be briefly addressed in the following. The list is by no means comprehensive.

The g_A problem

The difficulty with the axial coupling originates from the chiral charge commutator,

$$[Q_5^a, Q_5^b] = i\epsilon_{abc} Q^c, \qquad (9)$$

which mixes different N_C orders. In order to obey this relation the next-to-next to leading order is needed which in soliton models is carried by *non-adiabatic* fluctuations suppressed further by a collective angular momentum [10]. In scattering calculations non–adiabatic fluctuations are responsible for the Tomozawa–Weinberg split and also for the Δ-resonance to appear in the continuum. From the above current algebra (CA) derives the Adler–Weisberger sum rule

$$g_A^2 = 1 + R, \qquad (10)$$

where the "1" produced by the commutation relation is of order N_C^0 and R (i.e. an integral over the πN charge exchange cross–section) is positive. It is then immediately clear that the $1/N_C$ expansion cannot converge reasonably for g_A^2 and R which both are leading order N_C^2. The reason is that the empirical values $g_A^2 = 1.56$ and $R = 0.56$ are not large compared to the CA "1", which is two orders down. This is a very general problem and should appear in all approaches which rely on N_C counting.

In the soliton approach with (adiabatic) loop corrections it appears that R rather than g_A^2 is approached, and if the CA "1" is added by hand the satisfactory result $g_A = 1.20$ may be obtained from Table 1. With non–adiabatic fluctuations taken into account, *both* sides of the Adler–Weissberger sum rule could be checked.

Hot nucleons

It is generally believed that with increasing temperature hadronic matter undergoes a phase transition to a quark gluon plasma searched for in heavy ion collisions.

Already at lower temperatures baryon properties, as e.g. mass and couplings, are expected to change. The soliton model is predestined to describe nucleons in a hot pion gas because it contains the pions *and* the baryons as solitons of one and the same effective action. The temperature dependence of the soliton mass and of the axial coupling constant due to pion loops is shown in Fig. 5 [11]. It is noticed that the soliton mass is practically constant in the low–temperature region where the approach becomes reliable, while the axial coupling decreases with increasing temperature. This finding is in striking contrast to Brown–Rho scaling $M \sim f_\pi$ (pion decay constant) and $g_A \sim 1$ [12,13]. Of course this is no surprise, the temperature dependence is carried by the 1-loop contribution and does not scale like the tree approximation.

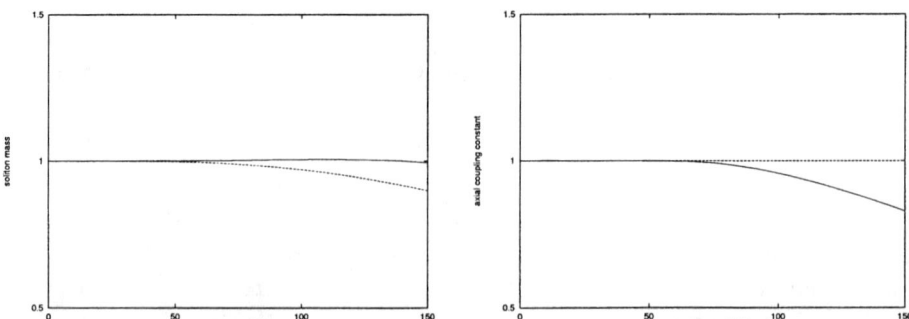

FIGURE 5. Temperature dependence of the soliton mass and of the axial coupling constant (solid lines). There is no Brown–Rho scaling $M \sim f_\pi$ and $g_A \sim 1$ (dashed lines).

SU(3) extension

The extension of the present program to SU(3) is essential for the description of strange baryons and also for nucleon properties sensitive to its strangeness content. The most appearent problem are the tree masses of the SU(3) chiral soliton which turn out at much too high energies, typically around 2GeV, because a large kaonic rotational energy is added. It was shown that kaon loops nicely compensate for this effect without violating N_C counting. It is convincing that the empirical nucleon mass lies in between the SU(2) and SU(3) symmetric cases [14]. Broken SU(3) was considered in Ref. [15].

Non-spherical solitons

The calculation of some baryon observables as e.g. the magnetic polarizability reguires the consideration of deformed solitons in leading order N_C. Another example is the famous static $B = 2$ configuration which possesses a torus like structure.

The application of the present program to the mass of this object is of particular interest, because it is conjectured that 1-loop corrections may destabilize the torus and all higher B configurations, which then decay into individual baryons (cf. also the section about the O(3) model).

So far the hedgehog symmetry of the soliton facilitated calculations enormously quite as does a spherical potential in ordinary potential scattering. Scattering e.g. at the $B = 2$ torus corresponds to non-spherical potential scattering and is technically a formidable task.

Non-adiabatic observables

Non-adiabatic baryon observables such as the nucleon-Δ split and the isoscalar magnetic moment connected with an angular velocity of the collective rotation are suppressed by $1/N_C$ already in tree approximation. The isoscalar magnetic moment $\mu^S = <r^2>_B /12\Theta_\pi$ related to the baryon radius (Θ_π is the pionic moment of inertia) turns out too small by more than a factor of 2 in all soliton models, e.g. $\mu^S = 0.18$ n.m. in our model. Here we encounter the peculiar situation that tree and 1-loop contributions are of the *same* N_C order.

Before doing the actual calculation, we may speculate about the pionic and kaonic 1-loop contributions aiming at the empirical value of the isoscalar magnetic moment $\mu^S = (\mu_p + \mu_n)/2 = 0.44$ n.m. and the corresponding SU(3) singlet magnetic moment $\mu^0 = 0.62$ n.m. obtained from a fit to the experimental hyperon magnetic moments. This would also solve another longstanding problem: the strange magnetic moment of the proton, which in tree approximation quite generally comes with the wrong sign. With the empirical value of μ^0 a *positive* strange magnetic moment $F_{2p}^{strange}(0) = 2\mu^0 - (\mu_p + \mu_n) = +0.36$ n.m. is predicted.

O(3) model in 2+1 dimensions

The present program has been applied also to other effective field theories such as the 2+1 dimenional O(3) model which plays an important role in the description of spin-textures in planar ferro and antiferromagnets. Similar to the hadronic O(4) solitons the spin-texture with topological charge $B = 2$ is classically stable. However with the 1-loop corrections included, the $B = 2$ soliton becomes unstable [16] and decays into two individual $B = 1$ solitons (which may be still weakly bound by a dipole force). Although different dimensionalities may cause qualitative differences a numerical example using the scale of hadron physics is considered in Table 2. We conjecture that exactly this situation might occur for the $B = 2$ torus in hadron physics.

TABLE 2. With the 1-loop corrections included the $B = 2$ soliton becomes unstable and decays into two $B = 1$ solitons. All energies are given in MeV.

	$E_{B=1}$	$E_{B=2}$	$E_{B=2} - 2E_{B=1}$
tree	1390	2640	-140
1-loop	-450	-530	$+370$
	940	2110	$+230$

CONCLUSIONS

The soliton picture of baryons may be pushed towards a more quantitative model by consideration of 1-loop contributions which are sizeable, actually typically of the order 1/3. The sign of the correction depends on the observable considered and generally improves the result towards the experimental datum (with one well understood exception: the axial quantities). Surprisingly, the model seems to work without quarks, so far we did not encounter baryon properties where we had to conclude that missing explicit quark degrees of freedom were important. One may argue that eventually in deep inelastic scattering quarks should be seen explicitly, although the relevant structure functions may in principle also be obtained from purly mesonic soliton models by making use of gradient expansion techniques.

The bad news is that the $1/N_C$ expansion converges rather slowly, in particular of course for the axial quantities. This statement is quite general and applies to all approaches (chiral quark models, NJL soliton models, etc.) which rely on the $1/N_C$ expansion. A rigorous calculation of the next-to-next to leading order, which would include 2-loop contributions, seems to be intractable.

REFERENCES

1. Weinberg S., *Physica* **96A**, 327 (1979).
2. Gasser J., and Leutwyler H., *Ann. Phys.* **158**, 142 (1984).
3. Gasser J., Sainio M.E., and Svarc A., *Nucl. Phys.* **B307**, 779 (1988).
4. Skyrme T.H.R., *Proc. Roy. Soc.* **A260**, 127 (1961).
5. Witten E., *Nucl. Phys.* **B160**, 57 (1979).
6. Dorey N., and Mattis M.P., *Phys. Rev.* **D52**, 2871 (1995).
7. Moussallam B., *Ann. Phys.* **225**, 264 (1993).
8. Meier F., and Walliser H., *Phys. Rep.* **289**, 383 (1997).
9. Rajaraman R., *Solitons and Instantons*, Amsterdam: North Holland, 1982.
10. Schwesinger B., and Walliser H., *Nucl. Phys.* **A490**, 602 (1988).
11. Walliser H., *Phys. Rev.* **D56**, 3866 (1997).
12. Bernard V., and Meissner U.-G., *Ann. Phys.* **206**, 50 (1991).
13. Brown G.E., and Rho M., *Phys. Rev. Lett.* **66**, 2720 (1991).
14. Walliser H., *Phys. Lett.* **B432**, 15 (1998).
15. Scoccola N.N., and Walliser H., *Phys. Rev.* **D58**, 094037 (1998).
16. Walliser H., and Holzwarth G., hep-ph/9907492.

Discriminating between chiral models of the nucleon via hard semi-inclusive $\vec{e}\vec{N}$ and $\vec{N}\vec{N}$ scattering

B. Dressler, K. Goeke, M.V. Polyakov[†], P. Schweitzer,
M. Strikman[†*], and C. Weiss

*Institut für Theoretische Physik II, Ruhr–Universität Bochum,
D–44780 Bochum, Germany*

[†]*Petersburg Nuclear Physics Institute, Gatchina,
St.Petersburg 188350, Russia*

[*]*Pennsylvania State University, University Park, PA 16802, U.S.A.*

Abstract. Measurements of the flavor asymmetry of the polarized antiquark distributions in the nucleon, $\Delta \bar{u}(x) - \Delta \bar{d}(x)$, would provide a unique opportunity to discriminate between different models of the role of the chiral structure of the nucleon in deep–inelastic scattering. In the pion cloud model of the nucleon's sea quark distributions the polarized flavor asymmetry is zero, while the large–N_c picture of the nucleon as a chiral soliton predicts large polarized flavor asymmetries. We discuss the prospects for measuring the polarized antiquark flavor asymmetries in two types of semi-inclusive scattering experiments: *i)* Hadron production in semi-inclusive polarized eN scattering (SMC, HERMES); *ii)* Drell–Yan lepton pair production in polarized NN scattering at HERA or RHIC energies. In particular, the Drell–Yan spin asymmetries are shown to be remarkably sensitive to $\Delta \bar{u}(x) - \Delta \bar{d}(x)$.

It can now be considered as established that non-perturbative effects play an essential role in generating the sea quark distributions in the nucleon, measured in scattering experiments at large momentum transfers. In particular, the violation of the Gottfried sum rule observed in deep–inelastic lepton–nucleon scattering [1], and the recent direct measurements of $\bar{u}(x) - \bar{d}(x)$ through Drell–Yan pair production in nucleon–nucleon collisions [2,3], have shown that there is a significant flavor asymmetry of the unpolarized antiquark distributions, which cannot be explained perturbatively. Other evidence is the large suppression of the strange sea compared to the nonstrange one at scales of the order of a few GeV^2.

It is widely believed that an explanation for the mentioned effects lies in the chiral structure of the nucleon. Two different approaches have been proposed to account for the role of the chiral degrees of freedom in DIS. In the so-called pion

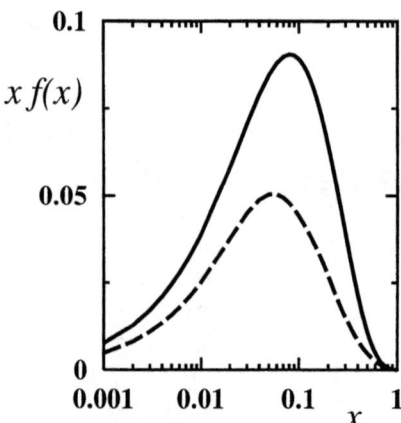

FIGURE 1. The polarized and unpolarized antiquark flavor asymmetries obtained in model calculations in the large-N_c limit (chiral quark–soliton model), evolved (LO) from the low normalization point of $\mu^2 = (600\,\mathrm{MeV})^2$ to a scale of $\mu^2 = (5\,\mathrm{GeV})^2$. Dashed line: Unpolarized flavor asymmetry, $x[\bar{d}(x) - \bar{u}(x)]$, see Ref. [7]. Solid line: Polarized flavor asymmetry, $x[\Delta\bar{u}(x) - \Delta\bar{d}(x)]$, see Refs. [6,9].

cloud model [4] one considers scattering off pions created via virtual processes $N \to N + \pi$, $N \to \Delta + \pi$, or $q \to q + \pi$. This mechanism can in principle generate a significant value of $\bar{u}(x) - \bar{d}(x)$, although this requires one to consider virtual pion momenta up to $\sim 1\,\mathrm{GeV}$ and relies on fine-tuning of the parameters of the model; see Ref. [5] for a discussion. Another approach is the description of the nucleon as a chiral soliton, which can be justified by the large–N_c limit of QCD. It has been shown in Refs. [6,7] that this picture contains all the ingredients for a fully quantitative description of the antiquark distributions in the nucleon, which preserves all fundamental qualitative properties of the distribution functions, such as positivity, sum rules *etc.* In particular, the results for $\bar{u}(x) - \bar{d}(x)$ obtained in this approach describe well the E866 Drell-Yan data [8].

Clearly, it would be of great interest to have at hands an observable effect which would allow one to discriminate between the two pictures. We have recently pointed out that a qualitative difference bewteen the two pictures lies in the predictions for the polarized antiquark flavor asymmetry, $\Delta\bar{u}(x) - \Delta\bar{d}(x)$ [9]. In the pion cloud model the polarization of the flavor asymmetry is zero [10]. There have been some attempts to generate polarization by including spin–1 resonances in this picture [11], which, however, presents severe conceptual difficulties; see Refs. [9,12] for a detailed discussion. In contrast, the large–N_c picture of the nucleon predicts that $\Delta\bar{u}(x) - \Delta\bar{d}(x)$ is much larger than the unpolarized $\bar{u}(x) - \bar{d}(x)$; the polarized asymmetry is parametrically enhanced by a factor of N_c. [The numerical results for the polarized [6,9] and unpolarized [7] antiquark flavor asymmetries obtained in this approach are shown in Fig.1 at a scale of $\mu^2 = (5\,\text{GeV})^2$.] Thus, measurements of $\Delta\bar{u}(x) - \Delta\bar{d}(x)$ would provide a unique opportunity to discriminate between the two approaches.

The flavor asymmetries of the polarized antiquark distributions are practically not constrained by the present DIS data. Inclusive polarized eN scattering, $\vec{e}+\vec{N} \to e + X$, which traditionally has been the main source of information about the spin structure of the nucleon, measures only the sum of the quark and antiquark distributions in the target. The observable here is the double spin asymmetry of the total cross section, which in the asymptotic regime $Q^2 \to \infty$ and in leading–order QCD is given by[1]

$$A_1(x, Q^2) \equiv \frac{\sigma^{tot}_{++} - \sigma^{tot}_{+-}}{\sigma^{tot}_{++} + \sigma^{tot}_{+-}} = \frac{\sum_a e_a^2 \Delta q_a(x, Q^2)}{\sum_b e_b^2 q_b(x, Q^2)}, \quad (1)$$

where the subscripts $\pm$ denote, respectively, the electron and nucleon helicities. Here Δq_a and q_a are the polarized and unpolarized quark distributions in the target, and the sums run over all light quark and antiquark flavors, $a, b = \{u, \bar{u}, d, \bar{d}, s, \bar{s}\}$. While the flavor decomposition can in principle be studied by comparing measurements with proton and neutron (i.e., nuclear) targets and making use of $SU(3)$ symmetry arguments, the separation of quark and antiquark contributions is not possible from inclusive DIS measurements alone. In particular, the flavor asymmetries of the polarized antiquark distributions, $\Delta\bar{u}(x) - \Delta\bar{d}(x)$ and $\Delta\bar{u}(x) + \Delta\bar{d}(x) - 2\Delta\bar{s}$, are practically unknown at present; these distributions were assumed to be zero in the GRSV95 parametrization [13]. Similar is the situation with the polarized gluon distribution, $\Delta G(x)$, which can in principle be deduced from logarithmic scaling violations.

Measurements of the individual quark– and antiquark distributions are, however, possible in semi-inclusive scattering experiments, where the detection of a particle produced in a hard partonic reaction offers a handle on the quantum numbers of the parent partons. In these Proceedings, following Refs. [9] and [12], we discuss the prospects for extracting the polarized antiquark flavor asymmetry from two types of polarized semi-inclusive scattering experiments:

i) Hadron production in semi-inclusive polarized eN scattering:
$\vec{e} + \vec{N} \to e + h + X$.

ii) Drell–Yan Lepton pair production in polarized NN scattering:
$\vec{N} + \vec{N} \to l^+l^- + X$.

For a more detailed description of the meson cloud picture and the large–N_c approach, and a discussion of their underlying assumptions, we refer to Refs. [9] and [12], and the references therein.

---

[1] At finite Q^2, the complete theoretical expressions for the spin asymmetry contain also kinematical factors involving inverse powers of Q^2, which are not shown here. See Ref. [9] for details.

i) *Hadron production in polarized semi-inclusive eN scattering.* The relevant observable here is the double spin asymmetry of the semi-inclusive cross section for producing a fast hadron of type h. In leading–order QCD this process is described as the fragmentation of the struck quark in the target, and the asymmetry is given by

$$A_1(x,Q^2) \equiv \frac{\sigma^h_{++} - \sigma^h_{+-}}{\sigma^h_{++} + \sigma^h_{+-}} = \frac{\sum_a e_a^2 \,\Delta q_a(x,Q^2)\, D_a^h(z,Q^2)}{\sum_b e_b^2 \, q_b(x,Q^2)\, D_b^h(z,Q^2)} \tag{2}$$

where the $D_a^h(z,Q^2)$ denote the unpolarized fragmentation functions for an (anti–)quark of type a to fragment in a hadron h carrying fraction z of its longitudinal momentum. Instead of the fixed–z asymmetry one usually considers the integrated asymmetry, where one replaces in Eq.(2) everywhere

$$D_a^h(z,Q^2) \;\rightarrow\; \int_{z_{\min}}^1 dz\, D_a^h(z;Q^2). \tag{3}$$

Here $z_{\min} > 0$ is a cutoff, which must be chosen sufficiently large such as to suppress so-called target fragmentation contributions [14].

Due to the presence of the fragmentation functions the semi-inclusive asymmetry, Eq.(2), contrary to the inclusive one, Eq.(1), allows to separate contributions from quarks and antiquarks in the target. Their extraction requires, however, precise knowledge of the fragmentation functions, both favored and unfavored ones. The fragmentation functions for a variety of charged hadrons have been measured some time ago by the EMC experiment [15]; for π^+ and π^- they have also recently been extracted from the HERMES data [14].

In order to investigate the effect of the polarized antiquark flavor asymmetries obtained in Refs. [6,9] on the observable semi-inclusive spin asymmetries one can proceed as follows (for details, see Ref. [9]). The individual polarized light quark and antiquark distributions $\Delta u(x)$, $\Delta \bar{u}(x)$, $\Delta d(x)$, $\Delta \bar{d}(x)$, $\Delta s(x)$, and $\Delta \bar{s}(x)$, figuring in the numerators in Eq.(2), can be expressed in terms of the six combinations

$$\Delta u(x) + \Delta \bar{u}(x), \quad \Delta d(x) + \Delta \bar{d}(x), \quad \Delta s(x) + \Delta \bar{s}(x), \tag{4}$$

$$\Delta \bar{u}(x) + \Delta \bar{d}(x) + \Delta \bar{s}(x), \tag{5}$$

$$\Delta \bar{u}(x) - \Delta \bar{d}(x), \tag{6}$$

$$\Delta \bar{u}(x) + \Delta \bar{d}(x) - 2\Delta \bar{s}(x). \tag{7}$$

Of these the first three, Eqs.(4), are measured directly in inclusive polarized DIS, so we evaluate them using the GRSV95 leading–order (LO) parametrization ("standard scenario"), which was obtained by fits to inclusive DIS data [13].[2] The flavor-singlet antiquark distribution, Eq.(5), we also take from the GRSV95 parametrization; this distribution is known only from the study of scaling violations in inclusive

---

[2] Actually, in DIS with proton or nuclear targets one is able to measure directly only two flavor combinations of these three distributions; however, the third one can be inferred using $SU(3)$ symmetry arguments.

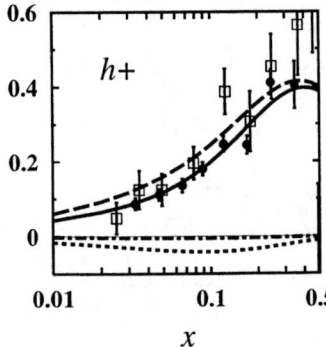
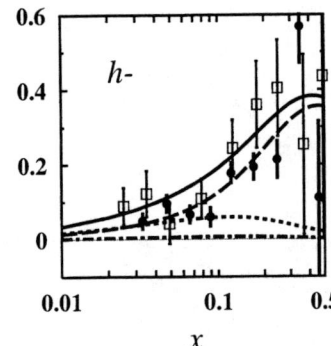

FIGURE 2. *The spin asymmetries for h^+ and h^- production in semi-inclusive DIS off the proton, at a scale of $Q^2 = 2.5\,\text{GeV}^2$, corresponding to the HERMES experiment (for details concerning the fragmentation functions see Ref. [9]).* <u>Dashed lines:</u> *The ratio Eq.(2) evaluated with the GRSV95 LO parametrization, assuming zero polarized antiquark flavor asymmetry.* <u>Solid lines:</u> *Results obtained including the polarized antiquark calculated at a low scale in the chiral quark–soliton model, see Fig.1.* <u>Dotted and dashed–dotted lines:</u> *Contributions proportional to $\Delta\bar{u}(x) - \Delta\bar{d}(x)$ and $\Delta\bar{u}(x) + \Delta\bar{d}(x) - 2\Delta\bar{s}(x)$ in the target, respectively.* <u>Open Squares:</u> *SMC data [18].* <u>Filled Circles:</u> *HERMES data [19].*

DIS and depends to some extent on the assumptions made about the polarized gluon distribution; however, the GRSV95 parametrization is in good agreement with the result of model calculations in the large–N_c limit [16]. For the polarized flavor asymmetries of the antiquark distribution, Eqs.(6) and (7), which are not constrained by DIS data, we use the results of the model calculation in the large–N_c limit of Refs. [6,9], evolved in LO from the low normalization point of $\mu^2 = (600\,\text{MeV})^2$ to the experimental scale. [In the chiral quark–soliton model in leading order of $1/N_c$ the $SU(3)$–octet asymmetry, Eq.(7), is proportional to the $SU(3)$–triplet one, Eq.(6); the proportionality factor can be expressed in terms of the F/D ratio; see Ref. [9].] Note that $\Delta\bar{u}(x) - \Delta\bar{d}(x)$ and $\Delta\bar{u}(x) + \Delta\bar{d}(x) - 2\Delta\bar{s}(x)$ do not mix with the other distributions under LO evolution. The "hybrid" polarized quark and antiquark distributions thus obtained, by construction, fit all the inclusive polarized DIS data in LO, while at the same time incorporating the polarized antiquark flavor asymmetry obtained in the model calculation in the large–N_c limit. Finally, to evaluate the denominator in Eq.(2) we use the GRV94 parametrization of the unpolarized parton distributions [17].

In Fig.2 we compare the theoretical predictions for the spin asymmetry in charged particle production ($h^+ \equiv \pi^+ + K^+ + p$, $h^- \equiv \pi^- + K^- + \bar{p}$) obtained including the polarized antiquark flavor asymmetries from the large–N_c model calculation (solid line), and with zero flavor asymmetries (dashed line). The results are shown for a scale of $Q^2 = 2.5\,\text{GeV}^2$ typical for the HERMES experiment, and with a cutoff $z_{\min} = 0.2$. One sees that the effect of $\Delta\bar{u}(x) - \Delta\bar{d}(x)$ in the target is rather sizable.

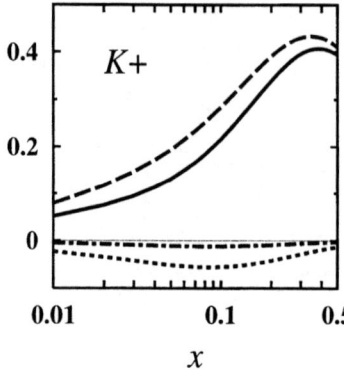 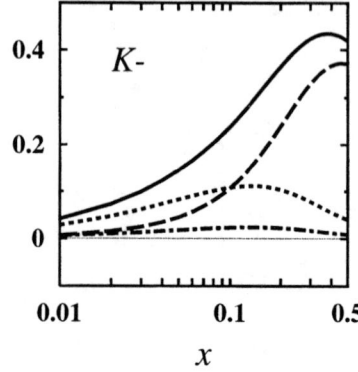

FIGURE 3. *Same as Fig.2, but for production of charged kaons, K^+ and K^-.*

Also, one sees that the large flavor asymmetry obtained in the model calculation is consistent with both the SMC [18] and the recent HERMES [19] data. Note that in the case of h^- production the spin asymmetries are very sensitive to the values of the unfavored fragmentation functions, so that the accuracy of the theoretical estimate is rather poor. The difference between the solid and dashed lines in Fig.2 gives an idea of the accuracy of the data which would be required in order to significantly discriminate between the case of a large and no flavor asymmetry. We conclude that the present level of accuracy is not sufficient for a significant comparison. However, improving statistics of the HERMES data could eventually change this situation.

Also of interest are the spin asymmetries in the semi-inclusive production of individual charged particles (π^+, π^-, K^+, K^-). Since K^- cannot be produced by favored fragmentation of either u or d quarks in the target, which is the dominant contribution to the semi-inclusive spin asymmetry in $\pi^\pm$ or $h^\pm$ production, one might expect a comparatively large effect from the flavor asymmetries of the polarized antiquark distributions. The LO theoretical predictions for K^+ and K^- production are shown in Fig.3.

The way to extract the flavor asymmetries discussed here requires accurate parametrizations for $\Delta u(x) + \Delta \bar{u}(x)$, $\Delta d(x) + \Delta \bar{d}(x)$ and $\Delta s(x) + \Delta \bar{s}(x)$, in order to be able to distinguish between the theoretical predictions of the asymmetry calculated with and without flavor asymmetry. Comparison between the results obtained with different parametrizations derived from fits to inclusive data (GRSV95 [13], Gehrmann–Stirling 96 [22]) show that the uncertainties in the theoretical predictions for A_1^h due to the variations in the present parametrizations of the inclusive data are comparable to the effect of the flavor asymmetries. Thus, at present it seems not possible to unambiguously extract the flavor asymmetry from the semi-inclusive data, due to the uncertainties in both inclusive and semi-inclusive data. One may hope that this situation changes as the HERMES experiment keeps taking data. We remark that measurements of semi-inclusive hadron production of the kind discussed here could also be performed at higher scales at the COMPASS experiment [20], or at the proposed TESLA linear accelerator [21].

ii) Drell–Yan Lepton pair production in polarized NN scattering. Another class of experiments which allow to measure separately the polarized quark– and antiquark distributions in the nucleon is Drell–Yan lepton pair production in polarized NN collisions, $N\uparrow + N\uparrow \to l^+l^- + X$. In leading–order QCD, the lepton pair is produced by the annihilation of a quark and an antiquark originating from the two nucleons. If the production is by exchange of a (time-like) virtual photon, the cross section exhibits a double spin asymmetry,

$$A_{LL}^\gamma(y;s,M^2) = \frac{\sigma_{++}^\gamma - \sigma_{+-}^\gamma}{\sigma_{++}^\gamma + \sigma_{+-}^\gamma} = \frac{\sum_a e_a^2\,\Delta q_a(x_1,M^2)\,\Delta q_{\bar{a}}(x_2,M^2)}{\sum_a e_a^2\,q_a(x_1,M^2)\,q_{\bar{a}}(x_2,M^2)}, \qquad (8)$$

where now the subscripts $+,-$ denote the longitudinal polarization of nucleons 1 and 2. Again, the sums over a and b run over all species of light quarks and antiquarks in the two hadrons, and the distribution $\bar{a}$ is the charge–conjugate of that with a. Here, M^2 is the invariant–mass squared of the produced lepton pair, and x_1, x_2 denote the longitudinal momentum fractions of the annihilating partons, which can be parametrized in terms of a rapidity variable, y, as $x_1 = (M^2/s)^{1/2}e^y$, $x_2 = (M^2/s)^{1/2}e^{-y}$, where s is the invariant energy of the colliding nucleons.

Again, we are interested in the sensitivity of the double spin asymmetry to the polarized antiquark flavor asymmetry in the nucleon. As above, we compare the LO theoretical prediction for the double spin asymmetry, Eq.(8), obtained with standard parametrizations of the polarized parton distributions derived from inclusive DIS [13,22], which assume zero polarized antiquark flavor asymmetry, and the results obtained including the polarized antiquark flavor asymmetry obtained in the model calculation in the large–N_c limit, see Fig.1. Fig.4 shows the theoretical predictions at a scale of $M^2 = 25\,\mathrm{GeV}^2$, and $s = (40\,\mathrm{GeV})^2$, corresponding to a proposed fixed target experiment using the HERA proton beam [23]. This time we show the results obtained using the GRSV95 [13] and the Gehrmann–Stirling [22] parametrizations. One sees that in both cases the inclusion of the flavor asymmetry of the antiquark distribution has a dramatic effect on the observable spin asymmetry, reversing even its sign compared to the case with $\Delta\bar{u}(x) - \Delta\bar{d}(x) = 0$ and $\Delta\bar{u}(x) + \Delta\bar{d}(x) - 2\Delta\bar{s}(x) = 0$. In the case of the GS96 parametrization we have shown the results obtained with sets "A" and "C", both with and without flavor asymmetry. (The GS96 "A", "B", and "C" parametrizations provide fits to the inclusive data with widely different assumptions about the shape of the input polarized gluon distributions [22].) One sees that the changes of A_{LL}^γ due to the inclusion of the flavor asymmetry (differences between corresponding solid and dashed curves) are much larger than the differences due to changes of the input gluon distribution (differences between the two dashed curves). Again, NLO corrections are known to enhance the importance of the polarized gluon distribution [24]; however, it is unlikely that their inclusion will change this conclusion. Thus, it

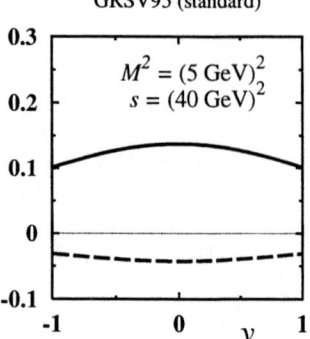 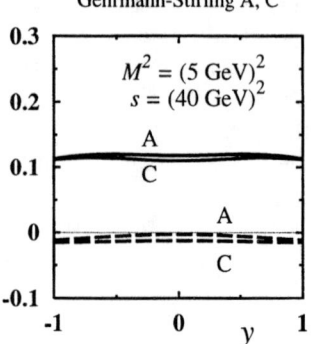

FIGURE 4. *Theoretical predictions for the double spin asymmetry for Drell–Yan lepton pair production in longitudinally polarized NN collisions, A^{γ}_{LL}, at a scale of $M^2 = 25\,\text{GeV}^2$, and for $s = (40\,\text{GeV})^2$.* <u>Dashed lines:</u> *The ratio Eq.(9) evaluated with the GRSV95 and GS96 "A" and "C" LO parametrizations, both assuming zero polarized antiquark flavor asymmetry.* <u>Solid lines:</u> *Results obtained including the polarized antiquark calculated at a low scale in the chiral quark–soliton model, see Fig.1.*

seems that polarized lepton pair production in NN collisions represents a promising venue for measuring the polarized antiquark flavor asymmetry in the nucleon.

The situation looks even simpler if a charged lepton pair is produced through an intermediate $W^{\pm}$ boson instead of a virtual photon. The observations of such processes will be possible at the RHIC collider, where one can reach values of $M^2 \approx M_W^2$. In this case, due to the parity–violating nature of the weak interaction the cross section exhibits already a single spin asymmetry,

$$A_L^{W^+} = \frac{\sigma_+^{W^+} - \sigma_-^{W^+}}{\sigma_+^{W^+} + \sigma_-^{W^+}} = \frac{\Delta u(x_1, M_W^2)\,\bar{d}(x_2, M_W^2) - \Delta\bar{d}(x_1, M_W^2)\,u(x_2, M_W^2)}{u(x_1, M_W^2)\,\bar{d}(x_2, M_W^2) + \bar{d}(x_1, M_W^2)\,u(x_2, M_W^2)}, \quad (9)$$

where now the subscripts $+, -$ denote the longitudinal polarization of nucleon 1; the polarization of nucleon 2 is averaged over. (For W^- one should exchange $u \leftrightarrow d, \bar{u} \leftrightarrow \bar{d}$ everywhere here.) The theoretical predictions for this single–spin asymmetry at $M^2 = M_W^2$, both with and without polarized antiquark flavor asymmetry, are shown in Fig.5, both for the GRSV95 and the GS96 parametrizations. One sees that the effect of the non-perturbative antiquark flavor asymmetry at the low normalization point is sizable even at this high scale.

To summarize, we have argued that a measurement of the polarized antiquark flavor asymmetry, $\Delta\bar{u}(x) - \Delta\bar{d}(x)$, would provide a unique opportunity to discriminate between different ways to describe the role of the chiral degrees of freedom in deep–inelastic scattering off the nucleon: the meson cloud picture, and the large–N_c

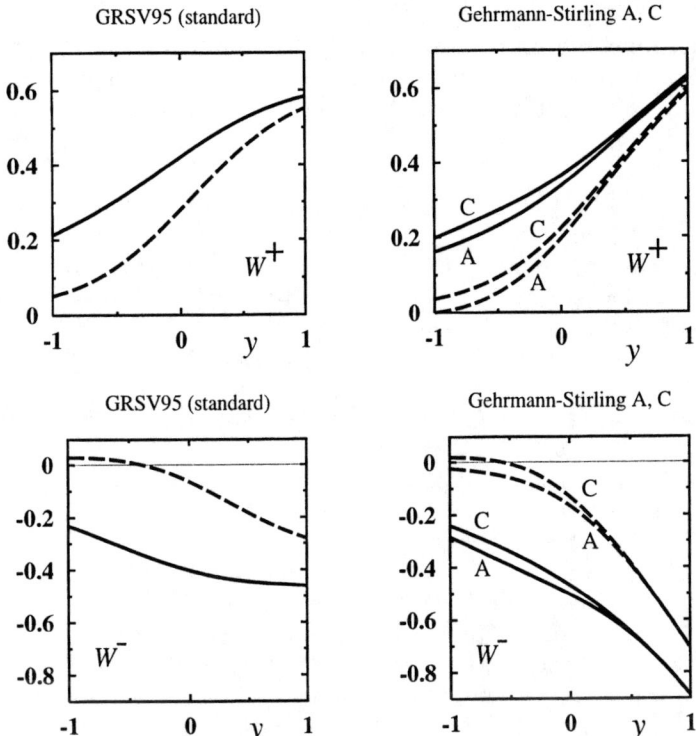

FIGURE 5. *The theoretical predictions for the single-spin asymmetry of the cross section for charged lepton pair production through W^+ (W^-) in polarized NN collisions, Eq.(9), at $M^2 = M_W^2$.* <u>Dashed lines:</u> *The ratio Eq.(9) evaluated with the GRSV95 [13] and GS96 LO [22] parametrizations, both assuming zero polarized antiquark flavor asymmetry.* <u>Solid lines:</u> *Results obtained including in addition the polarized antiquark flavor asymmetry calculated at a low scale in the chiral quark–soliton model, see Fig.1.*

description of the nucleon as a chiral soliton. We have reviewed semi-inclusive experiments as to their ability to detect a large flavor asymmetry as predicted by the large–N_c model calculations. Concerning hadron production in semi-inclusive eN scattering, while the large flavor asymmetry obtained in the chiral–quark soliton model of the nucleon is consistent with the recent HERMES data, the uncertainties in both inclusive and semi-inclusive data at present seem to preclude an unambiguous extraction of the flavor asymmetries. A promising candidate for measurement of the flavor asymmetries, however, is Drell–Yan lepton pair production in polarized NN collisions, where the polarized antiquark flavor asymmetry has a dramatic effect on the observables. One may thus expect that the future RHIC experiments will have a great impact on our understanding of the way in which the chiral degrees of freedom contribute to the structure of the nucleon seen in deep–inelastic scattering.

This work has been supported in part by the Deutsche Forschungsgemeinschaft (DFG), by a joint grant of the DFG and the Russian Foundation for Basic Research, by the German Ministry of Education and Research (BMBF), and by COSY, Jülich.

REFERENCES

1. New Muon Collaboration (P. Amaudruz et al.), Phys. Rev. Lett. **66** (1991) 2712; New Muon Collaboration (M. Arneodo et al.), Phys. Rev. **D 50** (1994) R1; Phys. Lett. **B 364** (1995) 107.
2. FNAL E866/NuSea Collaboration (E.A. Hawker et al.), Phys. Rev. Lett. **80**, 3715 (1998); FNAL E866/NuSea Collaboration (J.C. Peng et al.), Phys. Rev. **D 58**, 092004 (1998).
3. For a review, see: P.L. McGaughey, J.M. Moss, and J.C. Peng, Report LA-UR-99-850, hep-ph/9905409.
4. A.W. Thomas, Phys. Lett. **126 B** (1983) 97.
   L.L. Frankfurt, L. Mankiewicz, and M.I. Strikman, Z. Phys. **A 334** (1989) 343.
   E.M. Henley and G.A. Miller, Phys. Lett. **B 251** (1990) 453.
   W.Y.P. Hwang, J. Speth, and G.E. Brown, Z. Phys. **A 339** (1991) 383.
   W. Melnitchouk and A.W. Thomas, Phys. Rev. **D 47** (1993) 3794.
   H. Holtmann, N.N. Nikolaev, J. Speth, and A. Szczurek Z. Phys. **A 353** (1996) 411.
   H. Holtmann, A. Szczurek, and J. Speth, Nucl. Phys. **A 596** (1996) 397, *ibid.* 631.
5. W. Koepf, L.L. Frankfurt, and M. Strikman, Phys. Rev. **D 53** (1996) 2586.
6. D.I. Diakonov, V.Yu. Petrov, P.V. Pobylitsa, M.V. Polyakov and C. Weiss, Nucl. Phys. **B 480**, 341 (1996); Phys. Rev. **D 56**, 4069 (1997).
7. P.V. Pobylitsa, M.V. Polyakov, K. Goeke, T. Watabe, and C. Weiss, Phys. Rev. **D 59**, 034024 (1999).
8. B. Dressler et al., in: Proceedings of the 11th International Conference on Problems of Quantum Field Theory, Dubna, Russia, Jul. 13–17, 1998, hep-ph/9809487.
9. B. Dressler, K. Goeke, M.V. Polyakov, and C. Weiss, Bochum University Report RUB-TPII-12/99, hep-ph/9909541.
10. V.R. Zoller, Z. Phys. **C 53** (1992) 443; **C 60** (1993) 141.
11. R.J. Fries and A. Schäfer, Phys. Lett. **B 443**, 40 (1998); K.G. Boreskov, A.B. Kaidalov, Eur. Phys. J. **C 10**, 143 (1999).
12. B. Dressler, K. Goeke, M.V. Polyakov, P. Schweitzer, M. Strikman, and C. Weiss, Bochum University Report RUB-TPII-13/99, hep-ph/9910464.
13. M. Glück, E. Reya, M. Stratmann and W. Vogelsang, Phys. Rev. **D 53**, 4775 (1996).
14. Ph. Geiger, "Measurerment of Fragmentation Functions at HERMES", Doctoral Dissertation, Heidelberg University (1998).
15. European Muon Collaboration (M. Arneodo et al.), Nucl. Phys. **B 321**, 541 (1989).
16. M. Wakamatsu and T. Kubota, Phys. Rev. **D 60** (1999) 034020.
    P.V. Pobylitsa et al., in preparation.
17. M. Glück, E. Reya, and A. Vogt, Z. Phys. **C 67**, 433 (1995).

18. Spin Muon Collaboration (B. Adeva *et al.*), Phys. Lett. **B 420**, 180 (1998); Phys. Lett. **B 369**, 93 (1996).
19. HERMES Collaboration (K. Ackerstaff et al.), "Flavor decomposition of the polarized quark distributions in the nucleon from inclusive and semi-inclusive deep inelastic scattering", Report DESY-99-048, hep-ex/9906035.
20. COMPASS Collaboration (G. Baum *et al.*), "COMPASS: A proposal for a common muon and proton apparatus for structure and spectroscopy", CERN-SPSLC-96-14 (1996).
21. W.-D. Nowak, Report at the Workshop on polarized eN scattering at TESLA, Regensburg, Germany, Nov. 11–12, 1999.
22. T. Gehrmann and W.J. Stirling, Phys. Rev. **D 53** (1996) 6100.
23. M. Anselmino *et al.*, Proceedings of the Workshop "Future Physics at HERA", 1995/96, ed. G. Ingelman, A. DeRoeck and G. Klanner, DESY, Hamburg (1996), p.837.
    V.A. Korotkov and W.D. Nowak, Proceedings of the 2nd "ELFE Workshop", St. Malo, France, 1996.
24. T. Gehrmann, Nucl. Phys. **B 498** (1997) 245.

Stability of Multiquark Systems

Fl. Stancu [1]

*Institute of Physics, B. 5 University of Liege, Sart Tilman
B-4000 Liege 1, Belgium*

Abstract. We give a brief review of developments in the field of exotic hadrons formed of more than three quarks and/or antiquarks. In particular we discuss the stability of multiquark systems containing heavy flavours. We show that the gluon exchange model and the chiral constituent quark model based Goldstone boson (pseudoscalar meson) exchange give entirely different results.

INTRODUCTION

Here we consider systems formed of more than three quarks and/or antiquarks ($q^m \bar{q}^n$ with $m + n > 3$). These are a category of exotic hadrons. They are "exotic" with respect to "ordinary" hadrons which are either mesons ($q\bar{q}$) or baryons (q^3) systems. The existence or nonexistence of stable exotics against strong decays is crucial for understanding some aspects of the strong interactions. If discovered, their properties could be an important test for the validity of various quark models. Most theoretical and experimental effort has been devoted so far to systems described by the colour state $[222]_C$. These are the tetraquarks $q^2\bar{q}^2$ [1], the pentaquarks $q^4\bar{q}$ [2,3] and the hexaquarks q^6 [1]. They have a baryonic number B = 0, 1 and 2 respectively. Multibaryon systems with B up to 9 have also been studied within the SU(3) Skyrme model [4].

The most celebrated example of exotics is the H-particle (H- for hexaquark) predicted in the context of the MIT bag model more than 20 years ago [1]. It is a dibaryon with the flavor content of two Λ baryons, i. e. it contains light quarks only. Its existence remains controversial (see below).

From theoretical general arguments one expects an increase in stability of multiquark systems if they contain heavy flavours $Q = c$ or b. A recent review of the experimental efforts to search for the H-particle and charmed-strange pentaquarks can be found in [5]. In summary, no evidence for the production of a deeply bound H-particle has been observed, the production cross section being one order of magnitude than the theoretical estimates. By deeply bound state we understand a compact object, bound by about 80 MeV, as in Jaffe's picture [1]. However, a

---

[1] e-mail address: fstancu@ulg.ac.be.

molecular type structure, like that of the deuteron, is not excluded and there are some suggestive signals to be confirmed. Within the confidence level of the analyzed experiments, no convincing evidence for the production of pentaquarks with the flavour content $uuds\bar{c}$ and $udds\bar{c}$ has been observed either. However the existence of pentaquarks is not ruled out. The analysis done so far can provide a good starting point for future search in high statistics charm experiments at CERN [6] or Fermilab [7].

If experimentally discovered, the properties of multiquark systems would help to put constraints on phenomenological interquark forces. Indeed the theoretical predictions are model dependent. Here we are concerned with constituent quark models which simulate the low-energy limit of QCD and discuss theoretical predictions for compact objects. We compare results from constituent quark models where the spin-dependent term of the quark-quark interaction is described by the chromomagnetic part of the one gluon exchange (OGE) interaction [8] with results we obtained from a chiral model where the quarks interact via Goldstone boson exchange (GBE) [9,10], i.e. pseudoscalar mesons. In the latter model the hyperfine splitting in hadrons is due to the short-range part of the Goldstone boson exchange interaction between quarks, instead of the OGE interaction of conventional models. The GBE interaction is flavor dependent and its main merit is that it reproduces the correct ordering of positive and negative parity states in all parts of the considered spectrum. Moreover, the GBE interaction induces a strong short-range repulsion in the Λ-Λ system, which suggests that a deeply bound H-particle should not exist [11], in agreement with the high-sensitivity experiments at Brookhaven [12].

In the stability problem we are interested in the quantity

$$\Delta E = E(q^m \bar{q}^n) - E_T \qquad (1)$$

where $E(q^m \bar{q}^n)$ represents the multiquark energy and E_T is the lowest threshold energy for dissociation into two hadrons: two mesons for tetraquarks, a baryon + a meson for pentaquarks and two baryons for hexaquarks. In the right-hand side of (1) both terms are calculated within the same model. A negative ΔE suggests the possibility of a stable compact mutiquark system.

We first give a brief summary of the situation for tetraquarks, pentaquarks and hexaquarks. Next we outline the main characteristics of the GBE model, which is more recent and less known than the OGE model. In the last section we compare the results which we obtained in the frame of the GBE model with those from the literature, based on the OGE model.

THE TETRAQUARKS

The *light* tetraquarks are related to the study of meson-meson scattering as e.g. $\pi\pi$, $\pi\eta$, $\pi\eta'$, πK, etc. and to the identification of scalar mesons, i.e. mesons with quantum numbers $J^{PC} = 0^{++}$, having masses and decay properties which do not fit into a $q\bar{q}$ bound state. One expects a non-$q\bar{q}$ scalar component to play an important role in the mass range below 1800 MeV. The well observed isovector $a_0(980)$ and the isoscalar $f_0(980)$ mesons are interpreted as being $q^2\bar{q}^2$ states [1] or in a more realistic version as $K\overline{K}$ molecules [13] (see also [14]). It turns out that the $I=0$ states are the most complex both experimentally and theoretically [15]. Besides the $f_0(980)$ meson, there one has identified another three resonances namely $f_0(400-1200)$ or σ, $f_0(1370)$ and $f_0(1500)$. The $f_0(1370)$ is thought to be a $u\bar{u}+d\bar{d}$ state and the $f_0(1500)$ admits a glueball interpretation. The interpretation of σ remains open. A $q^2\bar{q}^2$ state is not excluded [16]. A recent reanalysis of the $\pi\pi$ scattering [17] reduces the interval of the σ mass to 400-800 MeV so that its central value returns to 600 MeV. The scalars $a_0(980)$, the $f_0(980)$, the $\sigma(600)$ and the $\kappa(900)$ (found in the analysis of πK scattering) are possible members of a scalar nonet (see e.g. [18]) and satisfy the Gell-Mann-Okubo mass formula [19].

The *heavy* tetraquarks have been studied in a variety of models and experimental search of double charmed tetraquarks ($cc\bar{u}\bar{d}$) are planned at CERN [6]. In the following let us denote by q a light quark u,d or s and by Q a heavy one c,b or t. Theoretical work has focussed on tetraquarks of type $QQ\bar{q}\bar{q}$ or equivalently $\overline{QQ}qq$ (see e.g. [20]). Note that one can also have $Q\overline{Q}q\bar{q}$ systems. These have two distinct thresholds $Q\overline{Q}+q\bar{q}$ and $Q\bar{q}+\overline{Q}q$. The latter is the same as for $\overline{QQ}qq$ and it can be shown [21] that $m_{Q\overline{Q}}+m_{q\bar{q}} \leq 2m_{Q\bar{q}}$ which means that $Q\overline{Q}+q\bar{q}$ is the lower threshold. Then assuming that the mass of $Q\overline{Q}q\bar{q}$ is the same as that of $QQ\bar{q}\bar{q}$, the latter has more chance to be bound. Also, it is more convenient for variational studies, where an upper bound is more conclusive about stability. Moreover $QQ\bar{q}\bar{q}$ has no meson-antimeson annihilation channels as $Q\overline{Q}q\bar{q}$ does have. As an example, detailed arguments in favour of the stability of $cc\bar{u}\bar{d}$ as compared to $c\bar{c}d\bar{u}$ are given in [6].

TABLE 1. Estimate of the size $R \sim (\alpha_s(m_Q)\,m_Q)^{-1}$ of a QQ pair

QQ	m_Q(GeV)	$\alpha_s(m_Q)$	R(fm)
cc	1.5	0.44	0.29
bb	5.0	0.28	0.14
tt	175.0	0.13	9.10^{-3}

The stability of $QQ\bar{q}\bar{q}$ relies on the fact that QQ brings a small kinetic energy into the system and forms a tightly bound pair of size $(\alpha_s(m_Q)m_Q)^{-1}$ (see Table 1). Then two heavy quarks act as an almost point-like heavy color antitriplet source. If Q is heavy enough, as it is the case of t, the short range Coulomb attraction plays an important role in the formation of the tetraquark system and leads to $tt\bar{q}\bar{q}$ stable states. The claim in Ref. [22] is that c and b are not heavy enough to enter such a mechanism. The alternative is the existence of weakly bound two heavy meson systems due to a potential determined at long distance by one pion exchange and calculable in chiral perturbation theory. One pion exchange meson-

meson interactions have also been discussed in Refs. [23] and [24].

Lattice gauge calculations became also recently available [25] and they may help in shedding more light into the intermeson potential and to isolate contributions of various mechanisms [26].

As mentioned above here we discuss results for stability looking at the quantity (1). We compare value of ΔE obtained in the literature from the OGE model [27] with those obtained in [28] from the GBE model. Ref. [28] considers only the most favourable configuration which is $\bar{3}3$ $S = 1, I = 0$. This means that QQ is in a $\bar{3}$ color state and $\bar{q}\bar{q}$ in a 3 color state. The mixing of the $6\bar{6}$ is neglected because one expects that this plays a negligible role in deeply bound heavy systems. Then the Pauli principle rquires $S_{12} = 1$ for QQ and $S_{34} = 0, I_{34} = 0$ for $\bar{q}\bar{q}$, if the relative angular momenta are zero for both subsystems. This gives a state of total spin $S = 1$ and isospin $I = 0$. In the channel of light quarks having $S^{light} = S_{34} = 0, I^{light} = I_{34} = 0$, as above, the lattice gauge calculations [25] produce a strong short-range attraction for $bb\bar{q}\bar{q}$, which is consistent with constituent quark model calculations [20,27,28].

THE HEXAQUARKS

We shall discuss hexaquarks before pentaquarks because they have been proposed first in historical order [1].

In the *light* sector the well known example is the H-particle which has been extensively studied in the literature. A recent and comprehensive review can be found in Ref. [29]. From the time it was proposed by Jaffe lots of theoretical studies have been performed within a variety of models as the bag model, the Skyrme model, constituent quark models, lattice calculations, QCD sum rules, etc. The results spread over a wide range of predictions depending on the model parameters and the approximations involved. In each model there are predictions for a bound state or for an unstable state. In the flavor singlet $uuddss$ system with $J^P = 0^+, I = 0$ the GBE model induces a strong repulsion of 847 MeV above the $\Lambda\Lambda$ threshold [11]. This implies that the H-particle should not exist as a compact object, in contrast to Jaffe's picture. A molecular type structure, as that of the deuteron, is not excluded however.

In the *heavy* sector attention has been focused on hexaquarks of type $uuddsQ$. These systems are like the H-particle where one of the s quarks has been replaced by a heavy one. Then when $Q = c$ the particle is denoted by H_c and when $Q = b$ the particle is denoted by H_b. In the context of a diquark model [30] the charmed hexaquark is found to be unstable but the bottom hexaquark is found to be stable by about 10 MeV. Calculations based on a chromomagnetic interaction give both H_c(I=0,J=3) and H_b(I=0,J=2 or 3) stable by 7.7 MeV up to 13.8 MeV [31]. In the GBE model both H_c and H_b turn out to be unstable, with about the same amount of repulsion above the respective thresholds, as for the H-particle [32]. This means that the heavy flavor has no effect on the stability in these cases.

THE PENTAQUARKS

The pentaquarks $P^0_{\bar{c}s} = uuds\bar{c}$ and $P^-_{\bar{c}s} = udds\bar{c}$ have been proposed as stable systems against strong decays nearly simultaneously in Refs. [2,3], about ten years after Jaffe's proposal for the H-particle. The more realistic calculations [33] which take into account the SU(3)-flavor breaking, etc. also lead to stability. In the OGE model the stable pentaquarks have negative parity (i.e. the parity of the antiquark) and require strangeness.

In the GBE model the best candidates to stability are not necessarily strange and have positive parity [34]. To understand these differences let us make the simplifying assumption that the heavy antiquark has an infinite mass. Then we need to care only about the light quarks, which we assume identical. The wave function of the light subsystem, containing radial, spin, flavor and color parts must be antisymmetric. The color part has necessarily the symmetry $[211]_C$ in order to form a color singlet together with the antiquark. Let us consider the spin $S = 0$ state of q^4. There are two ways to construct a totally antisymmetric state:

1) assume that the orbital part is symmetric i. e. has symmetry $[4]_O$. Then the flavor-spin part must have the symmetry $[31]_{FS}$. The inner product rules of the permutation group [35] require that the flavor part must be $[211]_F$. In the FS coupling this state reads

$$|1\rangle = \left([4]_O[211]_C[211]_{OC} \,;\, [211]_F[22]_S[31]_{FS}\right) \quad (2)$$

This state has $L = 0$, thus positive parity. Together with the antiquark this leads to a *negative* parity pentaquark. Obviously its flavor part $[211]_F$ requires strangeness. In the GBE model these pentaquarks are unbound [36]. In the CS coupling the state (2) has the same form but with C and F indices interchanged. It is the most favourable state in the OGE model [2,3], because it has the lowest allowed symmetry in the CS space.

2) assume that the flavor-spin part is symmetric. Then the Pauli principle requires that the orbital part should have the symmetry $[31]_O$. The spin state is $[22]_S$, as before, so that inner product rules require the symmetry $[22]_F$ for the flavor part in order to get $[4]_{FS}$. In the FS coupling this state reads

$$|2\rangle = \left([31]_O[211]_C[1^4]_{OC} \,;\, [22]_F[22]_S[4]_{FS}\right) \quad (3)$$

The lowest angular momentum associated to $[31]_O$ is $L = 1$ so that this state has negative parity and together with the antiquark gives a *positive* parity pentaquark. The $[22]_F$ symmetry indicates that strangeness is not required. The state $[4]_{FS}$ is the most favourable in the GBE model because it has the lowest symmetry in the FS space and, as will be shown in the next section, the GBE hyperfine interaction has a flavor-spin operator which, of course, takes the lowest expectation value for the most symmetric FS state.

The extension to the heavy flavor sectors of the Skyrmion approach [37] allowed to calculate the spectra of the lowest lying pentaquarks containing charm and bottom antiquarks [38]. Interestingly, the conclusions are similar to those of the GBE model: 1) the lowest pentaquarks have positive parity for any flavor content and 2) strangeness is not necessary in order to gain stability.

Finally, binding due to the long range one pion exchange has also been considered [39] and leads to a molecular type structure.

THE GBE MODEL

Here we refer to the GBE model as originally proposed by Glozman and Riska [9]. Its present status can be found in Ref. [40]. Besides the pseudoscalar meson exchange, both the vector and scalar meson exchanges are now incorporated in the model. The calculations presented here are based on the nonrelativistic version of Ref. [10].

The origin of the model lies in the spontaneous breaking of chiral symmetry in QCD which implies the existence of constituent quarks with a dynamical mass and Goldstone bosons (pseudoscalar mesons). Accordingly, it is assumed that the underlying dynamics in the low energy regime is due to Goldstone boson exchange between constituent quarks. In a nonrelativistic reduction for the quark spinors the quark meson vertex is proportional to $\vec{\sigma} \cdot \vec{q} \, \lambda^F$, with λ^F the Gell-Mann matrices, $\vec{\sigma}$ the Pauli matrices, and $\vec{q}$ the momentum of the meson. This generates a meson exchange interaction which is spin and flavor dependent. In the coordinate space the corresponding interaction potential contains two terms. One is the Yukawa potential tail and the other is a contact δ-interaction. When regularized, this generates the short range part of the quark-quark interaction. It is this short range part which dominates over the Yukawa part in the description of baryon spectra and leads to a corect order of positive and negative parity states both in nonstrange and strange baryons.

The dominant interaction is reinforced by the short-range part of the vector meson exchange (two correlated pions) [41].

The model is supported by the independent phenomenological analysis of the $L = 1$ baryons [42], by the $1/N_c$ expansion studies of the $L = 1$ nonstrange baryons [43] and by lattice studies [44].

TABLE 2. Results for ΔE for charmed tetraquarks, pentaquarks and hexaquarks. Each case corresponds to the most favourable I, J^P state.

System	Parity	ΔE for OGE[a]	ΔE for GBE[b]
$uu\bar{c}\bar{c}$	+	19 MeV (I=0,J=1) [27]	-185 Mev (I=0,J=1) [28].
$uuds\bar{c}$	-	-51 MeV (I=$1/2$,J=$1/2$) [33]	488 MeV (I=$1/2$,J=$1/2$) [36]
$uudd\bar{c}$	+	unbound	-75.6 MeV (I=0,J=$1/2$) [34]
$uuddsc$	+	-7.7 MeV (I=0,J=3) [31]	625 MeV (I=0,J=0 or 1) [32]

[a] In all cases a nonrelativistic Hamiltonian is used. It contains a linear confinement and a chromomagnetic spin-spin interaction .
[b] We use the nonrelativistic version of Ref. [10]

A schematic GBE model

In a schematic model the dominant GBE interaction takes the form

$$V_\chi = -C_\chi \sum_{i<j} \lambda_i^F \cdot \lambda_j^F \, \vec{\sigma}_i \cdot \vec{\sigma}_j \qquad (4)$$

with $C_\chi \cong 30$ MeV, determined from the Δ-N splitting [9]. It is useful to give an estimate of ΔE, based on the interaction (4). As an example, let us consider pentaquarks containing heavy flavor, for which equation (1) becomes

$$\Delta E = E(q^4\overline{Q}) - E(q^3) - E(q\overline{Q}) \qquad (5)$$

First we suppose that the confinement energy roughly cancels out in (5). Next, as in the previous section, we suppose that $m_Q \to \infty$. As a consequence, the quark-antiquark interaction can be neglected in the expectation value of (4) both for the pentaquark and the emitted heavy meson. Using the Casimir operator technique one finds that the contribution of V_χ to $E(q^3)$ is $-14\,C_\chi$. Now we have to distinguish the two cases introduced in the previous section.

Negative parity pentaquarks. In this case, in a harmonic oscillator basis, the contribution of the kinetic energy to ΔE is $3/4\,\hbar\omega$. This difference is exactly the kinetic energy associated to the extra degree of freedom in the pentaquark, corresponding to the relative motion between the q^3 and $q\overline{Q}$ subsystems. Using again the Casimir operator technique one finds that the state (2) leads to $\langle V_\chi \rangle = -16 C_\chi$. The separation energy becomes

$$\Delta E = 3/4\,\hbar\omega - (16-14)\,C_\chi = 128\ MeV. \qquad (6)$$

where the numerical value results from taking $\hbar\omega = 250$ MeV and $C_\chi = 30$ MeV [9]. The fact that ΔE is positive indicates that the GBE interaction leads to unbound negative parity pentaquarks of a compact type. This is confirmed by the more precise estimates [36] where ΔE obtained from a variational method is several hundred MeV for all strange or nonstrange pentaqurks under consideration, containing c or b antiquarks.

Positive parity pentaquarks. In this case the state (3) suggests that there is a unit of orbital excitation in the pentaquark due to the symmetry state [31]$_O$ of the q^4 subsystem. This leads to $\Delta K.E. = 5/4\,\hbar\omega$. But at the same time the contribution of the spin-flavor interaction becomes more attractive than for negative parity pentaquarks, giving $\langle V_\chi \rangle = -28 C_\chi$ due to the higher symmetry [4]$_{FS}$ present in (3). Then one has

$$\Delta E = 5/4\,\hbar\omega - (28-14)\,C_\chi = -100\ MeV. \qquad (7)$$

This proves that the attraction due to V_χ overcomes the excess in the kinetic energy due to the orbital excitation. This cannot happen for the OGE interaction which has a spin-color structure, thus is flavor-blind, and does not distinguish between the $[31]_{FS}$ and the $[4]_{FS}$ states. For this reason the positive parity pentaquarks are expected to be even more unbound than the negative parity ones. In particular the $uudd\bar{c}$ pentaquark will be unbound in any OGE model (see Table 2) inasmuch as the OGE interaction predicts unbound negative parity pentaquarks of the same flavor [33].

DISCUSSION

The main objective of this talk is to compare results for stability obtained from two constituent quark models: OGE and GBE. We illustrate the discussion with results for charmed multiquark systems as shown in Table 2. From this table and the results reported above for the H-particle one important conclusion can be drawn: when the GBE interaction stabilizes a system, the OGE interaction destabilizes it and vice-versa.

When the quark b is used instead of c the results are not so strickingly different but still show large differences in the predictions of the two models.

The challenging question of the existence of exotics remains unanswered so far. It is worthwhile to perform more elaborate calculations, based for example, on the resonating group method, in order to better understand the role played by various mechanisms and the dynamics of exotics. The experimental search would be of great help in putting constraints on the various effective quark-quark interactions.

Acknowledgements. I am most grateful to the organizing committee and in particular to Professor J. da Providencia for kindly inviting me to give this talk.

REFERENCES

1. Jaffe R. L., *Phys. Rev.* **D15**, 267 (1977).
2. Gignoux C., Silvestre-Brac B. and Richard J.-M., *Phys. Lett.* **B193**, 323 (1987).
3. Lipkin H. J., *Phys. Lett.* **B195**, 484 (1987); *Nucl. Phys.* **A625**, 207, (1997).
4. Scoccola N, these proceedings.
5. Ashery D., *Few-Body Systems Suppl.* **10**, 207 (1999).
6. Moinester M. A., *Z. Phys.* **A355**, 349 (1996); see also Baum G., et al., *COMPASS Collaboration* CERN-SPSLC-96-14, March 1996.
7. Aitala E. M., et al., *Phys. Rev. Lett.* **81**, 44 (1998).
8. de Rujula A., Georgi H. and Glashow L. S., *Phys. Rev.* **D12**, 147 (1975).
9. Glozman L. Ya. and Riska D. O., *Phys. Rep.* **268**, 263 (1996).
10. Glozman L. Ya., Papp Z. and Plessas W., *Phys. Lett.* **B381**, 311 (1996).
11. Stancu Fl., Pepin S., and Glozman L. Ya., *Phys. Rev.* **D57**, 4393 (1998).
12. Stotzer R. W., et al., *Phys. Rev. Lett.* **78**, 3646 (1997).
13. Weinstein J., and Isgur N., *Phys. Rev.* **D27**, 588 (1983); *Phys. Rev.* **D41**, 2236 (1990).

14. Brink D. M., and Stancu Fl., *Phys. Rev.* **D49**, 4665 (1994)
15. Review of Particle Physics, Particle Data Group, Spanier S., and Tornqvist N., *Eur. Phys. J.* **C3**, 390 (1998).
16. Schechter J., these proceedings; see also Achasov N. N., hep-ph/9904223.
17. Ishida S., hep-ph/9905260.
18. Black D., Fariborz A. H., Sannino F., and Schechter J., *Phys. Rev.* **D59**, 074026 (1999).
19. Ishida M., et al. hep-ph/9905261.
20. Brink D. M., and Stancu Fl., *Phys. Rev.* **D57**, 6778 (1998).
21. Richard J.-M., *Phys. Rev.* **A49**, 3573 (1994).
22. Manohar A. V., and Wise M. B., *Nucl. Phys.* **B399**, 17 (1993).
23. Tornqvist N. A., *Phys. Rev. Lett.* **67**, 556 (1991).
24. Ericson T. E. O., and Karl G., *Phys. Lett.* **B309**, 426 (1993).
25. Michael C., and Pennanen P., *Phys. Rev.* **D60**, 054012 (1999).
26. Barnes T., Black N., Dean D. J., and Swanson E. S., *Phys. Rev.* **C60**,045202 (1999).
27. Silvestre-Brac B., and Semay C., *Z. Phys.* **C57**, 273 (1993); **C59**, 457 (1993); **C61**, 271 (1994).
28. Pepin S., Stancu Fl., Genovese M., and Richard J.-M., *Phys. Lett.* **393**, 119 (1997).
29. Sakai T., Shimizu K., and Yazaki K., *Prog. Theor. Phys. Suppl.*, to be published.
30. Lichtenberg D., Roncaglia R., and Predazzi E., *J. Phys.* **G23**, 865 (1997).
31. Leandri J., and Silvestre-Brac B., *Phys. Rev.* **D47**, 5083 (1993).
32. Pepin S., and Stancu Fl., *Phys. Rev.* **D57**, 4475 (1998).
33. Leandri J., and Silvestre-Brac B., *Phys. Rev.* **D40**, 2340 (1989).
34. Stancu Fl., *Phys. Rev.* **D58**, 111501 (1998).
35. Stancu Fl., *Group Theory in Subnuclear Physics*, Oxford University Press, 1996, ch. 4 and 10.
36. Genovese M., Richard J.-M., Stancu Fl., and Pepin S., *Phys. Lett.* **B425**, 171 (1998).
37. Callan C. G., and Klebanov I., *Nucl. Phys.* **B202**, 365 (1985); Callan C. G., Hornbostel K., and Klebanov I., *Phys. Lett.* **B202**, 269 (1988).
38. Riska D. O., and Scoccola N. N., *Phys. Lett.* **B299**, 338 (1993).
39. Shmatikov M., *Phys. Lett.* **B349**, 411 (1995); *Nucl. Phys.*, **A612**, 449 (1997).
40. Glozman L. Ya., *Origins of the Baryon Spectrum*, hep-ph/9908423, to be published in *Nucl. Phys. A*.
41. Riska D. O., and G. E. Brown., *Nucl. Phys.* **A653**, 251 (1999).
42. Collins H., and Georgi H., *Phys. Rev.*, **D59**, 094010 (1999).
43. Carlson C. E., et al., *Phys. Rev.*, **D59**,114008 (1999); Carone C. D., hep-ph/9907412, to be published in *Nucl. Phys.*, A.
44. Liu K. F., et al., *Phys. Rev.*, **D59**, 112001 (1999).

Quantum Monte Carlo Studies of Relativistic Effects in ^{3}H and ^{4}He

A. Arriaga

Centro de Física Nuclear da Universidade de Lisboa
Departamento de Física da Faculdade de Ciências da Universidade de Lisboa

Abstract. Relativistic effects in ^{3}H and ^{4}He have been studied in the context of Relativistic Hamiltonian Dynamics, using Variational Monte Carlo Methods. Relativistic invariance is achieved through Poincaré group algebra, which introduces a boost interaction term defining the first relativistic effect considered. The second consists in the nonlocalities associated with the relativistic kinetic energy operator and with the relativistic one-pion exchange potential (OPEP). These nonlocalities tend to cancel, being the total effect on the binding energy attractive and very small, of the order of 1%. The dominant relativistic effect is due to the boost interaction, whose contribution is repulsive and of the order of 5%. The repulsive term of the nonrelativistic 3-body interaction has to be reduced by 37% so that the optimal triton binding energy is recovered, meaning that around 1/3 of this phenomenological term accounts for relativisitic effects. The changes induced on the wave functions of nuclei by these relativistic effetcs are very small and short ranged. Although the nonlocalities of OPEP, resulting in a reduction of 15%, are cancelled by other relativistic contributions, they may have significant effects on pion exchange currents in nuclei.

INTRODUCTION

At present QCD is vue as the fundamental theory to describe strong interactions. However, its nonperturbative character at low energies have been preventing its ability to quantitatively describe nuclear structure. This fact makes the development of effective calculable methods very important, for they may bring insights and estimates useful to progress the fundamental theory. Furthermore, due to quark confinement, the genuine QCD degrees of freedom, quarks and gluons, are not explicit at these regimes of energy, making effective hadronic degrees of freedom the appropriate to work with.

Conventional effective nonrelativistic models are formulated in the Schrödinger equation framework and consider an Hamiltonian containing a kinetic energy operator and effective two and three-body potentials. Due the remarkably small mass of the pion, the long range interaction between nucleons is believed to be mediated only by pion exchange. Therefore the two- and three-body interactions comprise asymptotic regions which are nonrelativistic limits of one- and two-pion exchanges diagrams respectively. Different models have been designed to describe their intermediate and short range parts, which are meant to account for effective representations of multiple meson exchange, relativistic effects and subnucleonic degrees of freedom. The values of the parameters of the NN interaction are adjusted through fits to the experimental data of the $A = 2$ system, both NN phase shifts and deuteron binding energy, and those of the NNN interaction to the experimental values of the binding energy of ^{3}H and density of nuclear matter. With these interactions, which exhibit an intricate spin-isospin structure, the Schrödinger equation is very hard to solve. Several thecniques have been developed, namely Faddeev-Yakubovsky [1], hyperspherical-harmonics [2] and Quantum Monte Carlo (QMC) methods [3,4]. While the first two are restricted to systems with $A \leq 4$, results for the binding energies of nuclei with $A \leq 8$ have been obtained by Pudliner et al. [5,6] with QMC methods, and they seem to exhibt a systematic underbinding for $A = 6, 7, 8$ [7].

The above discrepancy between theoretical and experimental results should be investigated, and one may start by asking the question whether a more explicit and correct treatment of relativistic effects allows to resolve it. Various relativistic approaches to light nuclei have been pursued lately, and they can be classified in two major categories: effective field theories and relativistic Hamiltonian dynamics. In the first category we refer to the Bethe-Salpeter equations for the two and three-body systems with a separable kernel [8], or with three-dimensional reductions such as minimal relativity [9] and the spectator formalism [10]. All these calculations have been restricted to nuclei with $A \leq 3$. Within the second category, QMC studies of relativistic effects in few-body nuclei have been carried out very recently [7], and in the following sections we discuss the results that have been obtained and draw some conclusions.

THE RELATIVISTIC HAMILTONIAN

The relativistic Hamiltonian dynamics is a formalism consistent with the requirements of special relativity and quantum mechanics [11], being therefore a linear theory satisfying Poincaré invariance. The dynamical equations are in Hamiltonian form which accommodate, in principle, either quark or hadronic degrees of freedom, and admit a large class of interactions [11].

The Poincaré group is the largest symmetry group of space-time transformations under which quantum mechanics matrix elements must be invariant. It is a Lie group with ten generators: the Hamiltonian for time translations, the three linear momentum operators for space translations, the three angular momentum operators for space rotations, and the three boost operators for the boosts transformations. These generators must satisfy a set of commutation relations [12], the group algebra, and cluster separabilty [13]. As Dirac pointed out [14], there are basically three ways of realizing the Poincaré algebra, depending upon the way the subsidiary condition, that relates time and space variables, is implemented: the point form, the light form and the instant form. In the present article we refer to the latter, where time is treated as a parameter and the physical conditions of the system are studied at each fixed value of time. With this choice all generators become independent of time and its conjugate variable, and the ones corresponding to space translations and rotations are interaction free [14]. Some advantages result: i) the orthogonality of the hamiltonian eigenstates is guaranted; ii) the probabilistic interpretation of wave functions is allowed; iii) the linear and angular momentum operators assume the conventional expressions; iv) the dynamical equation. the Schrödinger equation, becomes an Hamiltonian eigenvalue equation, allowing a resolution through methods developed in the nonrelativistic framework, namely QMC methods. On the other hand, the inclusion of retardation through energy dependent interactions is not possible, and this may be considered as a disadvantage.

Following the work of Bakamjian and Thomas [15] and Foldy [16], the relativistic Hamiltonian, a generator of the Poincaré group, can be written as [7]

$$H_R = \sum_i \left(\sqrt{m_i^2 + p_i^2} - m_i\right) + \sum_{i<j}[v_{ij} + \delta v_{ij}(\mathbf{P}_{ij})] + \sum_{i<j<k} V_{ijk}, \qquad (1)$$

where the first term is relativistic kinetic energy, v_{ij} the two-body potential in the Center of Mass (CM) frame of particles i and j (the frame in which $\mathbf{P}_{ij} = \mathbf{p}_i + \mathbf{p}_j = 0$) and V_{ijk} the three-body potential in CM frame of particles i, j and k ($\mathbf{P}_{ijk} = \mathbf{p}_i + \mathbf{p}_j + \mathbf{p}_k = 0$). Finally $\delta v_{ij}(\mathbf{P}_{ij})$ is the two-body boost interaction, depends on $\mathbf{P}_{ij}$ and is present by Poincaré invariance requirements. The boost interaction corresponding to the three-body force is neglected, as well as N body interactions with $N \geq 4$, which are both believed to be much smaller than the terms considered.

The boost interaction $\delta v_{ij}(\mathbf{P}_{ij})$ is determined from the CM potential v_{ij} through relativistic covariance [17,18]. It is expanded in powers of $P_{ij}^2/4m^2$ and, keeping only the leading term, is written as [12,19]

$$\delta v(\mathbf{P}) = -\frac{P^2}{8m^2}\tilde{v} + \frac{i}{8m^2}[\,\mathbf{P}\cdot\mathbf{r}\mathbf{P}\cdot\mathbf{p},\tilde{v}\,] + \frac{i}{8m^2}[\,(\boldsymbol{\sigma}_i - \boldsymbol{\sigma}_j)\times\mathbf{P}\cdot\mathbf{p},\tilde{v}\,], \qquad (2)$$

where $\mathbf{r}$ is the relative coordinate, $\mathbf{p} = (\mathbf{p}_i - \mathbf{p}_j)/2$ the relative momentum operator and $\sigma = 2\mathbf{S}$ the Pauli matrices for spin 1/2 particles. The subscripts ij of $v, \mathbf{P}, \mathbf{p}$ and $\mathbf{r}$ have been ommited for simplicity.

The various contributions of the boost interaction have been analysed and calculated for the three-and four-body system in Ref. [12]. The first and second terms of Eq. (2), denoted respectively by δv_{RE} and δv_{LC}, have classical origins in the relativistic energy-momentum relation and Lorentz contraction. The last term can be splitted in to two contributions, Thomas precession and quantum mechanics effects, which are denoted respectively by δv_{TP} and δv_{QM}. For example the contributions for the triton binding energy were found to be [12]: $\delta v_{RE} = 0.23(2)$MeV,

$\delta v_{LC} = 0.10(1)$MeV, $\delta v_{TP} = 0.016(2)$MeV and $\delta v_{QM} = -0.004(2)$MeV. Since the main contribution comes from the first two terms, the other have been neglected in the calculations of Ref. [7] that are going to discuss here.

The relation between δv and v, defined in Eq. (2), is determined by Poincaré invariance and is therefore model independent. All model dependence of H_R is contained in the v_{ij} and V_{ijk} potentials. The two-body interaction used is the isoscalar part of the Argonne v_{18} [20], with relativistic corrections in the asymptotic region. More precisely the interaction is expressed as:

$$v_{ij} = v_{ij}^{1\pi,R} + v_{ij}^{fen} \qquad (3)$$

where v_{ij}^{fen} stands for the intermediate and short range pieces and are phenomenological. $v_{ij}^{1\pi,R}$ denotes the relativistic OPEP, contains the relativistic corrections explained below, and replaces the static limit present in the nonrelativistic Argonne v_{18} interaction. In order to obtained a relativistic potential which is phase equivalent to the latter, the parameters of the phenomenological part had to be refitted to the data [7].

The OPEP is calculated from the one-pion-exchange diagram within effective pion-nucleon relativistic field theories. If this diagram is calculated on-the-energy-shell, then the OPEP is independent of the pseudoscalar or pseudovector nature of the $\pi - N$ coupling, and contains no energy dependence in the denominator of the pion propagator. Yet, as well known, in many-body calculations one needs this interaction off-the-energy-shell, and if energy conservation is released several models can be considered, which differ among themselves in the off-shell behavior and in the nature of the $\pi - N$ coupling. However, it has been shown that the different models, classified in Friar's notation with parameters $\tilde{\mu}$ and ν [21], are related by unitary transformations up to order p^2/m^2 [22,23]. The model of Ref. [7] corresponds to $\tilde{\mu} = 0$ and $\nu = 1/2$, is nonstatic, contains no energy dependence, thus no explicit retardation, and reads

$$v_{ij}^{1\pi,R}(\mathbf{p'},\mathbf{p}) = \frac{m}{\sqrt{m^2 + p'^2}} v_{ij}^{1\pi,NR}(\mathbf{q}) \frac{m}{\sqrt{m^2 + p^2}}, \qquad (4)$$

where $\mathbf{p}$ and $\mathbf{p'}$ are the initial and final momenta of nucleon i in the CM frame, $v_{ij}^{1\pi,NR}$ is the nonrelativistic OPEP and given by:

$$v_{ij}^{1\pi,NR}(\mathbf{q}) = -\frac{f_{\pi NN}^2}{\mu^2} \frac{\sigma_i \cdot \mathbf{q}\sigma_j \cdot \mathbf{q}\tau_i \cdot \tau_j}{\mu^2 + q^2}, \qquad (5)$$

with $f_{\pi NN}$ the pion–nucleon coupling constant, μ the pion mass and $\mathbf{q} = \mathbf{p} - \mathbf{p'}$ the momentum transfer.

Although calculations for $A = 2$ can be carried out in momentum space, for bigger systems the Monte Carlo calculations have to be performed in configuration space. In this space the nonrelativistic OPEP is a local potential, since $v^{1\pi,NR}$ depends only on $\mathbf{q}$. In contrast, $v^{1\pi,R}$ depends explicitly on p and p', and the Fourier transform yields a nonlocal relativistic OPEP in r-space. This Fourier transform is non trivial and requires an expansion of $v^{1\pi,R}$ in powers of $y = \mathbf{Q} \cdot \mathbf{q}^2/(m^2 + Q^2)$, where $\mathbf{Q} = (\mathbf{p} + \mathbf{p}')/2$. Exact momentum calculations for the deuteron show that this series converges very rapidly. In fact, the expectation value of the exact potential is $\langle v^{1\pi,R} \rangle = -18.797$ MeV, that of the first term (leading and independent of y) $\langle v^{1\pi,R}_{1st} \rangle = -18.644$ MeV, and that of the second term ($\propto y^2$) $\langle v^{1\pi,R}_{2nd} \rangle = -0.183$ MeV [7]. Keeping only the leading term the relativistic OPEP is approximated by

$$v^{1\pi,Rel}_{ij} = \frac{m^2}{m^2 + Q^2 + \frac{q^2}{4}} v_{\pi,NR}(\mathbf{q}). \tag{6}$$

With this new relativistic OPEP, the parameters of the phenomenological part are reffited to the data and, again, a phase equivalent potential to the Argonne v_{18} is obtained [7]. The corresponding r-space potential is given by [7]

$$v^{1\pi,Rel}_{ij}(\mathbf{r},r') = \frac{1}{3}\mu\frac{f^2_{\pi NN}}{4\pi} f(r') \left[F_{\sigma\tau}(r,r') \, \sigma_i \cdot \sigma_j + F_{t\tau}(r,r') \, S_{ij}(\hat{r},\hat{r}) \right] \tau_i \cdot \tau_j, \tag{7}$$

In the above equation we have:

$$F_{\sigma\tau}(r,r') = \frac{2}{\pi} \int q^2 dq \, \mathcal{Y}_\pi(q) \, j_0(qr) \, e^{-(\sqrt{m^2+q^2/4}-m)\,r'}, \tag{8}$$

$$F_{t\tau}(r,r') = \frac{2}{\pi} \int q^2 dq \, \mathcal{T}_\pi(q) \, j_2(qr) \, e^{-(\sqrt{m^2+q^2/4}-m)\,r'}, \tag{9}$$

$$f(r') = \frac{m^2}{4\pi} \frac{e^{-mr'}}{r'}, \tag{10}$$

$$S_{ij}(\hat{r},\hat{r}) = 3\,\sigma_i \cdot \hat{\mathbf{r}}\,\sigma_j \cdot \hat{\mathbf{r}} - \sigma_i \cdot \sigma_j, \tag{11}$$

$$\tag{12}$$

with

$$\mathcal{Y}_\pi(q) = \int Y_\pi(r) j_0(qr) r^2 dr, \tag{13}$$

$$\mathcal{T}_\pi(q) = \int T_\pi(r) j_2(qr) r^2 dr. \tag{14}$$

and finally

$$Y_\pi(r) = \frac{e^{-\mu r}}{\mu r}\left(1 - e^{-cr^2}\right), \tag{15}$$

$$T_\pi(r) = \left(1 + \frac{3}{\mu r} + \frac{3}{(\mu r)^2}\right)\left(1 - e^{-cr^2}\right)^2, \tag{16}$$

$$\tag{17}$$

where c is a cutoff parameter. In the limit $m \to \infty$, $f(r')$ becomes $\delta^3(\mathbf{r}')$, $F_{\sigma\tau}(r,r') \to Y_\pi(r)$ and $F_{t\tau}(r,r') \to T_\pi(r)$, and the nonrelativistic potential is recovered [20].

THE RELATIVISTIC WAVE FUNCTION

State of the art nonrelativistic variational wave functions have the form [24]:

$$|\Psi_v\rangle = \left(1 + \sum_{i<j<k} F_{ijk}\right)\left(\mathcal{S}\prod_{i<j} F_{ij}\right)|\Phi\rangle, \qquad (18)$$

where a symmetrized product of the two-body correlation operators, F_{ij}, and a sum of three-body correlation operators, F_{ijk}, act on the fully antisymmetric uncorrelated wave function Φ. These two- and three-body correlations induce the effects of the interactions on the wave function. F_{ij} involves products of operators, of the same type of the ones present in the interaction, and correlation functions satisfying Schrödinger-like two-body equations with appropriate boundary conditions. Their solutions comprise asymptotic parts which differ very much from nucleus to nucleus, and short range behaviors that are very similar to all nuclei [4].

In the case of $A = 2$ the correlation functions, in the two existing channels $^3S_1 - ^3D_1$, are simply proportional to the radial components of the wave function. They can be exactly obtained in momentum space for both nonrelativistic and relativistic Hamiltonians. However, for $A > 2$ the equations have to be solved in configuration space, and although this is easily done in the nonrelativistic case, the problem is much more difficult in relativistic situation. Therefore, a model for the relativistic pair correlation functions has been searched for, and a good approximation has been obtained through the method explained below [7].

As can be seen from Fig. 1, the nonrelativistic and relativistic radial wave functions for the deuteron are very similar. The small differences are very short ranged and since, as mentioned earlier, the correlation functions are almost idependent of A for small r, it can be expected that the relativistic corrections are very much alike for all nuclei. Considering only the most important channels, $^3S_1 - ^3D_1$ and $^1S_0$, the relativistic corrections are defined as

$$f^c_{0,1} = f^c_{0,1,NR}\,(1 + \lambda\,\xi_{^1S_0}), \qquad (19)$$
$$f^c_{1,0} = f^c_{1,0,NR}\,(1 + \lambda\,\xi_{^3S_1}), \qquad (20)$$
$$f^t_{1,0} = f^t_{1,0,NR}\,(1 + \lambda\,\xi_{^3D_1}), \qquad (21)$$

where λ is a variational parameter, to be determined by energy minimization for each nuclei, and ξ_c is defined for each channel c as:

$$\xi_c(r) = \frac{\phi_{c,R}(r) - \phi_{c,NR}(r)}{\phi_{c,NR}(r)}. \qquad (22)$$

The $\xi_{^3S_1}$ and $\xi_{^3D_1}$ can be exactly calculated using the relativistic and nonrelativistic deuteron radial wave functions for $\phi_{c,R}$ and $\phi_{c,NR}$. As anticipated they are very short ranged and are display in Fig. 2. A similar calculation for $\xi_{^1S_0}$ is not possible, because there is no bound state in this channel. Nevertheless, an artificial $^1S_0$

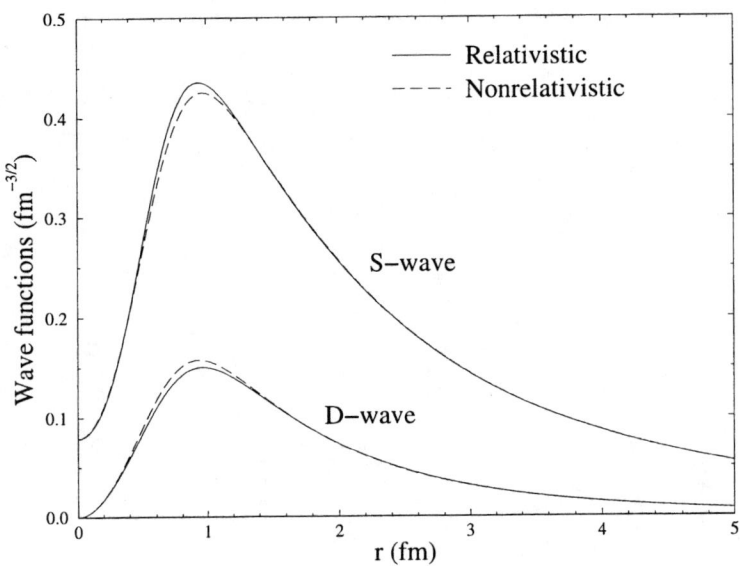

FIGURE 1. Deuteron radial wave functions.

bound state can be produced by slightly increasing the strength of intermediate range attraction in $v(^1S_0)$. Although the binding energy and the wave function for large r are very sensitive to small changes in $v(^1S_0)$, $\xi_{^1S_0}$ is relatively insentive as can be seen from the same figure, where this function is displayed for the cases corresponding to binding energies of 1 and 10 MeV.

RESULTS AND DISCUSSION

Variational Monte Carlo calculations for ^{3}H and ^{4}He were performed using techniques explained in detail in Ref. [7]. The results, reported in that reference, are here compiled in Table I, and correspond to the two-body interaction defined in the previous section and to the Urbana IX three-body interaction [5]. It is clear that the total relativistic effect on the binding energy is rather small, being ~0.3 MeV and ~1.8 MeV respectively for the triton and the α-particle.

The changes in $\langle T \rangle$ and $\langle v_{ij} \rangle$ cancel exactly, and by construction, in ^{2}H, since both H_{NR} and H_R are constrained to give the same deuteron binding energy, This cancellations seems to persist in ^{3}H and ^{4}He. In fact, the relativistic nonlocality corrections in the kinetic energy and in the two-body potential are -1.5(7) MeV and 1.3(7) MeV for the former, and 6(1) MeV and 6(1) MeV for the latter, leading to net attractive effects of $\sim -0.12 \pm 0.06$ MeV in and $\sim -0.17 \pm 0.10$ MeV respectively. This means that the main relativistic correction comes from the boost interaction, which is repulsive and 0.42 MeV for ^{3}H and 1.94 MeV for ^{4}He. On the other hand,

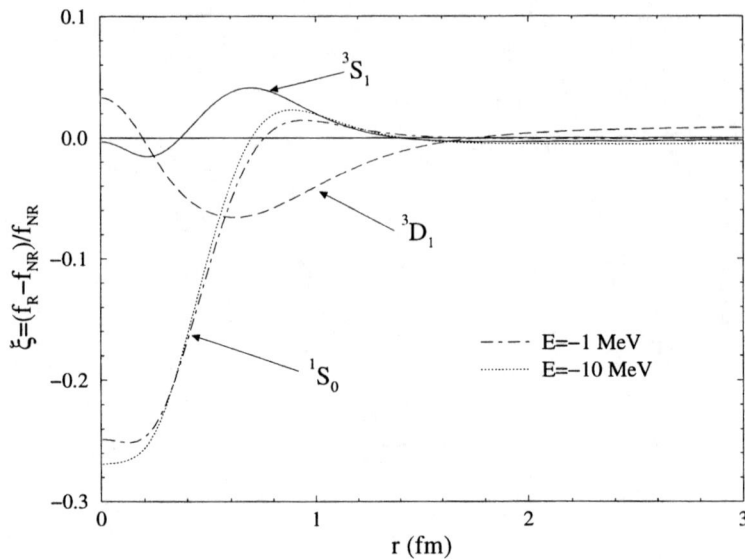

FIGURE 2. $\xi = \frac{f_R - f_{NR}}{f_{NR}}$ as a function of r.

TABLE 1. VMC results for ^{3}H and ^{4}He, calculated with 50,000 configurations. The results are in MeV.

	^{3}H			^{4}He		
	H_{NR}	H_R	$H_R - H_{NR}$	H_{NR}	H_R	$H_R - H_{NR}$
$\langle E \rangle^\dagger$	-8.24(3)	-7.94(4)	0.30(5)	-28.09(7)	-26.32(8)	1.8(1)
$\langle T \rangle$	50.1(5)	48.6(5)	-1.5(7)	104.8(9)	98.4(8)	-6(1)
$\langle \tilde{v}_{ij} \rangle$	-57.3(5)	-56.0(5)	1.3(7)	-127.6(9)	-121.5(9)	6(1)
$\langle \tilde{V}_{ijk} \rangle$	-1.06(3)	-1.03(3)		-5.29(9)	-5.20(8)	
$\langle \delta v_{ij} \rangle$		0.42(1)			1.94(3)	
$\langle \tilde{V}_{ijk}^R \rangle$	0.98(3)	1.01(3)		5.26(7)	5.38(8)	
$\langle \tilde{v}_\pi \rangle$	-44.0(2)	-38.3(2)	5.7(4)	-97.1(5)	-83.8(4)	13.3(1)

the changes induced on the wave functions of nuclei by these relativistic effetcs are very small and short ranged.

The parameters of the Urbana IX three-body interaction were adjusted in a nonrelativistic framework. In order to recover the optimal triton binding energy with the relativistic Hamiltonian, the repulsive term of this potential has to be reduced by 37%, meaning that around 1/3 of this phenomenological term accounts for relativisitic effects.

It is important to note that although the nonlocalities of OPEP, resulting in a reduction of around 15%, are cancelled by other relativistic contributions to the binding energy of the light nuclei considered, they may have significant effects in other situations, in particular on pion exchange currents in nuclei.

Finally, this relativistic approach to light nuclei should be pursued and applied to bigger systems, to try to answer the initial question whether a more explicit description of relativistic effects can resolve the discrepancy that appears to exist between theoretical and experimental binding energies.

REFERENCES

1. W. Glökle and H. Kamada, *Phys. Rev. Lett.* **71**, 971 (1993).
2. A. Kievsky, M. Viviani and S. Rosati, *Nucl. Phys.* **A551**, 241 (1993); **A577**, 511 (1994).
3. J. Carlson, in *Structure of Hadrons and Hadronic Matter* edited by O. Scholten and J. H. Koch, World Scientific, Singapore, 1991, p. 43
4. R. B. Wiringa, *Phys. Rev.* **C43**, 1585 (1991).
5. B. S. Pudliner, V. R. Pandharipande, J. Carlson and R. B. Wiringa, *Phys. Rev. Lett.* **74**, 4396 (1995).
6. B. S. Pudliner, V. R. Pandharipande, J. Carlson, Steven C. Pieper and R. B. Wiringa, *Phys. Rev.* **C56**, 1720 (1997).
7. J. L. Forest, V. R. Pandharipande and A. Arriaga, *Phys. Rev.* **C60**, 014002 (1999).
8. G. Rupp and J. A. Tjon, *Phys. Rev.* **C45**, 2133 (1992).
9. F. Sammaruca and R. Machleidt, *Few Body Syst.* **24**, 87 (1998).
10. Alfred Stadler and Franz Gross, *Phys. Rev. Lett.* **78**, 26 (1997).
11. B. D. Kiester and W. N. Polyzou, *Advances in Nucl. Phys.* **20**, 225 (1991).
12. J. L. Forest, V. R. Pandharipande and J. L. Friar, *Phys. Rev.* **C52**, 568 (1995).
13. F. Coester and W. N. Polyzou, *Phys. Rev.* **D26**, 1348 (1982).
14. P. A. M. Dirac *Rev. Mod. Phys.* **21**, 392 (1949).
15. B. Bakamjian and L. H. Thomas, *Phys. Rev.* **92**, 1300 (1952).
16. L. L. Foldy *Phys. Rev.* **122**, 275 (1961).
17. R. A. Krajcik and L. L. Foldy *Phys. Rev.* **D10**, 1777 (1974).
18. J. L. Friar *Phys. Rev.* **C12**, 695 (1975).
19. J. Carlson, V. R. Pandharipande and R. Schiavilla *Phys. Rev.* **C47**, 484 (1993).
20. R. B. Wiringa, V. G. J. Stoks and R. Schiavilla, *Phys. Rev.* **C51**, 38 (1995); **C54**, 646 (1996).
21. J. L. Friar, *Phys. Rev.* **C22**, 796 (1980).

22. J. Adam,Jr., H. Goller and H. Arenhovel, *Phys. Rev.* **C48**, 370 (1993).
23. J. L. Forest, preprint nucl-th/9905063
24. A. Arriaga, V. R. Pandharipande and R. B. Wiringa *Phys. Rev.* **C52**, 2362 (1995).

Computation of the nucleon properties in the chiral quark model[1]

Jürgen Baacke[2] and Hendrik Sprenger[3],[4]

Institut für Physik, Universität Dortmund, D - 44221 Dortmund, Germany

Abstract. Up to now the quark sea contributions to the energy and various physical observables in the chiral quark model have been formulated with the summation of levels. We will present a different technique for computing various physical observables, based on using the Euclidean Green functions and their K-spin partial wave reduction. In this framework it is not necessary to discretize the continuous spectrum by introducing a finite boundary. Besides this technical advantage such an alternative computation scheme makes it possible to obtain the numerical predictions of the model in an entirely independent way.

Using this formulation we perform a new self-consistent computation of the nucleon state in the Nambu-Jona-Lasinio model. We use the Pauli-Villars cutoff to define the model. Our results for the nucleon energy, the mesonic profile, and various observables essentially confirm results obtained by other groups using this regularization.

Similar techniques applied to Minkowskian Green functions can be used to compute the nucleon structure functions.

I INTRODUCTION

The Nambu-Jona-Lasinio (NJL) model [1] has received a wide attention during the last decade [2], following previous attempts [3,4] to describe the nucleon in the large-N_c limit on the basis of mesonic degrees of freedom. In the NJL model, the mesons appear as effective degrees of freedom, parametrizing condensates of the basic fermion fields. The basic model is a quark model with a four-fermion interaction, and therefore nonrenormalizable. The ultraviolet divergences are handled by introducing a cutoff which stays finite. Its functional form and numerical value, therefore, are relevant for the predictions of the model.

The use of the Pauli-Villars regularization opens the possibility to introduce a technique for computing the effective action which has been developed previously

---

[1] talk presented by H. Sp.
[2] Email address: baacke@physik.uni-dortmund.de
[3] Email address: sprenger@hal1.physik.uni-dortmund.de
[4] supported by the Graduiertenkolleg "Erzeugung und Zerfälle von Elementarteilchen"

[5] and has been applied to a various physical problems involving fluctuation determinants [6]. It is based on using the Euclidean Green functions instead of summation of levels. Computing functional determinants and other expectation values involving the quark continuum by summing over levels and Minkowski space wave functions requires the introduction of space boundaries in order to discretize the spectrum. The associated space cutoff has to be removed, introducing a numerical limiting procedure. This seems to be well under control. Nevertheless, level summation is certainly not a very economical technique. Thus the new technique can used for non-local cut-offs [7]. At the same time an alternative technique presents the possibility to obtain the predictions of the model in an independent way.

There are various methods to perform self-consistent computations [8]. As the equation of motion for the meson profiles requires to solve the equation of motion for the meson field, or, equivalently, the chiral angle, it is advantageous to be able to compute the functional derivative of the effective action with respect to the meson field. A technique for computing such derivatives using Euclidean Green functions has been set up recently [9]. It is the purpose of this work to transfer these techniques, **mutatis mutandis**, to the NJL model. Apart from the self-consistent computation of the mesonic profiles of the nucleon we also formulate the sea quark contributions to various other observables, such as the moment of inertia, in terms of Euclidean Green functions. This latter aspect of our work is of course important; without it, one would have to go back to level summation in computing these observables and our technique would loose much of its attractiveness.

The NJL model, as considered here, is not a unique theory; even within the restricted class of models in which only the $\pi-\sigma$ fields are taken into account, there is a wide variety. These models differ by the kind of regularization, as mentioned above, but also by applying it to various parts of the spectrum. In the NJL model it is used for the finite parts as well. Furthermore, it was usually only applied to the quark sea, and not to the valence contribution. We will use that version of the model recently used by Pobylitsa et al. [10] for computing the parton distributions. It only takes into account the π and σ fields, which are varied on the chiral circle only, and Paul-Villars regularization is applied to the quark sea, not to the filled bound state.

II THE MODEL

Starting with the NJL-Lagrangian [1]

$$\mathcal{L}_{\text{NJL}} = \bar{\psi}(i\gamma^\mu \partial_\mu - m)\psi + \frac{G}{2}\left[(\bar{\psi}\psi)^2 + (\bar{\psi}i\gamma_5\boldsymbol{\tau}\psi)^2\right] \quad (1)$$

one obtains after the standard bosonization procedure

$$S_{\text{NJL}} = \int d^4x \left\{ \bar{\psi}\left[i\gamma^\mu \partial_\mu - g(\sigma + i\boldsymbol{\pi}\cdot\boldsymbol{\tau}\gamma_5)\right]\psi - \frac{\mu^2}{2}\left(\sigma^2 + \boldsymbol{\pi}^2\right) + \frac{m\mu^2}{g}\sigma \right\}. \quad (2)$$

The parameters of the model are the fermion self-coupling G, the quark mass m, and the cutoff scale Λ. In the bosonized version these parameters appear as the quark-meson coupling g and the symmetry breaking mass parameter μ. They are related to the basic parameters as $g = \mu\sqrt{G}$ and $m\mu^2 = gf_\pi m_\pi^2$. The latter equation expresses μ in terms of physical constants and of the coupling g which remains a free parameter. A further relation is obtained from the gradient expansion of the effective action. The resulting kinetic term of the pion field is normalized correctly if the Pauli-Villars cutoff is fixed as $\Lambda = m_0\sqrt{\exp\left(\frac{4\pi^2}{N_c g^2}\right)}$. If the $\sigma - \pi$ field is varied only on the chiral circle, the second term in Eq. (2) is absent. For static pion fields the action is proportional to the time τ; the two remaining parts of the effective action then contribute to the energy as

$$E_{\text{fer}} = \frac{1}{\tau}\text{Tr}\log\left[-i\gamma^\mu\partial_\mu + \boldsymbol{m}(\boldsymbol{x})\right], \tag{3}$$

$$E_{\text{br}} = -m_\pi^2 f_\pi \int d^3x\,(\sigma - f_\pi). \tag{4}$$

Here the trace is taken over the quark sea and - in the case of baryons - over the filled bound states. $\boldsymbol{m}(\boldsymbol{x})$ is given - on the chiral circle - by $\boldsymbol{m}(\boldsymbol{x}) = m_0\bigl[\cos(\vartheta(r)) + i\gamma_5 \boldsymbol{\tau}\cdot\hat{\boldsymbol{x}}\sin(\vartheta(r))\bigr]$, where we have introduced the "dynamical quark mass" $m_0 = gf_\pi$ and used the hedgehog ansatz $\boldsymbol{\phi}(\boldsymbol{x}) = \hat{\boldsymbol{x}}\vartheta(r)$.

III BASIC RELATIONS

Given the profile $\vartheta(r)$, the energy of the corresponding nucleon state consists of the symmetry breaking part that can be evaluated trivially, and the contributions of valence and sea quarks. In order to evaluate the valence quark contribution we have to find the bound state energy by solving, with appropriate boundary conditions, the Dirac equation $(i\nu - H)\psi_0(\boldsymbol{x}) = 0$. Here we have introduced the Dirac Hamiltonian

$$H = -i\boldsymbol{\alpha}\cdot\boldsymbol{\nabla} + \gamma_0 \boldsymbol{m}(\boldsymbol{x}). \tag{5}$$

The computation of the quark sea contribution is more involved. We will recall here a method introduced previously [5] in which the computation of the zero point energy is related to the Euclidean Green function, that satisfies the equation

$$(i\nu - H)S_{\text{E}}(\boldsymbol{x},\boldsymbol{x}',\nu) = -\delta^3(\boldsymbol{x}-\boldsymbol{x}'). \tag{6}$$

The zero point energy E_{sea} can be computed as a contour integral around the positive imaginary axis in the complex ν-plane, see Fig. 1, as

$$E_{\text{sea}} = \int_{C_-} \frac{d\nu}{2\pi i}\nu\text{Tr}\int d^3x\,S_{\text{E}}(\boldsymbol{x},\boldsymbol{x},\nu). \tag{7}$$

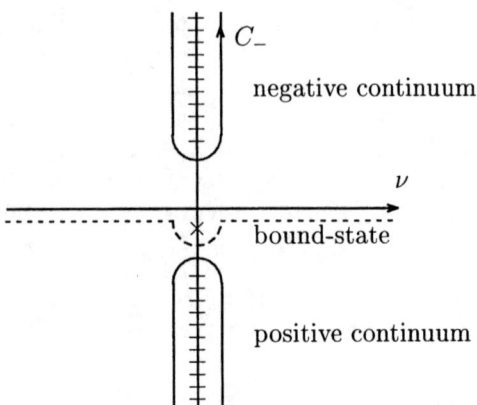

FIGURE 1. The complex ν-plane: Solid line: the contour C_- around the negative continuum states; dashed line: the deformed contour along the real ν axis, including, in addition, the bound state.

Deforming the contour to run along the real ν axis, and subtracting the zero point energy of the free Dirac operator $H_0 = -i\boldsymbol{\alpha}\cdot\boldsymbol{\nabla} + \gamma_0 m_0$, the integral takes the form

$$E_{\text{sea}} = -i \int_{-\infty}^{\infty} \frac{d\nu}{2\pi} \nu \text{Tr} \int d^3x \left[S_{\text{E}}(\boldsymbol{x},\boldsymbol{x},\nu) - S_{\text{E},0}(\boldsymbol{x},\boldsymbol{x},\nu)\right] . \tag{8}$$

The bound state can be included by deforming the contour to the one presented in Fig. 1 as a dashed line, however, we will consider the bound state separately here and below. It is convenient to introduce the bosonic Green function G_{E} via $S_{\text{E}} = (i\nu + H)G_{\text{E}}$ which satisfies

$$\left(\nu^2 + H^2\right) G_{\text{E}}(\boldsymbol{x},\boldsymbol{x}',\nu) = \left[\nu^2 - \Delta + m_0^2 + \mathcal{V}(\boldsymbol{x})\right] G_{\text{E}}(\boldsymbol{x},\boldsymbol{x}',\nu) = \delta^3(\boldsymbol{x}-\boldsymbol{x}') \tag{9}$$

with the potential or vertex operator $\mathcal{V}(\boldsymbol{x}) = i\boldsymbol{\gamma}\cdot\boldsymbol{\nabla}m(\boldsymbol{x})$. In terms of G_{E} the energy can be written as

$$E_0 = \int_0^\infty \frac{d\nu}{\pi} \nu^2 \int d^3x \text{Tr}\left[G_{\text{E}}(\boldsymbol{x},\boldsymbol{x},\nu) - G_{\text{E},0}(\boldsymbol{x},\boldsymbol{x},\nu)\right] . \tag{10}$$

The expressions for the zero point energy and the subsequent manipulations are formal. Even after subtracting the free zero point energy, they only make sense if properly regularized. The regularized zero point energy can be written as

$$E_{0,\text{reg}} = E_0(m_0) - \frac{m_0^2}{\Lambda^2} E_0(\Lambda) , \tag{11}$$

where the subtraction is to be understood to be done below the $\boldsymbol{x}$ and ν integrals, respectively (see below), before the partial wave summation and ν integration.

These prefactors have to be compensated in order to ensure the cancellation of the divergent integrals.

One obtains as an expression for the derivative of the zero point energy

$$\frac{\delta E_0^{(2)}}{\delta \phi_a(z)} = \text{tr} \sum_{K,P} \int d^3x \frac{\delta \mathcal{V}(x)}{\delta \phi_a(z)} \int_0^\infty \frac{d\nu}{2\pi} \overline{G_E^{(1)}}(x,x,\nu). \qquad (12)$$

The Euclidean Green function can be expanded [11] with respect to K-spin harmonics Ξ_n^{K,K_z} (for details see Appendix A or [12]). The radial Green functions can be written in terms of mode functions[5] $f_n^{\alpha+}(r)$ and $f_n^{\alpha-}(r)$ which are solutions regular at $r=0$ and as $r \to \infty$, respectively, of a system of radial differential equations given in Appendix A. Explicitly, they are given by

$$g_{mn}(r,r',\nu) = \kappa \left[\theta(r-r') f_m^{\alpha+}(r,\nu) f_n^{\alpha-}(r',\nu) + \theta(r'-r) f_m^{\alpha-}(r,\nu) f_n^{\alpha+}(r',\nu) \right]. \qquad (13)$$

The superscript α labels 4 linearly independent solutions.

In this basis, and using the reduced Green functions, the zero point energy takes the form as given in [5]. Analogously, the functional derivative of the energy can be written in terms of $g_{ij}^{(1)}$. Both expressions have to be regulated as implied by Eq. (11). The NJL-soliton is a system with baryon number equal to one. Therefore, one has to add the bound state part of the fermionic energy $E_0^{\text{comp}} = N_c E^{\text{bou}} + E_0^{(2)}$. Differentiating the bound state eigenvalue equation with respect to $\phi_a(z)$ and projecting with $\psi_0^\dagger$ one finds

$$\frac{\delta}{\delta \phi_a(z)} \omega_0 = \frac{1}{2\omega_0} \int d^3x \psi_0^\dagger(x) \frac{\delta \mathcal{V}^{0+}(x)}{\delta \phi_a(z)} \psi_0(x). \qquad (14)$$

The bound state energy is convergent. Thus it does not need to be regularized. Finally, the mesonic part of the energy and its derivative have to be evaluated.

IV PERTURBATIVE EXPANSION OF THE GREEN FUNCTION

After reduction of Eq. (9) to K-spin partial waves (see Appendix A) the differential equation for the partial wave Green functions $g_{mn}(r,r',\nu)$ becomes

$$\left[\delta_{nk} \left(\frac{d^2}{dr^2} + \frac{2}{r}\frac{d}{dr} - \frac{K_n(K_n+1)}{r^2} - \kappa^2 \right) - \mathcal{V}_{nk}(r) \right] g_{km}(r,r',\nu)$$
$$= -\delta_{nm} \delta(r-r')/r^2, \qquad (15)$$

---

[5] We omit the K-spin and parity superscripts here.

where $\kappa = \sqrt{\nu^2 + m_0^2}$ and where, again, we suppress the partial wave indices K and P. The potential $\mathcal{V}_{mn}$ depends on the K-spin, its explicit form is given in Appendix A. As already mentioned, we use for the numerical computation of the Green functions their standard expression (13) in terms of mode functions $f_n^{\alpha\pm}(r)$. The functions $f_n^{\alpha\pm}$ form 4×2 linearly independent systems (index $\alpha\pm$) of 4-component solutions (subscript n). A form independent of the choice of basis is given in [5,9]. Here we use a special convenient basis; it is defined by splitting off the free solutions, i.e. the modified Bessel functions $b_{K_n}^+(\kappa r) \equiv k_{K_n}(\kappa r)$ and $b_{K_n}^-(\kappa r) \equiv i_{K_n}(\kappa r)$ via

$$f_n^{\alpha\pm}(r) = \left[\delta_n^{\alpha\pm} + h_n^{\alpha\pm}(r)\right] b_{K_n}^{\pm}(\kappa r) \tag{16}$$

and by imposing the boundary condition $\lim_{r\to\infty} h_n^{\alpha\pm}(r) = 0$. The truncated mode-functions are obtained by solving the equations

$$\left[\frac{d^2}{dr^2} + 2\left(\frac{1}{r} + \kappa\frac{b_{K_n}'^{\pm}(\kappa r)}{b_{K_n}^{\pm}(\kappa r)}\right)\frac{d}{dr}\right] h_n^{\alpha\pm}(r) = \mathcal{V}_{nm}^K(r)\left[\delta_m^{\alpha} + h_m^{\alpha\pm}(r)\right]\frac{b_{K_m}^{\pm}(\kappa r)}{b_{K_n}^{\pm}(\kappa r)}. \tag{17}$$

This equation can be used for a perturbative expansion. Obviously, the functions $h_n^{\alpha\pm}(r)$ vanish to zeroth order in $\mathcal{V}$, so they are of order $\overline{(1)}$, which means of first and all higher orders in the potential. Once these solutions are known, the differential equation may be iterated to obtain the contribution of order $\overline{(2)}$. These radial Green functions are ready for being inserted into Eqs. (10) and (12).

V OBSERVABLES

One also has to write the observables in terms of Green Functions; without it, one would have to go back to level summation in computing these observables and our technique would loose much of its attractiveness. For the momentum of inertia one obtains

$$\left.\frac{\delta^2 E_0}{\delta\Omega_a\delta\Omega_b}\right|_{\Omega=0} = N_c \int_{-\infty}^{\infty}\frac{d\nu}{8\pi}\int d^3x \int d^3x' \text{tr}\left[\tau_a S_E^0(\boldsymbol{x},\boldsymbol{x}',\nu)\tau_b S_E^0(\boldsymbol{x}',\boldsymbol{x},\nu)\right],$$
$$+\frac{N_c}{2}\text{tr}\int d^3x d^3x' \psi_0(\boldsymbol{x})\psi_0^\dagger(\boldsymbol{x}')\tau_a S(\boldsymbol{x}',\boldsymbol{x},-iE^{\text{bou}})\tau_b. \tag{18}$$

The Euclidean Green function at any imaginary argument can again be related to the bosonic Green function via

$$S_E(\boldsymbol{x},\boldsymbol{x}',-iE^{\text{bou}}) = (H + E^{\text{bou}})G_E(\boldsymbol{x},\boldsymbol{x}',iE^{\text{bou}}). \tag{19}$$

and can be calculated with the same method.

The axial vector coupling constant is given [13,14] by the expectation value $g_A = \langle p\uparrow|\gamma_0\tau_3\gamma_3\gamma_5|p\uparrow\rangle$. It can be written as

$$g_A = -\frac{N_c}{3} \operatorname{tr} \int d^3x \, (\gamma_0\gamma_3\gamma_5\tau_3) \, \psi_0(\boldsymbol{x})\psi_0^\dagger(\boldsymbol{x})$$
$$-\frac{N_c}{3} \operatorname{tr} \int d^3x \int_{-\infty}^{\infty} \frac{d\nu}{2\pi} (\gamma_0\gamma_3\gamma_5\tau_3) \, S_E(\boldsymbol{x},\boldsymbol{x},\kappa) \,. \tag{20}$$

Pauli-Villars subtraction is implied. The quadratic radius is given by

$$\langle R^2 \rangle = \int_0^\infty r^4 dr \left\{ h^2(r) + j^2(r) \right\} \tag{21}$$
$$-3 \operatorname{tr} \int d^3x \int_0^\infty \frac{d\nu}{2\pi} r^2 S_E(\boldsymbol{x},\boldsymbol{x},\kappa) \,. \tag{22}$$

VI NUMERICS, RESULTS AND CONCLUSIONS

We have numerically implemented the expressions for the energy and its functional derivative presented in section 3 in the way described in [5] for the energy, and in [9] for its functional derivative.

The iteration proceeds as follows: For a given meson profile $\vartheta(r)$ one computes the mode functions and evaluates the functional derivative of the energy. One then requires the vanishing of the functional derivative. Extremizing the energy by requiring the functional derivative to vanish fixes $\vartheta(r)$. This profile is used as the input for the next iteration. The functions $h^{\alpha\pm}(r)$ have been computed in order $\overline{(1)}$ and $\overline{(2)}$ by solving (17) using a Runge-Kutta scheme. The accuracy of these solutions was checked by using the Wronskian relation.

The numerical results for the energy and other static parameters are presented in Table 1 and in Fig. 2.

We have presented here a self-consistent computation of the nucleon ground state in the Nambu-Jona-Lasinio model. In contrast to most previous calculations we have used a Pauli-Villars cutoff. It has been shown recently [15] that such a cutoff is favored by parton sum rules. Our results essentially confirm those of other groups, but this agreement is by no means guaranteed.

Besides presenting the analytical framework for the computation of self-consistent profiles, based on explicit expressions for the energy and its functional derivative, we have also derived explicit expressions for other observables.

Our numerical results are presented in Table 1. In Table 1 we also give some results obtained for $g = 4$ in Ref. [16], when using the same regularization. In view of the difference of the numerical approaches the agreement is very satisfactory. In Fig. 2 we plot the various parts of the energy. The nucleon mass is still too high, but lower than the one obtained with Schwinger proper-time cutoff.

In conclusion we have presented here a new approach for computing self-consistent meson profiles and static observables of the nucleon in the Nambu-Jona-Lasinio model, using a Pauli-Villars cutoff. In view of the fact that our numerical procedure is rather economical we think that it is worthwhile to pursue its application, e.g., to alternative versions of the model or to similar self-consistency problems.

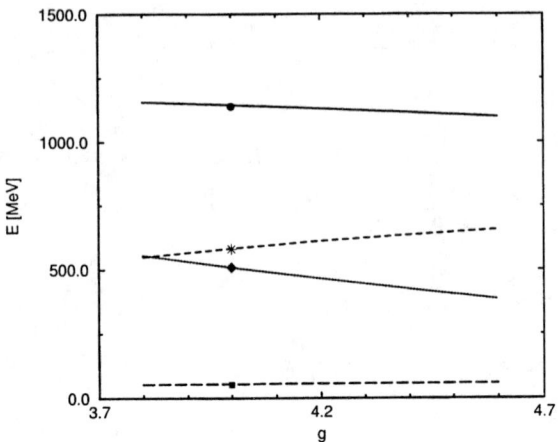

FIGURE 2. The energy of the nucleon. We display the energy as a function of g (solid line), the bound state energy (dotted line), the zero point energy (dashed line) and the energy of the symmetry breaking term (long dashed line). For comparison we also plot the total energy and its parts obtained previously by Döring et al. [5] for $g = 4$ (dot, square, diamond and cross).

g		3.8	4.0	4.2	4.4	4.6	4.0 [16]	Exp.
E_{bou}	[MeV]	555	508	464	424	386	507	
E_{con}	[MeV]	549	582	610	634	655	579	
Σ	[MeV]	53	55	57	58	58	52	45±9 [17]
E_{total}	[MeV]	1157	1145	1131	1116	1099	1139	
$1/\theta^{b-c}$	[MeV]	199	218	233	247	259		
$1/\theta^{c-c}$	[MeV]	1404	1264	1141	711	575		
$1/\theta_{total}$	[MeV]	174	186	193	183	178		
$M_\Delta - M_N$	[MeV]	261	279	290	274	267		294
M_N	[MeV]	1222	1214	1203	1184	1166		939
$\langle R^2 \rangle_{bou}$	[fm]	0.56	0.56	0.58	0.60	0.62		
$\langle R^2 \rangle_{con}$	[fm]	0.001	0.002	0.003	0.004	0.005		
$\langle R^2 \rangle_{total}$	[fm]	0.56	0.56	0.58	0.60	0.62	0.58	0.62
g_A^{bou}		0.72	0.71	0.71	0.70	0.70	0.71	
g_A^{con}		0.43	0.49	0.55	0.61	0.67		
g_A^{total}		1.15	1.20	1.25	1.31	1.37		1.23
$g_{\pi NN}$		11.6	11.7	11.8	11.8	11.8		13.1 [18]
g_A^{G-T}		1.15	1.16	1.16	1.16	1.17	0.96	1.23

TABLE 1. Static properties of the nucleon

VII MINKOWSKI SPACE GREEN FUNCTIONS

In order to calculate the nucleon structure functions one gets the following expression [15]

$$D = -\sum_{KP} \frac{N_C M_N}{4\pi^2} \int_M^\infty \frac{dk_0}{2\pi} \int_{|k_0 - x_{Bj} M_N|}^\infty dk\, k\, \mathrm{tr\,Im}\, s(k, k_3, k_0) \bigg|_{k_3 = x_{Bj} M_N - k_0}. \tag{23}$$

The function $\mathrm{Im}\, s$ is given by

$$\mathrm{tr\,Im}\, s(k, k_3, k_0) = \frac{2K+1}{4\pi} \mathrm{Im} \left[F_{11}^{KP} + F_{22}^{KP} + F_{33}^{KP} + F_{44}^{KP} \right.$$
$$\left. - \frac{k_3}{k} \left(F_{12}^{KP} + F_{21}^{KP} + F_{34}^{KP} + F_{43}^{KP} \right) \right], \tag{24}$$

where the Fourier-transformed radial fermionic Green functions can be written as

$$\mathrm{Im}\, F_{ij}^{KP}(k, \kappa) \sim \int_0^\infty dr\, r^2 \int_0^\infty dr'\, r'^2\, j_{K_i}(kr) j_{K_j}(kr') f_i^{\alpha-}(\kappa, r) f_j^{\alpha-}(\kappa, r'). \tag{25}$$

Here the functions $f_i^{\alpha-}$ are mode functions in Minkowski space, that can be decomposed as

$$f_n^{\alpha\pm}(r) = \left[\delta_n^{\alpha\pm} + c_n^{\alpha\pm}(r) \right] b_{K_n}^\pm(\kappa r), \tag{26}$$

where $b_{K_n}^\pm(\kappa r)$ are the Bessel and Hankel functions. The first part represents the S-Matrix. The $c^{\alpha\pm}$ are the solutions of the differential equations (17) by replacing the modified with the usual Bessel functions and $\kappa = \sqrt{k_0^2 - m_0^2}$. All further calculations are analogous to the Euclidian computations.

A PARTIAL WAVES AND RELATIONS OF THE SPINOR-ISOSPINORS

The expansion with respect to K-spin harmonics Ξ_{ij}

$$\psi_{K,K_z,P}(\boldsymbol{x}) = \begin{pmatrix} f_1^{K,P}(r) \Xi_1^{K,K_z}(\hat{\boldsymbol{x}}) + f_4^{K,P}(r) \Xi_4^{K,K_z}(\hat{\boldsymbol{x}}) \\ f_2^{K,P}(r) \Xi_2^{K,K_z}(\hat{\boldsymbol{x}}) + f_3^{K,P}(r) \Xi_3^{K,K_z}(\hat{\boldsymbol{x}}) \end{pmatrix} \text{ for parity}(-1)^{(K+1)} \tag{A1}$$

reduces the Dirac equation to radial equations for four coupled partial waves f_i. The Hamiltonian $H_{ij}^{KP} f_j = E f_i$ is squared in order to obtain a bosonic equation. One yields

$$\left(-\Delta_{ij}^K + m_0^2 \delta_{ij} + \mathcal{V}_{ij}^K \right) f_j = E^2 f_i, \tag{A2}$$

where

$$\Delta_{ij}^K = \delta_{ij}\frac{1}{r^2}\frac{d}{dr}r^2\frac{d}{dr} - \frac{K_i(K_i+1)}{r^2}. \tag{A3}$$

The notation K_i means $K_1 = K-1$, $K_2 = K_3 = K$ and $K_4 = K+1$. The potential for the parity $(-1)^{K+1}$ is given by

$$\mathcal{V}^K = \left\{ \begin{array}{cccc} c\left(S' + \frac{2KS}{r}\right) & C' & 0 & s\left(S' - \frac{S}{r}\right) \\ C' & c\left(-S' + \frac{2KS}{r}\right) & s\left(S' + \frac{S}{r}\right) & 0 \\ 0 & s\left(S' + \frac{S}{r}\right) & c\left(S' + \frac{2(K+1)S}{r}\right) & -C' \\ s\left(S' - \frac{S}{r}\right) & 0 & -C' & c\left(-S' + \frac{2(K+1)S}{r}\right) \end{array} \right\}, \tag{A4}$$

where S, C, S' and C' are functions of the profile $\vartheta(r)$ and s and c of the Grand-Spin K as given in [12]. For the parity $(-1)^K$ the sign of the mass has to be changed.

REFERENCES

1. Y. Nambu and G. Jona-Lasinio, Phys. Rev. **122**, 345 (1961); *ibid.* **124**, 246 (1961).
2. For recent reviews see: R. Alkofer, H. Reinhardt, and H. Weigel, Phys. Rept. **265**, 139 (1996); Christov et al., Prog. Part. Nucl. Phys. **A5**, 1 (1996).
3. T. Skyrme, Proc. Roy. Soc. Lond. **A260**, 17 (1961).
4. G. Adkins, C. Nappi, and E. Witten, Nucl. Phys. **B228**, 552 (1983).
5. J. Baacke, Z. Phys. **C53**, 402 (1992).
6. J. Baacke and S. Junker, Phys. Rev. **D49**, 2055 (1994).
7. B. Golli, W. Broniowski and G. Ripka, Phys. Lett. **B437** 24 (1998).
8. H. Reinhardt and R. Wünsch, Phys. Lett. **B215**, 577 (1988); T. Meissner, F. Grümmer, and K. Goeke, Phys. Lett. **B227**, 296 (1989).
9. J. Baacke and A. Sürig, Z. Phys. **C73**, 369 (1997).
10. P.V. Pobylitsa, M.V. Polyakov, K. Goeke, T. Watabe and C. Weiss, Phys. Rev. **D59** (1999) 034024.
11. S. Kahana and G. Ripka, Nucl. Phys. **A429**, 462 (1984).
12. J. Baacke and H. Sprenger, Phys. Rev. **D60** (1999) 054017.
13. M. Wakamatsu and H. Yoshiki, Nucl. Phys. **A524**, 561 (1991).
14. T. Meissner and K. Goeke, Nucl. Phys. **A524**, 719 (1991).
15. D. Diakonov, V. Petrov, P. Pobylitsa, M. Polyakov, and C. Weiss, Nucl. Phys. **B480**, 341 (1996).
16. F. Döring et al., Nucl. Phys. **A536**, 548 (1992).
17. J. Gasser, H. Leutwyler, and M. E. Sainio, Phys. Lett. **B253**, 252 (1991).
18. E. Matsinos, (1998), hep-ph/9807395.

Multibaryons in the Skyrme model

Norberto N. Scoccola

Physics Dept., Comisión Nac. Energía Atómica, Libertador 8250, (1429) Bs.As., Argentina
and
Universidad Favaloro, Solís 453, (1078) Buenos Aires, Argentina.

Abstract. Low-lying multibaryon configurations are studied within the bound state approach to the $SU(3)$ Skyrme model. We use approximate ansätze for the static background fields based on rational maps which have the same symmetries of the exact solutions. To determine the explicit form of the collective Hamiltonians and wave functions we only make use of these symmetries. Thus, the expressions obtained are also valid in the exact case. On the other hand, the meson bindings, inertia parameters and hyperfine splitting constants we calculate do depend on the detailed form of the ansätze and are, therefore, approximate. Using these values we compute the low-lying spectra of multibaryons with $B \leq 9$ and strangeness 0, -1 and $-B$. With these results the stability of some multilambda configurations is discussed.

I INTRODUCTION

In the last few years there have been several important developments in the determination of the lowest energy multiskyrmion configurations [1–3]. This type of solutions are essential for the understanding of multibaryons and, perhaps, nuclei in the framework of the topological chiral soliton models. So far, these models have proven to be useful for the description of quantities such as the masses, strong and electromagnetic properties of the octet and decuplet baryons, baryon-baryon interactions, etc. (see e.g. Refs. [4,5] and references therein). The knowledge of the properties of the multiskyrmion configurations opens the possibility of studying more complex baryonic objects. In fact, several investigations concerning non-strange multiskyrmion systems have been reported in the literature (see, e.g., Refs. [6–10]). Of particular interest are, however, the strange multibaryons. Perhaps the most celebrated example is the H dibaryon predicted in the context of the MIT bag model more than twenty years ago [11]. This exotic has been studied in various other models, including the Skyrme model [12–15], but its existence remains controversial both theoretically and experimentally. It has also been speculated that strange matter could be stable [16]. This has lead to numerous investigations of the properties of strange matter in bulk and in finite lumps (for a recent review see Ref. [17]). Moreover, with the new heavy ion colliders there is now the possibility of producing strange multibaryons in the laboratory [18]. In this situation the study of multibaryon systems within the $SU(3)$ Skyrme model appears to be very interesting. For general soliton configurations this is a quite hard numerical task since one has to deal with several coupled partial differential equations. However, the problem is greatly simplified if one introduces the (approximate) rational maps ansätze [19] for the multiskyrmion configurations. The construction of these ansätze is based on the analogy between BPS monopoles and skyrmions and requires that the approximate solutions have the same symmetries than the exact numerical ones. In fact, it is now known [1–3] that up to $B = 9$ these configurations are very symmetric. Namely, for $B = 2$ the solution corresponds to an axially symmetric torus while configurations with $B = 3 - 9$ possess the symmetries of the platonic polyhedra (e.g. tetrahedron for $B = 3$, etc) . In contrast with the exact solution, however, the rational map approximation assumes that the modulus of the static pionic field is radially symmetric while its direction depends only on the polar coordinates. In this contribution we will report on how to describe multibaryon states in the $SU(3)$ Skyrme model using these approximate ansätze.

II SYMMETRIC MULTISKYRMIONS AND RATIONAL MAPS

A rational map of order N is a map of $S^2 \to S^2$ of the form

$$R_N(z) = \frac{p(z)}{q(z)}$$

where p, q are polynomials of degree at most N in the stereographic coordinate $z = \tan(\theta/2)\,\exp(i\phi)$. It was shown by Donaldson [20] that there is a one-to-one correspondence between BPS monopoles of order k and rational maps of degree $N = k$. Using the analogy between this type of monopoles and the skyrmions, the authors of Ref. [19] proposed the following ansätze for the static soliton chiral field

$$U_N^{rat.}(\vec{r}) = \exp\left[i\vec{\tau}\cdot\hat{n}_N\;F(r)\right] \qquad (1)$$

where

$$\hat{n}_N = \left(\frac{2\Re(R_N)}{1+|R_N|^2},\frac{2\Im(R_N)}{1+|R_N|^2},\frac{1-|R_N|^2}{1+|R_N|^2}\right) \qquad (2)$$

Replacing Eq.(1) in the Skyrme model effective action

$$\Gamma_{eff} = \frac{f_\pi^2}{4}\int d^4x\;\mathrm{Tr}\;\partial_\mu U\partial^\mu U^\dagger + \frac{1}{32e^2}\int d^4x\;\mathrm{Tr}\left[U^\dagger\partial_\mu U, U^\dagger\partial_\nu U\right]^2 \qquad (3)$$

one gets the following expression for the soliton mass

$$M_{sol} = \frac{f_\pi^2}{2} \int d^3r \left[F'^2 + 2N \frac{\sin^2 F}{r^2} \left(1 + \frac{F'^2}{e^2 f_\pi^2} \right) + \frac{\mathcal{I}_N}{e^2 f_\pi^2} \frac{\sin^4 F}{r^2} \right] \quad (4)$$

where

$$\mathcal{I}_N = \frac{1}{4\pi} \int \frac{2i\, dz d\bar{z}}{(1+|z|^2)^2} \left(\frac{1+|z|^2}{1+|R_N|^2} \left| \frac{dR_N}{dz} \right| \right)^4 \quad (5)$$

To obtain the ansatz for a given baryon number $B = N$ one should proceed as follows. First, one constructs the most general map of degree N that has the symmetries of the exact solutions. Then, the resulting $\mathcal{I}_N$ has to be minimized with respect to the remaining free parameters. To perform the first step it is useful to recall that under a general $SO(3)$ transformation the stereographic coordinate z transforms as

$$z \to \frac{\alpha z + \beta}{-\bar{\beta} z + \bar{\alpha}} \quad (6)$$

where α, β are entries of the $J = 1/2$ representation of the corresponding rotation operator. We illustrate the method by considering the case $B = 2$. The most general map of degree $N = 2$ is

$$R_2 = \frac{\mu z^2 + \nu z + \lambda}{\delta z^2 + \gamma z + \xi} \quad (7)$$

If we impose the symmetries of the exact torus configuration (axial symmetry plus π rotations around the three cartesian axes) such general form reduces to

$$R_2 = \frac{z^2 - a}{-a z^2 + 1} \quad (8)$$

The value of a can be now determined by requiring that it should minimize M_{sol} (that is, $\mathcal{I}_2$). In this way one finds $a = 0$. Thus, the appropriate ansatz is

$$R_2 = z^2 \quad (9)$$

The explicit expressions of the rational maps corresponding to the other baryon numbers have been given in Ref. [19]. Once such maps are determined, the Euler-Lagrange equation for the soliton profile $F(r)$ can be numerically solved for each baryon number and the multiskyrmion masses M_{sol} evaluated.

The values of the soliton masses (per baryon number) for the different baryon numbers as calculated using the rational map ansätze are given in Table 1. For reference, the results corresponding to the skyrmion configurations which fully minimize the static energies [1–3] and the associate symmetry groups are also given. From this table one observes that the rational map ansätze indeed provide a very good approximation to the exact numerical solutions.

TABLE 1. Soliton mass per baryon number (in natural units $= 6\pi^2 f_\pi/e$) obtained by using the rational map ansatz (APPROX) as compared with the (EXACT) numerical minimization. The corresponding symmetry group G is also listed.

B	$\mathcal{I}$	APPROX	EXACT	G
1	1	1.232	1.232	$O(3)$
2	5.81	1.208	1.171	$D_{\infty,h}$
3	13.58	1.184	1.143	T_d
4	20.65	1.137	1.116	O_h
5	35.75	1.147	1.116	D_{2d}
6	50.76	1.137	1.109	D_{4d}
7	60.87	1.107	1.099	Y_h
8	85.63	1.118	1.100	D_{6d}
9	112.83	1.123	1.099	T_d

III STRANGE MULTIBARYONS

We turn now to the study of the strange multibaryons within the $SU(3)$ Skyrme model using the rational map ansätze described in the previous section. For this purpose, the effective action Eq.(3) has to be supplemented with the Wess-Zumino term and some suitable flavor symmetry breaking terms. In the calculations described below we have included terms that account for the different pseudoscalar meson masses and also for the difference between their decay constants.

To extend the model to $SU(3)$ flavor space we use the bound state approach, in which strange baryons appear as bound kaon-soliton systems [21]. Thus, we introduce a generalized Callan-Klebanov ansatz

$$U = \sqrt{U_N}\, U_K\, \sqrt{U_N} \qquad (10)$$

where U_N is the $SU(2)$ multiskyrmion field properly embedded into $SU(3)$ and U_K is the field that carries the strangeness. Its form is

$$U_K = \exp\left[i\frac{\sqrt{2}}{f_K}\begin{pmatrix} 0 & K \\ K^\dagger & 0 \end{pmatrix}\right] \qquad (11)$$

where K is the usual kaon isodoublet.

In the spirit of the bound state approach we consider first the problem of a kaon field in the background of a static multiskyrmion configuration. To describe such configuration we use the rational map ansatz approximation Eq.(1). Consequently, the ansatz for the kaon field should be

$$K = k_N(r,t)\, \vec{\tau}\cdot\hat{n}\, \chi \qquad (12)$$

where χ is a 1/2 spinor. Replacing Eqs.(10-12) in the effective action and performing the corresponding canonical transformations we obtain a quadratic Hamiltonian whose diagonalization leads to [22]

$$\left[-\frac{1}{r^2}\partial_r\left(r^2 h \partial_r\right) + m_K^2 + V - f\epsilon_N^2 - 2\lambda\epsilon_N\right] k(r) = 0 \qquad (13)$$

The radial functions f, h, λ and V depend on the baryon number B only through the integral $\mathcal{I}_N$. Their explicit expressions can be found in Ref. [22].

Eq.(13) has been solved numerically for different values of B using the values of Ref. [23] for f_π and e and setting m_K and f_K/f_π to their corresponding empirical values. The resulting eigenenergies are listed in Table 2. Also listed are the masses (per baryon number) of the corresponding $Y = 0$ states in the adiabatic approximation, $M_{Y=0}^{adiab}/B = M_{sol} + \epsilon$. These states are of particular interest since it has been claimed [11,24] that some of them can be stable against strong decays. As a general trend we see that the kaon binding energies $D_N^K = m_K - \epsilon_N$ decrease with increasing baryon number. However, as in the case of the energy required to liberate a single $B = 1$ skyrmion from the multisoliton background [2,3], we observe some deviation from a smooth behaviour, namely, $D_4^K > D_3^K$ and $D_7^K > D_6^K$. Consequently, such deviations will be also present in the multiskyrmion mass per baryon. Interestingly, this kind of phenomena has been also observed in some MIT bag model calculations [17]. There they are due to shell effects.

TABLE 2. Kaon eigenenergy ϵ_N and adiabatic approximation to the mass of the zero-hypercharge states (all in MeV) as a function of the baryon number B.

B	ϵ_N	$M_{Y=0}^{adiab.}/B$
1	222	1085
2	244	1091
3	255	1085
4	250	1047
5	263	1067
6	267	1064
7	262	1038
8	271	1055
9	276	1063

Using the values given in Table 2 we obtain

$$\begin{aligned} M_{2\Lambda} - 2M_\Lambda &= 12 \ MeV \\ M_{4\Lambda} - 2M_{2\Lambda} &= -176 \ MeV \\ M_{7\Lambda} - (M_{3\Lambda} - M_{4\Lambda}) &= -177 \ MeV \end{aligned} \qquad (14)$$

in the static soliton approximation (i.e. to $\mathcal{O}(N_c^0)$). These results seem to confirm previous speculations about the stability of the tetralambda in the Skyrme model

[24] and opens up the possibility of a stable heptalambda. On the other hand, they indicate that the H-particle, although very close to threshold, is not stable.

Within the static multiskyrmion approximation considered so far the spin and isospin quantum numbers of the bound kaon-multiskyrmion systems are not well defined. To recover good spin and isospin quantum numbers we proceed with the standard semi-classical collective quantization [23]. For $B > 1$, however, we should introduce independent spin and isospin rotations. The collective Lagrangian reads

$$L_{coll} = \tfrac{1}{2}\left[\Theta^J_{ab}\,\Omega_a\Omega_b + \Theta^I_{ab}\,\omega_a\omega_b + 2\,\Theta^M_{ab}\,\Omega_a\omega_b\right] - \left(c^J_{ab}\Omega_a + c^I_{ab}\omega_a\right)T_b \qquad (15)$$

Here, $\vec{\Omega}$ is the angular velocity corresponding to the spin rotation, $\vec{\omega}$ that of the isospin rotation and T_b is the kaon spin. Θ^J_{ab} and Θ^I_{ab} are the corresponding moments of inertia while Θ^M_{ab} is an inertia that mixes spin and isospin. The constants c^J_{ab} and c^I_{ab} are the hyperfine splitting constants which for $B = 1$ provide the Λ-Σ mass splitting. The explicit expressions of these inertia and hyperfine splitting tensors in terms of the soliton profile function $F(r)$ and the rational map $R_N(z)$ can be found in Ref. [25].

Using the standard definitions for the canonical conjugate momenta

$$\begin{aligned} J_a &= \tfrac{\partial L_{coll}}{\partial \Omega_a} = \Theta^J_{ab}\,\Omega_b + \Theta^M_{ab}\,\omega_b - c^J_{ab}\,T_b \\ I_a &= \tfrac{\partial L_{coll}}{\partial \omega_a} = \Theta^M_{ab}\,\Omega_b + \Theta^I_{ab}\,\omega_b - c^I_{ab}\,T_b \end{aligned} \qquad (16)$$

it is rather simply to find the general form of the collective Hamiltonian H_{coll}. Details are given in Ref. [25].

It is important to stress that the structure of the inertia and hyperfine splitting tensors appearing in Eq.(15) is strongly determined by the multiskyrmion symmetries. Using group theory arguments, it can be shown that (for symmetric skyrmions) such tensors are always diagonal. The number of independent diagonal entries, as well as whether the mixing inertias vanish or not, is also fixed by the properties of the corresponding symmetry group G. For example, for $B = 3$ the three components of the spin and isospin operators transform as the 3-dim irrep F_2 of the group T_d. Therefore, there is only one independent component for the spin inertia, one for isospin inertia and one for the mixing inertia. Similar analysis can be done for the hyperfine splittings. For $B = 4$, however, I_1, I_2 transform as the 2-dim irrep E_g of the group O_h while I_3 as the 1-dim irrep A_{2g} and the three components of $\vec{J}$ as the 3-dim irrep T_{1g}. Thus, for $B = 4$ we should have

$$\Theta^I_{11} = \Theta^I_{22} \neq \Theta^I_{33}\,; \qquad \Theta^J_{11} = \Theta^J_{22} = \Theta^J_{33}\,; \qquad \Theta^M_{aa} = 0 \qquad (17)$$

Finally, we have to determine the collective wave-functions. Their general form must be

$$|JJ_z, II_z, S\rangle = \sum_{J_3 I_3 T_3} \beta^{JIT}_{J_3 I_3 T_3}\, D^J_{J_z J_3}\, D^I_{I_z I_3}\, K^T_{T_3}$$

where $D^J_{J_z J_3}$ and $D^I_{I_z I_3}$ are $SU(2)$ Wigner functions and $\beta^{JIT}_{J_3 I_3 T_3}$ are some numerical coefficients that have to be fixed by requiring that these wave-functions transform as a 1-dim irrep of G. It is very important to notice that such irrep may not coincide with the trivial irrep. As well known when one performs an adiabatic symmetry operation on a skyrmion configuration one can pick a non-trivial phase. These are the so-called Filkenstein-Rubinstein phases. A detailed analysis of these phases for the configurations we are dealing with has been done by Irwin [10]. Using these phases one gets that, except for $B = 5, 6$, the wavefunctions should transform as the trivial irrep of G. For $B = 5$ they should transform as the A_2 irrep of D_{2d} and for $B = 6$ as the A_2 irrep of D_{4d}.

Having obtained the explicit form of the collective Hamiltonians and wavefunctions, the $\mathcal{O}(N_c^{-1})$ rotational contribution E_{rot} to the multiskyrmion masses can be calculated using first order perturbation theory. The numerical values of such contributions to the masses of the lowest lying non-strange baryons are given in Table 3 while those corresponding to the zero-hypercharge multibaryons are listed in Table 4.

TABLE 3. Quantum numbers and rotational energies of the lowest lying $S = 0$ states.

B	J^P	I	$E_{rot}[MeV]$
3	$1/2^+$	1/2	64
	$5/2^-$	1/2	147
4	0^+	0	0
	4^+	0	173
5	$1/2^+$	1/2	28
	$3/2^+$	1/2	40
6	1^+	0	7
	3^+	0	44
7	$7/2^+$	1/2	66
	$3/2^+$	3/2	98
8	0^+	0	0
	2^+	0	14
9	$1/2^+$	1/2	14
	$5/2^-$	1/2	30

From Table 3 we note that the quantum numbers of the ground states are consistent with those known for light nuclei with the exception of the odd values $B = 5, 7, 9$. We also observe that the lowest lying state has the lowest possible value of isospin and on the average mass splittings decrease for increasing baryon number. This is a consequence of the fact that, although all the moments of inertia increase with increasing baryon number, the increase of the spin inertia is much

TABLE 4. Quantum numbers and rotational energies of the lowest lying $Y = 0$ states.

B	J^P	I	$E_{rot}[MeV]$
3	$1/2^+$	1	50
	$3/2^-$	0	77
4	0^+	2	51
	0^+	0	72
5	$1/2^+$	1	29
	$1/2^-$	1	32
6	0^+	2	24
	0^-	1	26
7	$3/2^+$	2	32
	$5/2^+$	1	65
8	0^+	2	19
	2^+	2	31
9	$1/2^-$	2	25
	$3/2^-$	2	29

faster than that of the isospin one.

Using the values of the rotational corrections to the lowest lying $Y = I = 0$ states (some of which are listed in Table 4) one can see that stability of the 4Λ and the 7Λ is not affected by these corrections. For example, for the 4Λ there is a decrease of 36 MeV in the binding energy while that of the 7Λ is increased by 45 MeV.

IV   CONCLUSIONS

In this contribution we have reported on the description of multibaryons within the bound state approach to the $SU(3)$ Skyrme model. To describe the multiskyrmion backgrounds we have used ansätze based on rational maps. Such configurations are known to provide a good approximation to the exact numerical ones, and lead to a great simplification in the treatment of the kaon-soliton system. An important property of these approximate configurations is that they have the same symmetries as the exact ones. We have shown that the properties of the associated symmetry groups completely determine the explicit form of the collective Hamiltonians (namely, the detailed structure of the inertia and hyperfine splitting tensors). The same happens for the collective wavefunctions. In particular, we have shown how the Filkenstein-Rubinstein phases fix, in a unique way, the one dimensional irreducible representations as which each wave function should transform. Thus, the method to obtain the collective Hamiltonians and wave functions described here is also valid in the exact case. On the other hand, the numerical values of the meson bindings and of the independent inertia parameters and hyperfine splitting constants will depend on the detailed form of the ansätze and will be, therefore, approximate.

Using an effective action that provides a good description of the hyperon static properties we have studied the spectra of non-strange and strange multibaryons. In the case of non-strange baryons we found that, for even baryon number, the ground state quantum numbers coincide with those of known stable nuclei. It should be stressed, however, that in our opinion these quite compact multiskyrmion configurations should be interpreted as "multiquark bags" rather than normal nuclei. How these configurations are related with them is not yet clear. Another feature of the predicted spectra is that the low lying non-strange multibaryons always have the lowest possible value of isospin. This can be understood in terms of the behaviour of the inertia tensors as a function of the baryon number. The situation is more complicated in the case of strange particles for which there is a quite delicate interplay between the different terms contributing to the rotational energies. From the calculated spectra of strange multibaryon it results that some $Y = 0$ configurations could be stable against strong decays. Such configurations, usually called strangelets, are expected to be seen in RHIC [17,18].

Many of the ideas discussed in the present contribution can be extended to case of heavy flavor (e.g. charmed) multibaryons [26]. In such case, however, a proper treatment requires the use of an effective Lagrangian that accounts for both chiral symmetry and heavy quark symmetry. The present model has been also applied to the study of the binding of the η meson to few non-strange baryon systems [27].

We finish with a comment on the Casimir corrections to the multibaryon masses. Although these corrections are not expected to affect in any significant way the kaon eigenvalues and the rotational energies shown here, they might play some role in the determination of the multibaryon binding energies. Within the $SU(2)$ Skyrme model it has been shown [28] that they are responsible for the reduction of the otherwise large $B = 1$ soliton mass to a reasonable value when the empirical value of f_π is used. Here, we have avoided the $B = 1$ large mass problem by using the customary method of fitting f_π to reproduce the nucleon mass [23]. A more consistent approach should certainly use the empirical f_π and include the Casimir corrections. In this respect, there have been recently some efforts [29] to evaluate the corrections to the $B = 1$ mass in the $SU(3)$ Skyrme model. Unfortunately, even in the $SU(2)$ sector, almost nothing is known for $B > 1$. This is, of course, a very difficult task since it requires the knowledge of the meson excitation spectrum around the non-trivial multiskyrmion up to rather large energies.

ACKNOWLEDGEMENTS

The material presented here is based on work done with J.P. Garrahan and M. Schvellinger. Support provided by the grant PICT 03-00000-00133 from ANPCYT, Argentina is acknowledged. The author is fellow of the CONICET, Argentina. He would like to thank the members of the Organizing Committee for their warm hospitality during the workshop.

REFERENCES

1. V.B. Kopeliovich and B.E. Stern, JETP Lett. **45**, 203 (1987); J.J.M. Verbaarschot, Phys. Lett. **B195**, 235 (1987); N.S. Manton, Phys. Lett. **B192**, 177 (1987).
2. E. Braaten, S. Townsend and L. Carson, Phys. Lett. **B235**, 147 (1990).
3. R.A. Battye and P.M. Sutcliffe, Phys. Rev. Lett. **79**, 363 (1997).
4. I. Zahed and G.E. Brown, Phys. Rep. **142**, 1 (1986).
5. H. Weigel, Int J. Mod. Phys. **A11**, 2419 (1996).
6. E. Braaten and L. Carson, Phys. Rev. Lett. **56**, 1897 (1986).
7. L. Carson, Phys. Rev. Lett. **66**, 1406 (1991); Nucl. Phys. **A535**, 479 (1991).
8. T.S. Walhout, Nucl. Phys. **A531**, 596 (1991); Nucl. Phys. **A547**, 423 (1992).
9. N.R. Walet, Nucl. Phys. **A606**, 429 (1996).
10. P. Irwin, hep-th/9804142.
11. R.L. Jaffe, Phys. Rev. Lett. **38**, 195 (1977).
12. A.P. Balachandran et al, Phys. Rev. Lett. **52**, 887 (1984).
13. V.B. Kopeliovich, B. Schwesinger and B. Stern, Nucl. Phys. **A549**, 485 (1992).
14. J. Kunz and P.J. Mulders, Phys. Lett. **B215**, 449 (1988).
15. G.L. Thomas, N.N. Scoccola and A. Wirzba, Nucl. Phys. **A575**, 623 (1994).
16. E. Witten, Phys. Rev. **D30**, 272 (1984).
17. C. Greiner and J. Schaffner-Bielich, in *Heavy Elements and Related New Phenomena*, ed. R.K. Gupta and W. Greiner, (World Sci., Singapore); nucl-th/9801062.
18. K. Borer et al. (NA52 experiment), Phys. Rev. Lett. **72**, 1415 (1994); T.A. Armstrong et al. (E864 collaboration), Phys. Rev. Lett. **79**, 3612 (1997).
19. C.J. Houghton, N.S. Manton and P.M. Sutcliffe, Nucl. Phys. **B510**, 507 (1998).
20. S.K. Donalson, Com. Math. Phys. **96**, 387 (1984).
21. C. G. Callan and I. Klebanov, Nucl. Phys. **B262**, 365 (1985); N. N. Scoccola, H. Nadeau, M. Nowak and M. Rho, Phys. Lett. **B201**, 425 (1988); C. G. Callan, K. Hornbostel and I. Klebanov, Phys. Lett. **B202**, 269 (1988).
22. M. Schvellinger and N.N. Scoccola, Phys. Lett. **B430**, 32 (1998).
23. G.S. Adkins, C.R. Nappi and E. Witten, Nucl. Phys. **B228**, 552 (1983).
24. A.I. Issinskii, B.V. Kopeliovich and B.E. Shtern, Sov. J. Nucl. Phys. **48**, 133 (1988).
25. J.P. Garrahan, M. Schvellinger and N.N. Scoccola, Phys. Rev. **D**, in print (hep-ph/9906432).
26. B.V. Kopeliovich and W.J. Zakrzewski, JETP Lett **69**, 721 (1999); C.L. Schat and N.N. Scoccola, Phys. Rev. **D**, in print (hep-ph/9907271); B.V. Kopeliovich and W.J. Zakrzewski, hep-ph/9909365.
27. D.O. Riska and N.N. Scoccola, Phys. lett. **B444**, 21 (1998).
28. B. Moussallam, Ann. Phys. **225**, 264 (1993); F. Meier and H. Walliser, Phys.Rep. **289**, 383 (1997).
29. H. Walliser, Phys. Lett. **B432**, 15 (1998); N.N. Scoccola and H. Walliser, Phys. Rev. **D58**, 094037 (1998).

Form Factors of Baryons in a Confining and Covariant Diquark-Quark Model[1]

M. Oettel, S. Ahlig, R. Alkofer, and C. Fischer

Institute for Theoretical Physics, Tübingen University,
Auf der Morgenstelle 14, D-72076 Tübingen

Abstract. We treat baryons as bound states of scalar or axialvector diquarks and a constituent quark which interact through quark exchange. This description results as an approximation to the relativistic Faddeev equation for three quarks which yields an effective Bethe-Salpeter equation. Octet and decuplet masses and fully four-dimensional wave functions have been computed for two cases: assuming an essentially pointlike diquark on the one hand, and a diquark with internal structure on the other hand. Whereas the differences in the mass spectrum are fairly small, the nucleon electromagnetic form factors are greatly improved assuming a diquark with structure. First calculations to the pion-nucleon form factor also suggest improvements.

I MOTIVATION

Two approaches to the rich structure of strong interaction phenomena have been the topic of this workshop. The first one, effective theories like Chiral Perturbation Theory, resorts to including only physical fields with a suitable expansion parameter. The second approach, the building of effective models, often tries to interpolate between QCD and observable degrees of freedom by taking loans from the latter in terms of the assumed relevant degrees of freedom, such as (constituent) quarks. Different types of these models describe various aspects of baryon physics. Among them are nonrelativistic quark models, various sorts of bag models and approaches describing baryons by means of collective variables like topological or non-topological solitons [1]. Most of these models are designed to work in the low-energy region and generally do not match the calculations within perturbative QCD. Considering the great experimental progress in the medium energy range with momentum transfers between 1 and 5 GeV^2, there is a high demand for models describing baryon physics in this region that connects the low and high energy regimes.

---

[1] Supported by the BMBF (06–TU–888) and by the DFG (We 1254/4-1).

To describe this kind of physics, a fully covariant approach seems indispensable. Furthermore, the effects of quark confinement should be incorporated into a reliable description to avoid unphysical break-ups of baryons into their constituents. This is in sharp contrast to low-energy or static observables: baryon masses and magnetic moments, *e.g.*, can be understood in terms of a dynamically generated constituent quark mass through chiral symmetry breaking. Confinement plays seemingly an unimportant role.

The Nambu-Jona-Lasinio model in its various guises shows this feature of a dynamically generated quark mass and has thus been utilized to describe mesonic properties quite successfully [2]. The description of baryons within this model allows for two possibilities: They may appear as non-topological solitons [3,4] or as bound states of quark and diquark [5]. In ladder approximation, diquarks appear as poles in quark-quark scattering and therefore as physical particles. They are confined when going beyond ladder approximation [6]. A study which incorporates both, solitons and diquark-quark bound states [7], shows that the mesonic cloud and the quark-diquark interaction contribute about equally to the binding energy of the baryon.

On the other hand, the relativistic three-body problem can be simplified when discarding three-body irreducible interactions. The resulting Faddeev-type problem can be reduced further by assuming separable two-quark correlations which are usually called diquarks [8,9]. The Faddeev equations then collapse to a Bethe-Salpeter equation whose solutions describe the baryons. Quark and diquark hereby interact through quark exchange which restores full antisymmetry between the three quarks[2]. It is interesting to note that within the NJL model the two-quark correlations (or 4-point quark Green function) are separable in first order to yield a sum over poles of diquarks with different quantum numbers. In analogy to the meson spectrum[3], scalar and axialvector diquarks are assumed to be the lowest-lying and thus the most important particles. This line of approach has been taken in [9].

II THE MODEL

In the subsequent sections, we will follow this approach and derive an effective baryon Bethe-Salpeter equation with quark and diquark as constituents. However, to mimic confinement, we will avoid the diquark poles which would correspond to unphysical thresholds. To this end, consider the 4-point quark Green function in coordinate space,

$$G_{\alpha\beta\gamma\delta}(x_1, x_2, x_3, x_4) = \langle T(q_\gamma(x_3) q_\alpha(x_1) \bar{q}_\beta(x_2) \bar{q}_\delta(x_4)) \rangle , \qquad (1)$$

---

[2] Due to antisymmetry in the color indices and the related symmetrization of all other quantum numbers the Pauli principle leads to an attractive interaction in contrast to "Pauli repulsion" known in conventional few-fermion systems.

[3] Scalar diquarks correspond to pseudoscalar mesons and axialvector diquarks to vector mesons due to the intrinsically different parity of a fermion-antifermion pair compared to a fermion pair.

where α, β, γ, and δ denote the Dirac indices of the quarks. Assuming this 4-point function to be separable, we will parameterize scalar and axialvector diquark correlations as:

$$G^{\text{sep}}_{\alpha\gamma,\beta\delta}(p,q,P) := e^{-iPY}\int d^4X\, d^4y\, d^4z\; e^{iqz}e^{-ipy}e^{iPX} G^{\text{sep}}_{\alpha\beta\gamma\delta}(x_1,x_2,x_3,x_4) \quad (2)$$
$$= \chi_{\gamma\alpha}(p)\, D(P)\, \bar{\chi}_{\beta\delta}(q) + \chi^{\mu}_{\gamma\alpha}(p)\, D^{\mu\nu}(P)\, \bar{\chi}^{\nu}_{\beta\delta}(q),$$

P is the total momentum of the incoming and the outgoing quark-quark pair, p and q are the relative momenta between the quarks in these channels as y and z are the relative coordinates.

$\chi_{\alpha\beta}(p)$ and $\chi^{\mu}_{\alpha\beta}(p)$ are vertex functions of quarks with a scalar and an axialvector diquark, respectively. They belong to a $\bar{3}$-representation in color space and are flavor antisymmetric (scalar diquark) or flavor symmetric (axialvector diquark). For their Dirac structure we will retain the dominant contribution only, and a scalar function $P(p)$ which depends only on the relative momentum p between the quarks parameterizes the extension of the vertex in momentum space[4]:

$$\chi_{\alpha\beta}(p) = g_s(\gamma^5 C)_{\alpha\beta}\, P(p), \quad (3)$$
$$\chi^{\mu}_{\alpha\beta}(p) = g_a(\gamma^\mu C)_{\alpha\beta}\, P(p). \quad (4)$$

C denotes hereby the charge conjugation matrix and g_a and g_s are normalization constants at this stage. The choice

$$P(p) = 1 \quad (5)$$

corresponds to a point-like diquark whereas extended diquarks can be modeled as

$$P(p) = \left(\frac{\gamma^2}{\gamma^2 + p^2}\right)^n. \quad (6)$$

This specific form with $n=2$ or $n=4$ proved to be quite successful in describing electromagnetic properties of the nucleon when using scalar diquarks only [10].

To parameterize confinement, the propagators of scalar and axialvector diquark, appearing in eq. (2) as $D(P)$ and $D^{\mu\nu}(P)$, ought to be modified. Our chosen form,

$$D(p) = -\frac{1}{p^2 + m_{sc}^2}\left(1 - e^{-\left(1 + \frac{p^2}{m_{sc}^2}\right)}\right), \quad (7)$$

$$D^{\mu\nu}(p) = -\frac{\delta^{\mu\nu}}{p^2 + m_{ax}^2}\left(1 - e^{-\left(1 + \frac{p^2}{m_{ax}^2}\right)}\right), \quad (8)$$

removes the free particle poles at the cost of an essential singularity for time-like infinitely large momenta. The constituent quark propagator is modified likewise:

$$S(p) = \frac{i\slashed{p} - m_q}{p^2 + m_q^2}\left(1 - e^{-\left(1 + \frac{p^2}{m_q^2}\right)}\right). \quad (9)$$

---

[4] The Pauli principle requires then the relative momentum to be defined $p = \frac{1}{2}(p_\alpha - p_\beta)$, where p_α and p_β are the quark momenta [10].

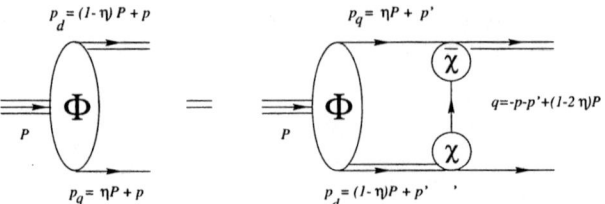

FIGURE 1. The baryon Bethe-Salpeter equation. The momentum partitioning parameter η distributes the relative momentum p' over quark and diquark.

With these ingredients, the Faddeev equations for the three quark system can be simplified enormously. To do this, one enters the Faddeev equations with an ansatz for the truncated, irreducible 3-quark correlation function (the 6-point quark Green function), which, *e.g.*, exhibits a pole from a spin-1/2 baryon:

$$G^{trunc}_{\alpha\beta\gamma,\delta\epsilon\zeta} \sim \frac{\Gamma_{\alpha\beta\gamma}(P;p,p_d,p_1)\,\bar{\Gamma}_{\delta\epsilon\zeta}(P;q,q_d,q_1)}{P^2+M^2}, \tag{10}$$

$$\Gamma_{\alpha\beta\gamma} = \chi_{\beta\gamma}(p_1)D(p_d)(\Phi^5(P,p)u)_\alpha + \chi^\mu_{\beta\gamma}(p_1)D^{\mu\nu}(p_d)(\Phi^\nu(P,p)u)_\alpha. \tag{11}$$

The flavor and color indices which have to be found after projection onto the baryon quantum numbers have been omitted here. The object of interest is now the nucleon vertex function $\Phi u = \begin{pmatrix} \Phi \\ \Phi^\mu \end{pmatrix} u$ (with u being a positive-energy Dirac spinor) which represents an effective spinor characterizing the scalar and the axialvector diquark correlations within the nucleon.

For this effective spinor, a coupled set of Bethe-Salpeter equations can be derived. Its complete derivation can be found in [9]. For spin-1/2 baryons in the flavor-symmetric case, the equation takes the form:

$$\begin{pmatrix} \Psi^5 \\ \Psi^{\mu'} \end{pmatrix}(p,P) := S(p_q) \begin{pmatrix} D & 0 \\ 0 & D^{\mu'\mu} \end{pmatrix}(p_d) \begin{pmatrix} \Phi^5 \\ \Phi^\mu \end{pmatrix}(p,P) \tag{12}$$

$$\begin{pmatrix} \Phi^5 \\ \Phi^\mu \end{pmatrix}(p,P) = \int \frac{d^4 p'}{(2\pi)^4} \frac{1}{2} \begin{pmatrix} -\chi S^T(q)\bar\chi & \sqrt{3}\chi^{\mu'}S^T(q)\bar\chi \\ \sqrt{3}\chi S^T(q)\bar\chi^\mu & \chi^{\mu'}S^T(q)\bar\chi^\mu \end{pmatrix} \begin{pmatrix} \Psi^5 \\ \Psi^{\mu'} \end{pmatrix}(p',P).$$

It is pictorially represented in Fig. 1. The attraction that leads to a bound state is the quark exchange between the two constituents. Note that we banned all unknown and possibly very complicated gluonic interactions between the quarks into the parameterization of the two-quark correlations. The quark exchange is a consequence of the structure of the Faddeev equations. The quark-diquark vertex from eqs. (3, 4) enters as the quark-diquark interaction vertex. This equation can be solved without any further approximation, especially without any non-relativistic reduction. First one decomposes the baryon vertex Φ (where each component is a 4×4-matrix) in Dirac space and projects onto positive parity and energy states. This procedure is described in detail in [11]. Choosing the rest frame of the nucleon, all independent components are regrouped as eigenstates of orbital angular momentum. As a final result, eight independent amplitudes, *i.e.* scalar functions

"non-relativistic" partial waves	$\begin{pmatrix} \chi \\ 0 \end{pmatrix}(\gamma_5 C)$	$\hat{P}^4 \begin{pmatrix} 0 \\ \chi \end{pmatrix}(\gamma^4 C)$	$\begin{pmatrix} i\sigma^i \chi \\ 0 \end{pmatrix}(\gamma^i C)$	$\begin{pmatrix} i\left(\hat{p}^i(\vec{\sigma}\hat{\vec{p}}) - \frac{\sigma^i}{3}\right)\chi \\ 0 \end{pmatrix}(\gamma^i C)$
spin	1/2	1/2	1/2	3/2
orbital angular momentum	s	s	s	d
"relativistic" partial waves	$\begin{pmatrix} 0 \\ \vec{\sigma}\hat{\vec{p}}\chi \end{pmatrix}(\gamma_5 C)$	$\hat{P}^4 \begin{pmatrix} (\vec{\sigma}\hat{\vec{p}})\chi \\ 0 \end{pmatrix}(\gamma^4 C)$	$\begin{pmatrix} 0 \\ i\sigma^i(\vec{\sigma}\hat{\vec{p}})\chi \end{pmatrix}(\gamma^i C)$	$\begin{pmatrix} 0 \\ i\left(\hat{p}^i - \frac{\sigma^i(\vec{\sigma}\hat{\vec{p}})}{3}\right)\chi \end{pmatrix}(\gamma^i C)$
spin	1/2	1/2	1/2	3/2
orbital angular momentum	p	p	p	p

TABLE 1. Components of the octet baryon vertex function with their respective spin and orbital angular momentum. $(\gamma_5 C)$ corresponds to scalar and $(\gamma^\mu C)$, $\mu = 1\ldots 4$, to axialvector diquark correlations. Note that the partial waves in the first row possess a non-relativistic limit.

which multiply the components, describe the spin-1/2 baryon as can be seen from Table 1. As the amplitudes still depend on two momenta (the relative momentum p and the total momentum P), an expansion in terms of Chebyshev polynomials for the variable $p \cdot P/(|p||P|)$ is performed. Thus the four-dimensional equation (12) can be reduced to a number of coupled one-dimensional integral equations [11,10] which we solved iteratively.

This procedure can be applied to spin-3/2 baryons as well [11]. Again eight independent amplitudes are found after spin and energy projection. Here, as a difference to spin-1/2 baryons, only one s partial wave exists which is found to dominate the expansion.

III RESULTS FOR OBSERVABLES

A Octet and Decuplet Masses

In our approach the strange quark constituent mass m_s is the only source of flavour symmetry breaking. Isospin is assumed to be conserved. The equations describing octet and decuplet baryons have been derived under the premises of flavour and spin conservation, i.e. only vertex function components with same spin and flavour content couple. Again the full set of equations can be found in [11]. The results for the cases of a pointlike diquark and an extended diquark are shown in Tab. 2. We chose scalar and axialvector diquark masses[5] to be equal and proportional to the sum of the two quark masses constituting the diquark. The proportionality constant is called ξ. The nucleon and the delta mass served as input

---

[5] The use of confining propagators renders the masses to be mere parameters which set the scale in the propagators, eqs. (7-9). They are of course unobservable.

	exp.	pointlike $P(p)=1$	extended diquark $P(p)=\left(\dfrac{\gamma^2}{\gamma^2+p^2}\right)^4$ $\gamma=0.5$ GeV
m_u (GeV)		0.5	0.56
m_s (GeV)		0.63	0.68
ξ		0.73	0.6
M_Λ (GeV)	1.116	1.133	1.098
M_Σ (GeV)	1.193	1.140	1.129
M_Ξ (GeV)	1.315	1.319	1.279
M_{Σ^*} (GeV)	1.384	1.380	1.396
M_{Ξ^*} (GeV)	1.530	1.516	1.572
M_Ω (GeV)	1.672	1.665	1.766

TABLE 2. Octet and decuplet masses.

to determine the normalization constants g_s and g_a appearing in eqs. (3,4). From the viewpoint of the effective quark-diquark theory, g_s and g_a reflect the coupling strengths in the two diquark channels.

As can be seen from the numbers, the mass splitting between octet and decuplet can be explained as a result of the relativistic dynamics only. In the case of extended diquarks, the splitting is even overestimated.

B Electromagnetic Form Factors

Calculation of observables within the Bethe-Salpeter framework proceeds along Mandelstam's formalism [12]. The two necessary ingredients are normalized nucleon-quark-diquark vertex functions and, in case of the electromagnetic form factors, the current operator. The vertex functions can be calculated as outlined in the previous section and their normalization is determined by the canonical normalization to the correct (fermionic) bound state residue, see, e.g., [13]. To this end, we define an object $G(p,p',P)$ involving the quark and diquark propagators and the exchange kernel appearing in the Bethe-Salpeter equation (12),

$$G(p,p',P) = (2\pi)^4\delta(p-p')S^{-1}(p_q)\begin{pmatrix} D^{-1} & 0 \\ 0 & (D^{\mu'\mu})^{-1} \end{pmatrix}(p_d) + \frac{1}{2}\begin{pmatrix} \chi S^T(q)\bar\chi & -\sqrt{3}\chi^{\mu'} S^T(q)\bar\chi \\ -\sqrt{3}\chi S^T(q)\bar\chi^\mu & -\chi^{\mu'} S^T(q)\bar\chi^\mu \end{pmatrix}. \tag{13}$$

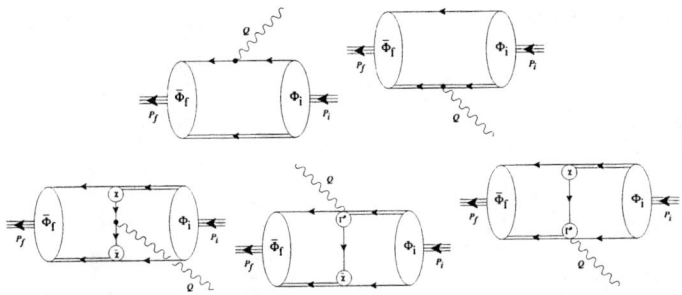

FIGURE 2. Diagrams that built up the baryon matrix elements of the electromagnetic current. The first row shows the diagrams of the impulse approximation, the second row the contributions of the exchange kernel.

With Λ^+ being the positive-energy projector, the normalization condition is:

$$-\int \frac{d^4 p}{(2\pi)^4} \int \frac{d^4 p'}{(2\pi)^4} \bar{\Psi}(p', P_n) \left[P^\mu \frac{\partial}{\partial P^\mu} G(p, p', P) \right]_{P=P_n} \Psi(p, P_n) \stackrel{!}{=} M\Lambda^+. \quad (14)$$

The current operator consists of the couplings of the photon to quark and diquark (impulse approximation) and to the exchange kernel G. For extended diquarks, it has been shown in [10] that the latter contribution encompasses two parts to make the total baryon current transversal and to reproduce the correct charge. These two parts are the interaction of the photon with the exchanged quark and its coupling to the diquark-quark vertex χ or χ^μ that can be described by a seagull like photon-quark-diquark vertex. In the case of pointlike diquarks, this seagull contribution vanishes.

To summarize, one has to calculate the diagrams given in Fig. 2. To ensure gauge invariance, the quark-photon and the diquark-photon vertices are of the Ball-Chiu type [14,15]. The seagull vertex is given by

$$\Gamma^\mu = e_\alpha \frac{4p^\mu - Q^\mu}{4pQ - Q^2} \left[\chi\left(p - \frac{Q}{2}\right) - \chi(p) \right] - \begin{pmatrix} \alpha \to \beta \\ Q \to -Q \end{pmatrix}. \quad (15)$$

An analogous relation is valid for the seagull involving the axialvector diquark vertex χ^ν. As before, p is the relative momentum between the two quarks, and e_α, e_β denote their respective charges.

We computed the Sachs form factors G_E and G_M for proton and neutron using the parameters given in Tab. 2. The results for the electric form factor are shown in Fig. 3. Clearly, the proton curve falls too weakly for a pointlike diquark which signals that the nucleon-quark-diquark vertex has too small a size in coordinate space. This is remedied by the introduction of the diquark structure. However, the neutron electric form factor seems to be quenched too strongly as compared to

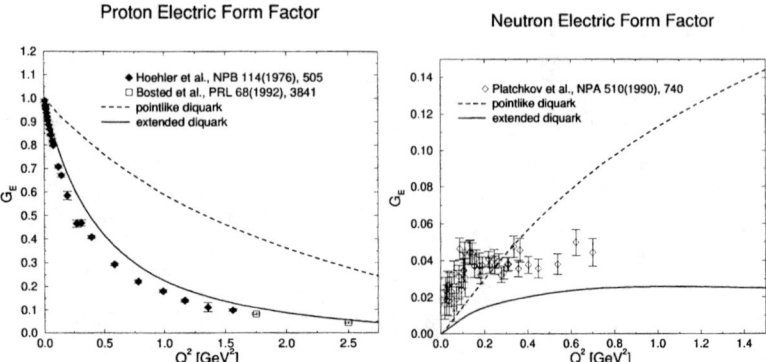

FIGURE 3. The electric form factor of proton and neutron.

the data[6]. Now this problem is probably due to overestimated axialvector diquark correlations within the nucleon. Retaining extended scalar diquarks only yields a very satisfactory description of the neutron G_E [10].

The nucleon magnetic moments have also improved with the introduction of the extended diquarks, see Fig. 4. Nevertheless, their absolute values are still about 13% too small in comparison with experiment although the ratio μ_p/μ_n is reproduced nicely. In our formalism, the diquarks have no anomalous magnetic moments since we do not properly resolve the diquark in the second impulse approximation diagram of Fig. 2. Performing Mandelstam's formalism for the diquark itself, *i.e.* coupling the photon to each of the quarks and letting them recombine to the diquark, would therefore certainly improve on the magnetic moments. In Fig. 4 we have also plotted separately the contributions of the impulse approximation and of the coupling to the exchange kernel. As the second contribution makes up more than 30 per cent of the total magnetic moment, the less involved impulse approximation is merely a rough guide to the behaviour of the magnetic form factor.

C Strong Form Factors

Among various strong processes that are candidates for closer scrutiny within our model, we have chosen first the pion-nucleon form factor $g_{\pi NN}(Q^2)$. Hereby we couple the pion to the quark only with its dominant Dirac amplitude $\sim \gamma_5$. This is certainly a good approximation as more detailed, microscopic calculations have shown [17]. The on-shell pion-quark vertex is dictated by PCAC and for the off-shell extrapolation we used a form proposed by ref. [18] and which has been

---

[6]) As has been pointed out in [16], these data should not be over-interpreted as systematic errors have been involved in extracting them from raw data. Nevertheless they give a feeling for the qualitative behaviour of the form factor.

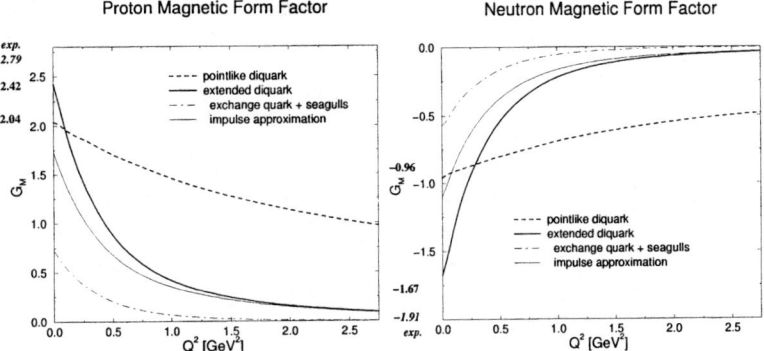

FIGURE 4. The magnetic form factor of proton and neutron.

applied in [15]. In our model, the diquark contributes nothing to $g_{\pi NN}$. This is a simple consequence of the Dirac algebra if one tries to couple the pion to each of the two quarks within the diquark. The results for the impulse approximation diagram only is shown in Fig. 5. Again, the fall-off in the case of the pointlike diquark is much slower than a monopole and appears to be unphysical. In contrast to this, $g_{\pi NN}$ for the extended diquark falls slightly stronger than a monopole with a width parameter of around 360 MeV. In the light of the results for the magnetic moments, the value of $g_{\pi NN}$ at $Q = 0$ may still be subject to sizeable corrections coming from the coupling to the exchange quark.

IV CONCLUSION

We have suggested a field theoretic model of baryons that makes use of diquarks which are a parameterization of the quark-quark correlations within baryons. Thereby we could retain full covariance. We parameterized confinement by a suitable modification of quark and diquark propagators to avoid unphysical thresholds.

Masses and four-dimensional vertex functions have been calculated for the baryon octet and decuplet. These vertex functions are the main ingredient for the calculation of observables such as the nucleon electromagnetic form factors. Whereas the mass spectrum is quite unsensitive to the extension of the diquarks, the form factors provide an effective mean to fix it. In these calculations gauge invariance was strictly maintained. However, the nucleon magnetic moments are still about 15 per cent to small. This we attribute to our incomplete handling of the electromagnetic structure of the diquark.

The computation of the pion-nucleon form factor is a necessary intermediate step to calculate production processes. As the pseudoscalar mesons do not couple to the diquarks, these processes are particularly transparent within the framework of our model. Additionally, a Λ hyperon in the final state renders the flavor algebra

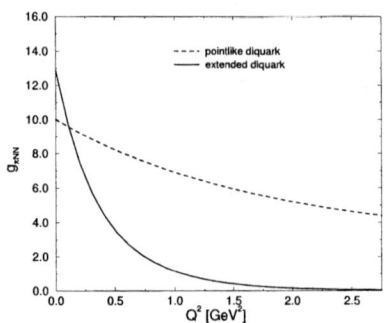

FIGURE 5. The strong form factor $g_{\pi NN}$.

simple, therefore we have chosen associated strangeness production ($pp \to pK\Lambda$) and kaon photoproduction ($\gamma p \to K\Lambda$) as further testing ground for our model [19].

Acknowledgement: M.O. thanks the organizers for the pleasant atmosphere at the workshop. The authors also want to express their gratitude to Hugo Reinhardt and Herbert Weigel for their support of this project.

REFERENCES

1. see e.g. Bhaduri, R., *Models of the Nucleon*, New York: Addison-Wesley, 1988.
2. see e.g. Ebert, D., et al., *Prog. Part. Nucl. Phys.* **33**, 1 (1994).
3. Alkofer, R., Reinhardt, H., and Weigel, H., *Phys. Rep.* **265**, 139 (1996).
4. Christov, C. V., et al., *Prog. Part. Nucl. Phys.* **37**, 1 (1996).
5. Reinhardt, H., *Phys. Lett. B* **244**, 316 (1990), see also Alkofer, R., and Reinhardt, H., *Chiral Quark Dynamics*, Heidelberg: Springer, 1995.
6. Hellstern, G., Alkofer, R., and Reinhardt, H., *Nucl. Phys. A* **625**, 697 (1997).
7. Zückert, U., et al., *Phys. Rev. C* **55** 2030, (1997).
8. Burden, C. J., Cahill, R. T., and Praschifka, J., *Aust. J. Phys.* **42**, 147 (1989).
9. Ishii, N., Bentz, W., and Yazaki, K., *Nucl. Phys. A* **51**, 617 (1995).
10. Oettel, M., Pichowsky, M. A., and von Smekal, L., *nucl-th/9909082*.
11. Oettel, M., et al., *Phys. Rev. C* **58** 2459, (1998).
12. Mandelstam, S., *Proc. Roy. Soc. A* **233**, 248 (1955).
13. Itzykson, C., and Zuber, B., *Quantum Field Theory*, New York: McGraw-Hill, 1985.
14. Ball, J. S., and Chiu, T.-W., *Phys. Rev. D* **22**, 2542 (1980).
15. Hellstern, G., et al., *Nucl. Phys. A* **627**, 679 (1997).
16. Passchier, I., et al., *Phys. Rev. Lett.* **82**, 4988 (1999).
17. see e.g. Tandy, P. C., *Prog. Part. Nucl. Phys.* **39**, 117 (1997).
18. Delbourgo, R., and Scadron, M. D., *J. Phys. G* **5**, 1621 (1979).
19. Alkofer, R., *Talk given at MENU99, to appear in the Proceedings*.

Nucleon-Nucleon interaction in a chiral constituent quark model

D. Bartz[1] and Fl. Stancu[2]

University of Liège, Institute of Physics, B.5, Sart Tilman, B-4000 Liège 1, Belgium

Abstract. We study the nucleon-nucleon (NN) problem as a six-quark system in a nonrelativistic chiral constituent quark model where the Hamiltonian contains a linear confinement and a pseudoscalar meson (Goldstone boson) exchange interaction between the quarks. This interaction has a long range Yukawa-type part, depending on the mass of the exchanged meson and a short range part, mainly responsible for the good description of the baryon spectra. We calculate the NN potential in the adiabatic approximation as a function of Z, the separation distance between the centres of the two three-quark clusters. The orbital part of the six-quark states is constructed either from the usual cluster model states or from molecular orbital single particle states. The latter are more realistic, having proper axially and reflectionally symmetries. In both cases the potential presents an important hard core at short distances, explained through the dominance of the $[51]_{FS}$ configuration. However in the molecular orbital basis the core is less repulsive, as a consequence of the fact that this basis gives a better upper bound for the energy of the six-quark system. We calculate the potential for the $^3S_1$ and $^3S_0$ channels with two different parametrizations. We find a small (few MeV) attractive pocket for one of these parametrizations. A middle range attraction is simulated by the addition of a σ-meson exchange interaction between quarks, of a form similar to that of the pseudoscalar meson exchange. The present study is an intermediate, useful step towards dynamical calculations based on the resonating group method.

INTRODUCTION

There have been many attempts to study the nucleon-nucleon interaction starting from a system of six interacting quarks described by a constituent quark model. These models explain the short range repulsion as due to the colour magnetic part of the one gluon exchange (OGE) interaction between quarks and due to quark interchanges between two $3q$ clusters [1,2]. To the OGE interaction it was necessary to add a scalar and a pseudoscalar meson exchange interaction between quarks of different $3q$ clusters in order to explain the intermediate- and long-range attraction between two nucleons [3–5].

---

[1] e-mail: d.bartz@ulg.ac.be
[2] e-mail: fstancu@ulg.ac.be

In [6] we have calculated the nucleon-nucleon (NN) interaction potential at zero-separation distance between two three-quark clusters in the frame of a constituent quark model [7–9] where the quarks interact via pseudoscalar meson i.e. Goldstone boson exchange (GBE) instead of OGE. An important motivation in using the GBE model is that it describes well the baryon spectra. In particular, it correctly reproduces the order of positive and negative parity states both for nonstrange [8] and strange [9] baryons where the OGE model has failed.

The underlying symmetry of the GBE model is related to the flavour-spin $SU_F(3) \times SU_S(2)$ group. Combining it with the S_3 symmetry, a thorough analysis performed for the $L = 1$ baryons [10] has shown that the chiral quark picture leads to more satisfactory fits to the observed baryon spectrum than the OGE models.

The one-pion exchange potential between quarks appears naturally as an iteration of the instanton induced interaction in the t-channel [11]. The meson exchange picture is also supported by explicit QCD latice calculations [12].

Another motivation in using the GBE model is that the exchange interaction contains the basic ingredients required by the NN problem. Its long-range part, required to provide the long-range NN interaction, is a Yukawa-type potential depending on the mass of the exchange meson. Its short-range part, of opposite sign to the long-range one, is mainly responsible for the good description of the baryon spectra [7–9] and also induces a short-range repulsion in the NN system, both in the $^3S_1$ and $^1S_0$ channels [13]. This study is an extention of [6] and we calculate here the interaction potential between two $3q$ clusters as a function of Z, the separation distance between the centres of the clusters. This separation distance is a good approximation of the Jacobi relative coordinate between the two clusters. Under this assumption, here we calculate the interaction potential in the adiabatic (Born-Oppenheimer) approximation, as explained below.

A common issue in solving the NN problem is the construction of adequate six-quark basis states. The usual choice is a cluster model basis [1,2,14]. In calculating the potential at zero-separation distance, in Ref. [6] we used molecular-type orbitals [15] and compared the results with those based on cluster model single-particle states. The molecular orbitals have the proper axially and reflectionally symmetries and can be constructed from appropriate combinations of two-centre Gaussians. By using molecular orbitals, in Ref. [6] we found that the height of the repulsion reduces by about 22 % and 25 % in the $^3S_1$ and $^1S_0$ channels respectively with respect to cluster model results. It is therefore useful to analyse the role of molecular orbitals at distances $Z \neq 0$. By construction, at $Z \to \infty$ the molecular orbital states are simple parity conserving linear combinations of cluster model states. Their role is expected to be important at short range at least. They also have the advantage of forming an orthogonal and complete basis while the cluster model (two-centre) states are not orthogonal and are overcomplete. For this reason we found that in practice they are more convenient to be used than the cluster model basis, where one must carefully [14] consider the limit $Z \to 0$. Here too, for the purpose of comparison we perform calculations both in the cluster and the molecular model.

In the following section we recall the procedure of constructing molecular orbital single-particle states starting from the two-centre Gaussians used in the cluster model calculations. Then the GBE Hamiltonian is presented. The subsequent section is devoted to results obtained for the NN potential. Next we introduce a middle range attraction through a scalar meson exchange interaction between quarks parametrized consistently with the pseudoscalar meson exchange. The last section is devoted to a summary and conclusions.

SINGLE-PARTICLE ORBITALS

In the cluster model one can define states which in the limit of large intercluster separation Z are right R and left L states

$$R = \psi\left(\vec{r} - \frac{\vec{Z}}{2}\right) \quad \text{and} \quad L = \psi\left(\vec{r} + \frac{\vec{Z}}{2}\right). \tag{1}$$

In the simplest cluster model basis these are ground state harmonic oscillator wave functions centered at $Z/2$ and $-Z/2$ respectively. They contain a parameter β which is fixed variationally to minimize the nucleon mass described as a $3q$ cluster within a given Hamiltonian. The states (1) are normalized but are not orthogonal at finite Z. They have good parity about their centers but not about their common center $\vec{r} = 0$.

From R and L one constructs six-quark states of given orbital symmetry $[f]_O$. The totally antisymmetric six-quark states also contain a flavour-spin part of symmetry $[f]_{FS}$ and a colour part of symmetry $[222]_C$. In the cluster model the most important basis states [13] for the Hamiltonian described in the following section are

$$\left| R^3 L^3 \, [6]_O \, [33]_{FS} \right\rangle \tag{2}$$

$$\left| R^3 L^3 \, [42]_O \, [33]_{FS} \right\rangle \tag{3}$$

$$\left| R^3 L^3 \, [42]_O \, [51]_{FS} \right\rangle \tag{4}$$

$$\left| R^3 L^3 \, [42]_O \, [411]_{FS} \right\rangle \tag{5}$$

Harvey [14] has shown that with a proper normalization the symmetry $[6]_O$ contains only s^6 and $[42]_O$ only $s^4 p^2$ configurations in the limit $Z \to 0$.

According to Ref. [15] let us consider also molecular orbital single-particle states. Most generally these are eigenstates of a Hamiltonian H_0 having axial and reflectional symmetries characteristic to the NN problem. These eigenstates have therefore good parity and good angular momentum projection. As in the cluster model basis where one uses the two lowest states R and L, in the molecular orbital basis we also consider the two lowest states, σ, of positive parity and π, of negative parity. From these we can construct pseudo-right r and pseudo-left l states as

$$\begin{bmatrix} r \\ l \end{bmatrix} = 2^{-1/2}(\sigma \pm \pi) \quad \text{for all } Z, \tag{6}$$

where

$$<r|r> = <l|l> = 1, \; <r|l> = 0. \tag{7}$$

In principle one can obtain molecular orbital single particle states from mean field calculations (see for example [16]). Here we approximate them by good parity, orthonormal states constructed from the cluster model states (1) as

$$\begin{bmatrix} \sigma \\ \pi \end{bmatrix} = [2(1 \pm <R|L>)]^{-1/2}(R \pm L), \tag{8}$$

Such molecular orbitals are a very good approximation to the exact eigenstates of a "two-centre" oscillator frequently used in nuclear physics or occasionally [17] in the calculation of the NN potential. They provide a convenient basis for the first step calculations based on the adiabatic approximation as described below.

Introduced in (6) they give

$$\begin{bmatrix} r \\ l \end{bmatrix} = \frac{1}{2}\left[\frac{R+L}{(1+<R|L>)^{1/2}} \pm \frac{R-L}{(1-<R|L>)^{1/2}}\right]. \tag{9}$$

At $Z \to 0$ one has $\sigma \to s$ and $\pi \to p$ (with $m = 0, \pm 1$) where s and p are harmonic oscillator states. Thus in the limit $Z \to 0$ one has

$$\begin{bmatrix} r \\ l \end{bmatrix} = 2^{1/2}(s \pm p), \tag{10}$$

and at $Z \to \infty$ one recovers the cluster model basis because $r \to R$ and $\ell \to L$.

From (r, l) as well as from (σ, π) orbitals one can construct six-quark states of required permutation symmetry, as shown in Ref. [15]. In the limit $Z \to 0$ six-quark states obtained from molecular orbitals contain configurations of type $s^n p^{6-n}$ with $n = 0, 1, ..., 6$. For example the $[6]_O$ state contains s^6, $s^6 p^4$, $s^2 p^4$ and p^6 configurations and the $[42]_O$ state associated to the S-channel contains $s^4 p^2$ and $s^2 p^4$ configurations. This is in contrast to the cluster model basis where $[6]_O$ contains only s^6 and $[42]_O$ only $s^4 p^2$ configurations, as mentioned above.

Besides being poorer in $s^n p^{6-n}$ configurations, the number of basis states is smaller in the cluster model although we deal with the same $[f]_O$ and $[f]_{FS}$ symmetries and the same harmonic oscillator states s and p in both cases. This is due to the existence of three-quark clusters only in the cluster model states, while the molecular basis also allows configurations with five quarks to the left and one to the right, or vice versa, or four quarks to the left and two to the right or vice versa (see Eqs (11)-(19) of [18]). At large separations these states act as "hidden colour" states but at short- and medium-range separation distances they are expected to bring a significant contribution, as we shall see below. The "hidden colour" are states where a $3q$ cluster in an s^3 configuration is a colour octet, in contrast to the nucleon which is a colour singlet. Their role is important at short separations but it vanishes at large ones (see e.g. [14]).

HAMILTONIAN

The GBE Hamiltonian considered in this study has the form [8,9] :

$$H = \sum_i m_i + \sum_i \frac{\vec{p}_i^2}{2m_i} - \frac{(\sum_i \vec{p}_i)^2}{2\sum_i m_i} + \sum_{i<j} V_{\text{conf}}(r_{ij}) + \sum_{i<j} V_\chi(r_{ij}), \qquad (11)$$

with the linear confining interaction :

$$V_{\text{conf}}(r_{ij}) = -\frac{3}{8}\lambda_i^c \cdot \lambda_j^c (V_0 + C\, r_{ij}), \qquad (12)$$

and the spin–spin component of the GBE interaction in its $SU_F(3)$ form :

$$V_\chi(r_{ij}) = \left\{ \sum_{F=1}^{3} V_\pi(r_{ij})\lambda_i^F \lambda_j^F \right. \\ \left. + \sum_{F=4}^{7} V_K(r_{ij})\lambda_i^F \lambda_j^F + V_\eta(r_{ij})\lambda_i^8 \lambda_j^8 + V_{\eta'}(r_{ij})\lambda_i^0 \lambda_j^0 \right\} \vec{\sigma}_i \cdot \vec{\sigma}_j, \qquad (13)$$

with $\lambda^0 = \sqrt{2/3}\,\mathbf{1}$, where $\mathbf{1}$ is the 3×3 unit matrix. The interaction (13) contains $\gamma = \pi, K, \eta$ and η' meson-exchange terms and the form of $V_\gamma(r_{ij})$ is given as the sum of two distinct contributions : a Yukawa-type potential containing the mass of the exchanged meson and a short-range contribution of opposite sign, the role of which is crucial in baryon spectroscopy.

In the parametrization of Ref. [8] the exchange potential due to a meson γ has the form

$$V_\gamma^I(r) = \frac{g_\gamma^2}{4\pi} \frac{1}{12 m_i m_j} \{\theta(r - r_0)\mu_\gamma^2 \frac{e^{-\mu_\gamma r}}{r} - \frac{4}{\sqrt{\pi}} \alpha^3 \exp(-\alpha^2(r - r_0)^2)\}. \qquad (14)$$

The shifted Gaussian of Eq. (14) results from a pure phenomenological fit (see below) of the baryon spectrum with

$$r_0 = 0.43\, fm, \quad \alpha = 2.91\, fm^{-1}, \qquad (15)$$

For the parametrization of Ref. [9], the potential has the form

$$V_\gamma^{II}(r) = \frac{g_\gamma^2}{4\pi} \frac{1}{12 m_i m_j} \{\mu_\gamma^2 \frac{e^{-\mu_\gamma r}}{r} - \Lambda_\gamma^2 \frac{e^{-\Lambda_\gamma r}}{r}\}. \qquad (16)$$

where

$$\Lambda_\gamma = \Lambda_0 + \kappa\, \mu_\gamma, \quad \Lambda_0 = 5.82\, fm^{-1}, \quad \kappa = 1.34. \qquad (17)$$

In the following, we shall call the form (14) Model I and the form (16) Model II. For a system of u and d quarks only, as it is the case here, the K-exchange does not contribute. The apriori determined parameters of the GBE model are the masses

TABLE 1. Parameters of the Hamiltonian (11-18)

Model	V_0 (MeV)	C (fm^{-2})	$g_8^2/4\pi$	$g_0^2/4\pi$	Ref.
I	0	0.474	0.67	1.206	[8]
II	-112	0.77	1.24	2.765	[9]

$$m_{u,d} = 340\, MeV,\ \mu_\pi = 139\, MeV,\ \mu_\eta = 547\, MeV,\ \mu_{\eta'} = 958\, MeV. \qquad (18)$$

The other parameters are given in Table I.

It is useful to comment on Eqs. (14) and (16). The coupling of pseudoscalar mesons to quarks (or nucleons) gives rise to a two-body interaction potential which contains a Yukawa-type term and a contact term of opposite sign (see e.g. [19]). The second term of (14) or (16) stems from the contact term, regularized with parameters fixed phenomenologically. Certainly more fundamental studies are required to understand this second term and attempts are being made in this direction. The instanton liquid model of the vacuum (for a review see [20]) implies point-like quark-quark interactions. To obtain a realistic description of the hyperfine interaction this interaction has to be iterated in the t-channel [11]. The t-channel iteration admits a meson exchange interpretation [21].

RESULTS

We diagonalize the Hamiltonian (11)-(18) in the six-quark cluster model basis and in the six-quark molecular orbital basis for values of the separation distance Z up to 2.5 fm. Using in each case the lowest eigenvalue, denoted by $\langle H \rangle_Z$ we define the NN interaction potential in the adiabatic (Born-Oppenheimer) approximation as

$$V_{NN}(Z) = \langle H \rangle_Z - 2m_N - K_{rel} \qquad (19)$$

Here m_N is the nucleon mass obtained as a variational s^3 solution for a $3q$ system described by the Hamiltonian (11). The wavefunction has the form $\phi \propto \exp\left[-(\rho^2 + \lambda^2)/2\beta^2\right]$ where $\rho = (\vec{r}_1 - \vec{r}_2)/\sqrt{2}$ and $\vec{\lambda} = (\vec{r}_1 + \vec{r}_2 - 2\vec{r}_3)/\sqrt{6}$. The minimum for $m_N = \langle H \rangle_{3q}$ is 970 MeV in the Model I and 1311 MeV in the Model II respectively, reached at the same β in both models. The same value of β is also used for the 6q system. This is equivalent with imposing the "stability condition" which is of crucial importance in resonating group method (RGM) calculations [1,2]. The quantity K_{rel} represents the relative kinetic energy of two $3q$ clusters separated at infinity

$$K_{rel} = \frac{3\hbar^2}{4m\beta^2} \qquad (20)$$

where m above and in the following designates the mass of the u or d quark. For our value of β this gives $K_{rel} = 0.448$ GeV.

Cluster model

At $Z \to \infty$ the symmetries corresponding to baryon-baryon channels, namely $[51]_{FS}$ and $[33]_{FS}$, must appear with proper coefficients, as given by Eq. (21). The contribution due to these symmetries must be identical to the contribution of V_χ to two nucleon masses also calculated with the Hamiltonian (11). This is indeed the case. In the total Hamiltonian the contribution of the $[411]_{FS}$ V_χ state tends to infinity when $Z \to \infty$. Then this state decouples from the rest which is natural because it does not correspond to an asymptotic baryon-baryon channel. It plays a role at small Z but at large Z its amplitude in the NN wavefunction vanishes, similarly to the "hidden colour" states. Actually, in diagonalizing the total Hamiltonian in the basis (2)-(5) we obtain an NN wavefunction which in the limit $Z \to \infty$ becomes [14]

$$\psi_{NN} = \frac{1}{3} \,|[6]_o[33]_{FS}\rangle + \frac{2}{3} \,|[42]_o[33]_{FS}\rangle - \frac{2}{3} \,|[42]_o[51]_{FS}\rangle \quad (21)$$

The adiabatic potential drawn in Figs. 1 and 2 is defined according to Eq. (19) where $\langle H \rangle_Z$ is the lowest eigenvalue resulting from the diagonalization. Fig. 1 corresponds to $S = 1$, $I = 0$ in the Model I and Fig. 2 to the same channel in the Model II. Note that from these curves one should subtract K_{rel} of Eq. (20) in order to obtain the asymptotic value zero for the potential. Similar results are obtained in the $SI = (01)$ channel. One can see that the potential is repulsive at any Z in the Model I, but a small attractive pocket appears in the potential of the Model II (see the zoom in the inside box of Fig. 2).

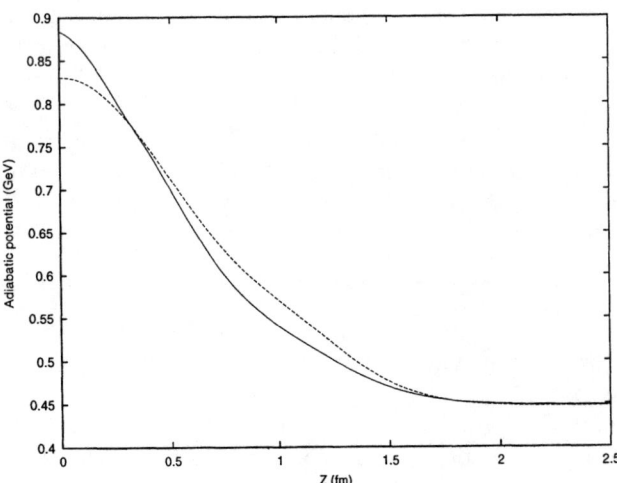

FIGURE 1. Comparison of the adiabatic potential in Model I, for $SI = (10)$, calculated in the cluster model (solid curve) and the molecular orbital basis (dashed curve).

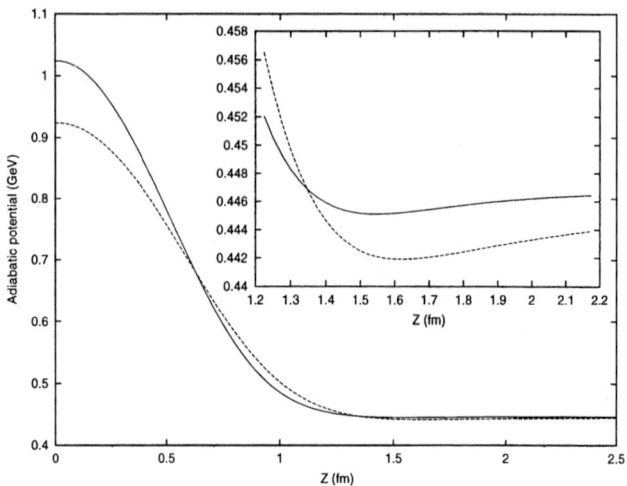

FIGURE 2. Same as Fig. 1 but for the Model II. The upper box represents a zoom of the small attractive pocket.

Molecular orbital basis

In the molecular orbital basis the asymptotic form of ψ_{NN} is also given by Eq. (21) inasmuch as $r \to R$ and $l \to L$ as indicated below Eq. (10).

We diagonalize also the Hamiltonian and use its lowest eigenvalue to obtain the NN potential according to the definition (19). The $S = 1$, $I = 0$ case is illustrated for Models I and II in Figs. 1 and 2 respectively, for a comparison with the cluster model basis. The repulsion reduces by about 15 % in the $^3S_1$ channels when passing from the cluster model basis to the molecular orbital basis. From Figs. 1 and 2 one can see that the molecular orbital basis has an important effect up to about $Z \approx 1.5$ fm giving a lower potential at small values of Z. For $Z \approx 1$ fm it gives a potential larger by few tens of MeV than the cluster model potential. However there is no attraction at all in the Model I (for details see Ref. [18]).

MIDDLE RANGE ATTRACTION

In principle we expected some attraction at large Z due to the presence of the Yukawa potential tail in Eq. (14). This is indeed the case in Model II. In Model I, we adopt a consistent procedure assuming that besides the pseudoscalar meson exchange interaction there exists an additional scalar, σ-meson exchange interaction between quarks. This is in the spirit of the spontaneous chiral symmetry breaking mechanism on which the GBE model is based. The σ-meson is the chiral partner of the pion and it should be considered explicitly.

Actually once the one-pion exchange interaction between quarks is admitted, one can inquire about the role of at least two-pion exchanges. Recently it was found [21] that the two-pion exchange also plays a significant role in the quark-quark interaction. It enhances the effect of the isospin dependent spin-spin component of the one-pion exchange interaction and cancels out its tensor component. Apart from that it gives rise to a spin independent central component, which averaged over the isospin wave function of the nucleon it produces an attractive spin independent interaction. These findings also support the introduction of a scalar (σ-meson) exchange interaction between quarks as an approximate description of the two-pion exchange loops.

For consistency with the parametrization [8] we consider here a scalar quark-quark interaction of the form

$$V_\sigma(r) = \frac{g_\sigma^2}{4\pi} \frac{1}{12 m_i m_j} \{\theta(r - r_0')\mu_\sigma^2 \frac{e^{-\mu_\sigma r}}{r} - \frac{4}{\sqrt{\pi}}\alpha'^3 \exp(-\alpha'^2(r - r_0')^2)\}. \qquad (22)$$

where $\mu_\sigma = 675$ MeV and r_0', α' and the coupling constant $g_\sigma^2/4\pi$ are arbitrary parameters. In order to be effective at medium-range separation between nucleons we expect this interaction to have $r_0' \neq r_0$ and $\alpha' \neq \alpha$. Note that the factor $1/m_i m_j$ has only been introduced for dimensional reasons.

We first looked at the baryon spectrum with the same variational parameters as before. The only modification is a shift of the whole spectrum which would correspond to taking $V_0 \approx -60$ MeV in Eq. (12).

For the $6q$ system we performed calculations in the molecular basis, which is more appropriate than the cluster model basis. We found that the resulting adiabatic

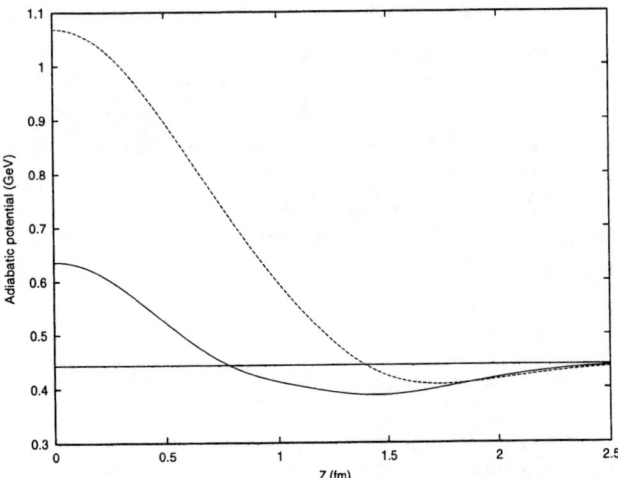

FIGURE 3. The adiabatic potential Model I in the molecular orbital basis for $SI = (10)$ (full curve) and $SI = (01)$ (dashed curve) with pseudoscalar + scalar quark-quark interaction.

potential is practically insensitive to changes in μ_σ and r_0' but very sensitive to α'. In Fig. 3 we show results for

$$r_0' = 0.86\,fm,\ \alpha' = 1.47\,fm^{-1},\ g_\sigma^2/4\pi = g_8^2/4\pi \qquad (23)$$

One can see that V_σ produces indeed an attractive pocket, deeper for $SI = (10)$ than for (01), as it should be for the NN problem. The depth of the attraction depends essentially on α'. The precise values of the parameters entering Eq. (22) should be determined in further RGM calculations. As mentioned above the Born-Oppenheimer potential is in fact the diagonal RGM kernel. It is interesting that an attractive pocket is seen in this kernel when a σ-meson exchange interaction is combined with pseudoscalar meson exchange and OGE interactions (hybrid model), the whole being fitted to the NN problem. The same interaction should also be introduced in Model II, but here, our purpose is to show how the σ-exchange interaction gives some middle-range attraction.

SUMMARY

We have calculated the NN potential in the adiabatic approximation as a function of Z, the separation distance between the centres of the two $3q$ clusters. We used two different parametrizations of a constituent quark model where quarks interact via pseudoscalar meson exchange. The orbital part of the six-quark states was constructed either from cluster model or molecular orbital single particle states. The latter are more realistic, having the proper axially and reflectionally symmetries. Also technically they are more convenient. We explicitly showed that they are important at small values of Z. In particular we found that the NN potential obtained in the molecular orbital basis has a less repulsive core than the one obtained in the cluster model basis. However none of the bases leads to an attractive pocket in one of the parametrizations considered here. We have simulated this attraction by introducing a σ-meson exchange interaction between quarks.

The present calculations give us an idea about the size and shape of the hard core produced by the GBE interaction. Except for small values of Z the two bases give rather similar potentials. Taking Z as a generator coordinate the following step is to perform a dynamical study based on the resonating group method which will provide phase-shifts to be compared to the experiment. These calculations are underway.

REFERENCES

1. Oka M. and Yazaki K., Int. Rev. Nucl. Phys., vol. 1 (Quarks and Nuclei, ed. W. Weise), World Scientific, Singapore, p. 490 (1984)
2. Shimizu K., Rep. Prog. Phys. **52** (1989) 1
3. Kusainov A. M., Neudatchin V. G. and Obukhovsky I. T., Phys. Rev. **C44** (1991) 1343

4. Zhang Z., Faessler A., Straub U. and Glozman L. Ya., Nucl. Phys. **A578** (1994) 573; Valcarce A., Buchman A., Fernandez F. and Faessler A., Phys. Rev. **C50** (1994) 2246
5. Fujiwara Y., Nakamoto C. and Suzuki Y., Phys. Rev. Lett. **76** (1996) 2242; Phys.Rev. **C54** (1996) 2180
6. Bartz D. and Stancu Fl., Phys. Rev. **C59** (1999) 1756
7. Glozman L. Ya. and Riska D. O., Phys. Rep. **268** (1996) 263
8. Glozman L. Ya., Papp Z. and Plessas W., Phys. Lett. **B381** (1996) 311
9. Glozman L. Ya., Papp Z., Plessas W., Varga K. and Wagenbrunn R. F., Nucl. Phys. **A623** (1997) 90c
10. Collins H. and Georgi H., Phys. Rev. **D59** (1999) 094010
11. Glozman L. Ya. and Varga K., e-print hep-ph/9901439, to appear in Phys. Rev. D
12. Chu M. C. et al., Phys. Rev. **D49** (1994) 6039; Negele J. W., Nucl. Phys. Proc. Suppl **73** (1999) 92; Liu K. F. et al., Phys. Rev. **D59** (1999) 112001; Aoki S. et al., Phys. Rev. Lett. **82** (1999) 4392
13. Stancu Fl., Pepin S. and Glozman L. Ya., Phys. Rev. **C56** (1997) 2779; Phys. Rev. **D57** (1998) 4393; Shimizu K. and Glozman L. Ya., e-print nucl-th/9906008
14. Harvey M., Nucl. Phys. **A352** (1981) 301; ibid. **A481** (1988) 834
15. Stancu Fl. and Wilets L., Phys. Rev. **C36** (1987) 726
16. Koepf W., Wilets L., Pepin S. and Stancu Fl., Phys. Rev. **C50** (1994) 614
17. Robson D., Phys. Rev. **D35** (1987) 1029
18. Bartz D. and Stancu Fl., Phys. Rev. **C60** (1999) 055207
19. Brown G. E. and Jackson A. D., *The nucleon-nucleon interaction*, North Holland, Amsterdam, 1976 p. 5
20. Schäfer T. and Shuryak E. V., Rev. Mod. Phys. **70** (1998) 323
21. Riska D. O. and Brown G. E., Nucl. Phys. **A653** (1999) 251
22. Glozman L. Ya., Plessas W., Varga K. and Wagenbrunn R. F., Phys. Rev. **D58** (1998) 094030
23. Stancu Fl., *Group Theory in Subnuclear Physics*, Clarendon Press, Oxford, 1996, chapters 4 and 10

Chiral NN interactions in nuclear matter

Boris Krippa

Department of Physics and Astronomy, Free University of Amsterdam,
De Boelelaan 1081, 1081 HV Amsterdam.

Abstract. We consider an effective field theory of NN system in nuclear medium. The shallow bound states, which complicate the effective field theory analysis and lead to the large scattering length in the vacuum case do not exist in matter. We show that the next-to-leading order terms in the chiral expansion of the effective NN potential can be interpreted as corrections so that the expansion is systematic. It is pointed out however that it is still useful to treat the problem nonperturbatively since it may allow for the consideration of the nuclear systems with the density smaller that the normal nuclear matter one. The potential energy per particle is calculated. The possible directions in constructing the chiral theory of nuclear matter are outlined.

Effective Field Theory (EFT) has become a popular tool for studying nuclear interactions. EFT is based on the idea to use the Lagrangian with the appropriate effective degrees of freedom instead of the fundamental ones in the low-energy region (for review of EFT see, for example [1]). This Lagrangian should include all possible terms allowed by the symmetries of the underlying QCD. The states which can be treated as heavy, compared to the typical energy scale involved, are integrated out. They are hidden in the Low Energy Effective Constants (LEC's) of the corresponding Lagrangian. The physical amplitudes can be represented as the sum of certain graphs, each of them being of a given order in Q/Λ, where Q is a typical momentum scale and Λ is a parameter reflecting the scale of the short range physics. The relative contribution of each graph can roughly be estimated using chiral counting rules [2]. The relevant degrees of freedom in the nuclear domain are nucleons and pions. In the case of meson-meson [3] and meson-nucleon [4] interactions the perturbative chiral expansion can be organized in a consistent way. However, being applied to the NN system EFT encounters serious problem which is due to existence of the bound states near threshold [5]. It results in the large nucleon-nucleon scattering length and makes the perturbative expansion divergent. Weinberg suggested [5] to apply chiral counting rules to the certain class of the irreducible diagrams which should then be summed up to infinite order by solving the Lippmann-Schwinger (LS) equation. The irreducible diagrams can be treated as the effective potential in this case. Different aspects of the chiral NN problem have been discussed since then [6]. The concept of EFT has also intensively been used to study nuclear matter [7–10]. In [7] the effective chiral Lagrangian was constructed and the "naturalness" of the effective coupling constants has been demonstrated. The possible counting rules for nuclear matter have been discussed in [9]. These two lines of development of the chiral nuclear physics are in some sense similar to the tendencies existed some time ago in conventional nuclear physics with the phenomenological two body forces. On the one hand, the phenomenological NN potentials were used to describe nucleon-nucleon cross sections and phase shifts. On the other hand, nuclear mean field approaches provided a reasonable description of the bulk properties of nuclear matter. The unification of these two approaches then led to the famous Bethe-Goldstone (BG) equation [11] for the G-matrix which is an analog of scattering T-matrix, satisfying the LS equation. It is therefore reasonable to follow the same strategy and, being equipped with the chiral theory of NN interaction in vacuum, try to construct the chiral G-matrix, describing the effective interactions of two nucleons in medium. One can easily see the qualitative difference between vacuum and medium cases. In nuclear medium because of Pauli blocking the intermediate states with the momenta less than Fermi momentum p_F are forbidden. Therefore, the nucleon propagator does not exhibit a pole. Moreover, the shallow bound or virtual NN states, which constitute the main difficulty of the problem in vacuum, simply do not exist in nuclear matter because of interaction of the NN pair with nuclear mean field. It means that the effective scattering length becomes considerably smaller compared to the vacuum one. The value of a scattering length is determined by the position of the singularity, nearest to the physical region. In the vacuum case, for example, the virtual deuteron bound state is very close to the NN threshold leading to the unnaturally large scattering length. The moderate value of the in-medium scattering length would indicate, in some sense, that the typical scale of the NN interactions gets "more natural" in nuclear matter.

We start from the standard nucleon-nucleon effective chiral Lagrangian which can be written as follows

$$\mathcal{L} = N^\dagger i \partial_t N - N^\dagger \frac{\nabla^2}{2M} N - \frac{1}{2} C_0 (N^\dagger N)^2 - \frac{1}{2} C_2 (N^\dagger \nabla^2 N)(N^\dagger N) + h.c. + \ldots . \quad (1)$$

We consider the simplest case of the NN scattering in the $^1S_0$ state and assume zero total 3-momentum of NN pair in the medium. The inclusion of the nonzero total 3-momentum does not really change anything qualitatively and only makes the calculations technically more involved. The G-matrix is given by

$$G(p',p) = V(p',p) + M \int \frac{dq\, q^2}{2\pi^2} V(p',q) \frac{\theta(q - p_F)}{M(\epsilon_1(p) + \epsilon_2(p')) - q^2} G(q,p), \quad (2)$$

Here ϵ_1 and ϵ_2 are the single-particle energies of the bound nucleons. They are affected by the nuclear mean field. In nuclear medium such corrections lead to the nucleon effective mass slightly different from that in free space. We used the value $M = 0.8 M_0$, where M_0 is the nucleon mass in vacuum. One notes that this value is close to one usually accepted in nuclear mean field theories. The standard strategy of treating the chiral NN problem in vacuum is the following. One computes amplitudes up to a given chiral order in the terms of the effective constants C_0 and C_2 which are then determined by comparing the calculated amplitude with some experimental data. Having these constants fixed one can calculate the other observables. We will follow the similar strategy in the nuclear matter case and proceed as follows. We choose exactly solvable separable potential with parameters adjusted to the value of the potential energy per particle in nuclear matter. Then we solve the BG equation with the effective constants C_0 and C_2. The numerical values of these constants are determined comparing the phenomenological and EFT G-matrix at some fixed kinematical points. The check of consistency we used is the difference between C_0's determined in the leading and subleading orders. If the difference between the values of C_0 needed to fit the data in leading and subleading order is of higher order then the procedure of truncation of the standard chiral expansion is justified. In the vacuum case the corresponding difference was found to be large [12]. Using a simple separable potential

$$V = -\lambda |\eta\rangle\langle\eta| \tag{3}$$

with the form factors

$$\eta(p) = \frac{1}{(p^2 + \beta^2)^{1/2}} \tag{4}$$

One can easily get

$$\frac{1}{T(k,k)} = V(k,k)^{-1}\left[1 - M_0 \int \frac{dq\, q^2}{4\pi^2} \frac{V(q,q)}{k^2 - q^2}\right] \tag{5}$$

The experimental values of scattering length a and effective radius r_e are

$$a = -23.71 \pm 0.013\,\text{fm} \qquad r_e = 2.73 \pm 0.03\,\text{fm}. \tag{6}$$

These values can be reproduced if we choose

$$\lambda = 1.95 \qquad \beta = 0.8\,\text{fm} \tag{7}$$

The solution of the BG equation for the separable potential is a simple generalization of the one for the LS equation

$$G(k,k) = -\eta^2(k)\left[\lambda^{-1} + \frac{M}{2\pi^2}\int dq\, q^2 \frac{\theta(q - p_F)\eta^2(q)}{k^2 - q^2}\right]^{-1} \tag{8}$$

However, the phenomenological G-matrix with the parameters determined from the effective range expansion fit leads to the somewhat lower the potential energy than the usually accepted value $\sim$ -16 MeV.

To get a better fit we choose

$$\lambda = 2.4 \qquad \beta = 1.1 \, \text{fm} \qquad (9)$$

These values are fairly close to the vacuum ones and provide the potential energy per particle in a good agreement with the empirical value. The parameters λ and β being substituted in the G-matrix lead to $a_m \simeq r_m \simeq 0(1)$, where a_m and r_m are the in-medium analogs of scattering length and effective radius. One notes that effective radius is much less affected by the medium effects. It is quite natural since the value of the effective radius is only weakly sensitive to the bound state at threshold and is of the "almost natural" size already in the vacuum case. The absolute value of the in-medium scattering length is considerably reduced compared to the vacuum one. It clearly indicates that, as expected, the shallow virtual nucleon-nucleon bound state is no longer present in nuclear medium. Thus, one can avoid significant part of the difficulties typical for the chiral NN problem in vacuum. Having determined the phenomenological G-matrix one can now solve the BG equation using leading and sub-leading orders of the NN effective chiral Lagrangian. The solution is similar to the vacuum case [12] and can be represented as follows

$$\frac{1}{G(k,k)} = \frac{(C_2 I_3(k,p_F) - 1)^2}{C_0 + C_2^2 I_5(k,p_F) + k^2 C_2(2 - C_2 I_3(k,p_F))} - I(k,p_F), \qquad (10)$$

where we defined

$$I_n \equiv -\frac{M}{(2\pi)^2} \int dq\, q^{n-1} \theta(q - p_F). \qquad (11)$$

and

$$I(k) \equiv \frac{M}{2\pi^2} \int dq\, \frac{q^2 \theta(q - p_F)}{k^2 - q^2}. \qquad (12)$$

These integrals are divergent so the renormilization should be carried out. The procedure used is similar to that adopted in Ref. [13] to study the EFT approach to the NN interaction in vacuum. We subtract the divergent integrals at some kinematical point $p^2 = -\mu^2$. After subtraction the renormalized G-matrix takes the form

$$\frac{1}{G^r(k,k)} = \frac{1}{C_0^r(\mu) + 2k^2 C_2^r(\mu)} + \frac{M}{4\pi}[p \log \frac{p_F - p}{p_F + p} - i\mu \log \frac{p_F - i\mu}{p_F + i\mu}], \qquad (13)$$

One notes that in the $p_F \to 0$ limit the vacuum chiral NN amplitude is recovered. We choose the value $\mu = 0$ as a subtraction point. The μ dependence of LEC's is governed by the renormalization group equation. Now one can determine the LEC's by equating the EFT and phenomenological G-matrices at some kinematical points. We used the values $p = \frac{p_F}{2}; \frac{p_F}{3}$ as such points. The assumed value of the Fermi-momentum is $p_F = 1.37$ fm. In the following we will omit the label "r" implying that we always deal with renormalized quantities. We found $C_0 = -1.86 fm^2$ in LO. In NLO one gets $C_0 = 2.64 fm^2$ and $C_2 = 0.84 fm^4$ so that the inclusion of the NLO corrections give rise to the approximately 40% change in the value of C_0. It indicates that the chiral expansion is systematic in a sense that adding of the NLO terms in the effective Lagrangian results in a "NLO change" of the coefficients which have already been determined at LO. The natural size of the in-medium scattering length and moderate changes experienced by the coupling constant C_0 might, in principle, indicate the possibility of the perturbative calculations. However, in spite of this, it is still more useful to treat this problem in the nonperturbative manner. There are few reasons for the nonperturbative treatment. Firstly, the corrections themselves are quite significant. Secondly, the overall (although distant) goal of the EFT description is to derive both nuclear matter and the vacuum NN amplitude from the same Lagrangian, However, it is hard to say at what densities the dynamics becomes intrinsically nonperturbative, so it is better to treat the problem nonperturbatively from the beginning. The nonperturbative treatment may also turn out important to get the correct saturation curve since at some density lower than the normal nuclear one the scattering length starts departing from its natural value and some sort of the nonperturbative approach becomes inevitable. Thirdly, in the processes involving both the nonzero density and temperature, such as heavy ion collisions, the value of the Fermi-momentum can effectively be lowered again making the nonperturbative treatment preferable.

Let's now calculate the potential energy per particle using the expression for the in-medium chiral NN scattering amplitude. The potential energy of nuclear matter can be evaluated from

$$U_{tot} = \frac{1}{2} \sum_{\mu,\nu} <\mu\nu|G(\epsilon_\mu + \epsilon_\nu)|\mu\nu - \nu\mu> \tag{14}$$

The summation goes over the states with momenta below p_F. Here it is seen that G amplitude plays the role of an effective chiral two-body interaction in nuclear medium. The calculations using the lowest order G-matrix result in the value $\frac{U(^1S_0)}{A} \simeq -17 MeV$. The inclusion of the next-to-leading order corrections gives rise to the value $\frac{U(^1S_0)}{A} \simeq -13.1 MeV$. One notes that both $\frac{U(^1S_0)}{A}$ and C_0 experience corrections of the same order when NLO terms are included in the effective Lagrangian. The similar calculations done in the triplet s-wave channel give rise to to the value $\frac{U(^3S_1)}{A} \simeq -17.3(-13.2)$ MeV in LO (NLO). The values of the potential energy obtained with chiral approach looks quite reasonable although they are somewhat smaller than the standard values usually obtained in the calculations with the phenomenological two-body forces [14]. One can therefore conclude that there is still a room for both pionic effects and three particle correlations which should be included in a chirally invariant manner.

The validity of the EFT description is restricted by some cutoff parameter reflecting the short range physics effects. Its value deserves some comments in the context of applying of the EFT methods to nuclear matter. The scale where the

EFT treatment ceases to be valid should approximately correspond to the scale of the short range correlations (SRC), that is, $\sim 2.5 fm^{-1}$. The description of SRC is hardly possible in the framework of EFT so the value $\Lambda \sim 2 fm^{-1}$ might put natural constraint on the EFT description of nuclear matter. To make the chiral expansion meaningful the chiral counting rules in nuclear matter must be established. This is still open problem. However, the above obtained results suggest that the relevant expansion parameter could be something like $\frac{<p>}{\Lambda} \sim \frac{<m_\pi>}{\Lambda} \sim 0.3 - 0.4$. Of course, until pion effects are taken into account this estimate can only be suggestive. Moreover, many other things remain to be done to make the qualitative description of nuclear matter possible. Beside pionic effects one needs to include many body forces and formulate chiral counting rules. One should also find a way to remove off-shell ambiguities order by order and calculate the nucleon self-energy up to a given chiral order to make EFT description of nuclear matter fully consistent.

ACKNOWLEDGMENTS

Author is very grateful for the support and warm hospitality from SRCSSM at the University of Adelaide where the initial part of this work was done.

REFERENCES

1. A. Manohar, *Effective Field Theories*, **hep-ph/9506222**; D. B. Kaplan, *Effective Field Theories*, **nucl-th/9506035**.
2. S. Weinberg, Physica A **96**, 32 (1979).
3. J Gasser and H. Leutwyler, Annals of Physics **158**, 142 (1984).
4. G. Ecker and M. Mojzis, Phys. Lett. B. **365** 312 (1996); Bernard, N. Kaiser, and U.-G. Meißner, Nucl. Phys. **A615**, 483 (1997).
5. S. Weinberg, Nucl. Phys. **B363**, 3 (1991).
6. D. B. Kaplan, M. Savage, and M. B. Wise, Nucl. Phys., **B478**, 629 (1996); D. B. Kaplan, M. Savage, and M. B. Wise, Phys. Lett., **B424**, 390 (1998), D. B. Kaplan, M. Savage, and M. B. Wise, Nucl. Phys., **B534**, 329 (1998) U. van Kolck, **nucl-th/9808007**; M. C. Birse, **nucl-th/9806038**; J. V. Steele and R. J. Furnstahl, Nucl. Phys., **A637**, 46(1998); T.-S. Park, K.Kubodera, D.-P. Min and M. Rho, Phys. Rev. C **58**, R637 (1998); G. P. Lepage, **nucl-th/9706029**; T. Mehen and I. W. Stewart, **nucl-th/9806038**.
7. R. Furnstahl, B. Serot and H.-B. Tang, Nucl. Phys., **A618**, 445 (1997).
8. B. Lynn, Nucl. Phys., **B402**, 281 (1993).
9. M. Lutz, Nucl. Phys., **A642**, 171 (1998).
10. J. Friar, **nucl-th/9804010**.
11. H. A. Bethe and J. Goldstone, Proc. Roy. Soc. (London) **A238**, 157 (1957).
12. S. Beane, T. D. Cohen and D. Phillips, Nucl. Phys., **A632**, 445 (1997).
13. J, Gegelia Phys. Lett., **B429**, 227 (1998).
14. M. I. Haftel and F. Tabakin, Nucl. Phys., **A158**, 1 (1970).

A Sketch of Two and Three Bodies

Harald W. Grießhammer[†1]

[†]*Nuclear Theory Group, Department of Physics, University of Washington,*
Box 351 560, Seattle, WA 98195-1560, USA
and
Institut für Theoretische Physik, Physik-Department der
Technischen Universität München, 85748 Garching, Germany (permanent address)

Abstract. A cartoon of the Effective Field Theory of many nucleon systems is drawn, concentrating on Compton scattering in the two nucleon system, and on nd scattering in the three body system.

The purpose of this presentation is to give a concise introduction into the Effective Field Theory (EFT, for a review see e.g. [1]) of two and three nucleon systems as it emerged in the last three years. However, I can only give a "teaser" with a lot of words and figures and a few cheats in details, referring to the literature, esp. to the excellent proceedings of the INT-Caltech Workshops 1998 and 1999 [2]. I concentrate on work undertaken with J.-W. Chen, R.P. Springer and M.J. Savage in [3,4], P.F. Bedaque in [5,6], and F. Gabbiani in [6]. M. Birse's talk at this Workshop provides a more formal investigation of the EFT of the two nucleon system, and U.-G. Meißner's alternative approach he presented here follows Weinberg's original suggestion [7] but needs to be studied further.

Effective Field Theory methods are largely used in many branches of physics where a separation of scales exists. In low energy nuclear systems, the two well separated scales are, on one side, the low scales of the typical momentum of the process considered and the pion mass, and on the other side the higher scales associated with chiral symmetry and confinement. This separation of scales was explored with great success in the mesonic sector (Chiral Perturbation Theory [8]) and in the one baryon sector (Heavy Baryon Chiral Perturbation Theory [9]), producing a low energy expansion of a variety of observables (see also the Chiral Perturbation Theory section of this Workshop). It provided for the first time a description of strongly interacting particles which is systematic, rigorous and model independent (meaning, independent of assumptions about the non-perturbative QCD dynamics).

[1)] Email: hgrie@physik.tu-muenchen.de

Three main ingredients enter the construction of an EFT: The Lagrangean, the power counting and a regularisation scheme. First, the relevant degrees of freedom have to be identified. In his original suggestion how to extend EFT methods to systems containing two or more nucleons, Weinberg [7] noticed that below the Δ production scale, only nucleons and pions need to be retained as the infrared relevant degrees of freedom of low energy QCD. Because at these scales the momenta of the nucleons are small compared to their rest mass, the theory becomes non-relativistic at leading order in the velocity expansion, with relativistic corrections systematically included at higher orders. The most general chirally (and iso-spin) invariant Lagrangean consists hence of contact interactions between non-relativistic nucleons, and between nucleons and pions, with the first few terms of the form

$$\mathcal{L}_{NN} = N^\dagger (i\partial_0 + \frac{\vec{\partial}^2}{2M})N + \frac{f_\pi^2}{8} \mathrm{tr}[(\partial_\mu \Sigma^\dagger)(\partial^\mu \Sigma)] + g_A N^\dagger \vec{A} \cdot \sigma N$$
$$- C_0 (N^T P^i N)^\dagger (N^T P^i N) +$$
$$+ \frac{C_2}{8}\left[(N^T P^i N)^\dagger (N^T P^i (\vec{\partial} - \overleftarrow{\partial})^2 N) + \mathrm{h.c.}\right] + \ldots,$$
(1)

where $N = \binom{p}{n}$ is the nucleon doublet of two-component spinors and P^i is the projector onto the iso-scalar-vector channel, $P^{i,b\beta}_{a\alpha} = \frac{1}{\sqrt{8}}(\sigma_2 \sigma^i)^\beta_\alpha (\tau_2)^b_a$. σ (τ) are the Pauli matrices acting in spin (iso-spin) space. The iso-vector-scalar part of the NN Lagrangean introduces more constants C_i and interactions and has not been displayed for convenience. The field ξ describes the pion, $\xi(x) = \sqrt{\Sigma} = e^{i\Pi/f_\pi}$, $f_\pi = 130$ MeV. D_μ is the chiral covariant derivative $D_\mu = \partial_\mu + V_\mu$, and the vector and axial currents are $V_\mu = \frac{1}{2}(\xi \partial_\mu \xi^\dagger + \xi^\dagger \partial_\mu \xi)$, $A_\mu = \frac{i}{2}(\xi \partial_\mu \xi^\dagger - \xi^\dagger \partial_\mu \xi)$. The interactions involving pions are severely restricted by chiral invariance. As such, the theory is an extension to the many nucleon system of Chiral Perturbation Theory and Heavy Baryon Chiral Perturbation Theory. Like in its cousins, all short distance physics – branes and strings, quarks and gluons, resonances like the Δ or σ – is integrated out into the coefficients of the low energy Lagrangean. In principle, these constants could be derived by solving QCD or via models of the short distance physics like resonance saturation. The most common and practical way to determine those constants, though, is by fitting them to experiment.

The EFT with pions integrated out (formally, $g_A = 0$ in (1)) is valid below the pion cut and was recently pushed to very high orders in the two-nucleon sector [10] where accuracies of the order of 1% were obtained. It can be viewed as a systematisation of Effective Range Theory with the inclusion of relativistic and short distance effects traditionally left out in that approach.

Because the Lagrangean (1) consists of infinitely many terms only restricted by symmetry, an EFT may at first sight suffer from lack of predictive power. Indeed, as the second part of its formulation, predictive power is ensured only by establishing a power counting scheme, i.e. a way to determine at which order in a momentum expansion different contributions will appear, and keeping only and all the terms up to a given order. The dimensionless, small parameter on which the expansion is based is the typical momentum Q of the process in units of the scale Λ at which the theory is expected to break down, with estimates ranging from $\Lambda_\pi \approx 300$ to 800 MeV [2] in the two body system for the theory with pions. The pionless theory should be in disagreement with experiment starting at the pion cut, $\Lambda_{no\pi} \approx 140$ MeV. Values for Λ and Q have to be determined from comparison to experiments and are a priori unknown. Assuming that all contributions are of natural size, i.e. ordered by powers of Q, the systematic power counting ensures that the sum of all terms left out when calculating to a certain order in Q is smaller than the last order retained, allowing for an error estimate of the final result.

Even if calculations of nuclear properties were possible starting from the underlying QCD Lagrangean, EFT simplifies the problem considerably by factorising it into a short distance part (subsumed into the coefficient of the Lagrangean) and a long distance part which contains the infrared-relevant physics and is dealt with by EFT methods. EFT provides an answer of finite accuracy because higher order corrections are systematically calculable and suppressed in powers of Q. Hence, the power counting allows for an error estimate of the final result, with the natural size of all neglected terms known to be of higher order. Relativistic effects, chiral dynamics and external currents are included systematically, and extensions to include e.g. parity violating effects are straightforward. Gauged interactions and exchange currents are unambiguous. Results obtained with EFT are easily dissected for the relative importance of the various terms. Because only S-matrix elements between on-shell states are observables, ambiguities nesting in "off-shell effects" are absent. On the other hand, because only symmetry considerations enter the construction of the Lagrangean, EFTs are less restrictive as no assumption about the underlying QCD dynamics is incorporated.

In systems involving two or more nucleons, establishing such a power counting is complicated by the fact that unnaturally large scales have to be accommodated, so that some coefficients in the Lagrangean may not be of natural size and hence possibly jeopardise power counting: Given that the typical low energy scale in the problem should be the mass of the pion as the lightest particle emerging from QCD, fine tuning seems to be required to produce the large scattering lengths in the $^1S_0$ and $^3S_1$ channels ($1/a^{^1S_0} = -8.3$ MeV, $1/a^{^3S_1} = 36$ MeV). Since there is a bound state in the $^3S_1$ channel with a binding energy $B = 2.225$ MeV and hence a typical binding momentum $\gamma = \sqrt{MB} \simeq 46$ MeV well below the scale Λ at which the theory should break down, it is also clear that at least some processes have to be treated non-perturbatively in order to accommodate the deuteron. Most likely, these small scales do not arise from the fact that the real world is close to the chiral limit: In the singlet channel, for instance, the one pion exchange potential vanishes in the chiral limit and thus cannot be the cause of the fine tuning. The fine tuning then must be a result of short distance physics.

A way to incorporate this fine tuning into the power counting was suggested by Kaplan, Savage and Wise [11]: At very low momenta, contact interactions with several derivatives – like $p^2 C_2$ and the pion-nucleon interactions – should become unimportant, and we are left only with the contact interactions proportional to C_0. The leading order contribution to nucleons scattering in an S wave comes hence from four nucleon contact interactions and is summed geometrically as in Fig. 1 to all orders to produce the shallow real bound state, i.e. the deuteron.

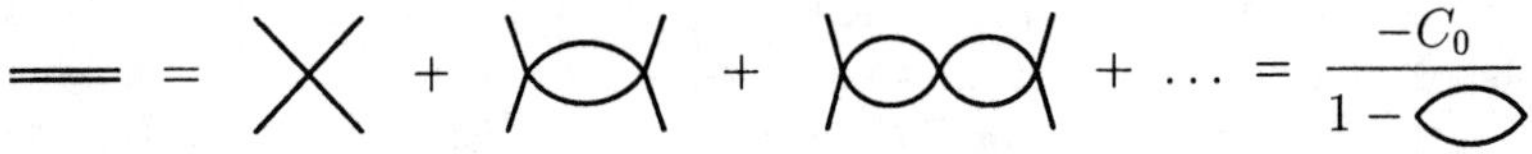

$$= \;=\; \times \;+\; \bowtie \;+\; \bowtie\!\bowtie \;+\; \ldots \;=\; \frac{-C_0}{1-\bigcirc}$$

FIGURE 1. Re-summation of the contact interactions into the deuteron propagator.

How to justify this? Any diagram can be estimated by scaling momenta by a factor of Q and non-relativistic kinetic energies by a factor of Q^2/M. The remaining integral includes no dimensions and is taken to be of the order Q^0 and of natural size. This scaling implies the rule that nucleon propagators contribute one power of M/Q^2 and each loop a power of Q^5/M. Assuming that

$$C_0 \sim \frac{1}{MQ}, \quad C_2 \sim \frac{1}{M\Lambda Q^2}, \tag{2}$$

the diagrams contributing at leading order to the deuteron propagator are indeed an infinite number as shown in Fig. 1, each one of the order $1/(MQ)$. The regulator dependent, linear divergence in each of the bubble diagrams does not show in dimensional regularisation as a pole in 4 dimensions, but it does appear as a pole in 3 dimensions which we subtract following the Power Divergence Subtraction scheme [11]. Dimensional regularisation is chosen to explicitly preserve the systematic power counting as well as all symmetries (esp. chiral invariance) at each order in every step of the calculation. At leading (LO), next-to-leading order (NLO) and often even NNLO in the two nucleon system, it also allows for simple, closed answers whose analytic structure is readily asserted. The deuteron propagator

$$\frac{4\pi}{M} \frac{-i}{\frac{4\pi}{MC_0} + \mu - \sqrt{\frac{\vec{p}^2}{4} - Mp_0 - i\varepsilon}} \tag{3}$$

has the correct pole position and cut structure when one chooses

$$C_0(\mu) = \frac{4\pi}{M} \frac{1}{\gamma - \mu}. \tag{4}$$

Indeed, when choosing $\mu \sim Q$, the leading order contact interaction scales as demanded in (2) and – as expected for a physical observable – the NN scattering amplitude becomes independent of μ, the renormalisation scale or cut-off chosen. The same can be shown for the higher order coefficients, so that the scheme is self-consistent. Power Divergence Subtraction moves hence a somewhat arbitrary amount of the short distance contributions from loops to counterterms and makes precise cancellations manifest which arise from fine tuning. Notice that the resummed deuteron propagator has the same order $1/(MQ)$ as each diagram in Fig. 1.

One surprising result arises from this analysis because chiral symmetry implies a derivative coupling of the pion to the nucleon at leading order. The contribution from one pion exchange includes a factor of Q^{-2} from the pion propagator and a factor of Q^2 coming from the pion-nucleon vertices, so that for momenta of the order of the pion mass, the instantaneous one pion exchange scales as Q^0 and is *smaller* than the contact piece C_0 which according to (2) scales as Q^{-1}. Iterated and radiative pion exchanges are suppressed even further. Pion exchange and higher derivative contact terms appear hence only as perturbations at higher orders. In contradistinction to iterative potential model approaches, each higher order contribution is inserted only once. In this scheme, the only non-perturbative physics responsible for nuclear binding is extremely simple, and the more complicated pion contributions are at each order given by a finite number of diagrams. For example, the NLO contributions to the deuteron are the one instantaneous pion exchange and the four nucleon interaction with two derivatives, Fig. 2. The constants are determined e.g. by demanding the correct deuteron pole position and residue [12].

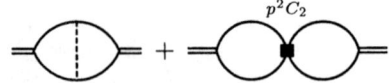

FIGURE 2. The NLO corrections to the deuteron.

In the two body sector, the theory thus emerging has been put to extensive tests at NLO and NNLO, giving for the first time analytic answers to many deuteron properties, see e.g. [2]. Although in general process dependent, the expansion parameter is found to be of the order of $\frac{1}{3}$ in most applications, so that NLO calculations can be expected to be accurate to about 10%, and NNLO calculations to about 4%. In all cases, experimental agreement is within the estimated theoretical uncertainties, and in some cases, previously unknown counterterms could be determined.

The elastic deuteron Compton scattering diagrams to NLO are partially obtained by gauging the Lagrangean (1), i.e. by replacing ordinary derivatives by covariant ones: At LO, a seagull-graph and one graph in which the incident and outgoing photon couple to the same nucleon are found. At NLO, the photons are attached in all possible ways to the corrections in Fig. 2, including to the pion, the $NN\pi$ vertex and the C_2 vertex. The Fermi interaction $\vec{\sigma} \cdot \vec{B}$ probes the $^1S_0$ intermediate NN state and enters at NLO, too. Finally, the iso-scalar electric nuclear polarisability was shown to come from relativistic ("radiative") pions in Chiral Perturbation Theory [13], $\alpha_{E,N} = \frac{5g_A^2 \alpha}{48\pi f_\pi^2 m_\pi}$, and is NLO. The cross section fits finally on less than one page with functions not more complicated than Logarithms and Arcustangentes [4] and contains no free parameters. Comparison with the Urbana experiment [14] in Fig. 3 shows good agreement, with the pion graphs that dominate the electric polarisability of the nucleon necessary to improve it. The deuteron scalar and tensor electric and magnetic polarisabilities are also easily extracted [3].

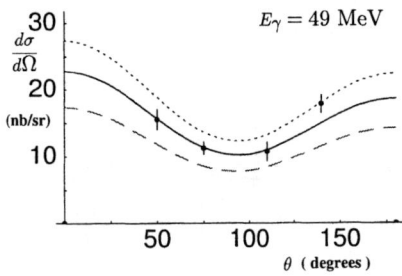 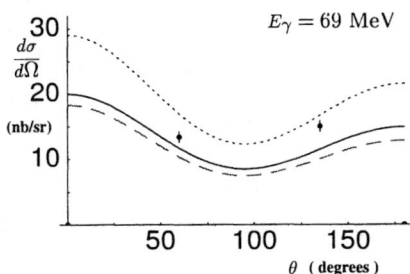

FIGURE 3. The differential cross section for elastic γ-deuteron Compton scattering at incident photon energies of $E_\gamma = 49$ MeV and 69 MeV in an EFT with explicit pions [4], no free parameters. Dashed: LO; dotted: NLO without the graphs that contribute to the nucleon polarisability; solid curve: complete NLO result.

In the three body sector, even the leading order calculation is too complex for a fully analytical solution. Still, the equations that need to be solved are computationally trivial and can furthermore be improved systematically by higher order corrections that involve only (partly analytical, partly numerical) integrations, as opposed to many-dimensional integral equations arising in other approaches. The nd system provides a laboratory in which many complications of the other channels are not encountered: The absence of Coulomb interactions ensures that only properties of the strong interactions are probed. In the quartet channel, the Pauli principle forbids three body forces [15] in the first few orders. Because the calculation is parameter-free, it allows one to determine the range of validity of the KSW scheme without a detailed analysis of the fitting procedure. Although e.g. the quartet scattering length is large, no extra fine tuning except the one for the deuteron is required. In the S wave, spin-doublet (triton) channel, the situation is more complicated. An unusual renormalisation of the three-body force makes it large and as important as the leading two-body forces [16]. More work is needed there.

A comparative study between the theory with explicit pions and the one with pions integrated out was performed in [5] for the spin quartet S wave. As seen above, the two theories are identical at LO: All graphs involving only C_0 interactions are of the same order and form a double series which is not geometrical and cannot be summed analytically. One is hence left with the task of summing all "pinball" diagrams (first line of Fig. 4). Summing all "bubble-chain" sub-graphs into the deuteron propagator, one can however obtain the solution numerically from the integral equation pictorially shown in the lower line of Fig 4. A code runs within seconds on a personal computer.

The power counting shows that at NLO, we have additional contributions from: $p^2 C_2$ insertions and pion exchange corrections to the deuteron propagator depicted in the first line of Fig. 5; pionic vertex corrections to C_0 (second line); and the

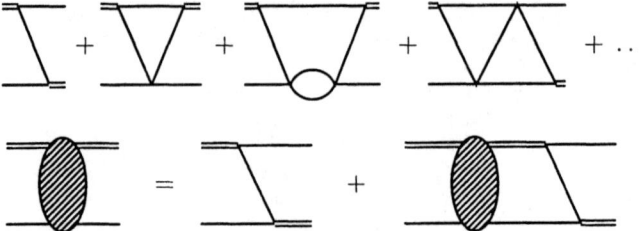

FIGURE 4. The double infinite series of LO "pinball" diagrams, some of which are shown in the first line, is equivalent to the solution of the Faddeev equation shown in the second line.

pion diagram of the last line which corrects the three particle intermediate state. Here, we used a re-formulation of the Lagrangean (1) in order not to have poorly convergent diagrams containing C_2 like the second one in the first line of Fig. 4, see [5] for details. The calculation without explicit pions was carried out to NNLO.

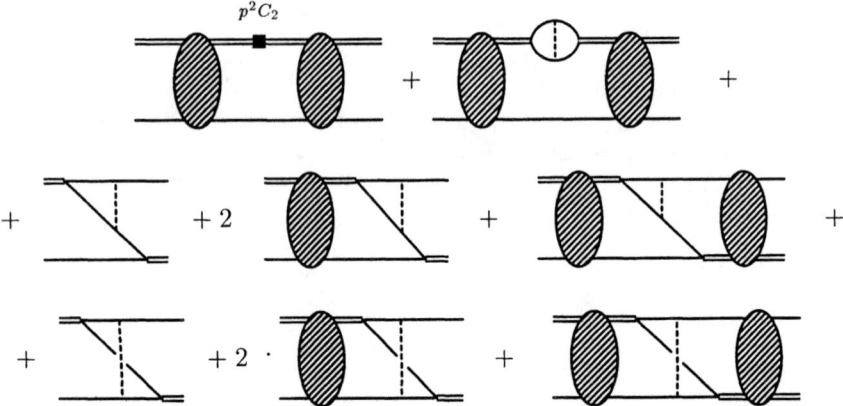

FIGURE 5. The NLO contributions to nd scattering in the quartet channel. First line: Corrections to the deuteron propagator; second line: pionic corrections to the C_0 vertex ; third line: pionic corrections to three particle breakup in the intermediate state. Permuted graphs left out.

All calculations demonstrate convergence. The scattering length is $a(^4S_{\frac{3}{2}}, \text{LO}) = (5.1 \pm 1.5)$ fm, and at NLO with (without) perturbative pions $a(^4S_{\frac{3}{2}}, \text{NLO}, \pi) = (6.8 \pm 0.7)$ fm ($a(^4S_{\frac{3}{2}}, \text{NLO}, \text{no}\pi) = (6.7 \pm 0.7)$ fm). At NNLO, [15] report $a(^4S_{\frac{3}{2}}, \text{NNLO}, \text{no}\pi) = (6.33 \pm 0.1)$ fm, and the experimental value is $a(^4S_{\frac{3}{2}}, \exp) = (6.35 \pm 0.02)$ fm [17]. Comparing the NLO correction to the LO scattering length provides one with the familiar error estimate at NLO: $(\frac{1}{3})^2 \approx 10\%$. The NLO calculations with and without pions lie within each other's error bar. The NNLO calculation is inside the error ascertained to the NLO calculation and carries itself an error of about $(\frac{1}{3})^3 \approx 4\%$. NLO and LO contributions become comparable for momenta of more than 200 MeV. In the imaginary part shown in Fig. 6, the same pattern emerges with a slightly more pronounced difference between the pion-less and pions-full theory. Because results obtained with EFT are easily dissected for the relative importance of the various terms, one concludes that pionic corrections to nd scattering in the quartet S wave channel – although formally NLO – are indeed much weaker: The calculation with perturbative pions and with pions integrated out do not differ significantly over a wide range of momenta. The difference should appear for momenta of the order of m_π and higher because of non-analytical contributions of the pion cut. However, it is very moderate for momenta of up to 300 MeV in the centre-of-mass frame ($E_{cm} \approx 70$ MeV), see Fig. 6. This and the lack of data makes it difficult to assess whether the KSW power counting scheme to include pions as perturbative increases the range of validity over the pion-less theory, but effects from the pion cut are seemingly weak.

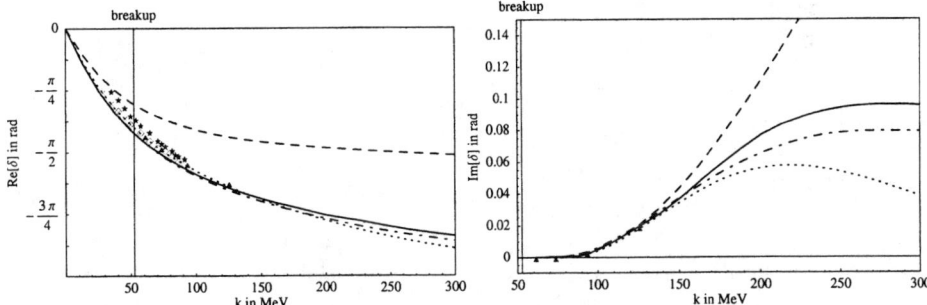

FIGURE 6. Real and imaginary parts in the quartet S wave phase shift of nd scattering versus the centre-of-mass momentum [5]. Dashed: LO; solid (dot-dashed) line: NLO with perturbative pions (pions integrated out); dotted: NNLO without pions [15,6]. Realistic potential models: squares from [18], crosses from [19], triangles from [20]. Stars: TUNL pd phase shift analysis [18].

Finally, the real and imaginary parts of the higher partial waves $l = 1, \ldots, 4$ in the spin quartet and doublet channel were presented in [6] in a papameter-free calculation. Figure 7 shows two examples. Comparison of the LO with the NLO and NNLO result demonstrates convergence of the EFT, with the expansion parameter again about $\frac{1}{3}$. It is interesting that the NLO correction is for high enough energies sometimes sizeable, while the NNLO correction is in general very small. Within the range of validity of this pion-less theory, convergence is good, and the results agree with potential model calculations (as available) within the theoretical uncertainty. That makes one optimistic about carrying out higher order calculations of problematic spin observables like the A_y problem where the EFT approach will differ from potential model calculations due to the inclusion of three-body forces.

Acknowledgements

It is a great pleasure to thank my collaborators – J.-W. Chen, R.P. Springer and

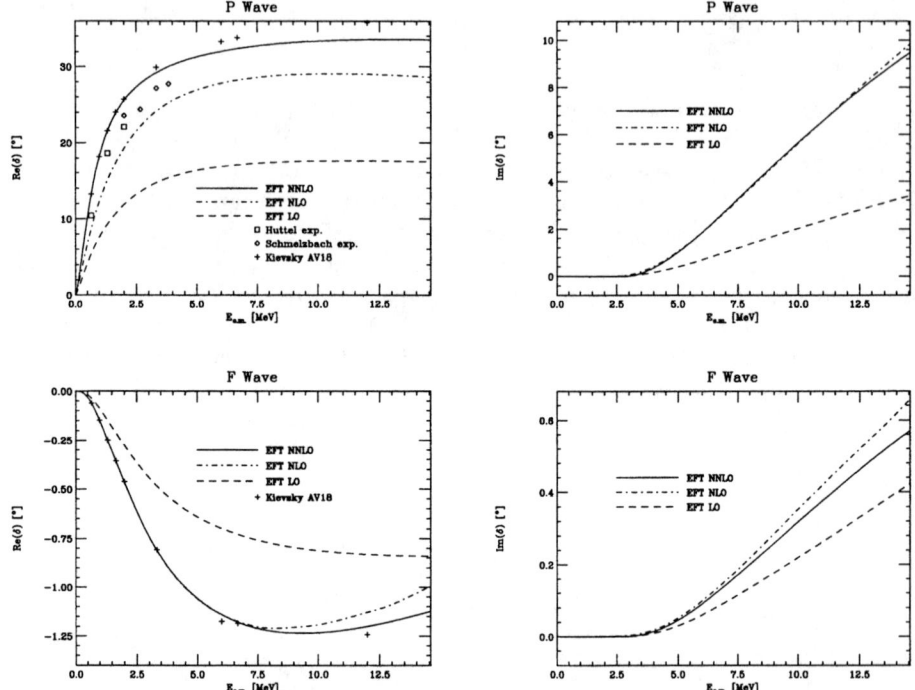

FIGURE 7. Real and imaginary parts of the quartet P (top) and doublet F (bottom) wave phase shift of nd scattering versus the centre-of-mass energy in the EFT without pions [6]. Dashed: LO; dot-dashed: NLO; solid line: NNLO. The experiments by Huttel et al. [21] and Schmelzbach et al. [22] are denoted by open squares and diamonds, respectively. The calculations of Kievsky et al. (crosses) are from Refs. [19] below breakup ($E_{cm} = B$) and [23] above breakup.

M.J. Savage in [3,4], P.F. Bedaque in [5,6], and F. Gabbiani in [6] – for a lot of fun, and the EFT group at the INT and the University of Washington in Seattle for a number of valuable discussions. The work was supported in part by the Department of Energy grant DE-FG03-97ER41014 and the Bundesministerium für Bildung und Forschung.

REFERENCES

1. G.P. Lepage, "What Is Renormalization?," *Invited lectures given at TASI'89 Summer School, Boulder, CO, Jun 4-30, 1989*, in: From Actions to Answers, T. DeGrand and D. Toussaint, eds., World Scientific 1990.
2. U. van Kolck, M. Savage and R. Seki, eds., "Nuclear Physics with Effective Field Theory", Proceedings of the INT-Caltech Workshop at Caltech (1998), World Scientific; P. Bedaque, U. van Kolck, M. Savage and R. Seki, eds., "Nuclear Physics

with Effective Field Theory II", Proceedings of the INT-Caltech Workshop at the INT (1999), World Scientific, in press.
3. J. Chen, H.W. Grießhammer, M.J. Savage and R.P. Springer, Nucl. Phys. **A644**, 221 (1998), nucl-th/9806080.
4. J. Chen, H.W. Grießhammer, M.J. Savage and R.P. Springer, Nucl. Phys. **A644**, 245 (1998), nucl-th/9809023.
5. P.F. Bedaque and H.W. Grießhammer, nucl-th/9907077 (to appear in Nucl. Phys. **A**).
6. F. Gabbiani, P.F. Bedaque and H.W. Grießhammer, *Higher Partial Waves in an Effective Field Theory Approach to nd Scattering*, preprint DOE/ER/40561-74-INT99, DUKE-TH-99-198, TUM-T39-99-23.
7. S. Weinberg, Nucl. Phys. **B363**, 3 (1991).
8. J. Gasser and H. Leutwyler, Ann. Phys. **158**, 142 (1984); S. Weinberg, Physica **96A**, 327 (1979).
9. J. Gasser, M.E. Sainio and A. Svarc, Nucl. Phys. **B307**, 779 (1988); E. Jenkins and A.V. Manohar, Phys. Lett. **B255**, 558 (1991).
10. J. Chen, G. Rupak and M.J. Savage, nucl-th/9902056.
11. D.B. Kaplan, M.J. Savage and M.B. Wise, Phys. Lett. **B424**, 390 (1998), nucl-th/9801034; D.B. Kaplan, M.J. Savage and M.B. Wise, Nucl. Phys. **B534**, 329 (1998), nucl-th/9802075.
12. D.R. Phillips, G. Rupak and M.J. Savage, nucl-th/9908054.
13. V. Bernard, N. Kaiser and U. Meissner, *Phys. Rev. Lett.* **67**, 1515 (1991); *Nucl. Phys.* B **373**, 364 (1992); *Phys. Lett.* B **319**, 269 (1993).
14. M.A. Lucas, Ph. D. thesis, University of Illinois at Urbana-Champaign (1994).
15. P.F. Bedaque, H.W. Hammer and U. van Kolck, Phys. Rev. **C58**, R641 (1998), nucl-th/9802057; P.F. Bedaque and U. van Kolck, Phys. Lett. **B428**, 221 (1998), nucl-th/9710073.
16. P.F. Bedaque, H.W. Hammer and U. van Kolck, Phys. Rev. Lett. **82**, 463 (1999), nucl-th/9809025; P.F. Bedaque, H.W. Hammer and U. van Kolck, Nucl. Phys. **A646**, 444 (1999), nucl-th/9811046; P.F. Bedaque, H.W. Hammer and U. van Kolck, nucl-th/9906032.
17. W. Dilg, L. Koester and W. Nistler, Phys. Lett. **B36**, 208 (1971).
18. W. Tornow and H. Witała, "Proton-Deuteron Phase-Shift Analysis Above the Deuteron Breakup Threshold", Technical Report TUNL XXXVI (1996-97).
19. A. Kievsky, S. Rosati, W. Tornow and M. Viviani, Nucl. Phys. **A607**, 402 (1996).
20. D. Hüber, J. Golak, H. Witała, W. Glöckle and H. Kamada, "Phase Shift and Mixing Parameters for Elastic nd Scattering above the Breakup Threshold".
21. E. Huttel, W. Arnold, H. Baumgart, H. Berg and G. Clausnitzer, *Nucl. Phys.* A **406**, 443 (1983).
22. P.A. Schmelzbach, W. Grübler, R.E. White, V. König, R. Risler and P. Marmier, *Nucl. Phys.* A **197**, 273 (1972).
23. A. Kievsky, private communication.

VARIOUS APPROACHES TO QCD

A renormalisation-group treatment of two-body scattering

Michael C. Birse, Judith A. McGovern and Keith G. Richardson

Theoretical Physics Group, Department of Physics and Astronomy
University of Manchester, Manchester, M13 9PL, UK

Abstract. A Wilsonian renormalisation group is used to study nonrelativistic two-body scattering by a short-ranged potential. We identify two fixed points: a trivial one and one describing systems with a bound state at zero energy. The eigenvalues of the linearised renormalisation group are used to assign a systematic power-counting to terms in the potential near each of these fixed points. The expansion around the nontrivial fixed point is shown to be equivalent to the effective-range expansion.

INTRODUCTION

Recently there has been much interest in the possibility of developing a systematic treatment of low-energy nucleon-nucleon scattering using the techniques of effective field theory [1–3]. Here we approach the problem using Wilson's continuous renormalisation group [4] to examine the low-energy scattering of nonrelativistic particles interacting through short-range forces [5].

The starting point for the renormalisation group (RG) is the imposition of a momentum cut-off, $|\mathbf{k}| < \Lambda$, separating the low-momentum physics which we are interested in from the high-momentum physics which we wish to "integrate out". Provided that there is a separation of scales between these two regimes, we may demand that low-momentum physics should be independent of Λ.

The second step is to rescale the theory, expressing all dimensioned quantities in units of Λ. As the cut-off Λ approaches zero, all physics is integrated out until only Λ itself is left to set the scale. In units of Λ any couplings that survive are just numbers, and these define a "fixed point". Such fixed points correspond to systems with no natural momentum scale. Examples include the trivial case of a zero scattering amplitude and the more interesting one of a bound state at exactly zero energy.

Real systems can then be described in terms of perturbations away from one of these fixed points. For perturbations that scale as definite powers of Λ, we can set up a power-counting scheme: a systematic way to organise the terms in an effective potential or an effective field theory. A fixed point is said to be stable if all perturbations vanish like positive powers of Λ as $\Lambda \to 0$ and unstable if one or more of them grows with a negative power of Λ.

TWO-BODY SCATTERING

We consider s-wave scattering by a potential that consists of contact interactions only. Expanded in powers of energy and momentum this has the form

$$V(k',k,p) = C_{00} + C_{20}(k^2 + k'^2) + C_{02}\, p^2 \cdots, \tag{1}$$

where k and k' denote momenta and energy-dependence is expressed in terms of the on-shell momentum $p = \sqrt{ME}$. Below all thresholds for production of other particles, this potential should be an analytic function of k^2, k'^2 and p^2.

Low-energy scattering is conveniently described in terms of the reactance matrix, K. This is similar to the scattering matrix T, except for the use of standing-wave boundary conditions. It satisfies the Lippmann-Schwinger (LS) equation (see [6])

$$K(k',k,p) = V(k',k,p) + \frac{M}{2\pi^2}\mathcal{P}\int q^2 dq \frac{V(k',q,p)K(q,k,p)}{p^2 - q^2}, \tag{2}$$

where $\mathcal{P}$ denotes the principal value.

On-shell, with $k = k' = p$, the K-matrix is related to the phase-shift by

$$\frac{1}{K(p,p,p)} = -\frac{M}{4\pi} p \cot \delta(p), \tag{3}$$

which means it has a simple relation to the effective-range expansion [7],

$$p \cot \delta(p) - \frac{1}{a} + \frac{1}{2} r_e p^2 + \cdots, \tag{4}$$

where a is the scattering length and r_e is the effective range. We shall see that this turns out to be equivalent to an expansion around a nontrivial fixed point of the RG.

RENORMALISATION GROUP

To set up the RG we first impose a momentum cut-off on the intermediate states in the LS equation (2). This can be written

$$K = V(\Lambda) + V(\Lambda) G_0(\Lambda) K, \tag{5}$$

where we have included a sharp cut-off in the free Green's function,

$$G_0 = \frac{M\theta(\Lambda - q)}{p^2 - q^2}. \tag{6}$$

We now demand that $V(k',k,p,\Lambda)$ varies with Λ in order to keep the off-shell K-matrix independent of Λ:

$$\frac{\partial K}{\partial \Lambda} = 0. \tag{7}$$

This is sufficient to ensure that all scattering observables do not depend on Λ. Differentiating the LS equation (5) with respect to Λ and then operating from the right with $(1+G_0 K)^{-1}$, we get

$$\frac{\partial V}{\partial \Lambda} = \frac{M}{2\pi^2} V(k',\Lambda,p,\Lambda) \frac{\Lambda^2}{\Lambda^2 - p^2} V(\Lambda,k,p,\Lambda). \tag{8}$$

We now introduce dimensionless momentum variables, $\hat{k} = k/\Lambda$ etc., and a rescaled potential,

$$\hat{V}(\hat{k}',\hat{k},\hat{p},\Lambda) = \frac{M\Lambda}{2\pi^2} V(\Lambda \hat{k}', \Lambda \hat{k}, \Lambda \hat{p}, \Lambda). \tag{9}$$

From the equation (8) satisfied by V we find that the rescaled potential satisfies the RG equation

$$\Lambda \frac{\partial \hat{V}}{\partial \Lambda} = \hat{k}' \frac{\partial \hat{V}}{\partial \hat{k}'} + \hat{k}\frac{\partial \hat{V}}{\partial \hat{k}} + \hat{p}\frac{\partial \hat{V}}{\partial \hat{p}} + \hat{V} + \hat{V}(\hat{k}',1,\hat{p},\Lambda)\frac{1}{1-\hat{p}^2}\hat{V}(1,\hat{k},\hat{p},\Lambda). \tag{10}$$

FIXED POINTS

We are now in a position to look for fixed points: solutions of (10) that are independent of Λ. These provide the possible low-energy limits of theories as $\Lambda \to 0$ and hence the starting points for systematic expansions of the potential.

The trivial fixed point

One obvious solution of (10) is the trivial fixed point,

$$\hat{V}(\hat{k}',\hat{k},\hat{p},\Lambda) = 0, \tag{11}$$

which describes a system with no scattering.

For systems described by potentials close to the fixed point we can expand in terms of eigenfunctions, $\hat{V} = \Lambda^\nu \phi(\hat{k}',\hat{k},\hat{p})$, of the linearised RG equation,

$$\hat{k}' \frac{\partial \phi}{\partial \hat{k}'} + \hat{k}\frac{\partial \phi}{\partial \hat{k}} + \hat{p}\frac{\partial \phi}{\partial \hat{p}} + \phi = \nu \phi. \tag{12}$$

These have the form

$$\hat{V}(\hat{k}',\hat{k},\hat{p},\Lambda) = C\Lambda^\nu \hat{k}'^l \hat{k}^m \hat{p}^n, \tag{13}$$

with eigenvalues $\nu = l+m+n+1$, where l, m and n are non-negative even integers. The eigenvalues are all positive and so the fixed point is a stable one: all nearby potentials flow towards it as $\Lambda \to 0$.

The corresponding unscaled potential has the expansion

$$V(\hat{k}',\hat{k},\hat{p},\Lambda) = \frac{2\pi^2}{M} \sum_{l,n,m} \hat{C}_{lmn}\Lambda_0^{-\nu} k'^l k^m p^n, \tag{14}$$

where we have written the coefficients in dimensionless form by taking out powers of Λ_0, the scale of the short-distance physics. The power counting in this expansion is just the one proposed by Weinberg [1] if we assign an order $d = \nu - 1$ to each term in the potential. This fixed point can be used to describe systems where the scattering at low energies is weak and can be treated perturbatively. It is not the appropriate starting point for s-wave nucleon-nucleon scattering, where the scattering length is large.

A nontrivial fixed point

The simplest nontrivial fixed point is one that depends on energy only, $\hat{V} = \hat{V}_0(\hat{p})$. It satisfies

$$\hat{p}\frac{\partial \hat{V}_0}{\partial \hat{p}} + \hat{V}_0(\hat{p}) + \frac{\hat{V}_0(\hat{p})^2}{1-\hat{p}^2} = 0. \tag{15}$$

The solution, which must be analytic in $\hat{p}^2$, is

$$\hat{V}_0(\hat{p}) = -\left[1 - \frac{\hat{p}}{2}\ln\frac{1+\hat{p}}{1-\hat{p}}\right]^{-1}. \tag{16}$$

Although the detailed form of this potential is specific to our particular choice of cut-off, the fact that it tends to a constant as $\hat{p} \to 0$ is a generic feature, which is present for any regulator.

The corresponding unscaled potential is

$$V_0(p,\Lambda) = -\frac{2\pi^2}{M}\left[\Lambda - \frac{p}{2}\ln\frac{\Lambda+p}{\Lambda-p}\right]^{-1}. \tag{17}$$

The solution to the LS equation for K with this potential is infinite, or rather $1/K = 0$. This corresponds to a system with infinite scattering length, or equivalently a bound state at exactly zero energy.

To study the behaviour near this fixed point we consider small perturbations about it that scale with definite powers of Λ:

$$\hat{V}(\hat{k}',\hat{k},\hat{p},\Lambda) = \hat{V}_0(\hat{p}) + C\Lambda^\nu \phi(\hat{k}',\hat{k},\hat{p}). \tag{18}$$

These satisfy the linearised RG equation

$$\hat{k}'\frac{\partial\phi}{\partial\hat{k}'} + \hat{k}\frac{\partial\phi}{\partial\hat{k}} + \hat{p}\frac{\partial\phi}{\partial\hat{p}} + \phi + \frac{\hat{V}_0(\hat{p})}{1-\hat{p}^2}\left[\phi(\hat{k}',1,\hat{p}) + \phi(1,\hat{k},\hat{p})\right] = \nu\phi. \tag{19}$$

Solutions to (19) that depend only on energy ($\hat{p}$) can be found straightforwardly by integrating the equation. They are

$$\phi(\hat{p}) = \hat{p}^{\nu+1}\hat{V}_0(\hat{p})^2. \tag{20}$$

Requiring that these be well-behaved as $\hat{p}^2 \to 0$, we find the RG eigenvalues $\nu = -1, 1, 3, \ldots$. The fixed point is unstable: it has one negative eigenvalue.

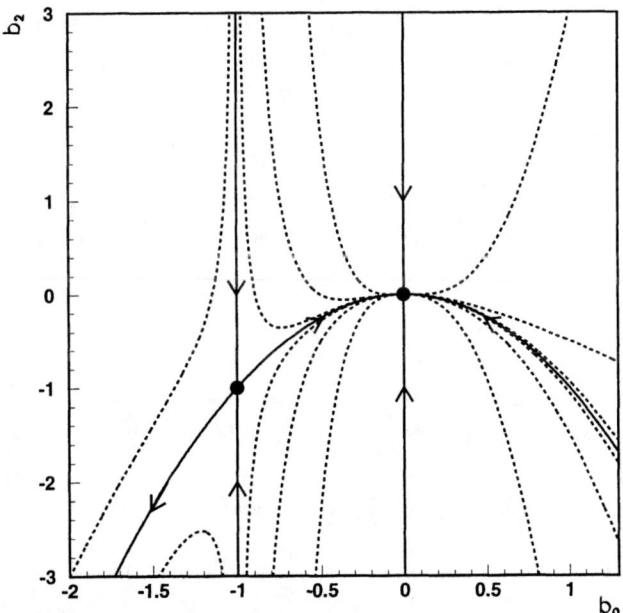

FIGURE 1. The RG flow of the first two terms in the expansion of the rescaled potential in powers of energy. The two fixed points are indicated by the black dots. The solid lines are flow lines that approach one of the fixed points along a direction corresponding to an RG eigenfunction; the dashed lines are more general flow lines. The arrows indicate the direction of flow as $\Lambda \to 0$.

The instability can be seen from the RG flow in Fig. 1. Only potentials that lie exactly on the "critical surface" flow into the nontrivial fixed point as $\Lambda \to 0$.

Any small perturbation away from this surface eventually builds up and drives the potential either to the trivial fixed point at the origin or to infinity.

The corresponding unscaled potential is

$$V(k',k,p,\Lambda) = V_0(p,\Lambda) + \frac{M}{2\pi^2}\left(C_{-1} + C_1 p^2 + \cdots\right) V_0(p,\Lambda)^2. \tag{21}$$

For perturbations around the nontrivial fixed point, we can assign an order $d = \nu - 1 = -2, 0, 2, \ldots$ to each term in the potential. This power counting for (energy-dependent) perturbations agrees with that found by Kaplan, Savage and Wise [2] using a "power divergence subtraction" scheme and also by van Kolck [3] in a more general subtractive renormalisation scheme. The equivalence can be seen by making the replacement

$$V_0 = -\frac{2\pi^2}{M\Lambda} + \cdots \to -\frac{4\pi}{M\mu}, \tag{22}$$

where μ is the renomalisation scale introduced by Kaplan, Savage and Wise in their subtraction scheme, and which plays an analogous role to the cut-off Λ in our approach.

The on-shell K-matrix for this potential is (to any order in the C's)

$$\frac{1}{K(p,p,p)} = -\frac{M}{2\pi^2}\left(C_{-1} + C_1 p^2 + \cdots\right). \tag{23}$$

This is just the effective-range expansion (4). There is a one-to-one correspondence between the perturbations in V and the terms in that expansion,

$$C_{-1} = -\frac{\pi}{2a}, \qquad C_1 = \frac{\pi r_e}{4}. \tag{24}$$

The expansion around the nontrivial fixed point is the relevant one for systems with large scattering lengths, such as s-wave nucleon-nucleon scattering.

WEAK LONG-RANGE FORCES

The treatment outlined above is only valid at very low momenta, where all pieces of the potential can be regarded as short-range. To extend it to describe nucleon-nucleon scattering at higher momenta, we would like to include pion-exchange forces explicitly. The longest-ranged of these is single pion exchange, which provides a central Yukawa potential,

$$V_{1\pi}(\mathbf{k}',\mathbf{k}) = -\frac{4\pi\alpha_\pi}{(\mathbf{k}-\mathbf{k}')^2 + m_\pi^2}, \tag{25}$$

where

$$\alpha_\pi = \frac{g_A^2 m_\pi^2}{16\pi f_\pi^2} \simeq 0.072. \qquad (26)$$

As in chiral perturbation theory, we want to treat the pion mass as a new low-energy scale (in addition to the momentum and energy variables). This can be done by defining a rescaled variable $\hat{m}_\pi = m_\pi/\Lambda$ and applying the RG as above. The corresponding term in the rescaled potential is

$$\hat{V}_{1\pi}(\hat{\mathbf{k}}', \hat{\mathbf{k}}, \hat{m}_\pi, \Lambda) = -\Lambda \frac{Mg_A^2}{8\pi^2 f_\pi^2} \frac{\hat{m}_\pi^2}{(\hat{\mathbf{k}} - \hat{\mathbf{k}}')^2 + \hat{m}_\pi^2}. \qquad (27)$$

It scales as Λ^1, like the effective-range term in the potential above. This suggests that one-pion exchange (OPE) can be treated as a perturbation. It would contribute at next-to-leading order (NLO) in the potential.

However questions remain about whether OPE is really weak enough for a perturbative treatment to be useful. A possible scale for nonperturbative long-range physics is the pionic "Bohr radius":

$$R = \frac{2}{\alpha_\pi M} \simeq 5.8 \text{ fm}. \qquad (28)$$

This should be compared with the range of the Yukawa potential, $r_\pi = 1/m_\pi = 1.4$ fm, which cuts off the potential at long distances, preventing the formation of a bound state. The ratio of these scales is

$$\frac{r_\pi}{R} \simeq 0.24, \qquad (29)$$

Although this is smaller than the critical value of 0.84, at which a bound state forms [6], one might expect relatively slow convergence of the perturbation series.

Further questions are raised when the contribution of OPE to the effective range is examined. A perturbative treatment (to NLO in an expansion in powers of momenta, m_π and $1/a$, as in [8]) gives a short-range contribution to the effective $^1S_0$ range of

$$r_e^0 = r_e - \frac{2\alpha_\pi M}{m_\pi^2} \qquad (30)$$
$$= 2.62 - 1.38 = 1.24 \text{ fm}. \qquad (31)$$

It is also possible to set up a distorted-wave effective-range expansion, in which the long-range interaction is treated all orders [9]. This is essentially an expansion in powers of energy of $p\cot(\delta - \delta_{1\pi})/|\mathcal{F}_{1\pi}(p)|^2$ where $\delta_{1\pi}$ is the OPE phase shift and $\mathcal{F}_{1\pi}(p)$ the corresponding Jost function [6]. The resulting purely short-range effective range is [10] (see also [11])

$$r_e^0 = 4.2 \text{ fm}. \qquad (32)$$

This is significantly different from the perturbatively corrected effective range (30). The difference may be an indication of either strong forces with two-pion range, or of strong short-range forces with a complicated structure [12].

SUMMARY

We have applied Wilson's renormalisation group to nonrelativistic two-body scattering and identified two important fixed points [5].

The first is the trivial fixed point. Perturbations around it can be used to describe systems with weak scattering. These perturbations can be organised according to Weinberg's power counting [1].

The second fixed point describes systems with a bound state at exactly zero energy. In this case the relevant power-counting is the one found by Kaplan, Savage and Wise [2] and van Kolck [3]. The expansion around this fixed point is exactly equivalent to the effective-range expansion.

These ideas can be extended in various ways. Short-range interactions in other numbers of spatial dimensions can be studied. The critical dimension for instability of the nontrivial fixed point is $D = 2$, which has been studied for some time in the context of anyons [13,14].

Three-body systems are also being studied from the point of view of effective field theory [15,16]. In some cases these display much more complicated behaviour under the RG than the two-body ones discussed above [17].

Various nucleon-nucleon scattering observables as well as deuteron properties have been calculated using the expansion around the nontrivial fixed point [2,18–20]. In this approach, pion-exchange forces are treated as perturbations. An alternative approach which is being explored by other groups is to use Weinberg's power counting in the expansion of the potential, but then to iterate that potential to all orders in the LS equation [21–25]. This may provide a way to evade the problems of slow convergence when OPE is included explicitly [8,12].

Finally, strong long-ranged interactions, such as the Coulomb force, lead to quite different behaviour from the examples discussed here. They can still be treated using similar techniques, as in NRQED [26] and NRQCD [27].

REFERENCES

1. S. Weinberg, *Phys. Lett.* **B251**, 288 (1990); *Nucl. Phys.* **B363**, 3 (1991).
2. D. B. Kaplan, M. J. Savage, and M. B. Wise, *Phys. Lett.* **B424**, 390 (1998); *Nucl. Phys.* **B534**, 329 (1998); *Phys. Rev.* **C59**, 617 (1999).
3. U. van Kolck, *Nucl. Phys.* **A645**, 273 (1999).
4. K. G. Wilson and J. G. Kogut, *Phys. Rep.* **12**, 75 (1974); J. Polchinski, *Nucl. Phys.* **B231**, 269 (1984).
5. M. C. Birse, J. A. McGovern and K. G. Richardson, hep-ph/9807302, *Phys. Lett.* **B** (in press).
6. R. G. Newton, *Scattering Theory of Waves and Particles*, New York: Springer-Verlag, 1982.
7. J. M. Blatt and J. D. Jackson, *Phys. Rev.* **76**, 18 (1949); H. A. Bethe, *Phys. Rev.* **76**, 38 (1949).
8. T. D. Cohen and J. M. Hansen, *Phys. Rev.* **C59**, 13 (1999).

9. M. van Haeringen and L. P. Kok, *Phys. Rev.* **A26**, 1218 (1982).
10. K. G. Richardson, Ph.D. thesis, University of Manchester, 1999.
11. J. V. Steele and R. V. Furnstahl, *Nucl. Phys.* **A645**, 439 (1999).
12. D. B. Kaplan and J. V. Steele, nucl-th/9905027.
13. R. Jackiw, in *M. A. B. Bég Memorial Volume*, edited by A. Ali and P. Hoodbhoy, (World Scientific, Singapore, 1991).
14. C. Manuel and R. Tarrach, *Phys. Lett.* **B328**, 113 (1994).
15. P. F. Bedaque and U. van Kolck, Phys. Lett. **B428**, 221 (1998); P. F. Bedaque, H.-W. Hammer and U. van Kolck, Phys. Rev. **C58**, R641 (1998).
16. P. F. Bedaque and H. W. Griesshammer, nucl-th/9907077.
17. P. F. Bedaque, H.-W. Hammer and U. van Kolck, *Phys. Rev. Lett.* **82**, 463 (1999); *Nucl. Phys.* **A646**, 444 (1999); nucl-th/9906032.
18. J.-W. Chen, H. W. Griesshammer, M. J. Savage and R. P. Springer, *Nucl. Phys.* **A644**, 221 (1998).
19. M. J. Savage, K. A. Scaldeferri and M. B. Wise, *Nucl. Phys.* **A652**, 273 (1999).
20. S. Fleming, T. Mehen and I. W. Stewart, nucl-th/9906056.
21. C. Ordonez, L. Ray and U. van Kolck, *Phys. Rev.* **C53**, 2086 (1996).
22. E. Epelbaum, W. Glöckle and U.-G. Meissner, *Nucl. Phys.* **A637**, 107 (1998); nucl-th/9910064.
23. U. van Kolck, nucl-th/9902015.
24. S. R. Beane, M. Malheiro, D. R. Phillips and U. van Kolck, *Nucl. Phys.* **A656**, 367 (1999).
25. D. R. Phillips and T. D. Cohen, nucl-th/9906091.
26. W. E. Caswell and G. P. Lepage, *Phys. Lett.* **B167**, 437 (1986); P. Labelle, *Phys. Rev.* **D58**, 093013 (1998).
27. G. T. Bodwin, E. Braaten and G. P. Lepage, *Phys. Rev.* **D51**, 1125 (1995) [erratum: *ibid.* **D55**, 5853 (1997)]; H. W. Griesshammer, *Phys. Rev.* **D58**, 094027 (1998).

QCD dynamics at $\theta \sim \pi$

Andrei V. Smilga*

*Université de Nantes, 2, Rue de la Houssinière, BP 92208, F-44322, Nantes CEDEX 3, France
and
ITEP, B. Cheremushkinskaya 25, Moscow 117218, Russia.

Abstract. Taking into account the terms $\sim m^2$ in the effective chiral lagrangian, we show that, at $\theta = \pi$, the theory with 2 light quarks of equal mass involves two degenerate vacuum states separated by a barrier. For $N_f = 3$, the energy barrier between two vacua appears already in the leading order in mass. This corresponds to the first order phase transition at $\theta = \pi$. The surface energy density of the domain wall separating two different vacua is calculated. In the immediate vicinity of the point $\theta = \pi$, two minima of the potential still exist, but one of them becomes metastable. The probability of the false vacuum decay is estimated.

It is very well known that in QCD with N_f massless quarks, chiral symmetry $SU(N_f) \otimes SU(N_f)$ is spontaneously broken down to $SU_V(N_f)$. (This is an experimental fact for $N_f = 2, 3$. Probably, no spontaneous chiral symmetry breaking occurs for large enough number of flavors $N_f \sim 8 - 10$ [1] which will be of no concern for us here). Spontaneous symmetry breaking means that the order parameter $< \bar{q}_R^i q_L^j >$ (i, j are flavor indices) can acquire an arbitrary direction in the flavor space. Massless Goldstone particles appear. If the free quark masses are not zero, the axial chiral symmetry is broken explicitly and the minimum of the energy functional corresponds to a particular flavor orientation of the condensate. Goldstones acquire small masses. In the real World with $m_u \approx 4$ MeV, $m_d \approx 7$ MeV, $m_s \approx 150$ MeV, and $\theta = 0$, the vacuum state is unique.

It is interesting to study also other variants of the theory with different values of masses and θ. It was known for a long time [2,3] that in the theory with equal light quark masses and $\theta = \pi$, there are two degenerate vacuum states. This is best seen in the framework of the effective chiral lagrangian describing only the light pseudogoldstone degrees of freedom. In the leading order in mass, the effective potential is

$$V = -\Sigma \, \text{Re} \left[\text{Tr} \left\{ \mathcal{M} e^{i\theta/N_f} U^\dagger \right\} \right] \qquad (1)$$

where $U = \exp\{2i\phi^a t^a / F_\pi\}$ (ϕ^a are pseudogoldstone fields), $\mathcal{M}$ is the quark mass matrix and Σ is the absolute value of the quark condensate.

Suppose $N_f = 3$, $\mathcal{M} = m\hat{1}$, and $\theta = 0$. The minimum of the energy is achieved at $U = \hat{1}$. For $\theta = \pi$, there are two different minima with $U = \hat{1}$ and $U = e^{2\pi i/3}\hat{1}$. They are separated by the energy barrier. The appearance of two vacuum states corresponds to spontaneous breaking of the CP–symmetry by the Dashen mechanism [4].

The situation is, however, more confusing for $N_f = 2$. The trace of a $SU(2)$ matrix is always real which means that, at $\mathcal{M} = m\hat{1}$ and $\theta = \pi$, the potential (1) does not depend on U at all. That would mean that the explicit breaking of the chiral symmetry is absent, pions are massless at this point and the phase transition would be of the second rather than of the first order. We understand, however, that the chiral symmetry of the original QCD lagrangian *is* broken explicitly by the quark mass term for all values of θ including $\theta = \pi$, and the situation looks paradoxical.

This question was left undiscussed in the original papers [2,3], and only comparatively recently it was shown how the paradox is resolved [5]. To this end, one should to take into account the terms of order $\sim m^2$ in the effective chiral lagrangian. If doing so, the continuous vacuum degeneracy at $\theta = \pi$ is lifted, and we obtain again only two vacuum states separated by a barrier. [1]

In what follows, we confirm this finding. The main aim of the paper is to bring the analysis of Ref. [5] into contact with standard chiral theory notations and wisdom and to perform some quantitative estimates both for $N_f = 3$ and $N_f = 2$ for the height of the energy barrier, the surface energy density of the domain walls interpolating between two degenerate vacua at $\theta = \pi$, and for the decay rate of metastable vacuum states at the vicinity of the phase transition point.

Let us discuss first in some details the case $N_f = 3$ where no complications due to higher-order terms arise. By a conjugation $U \to VUV^\dagger$, any unitary matrix U can be brought into the diagonal form $U = \text{diag}(e^{i\alpha}, e^{i\beta}, e^{-i(\alpha+\beta)})$. When $\mathcal{M} = m\hat{1}$, a conjugation does not change the potential (1). For diagonal U, the latter acquires the form

$$U(\alpha, \beta) = -m\Sigma \left[\cos\left(\alpha - \frac{\theta}{3}\right) + \cos\left(\beta - \frac{\theta}{3}\right) + \cos\left(\alpha + \beta + \frac{\theta}{3}\right)\right] \quad (2)$$

The function U has six stationary points:

$$\mathbf{I}: \alpha = \beta = 0, \quad \mathbf{II}: \alpha = \beta = -\frac{2\pi}{3}, \quad \mathbf{III}: \alpha = \beta = \frac{2\pi}{3} \quad \mathbf{IV}: \alpha = \beta = -\frac{2\theta}{3} + \pi,$$

$$\mathbf{IV}a: \alpha = -\alpha - \beta = -\frac{2\theta}{3} + \pi, \quad \mathbf{IV}b: \beta = -\alpha - \beta = -\frac{2\theta}{3} + \pi \quad (3)$$

---

[1] One could, of course, anticipate it. For $\theta = 0$ and in the leading order in quark mass, the mass of pions is given by the Gell-Mann – Oakes – Renner relation $F_\pi^2 M_\pi^2 = (m_u + m_d)\Sigma$. A theory with $m_u = m_d$ and $\theta = \pi$ is equivalent to the theory with $m_u = -m_d$ and $\theta = 0$ (only $\theta_{\text{phys}} = \theta + \arg(\det\mathcal{M})$ is relevant). In that case, the pions seem to stay massless. However, the Gell-Mann – Oakes – Renner relation is true only in the leading order in m_q. Higher order corrections bring about a nonzero mass to pions.

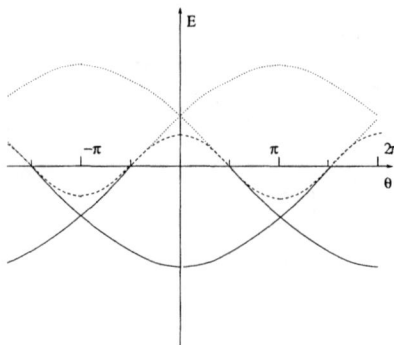

FIGURE 1. Stationary point of $E(\alpha,\beta)$ for different θ. Solid lines are the minima, dotted lines are the maxima, and dashed line are the saddle points

The points **IV**a and **IV**b are obtained from from **IV** by Weyl permutations and their physical properties are the same. Actually, we have here not 3 distinct stationary points, but the whole 4–dimensional manifold $SU(3)/[SU(2)\otimes U(1)]$ of the physically equivalent stationary points related to each other by conjugation. The values of the potential at the stationary points are

$$E_{\mathbf{I}} = -3m\Sigma\cos\frac{\theta}{3}, \quad E_{\mathbf{II}} = -3m\Sigma\cos\frac{\theta+2\pi}{3}, \quad E_{\mathbf{III}} = -3m\Sigma\cos\frac{\theta-2\pi}{3},$$
$$E_{\mathbf{IV}} = m\Sigma\cos\theta \quad (4)$$

Studying the expressions (4) and the matrix of the second derivatives at $\theta = \pi$, one can readily see that *i)* the points **I** and **III** are the degenerate minima; *ii)* the point **II** is the maximum, and *iii)* the points **IV** are saddle points. When θ is slightly less than π, **I** is a global minimum while **III** is still a minimum, but of a local variety. The latter coalesces with all three saddle points at $\theta = \pi/2$. At this point the eigenvalues of the second derivative matrix pass zero and, at still lower values of θ, a metastable minimum does not exist. When we instead make θ larger than π, the picture is symmetric, only the minima **I** and **III** change their roles. At $\theta = 0$, the picture is exactly reversed compared to what we had at $\theta = \pi$: there is one global minimum **I**, two degenerate maxima **II** and **III**, and a surface of saddle points **IV**.

Fig. 1 illustrates how the stationary points of the potential are moved when the vacuum angle is changed. One can show that metastable vacua are absent at $\theta = 0$ also with physical values of masses.

Two degenerate vacua at $\theta = \pi$ are separated by the domain wall. To find the profile of this wall, we have to restore the kinetic term $\frac{F_\pi^2}{4}\text{Tr}\{\partial_\mu U \partial_\mu U^\dagger\}$ in the effective lagrangian and seek for the field configurations depending only on one spatial coordinate x with the boundary conditions $U(-\infty) = 1$, $U(\infty) = e^{2i\pi/3}$ and realizing the minimum of the energy functional

$$E = \mathcal{A}\sigma = \mathcal{A}\int_{-\infty}^{\infty} dx \left(\frac{F_\pi^2}{4}\text{Tr}\{\partial_x U \partial_x U^\dagger\} - m\Sigma\text{Re}\left[e^{i\theta/3}\text{Tr}\{U^\dagger\}\right]\right) \quad (5)$$

($\mathcal{A}$ is the total area factor). In our case, it suffices to seek for the solutions in the class $U = \text{diag}(e^{i\alpha(x)}, e^{i\alpha(x)}, e^{-2i\alpha(x)})$. Introducing $\gamma = \alpha - \pi/3$, subtracting the vacuum energy and using the Gell-Mann – Oakes – Renner relation, the expression (5) is rewritten as

$$\sigma = 3F_\pi^2 \int_{-\infty}^{\infty} dx \left[\frac{1}{2}\gamma'^2 + \frac{M_\pi^2}{3}\left(\cos\gamma - \frac{1}{2}\right)^2 \right] \quad (6)$$

The corresponding equations of motion with the boundary conditions $\gamma(\pm\infty) = \pm\pi/3$ can be readily integrated. The first integral is $\gamma' = M_\pi\sqrt{2/3}(\cos\gamma - 1/2)$. Integrating it further, we obtain the solution

$$\cos\gamma = \frac{E^2 + 4E + 1}{2(E^2 + E + 1)} \quad (7)$$

where $E = \exp\{M_\pi x/\sqrt{2}\}$. The solution (7) is centered at $x = 0$ where it passes through the saddle point **IV**. There are, of course, other solutions obtained from Eq.(7) by a shift of x. Also we could have chosen the Ansätze $U = \text{diag}(e^{i\alpha(x)}, e^{-2i\alpha(x)}, e^{i\alpha(x)})$ or $U = \text{diag}(e^{-2i\alpha(x)}, e^{i\alpha(x)}, e^{i\alpha(x)})$ and obtain two other wall solutions (with the same properties) passing through the saddle points **IV**a and **IV**b. The wall surface tension is

$$\sigma = \frac{9}{\sqrt{2}}F_\pi^2 M_\pi \int_0^\infty \frac{E dE}{(E^2 + E + 1)^2} = 3\sqrt{2}\left(1 - \frac{\pi}{3\sqrt{3}}\right)M_\pi F_\pi^2 \quad (8)$$

Suppose now that $\theta = \pi + \phi$ with $0 < |\phi| \ll 1$. The energies of the vacua are not degenerate anymore but are splitted apart by the value

$$\Delta\mathcal{E} \approx m\Sigma\sqrt{3}|\phi| \quad (9)$$

A metastable vacuum should decay with the formation of bubbles of the stable phase. The quasiclassical formula for the decay rate per unit time per unit volume was derived in [6]:

$$\Gamma \propto \exp\left\{-\frac{27}{2}\pi^2\frac{\sigma^4}{(\Delta\mathcal{E})^3}\right\} \quad (10)$$

where σ is the surface tension the bubble. Substituting here Eqs.(8, 9) [Strictly speaking, at $\theta \neq \pi$, the bubble surface tension does not coincide with Eq.(8) but is somewhat less going to zero at $\theta = \pi/2$ or $\theta = 3\pi/2$. But at small $|\phi|$, the expression (8) is correct], we obtain [2]

---

[2] Note the difference with the rough estimate $\ln\Gamma \sim -F_\pi^2/(M_\pi^2|\phi|^2)$ for the same quantity in the Witten's paper [3]. First, one has $|\phi|^3$ rather than $|\phi|^2$ in the denominator and, second, a huge numerical factor pops up.

$$\Gamma \propto \exp\left\{-\frac{CF_\pi^2}{M_\pi^2|\phi|^3}\right\} \tag{11}$$

with $C = 2^4 \cdot 3^5 \sqrt{3}\pi^2[1 - \pi/(3\sqrt{3})]^4$. Note that the numerical factor C in the exponent is tremendously large, i.e. lifetime of metastable states would be tremendously large (much larger than the lifetime of the Universe) for almost all θ in the interval $(\frac{\pi}{2}, \frac{3\pi}{2})$ not too close to its boundaries (where metastable states disappear). It is a real pity that such a beautiful possibility is not realized in Nature [3].

As was mentioned before, we have to take into account here the terms of higher order in mass in the effective potential. For $N_f = 2$, it is convenient to make benefit of the fact that $SU(2) \otimes SU(2) \equiv O(4)$ and to use the 4–vector notations so that $U = U_\mu \sigma_\mu = U_0 + iU_i\sigma_i$, $U_\mu^2 = 1$. The 2×2 complex mass matrix involves 8 real parameters which is convenient to "organize" in two different isotopic 4– vectors [4]

$$\begin{aligned}\chi_\mu &= \frac{\Sigma}{F_\pi^2}\left(\text{Re Tr}\{\mathcal{M}e^{i\theta/2}\},\ \text{Im Tr}\{\mathcal{M}e^{i\theta/2}\sigma_i\}\right) \\ \tilde{\chi}_\mu &= \frac{\Sigma}{F_\pi^2}\left(\text{Im Tr}\{\mathcal{M}e^{i\theta/2}\},\ -\text{Re Tr}\{\mathcal{M}e^{i\theta/2}\sigma_i\}\right)\end{aligned} \tag{12}$$

In the second order in $\chi, \tilde{\chi}$, the most general form of the potential is [8]

$$V(U_\mu) = -F_\pi^2(\chi_\mu U_\mu) - l_3(\chi_\mu U_\mu)^2 - l_7(\tilde{\chi}_\mu U_\mu)^2 \tag{13}$$

where $l_{3,7}$ are some dimensionless coefficients (the coefficients $l_{1,2,4,5,6}$ multiply the structures involving the derivatives of the field U in the effective lagrangian). The term $\propto (\chi_\mu U_\mu)(\tilde{\chi}_\mu U_\mu)$ is not allowed because it would lead to CP breaking even at $\theta = 0$.

The term $\sim (\chi_\mu U_\mu)^2$ can always be neglected compared to the leading one for small masses and is not interesting. On the contrary, the term involving l_7 has a different θ – dependence and, for $\mathcal{M} = m\hat{1}$ and $\theta \sim \pi$ when the leading term vanishes, determines the whole dynamics.

Before proceeding further, let us try to extract an information on the numerical value of the constant l_7. The following relation belonging to the same class as the well–known Weinberg sum rule and derived in Ref. [8] is very useful:

$$-8\left(\frac{\Sigma}{F_\pi^2}\right)^2 l_7 \delta^{ik} = \int d^4x \left[<S^i(x)S^k(0)> - \delta^{ik} <P^0(x)P^0(0)>\right] \tag{14}$$

where $S^i = \bar{q}\sigma^i q$ and $P^0 = i\bar{q}\gamma^5 q$. Let us calculate the right–hand side of Eq. (14) as an Euclidean functional integral. Let us first calculate the correlators in

---

[3] Recently, Halperin and Zhitnitsky argued the existence of metastable states in the real QCD at $\theta = 0$ [7]. However, their arguments were based on a particular model form of the effective potential incorporating also glueball degrees of freedom and involving certain cusps. The status of this potential is not quite clear by now.
[4] We use the notations of Ref. [8].

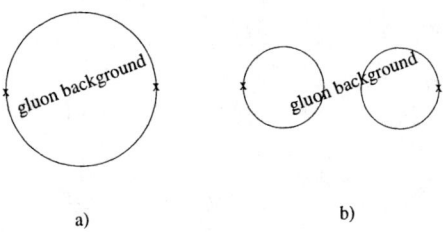

FIGURE 2. Connected and disconnected contributions to the quark current correlators.

a *particular gauge field background.* Only the connected quark diagram depicted in Fig. 2a contributes to the correlator of the scalar isovector densities. The pseudoscalar correlator receives also a contribution from the disconnected graph in Fig. 2b. The solid lines in Fig. 2 stand for the quark Green's functions in a particular gauge field background for which we use the spectral decomposition

$$G_A(x,y) = \sum_n \frac{\psi_n(x)\psi_n^\dagger(y)}{m - i\lambda_n} \qquad (15)$$

where λ_n are the eigenvalues of the *massless* Euclidean Dirac operator in an external gauge field and $\psi_n(x)$ are its eigenfunctions. All non–zero eigenvalues are paired: for any eigenfunction $\psi_n(x)$ with non-zero eigenvalue λ_n, $\tilde\psi_n(x) = \gamma^5\psi_n(x)$ is also an eigenfunction with $\tilde\lambda_n = -\lambda_n$. There are also zero modes. For each flavor, their number coincides with the topological charge $\nu = (1/32\pi^2)\int d^4x G_{\mu\nu}^a \tilde G_{\mu\nu}^a$ of the gauge field configuration.

Now, for large Euclidean volumes $mV\Sigma \gg 1$, zero mode contribution is irrelevant for the *connected* graphs (cf. the discussion in Ref. [9]). On the other hand, the *only* contribution in the disconnected graph for the pseudoscalar correlator is due to zero modes. This contribution is very large $\propto \nu^2/m^2$ and is of a paramount importance. Plugging in the Green's functions (15) in the correlators in Eq.(14), pairing together positive and negative λ, and integrating over gauge fields, we obtain for large volumes

$$2\left(\frac{\Sigma}{F_\pi^2}\right)^2 l_7 = \frac{1}{V}\left[\left\langle {\sum_n}' \frac{m^2 - \lambda_n^2}{(m^2+\lambda_n^2)^2}\right\rangle + \left\langle {\sum_n}' \frac{1}{m^2+\lambda_n^2}\right\rangle - \frac{<\nu^2>}{m^2}\right] \qquad (16)$$

where $\sum_n'$ means the summation over positive eigenvalues only and the symbol $<\ldots>$ stands for the gauge field averaging. The first term in the R.H.S. comes from the scalar isovector correlator, the second term is the contribution of the connected graph to the pseudoscalar correlator, and the last term is due to the disconnected graph.

Introducing the spectral density $\rho(\lambda) = \langle \sum_n \delta(\lambda - \lambda_n)\rangle$, the relation (16) is rewritten as

$$l_7 = \frac{F_\pi^4}{\Sigma^2}\left[m^2\int_0^\infty \frac{\rho(\lambda)d\lambda}{(\lambda^2+m^2)^2} - \frac{<\nu^2>}{2m^2 V}\right] \tag{17}$$

This relation belongs to the same class as the famous Banks and Casher relation for the fermion condensate

$$\Sigma = 2m\int_0^\infty \frac{\rho(\lambda)d\lambda}{\lambda^2+m^2} = \pi\rho(0) + O(m) \tag{18}$$

For some more examples, see Refs. [10,11].

Both terms in (17) are singular $\propto 1/m$ in the chiral limit. The singularity cancels out, however. Indeed, the leading infrared contribution in the first term in square brackets is $\pi\rho(0)/(4m) \sim \Sigma/(4m)$ which is the same as for the second term due to the known result for the topological succeptibility in a theory with light quarks of the same mass [12]

$$\frac{1}{V}<\nu^2> = \frac{m\Sigma}{N_f} + O(m^2) \tag{19}$$

The absense of singularity in the isoscalar pseudoscalar correlator means that the corresponding meson is massive: $U(1)$ problem is resolved by the 't Hooft mechanism due to fermion zero modes in topologically non–trivial gauge backgrounds. l_7 is given thereby by a constant term $\sim O(1)$ which is left out after the cancellation of singular terms. This constant is completely determined by the term $\propto m^2$ in $<\nu^2>$. Indeed, for $N_f = 2$, the spectral density is analytic at $\lambda = 0$: $\rho(\lambda) = \rho(0) + \mu\lambda^2 + \ldots$ [10]. The extra infrared contribution is of order $O(m)$ and can be neglected. [5]. We finally obtain

$$l_7 = \lim_{m\to 0}\frac{F_\pi^4}{4\Sigma^2}\left[\frac{\Sigma - 2<\nu^2>/(mV)}{m}\right] \tag{20}$$

What can be said about m–dependence of the topological succeptibility in the next–to–leading order in mass ? Let us first see what happens in a theory with two light quarks embedded in the theory involving also the third quark which is much more massive but still light enough (like it is the case in the real World). The topological succeptibility in the theory with 3 light quarks $m_u = m_d \equiv m \ll m_s < \mu_{\rm hadr}$ is given by the expression [2,3]

$$\frac{<\nu^2>}{V} = \frac{mm_s\Sigma}{m+2m_s} = \frac{m\Sigma}{2} - \frac{m^2\Sigma}{4m_s} + O(m^3) \tag{21}$$

i.e. the second term is *negative* [6] which means that l_7 as given by Eq.(20) is *positive*. We obtain

---

[5] Note also that the spectral integral in Eq.(17) involves a logarithmic *ultraviolet* singularity at large λ where the spectral density is the same as for free fermions $\rho(\lambda) \sim \lambda^3$. It is multiplied, however, by m^2 and can be dropped out by that reason.

[6] That is quite natural, of course. The presence of an extra light quark brings about the suppression of large ν due to the extra factor m_s^ν in the fermion determinant.

$$l_7 = \frac{F_\pi^4}{8m_s\Sigma} \equiv \frac{F_\pi^2}{6M_\eta^2} \approx 5\cdot 10^{-3} \qquad (22)$$

The same result could be obtained in a more direct way if saturating the pseudoscalar correlator in Eq.(14) by the η - meson pole [8].

The estimate (22) is quite good for the real QCD. What we are interested in here, however, is a hypothetical theory with $\theta \sim \pi$ and just two light flavors. Remarkably, an analytic result for l_7 can be obtained also in this case if the number of colors N_c is assumed to be large. For large N_c, the axial $U(1)$ symmetry is almost not affected by the anomaly which means that η' - meson is relatively light: $M_{\eta'}^2 \sim \mu_{\text{hadr}}^2/N_c$. We can saturate now the pseudoscalar correlator by the η' pole to obtain

$$l_7 = \frac{F_\pi^2}{2M_{\eta'}^2} \qquad (23)$$

The same result can be obtained via the relation (20). For large N_c and $N_f = 2$, the topological succeptibility is known to be [13,3]

$$\frac{1}{V}<\nu^2> = \frac{m\tau\Sigma}{2\tau+m\Sigma} = \frac{m\Sigma}{2} - \frac{m^2\Sigma^2}{4\tau} + O(m^3) \qquad (24)$$

where τ is the topoligical succeptibility in the pure Yang–Mills theory. Again, the term $\sim m^2$ in $<\nu^2>/V$ is negative which leads to the positive l_7 which coincides with (23) due to the relation $F_\pi^2 M_{\eta'}^2 = 4\tau$ which holds in the limit $m \to 0$. We assume that l_7 is positive also for small number of colors down to $N_c = 2$. Indeed, in the limit $m \to \infty$, the topological succeptibility should coincide with τ. It is natural that the series in m for small masses [the analog of Eq.(24)] should have alternating signs. We cannot, unfortunately, formulate this statement (the positiveness of l_7) as an exact theorem though our *suspicion* is that such a theorem can somehow be proven.

We are ready now to discuss the vacua dynamics in the region $\theta \sim \pi$. Assume $U = \text{diag}(e^{i\alpha}, e^{-i\alpha})$. The potential (13) (with $l_3 = 0$) is

$$V(\alpha) - -2m\Sigma\cos\frac{\theta}{2}\cos\alpha - 4l_7 m^2 \sin^2\frac{\theta}{2}\left(\frac{\Sigma}{F_\pi^2}\right)^2 \cos^2\alpha \qquad (25)$$

Defining $\phi = \theta - \pi$, it can be rewritten for small $|\phi|$ as

$$V(\alpha) = m\Sigma\phi\cos\alpha - 4l_7 m^2 \left(\frac{\Sigma}{F_\pi^2}\right)^2 \cos^2\alpha \qquad (26)$$

If

$$|\phi| < \phi_* = \frac{8l_7 m\Sigma}{F_\pi^4}, \qquad (27)$$

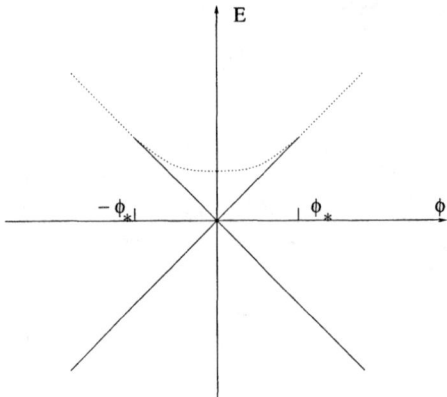

FIGURE 3. Stationary point of $E(\alpha)$ for $N_f = 2$ in the region of small $|\phi| = |\theta - \pi|$. Solid lines are minima and dotted lines are maxima.

the function (26) has four stationary points:

$$\begin{aligned}
&\textbf{I} \quad \alpha = 0 \quad \text{with} \quad E_\textbf{I} = m\Sigma\phi - 4l_7 m^2 \left(\frac{\Sigma}{F_\pi^2}\right)^2 \\
&\textbf{II} \quad \alpha = \pi \quad \text{with} \quad E_\textbf{II} = -m\Sigma\phi - 4l_7 m^2 \left(\frac{\Sigma}{F_\pi^2}\right)^2 \\
&\textbf{III, IIIa} \quad \alpha = \pm\arccos\frac{\phi F_\pi^4}{8l_7 m\Sigma} \quad \text{with} \quad E_\textbf{III} = \frac{F_\pi^4 \phi^2}{16 l_7}
\end{aligned} \qquad (28)$$

Again, the points **III, IIIa** are related to each other by Weyl symmetry and we have actually the whole surface $SU(2)/U(1) \equiv S^2$ of the equivalent stationary points. Studying the second derivatives $\partial^2 V/\partial\alpha^2$ for the branches (28), one readily sees that, in the region (27), the points **I** and **II** present local minima (for $\phi < 0$ or $\theta < \pi$ the absolute minimum is **I** while **II** is a metastable state; for $\theta > \pi$, it is the other way round). The points **III** are degenerate maxima. The picture is depicted in Fig. 3. Physically, it is exactly the same as in the case $N_f = 3$ and we have a first order phase transition. Only the width of the region in θ where two local minima coexist is much more narrow than in the case $N_f = 3$ and goes to zero in the chiral limit $m \to 0$.

Note that with negative l_7, the picture would be reversed compared to that in Fig. 3 so that we would have in the region (27) a surface of degenerate *minima* at the points **III**. That corresponds to spontaneous breaking of flavor symmetry $SU_V(2)$ [5] (when $\theta \neq 0$, the contributions to the partition function coming from topologically nontrivial sectors are not positively defined, and the Vafa–Witten theorem [14] which prohibits spontaneous breaking of vector flavor symmetry at $\theta = 0$ does not apply). Besides, at $\theta \neq \pi$ we would have the maxima of inequal height at the points **I** and **II**. Instead of one first order phase transition at $\theta = \pi$ we would have two consequent second order phase transitions at $\theta = \pi \pm \phi_*$. We find this picture rather unnatural and consider it as an additional argument why l_7 should be always positive.

It is not difficult now to perform the same program as for $N_f = 3$ and to find the surface energy density of the domain wall at $\theta = \pi$ and the decay rate of metastable vacua at the vicinity of $\theta = \pi$. The wall configuration and the energy density at $\theta = \pi$ are obtained by minimizing the functional

$$\sigma = \int_{-\infty}^{\infty} \left[\frac{F_\pi^2}{2} (\partial_x \alpha)^2 - 2l_7 m^2 \left(\frac{\Sigma}{F_\pi^2} \right)^2 (\cos 2\alpha - 1) \right] dx \tag{29}$$

with the boundary conditions $\alpha(-\infty) = 0$, $\alpha(\infty) = \pi$. The equations of motion have a simple solution

$$\alpha(z) = 2 \arctan \left[\exp \left\{ \sqrt{8l_7} \frac{m\Sigma}{F_\pi^3} x \right\} \right] \tag{30}$$

The surface energy density is $\sigma = m\Sigma/(F_\pi)\sqrt{32l_7}$. It is much lower numerically than in the case $N_f = 3$ and goes to zero in the chiral limit. The rate of metastable vacuum decay at $|\phi| \ll \phi_*$ is estimated as

$$\Gamma \propto \exp \left\{ -12^3 \pi^2 l_7^2 \frac{m\Sigma}{F_\pi^4 |\phi|^3} \right\} \tag{31}$$

Acknowledgements: It is a pleasure to thank the organizers for the inspired scientific atmosphere of the meeting. I am indebted to M. Creutz, H. Leutwyler and M. Shifman for illuminating discussions.

REFERENCES

1. T. Banks and A. Zaks, *Nucl. Phys.* **B82** (1982) 196; T. Appelquist and S.B. Selipsky, *Phys. Rev. Lett.* **79** (1997) 2767.
2. P. di Vecchia and G. Veneziano, *Nucl. Phys.* **B171** (1980) 253.
3. E. Witten, *Ann. Phys.* **128** (1980) 363.
4. R.F. Dashen, *Phys. Rev.* **D3** (1971) 1879.
5. M. Creutz, *Phys. Rev.* **D52** (1995) 2951.
6. M.B. Voloshin, I.Yu. Kobzarev, and L.B. Okun, *Sov. J. Nucl. Phys.* **20** (1975) 644.
7. I. Halperin and A. Zhitnitsky, hep-ph/9803301.
8. J. Gasser and H. Leutwyler, *Ann. Phys.* **158** (1984) 142.
9. H. Leutwyler and A.V. Smilga, *Phys. Rev.* **D46** (1992) 5607.
10. A.V. Smilga and J. Stern, *Phys. Lett.* **B318** (1993) 531.
11. A.V. Smilga and J.J.M. Verbaarschot, *Phys. Rev.* **D54** (1996) 1087.
12. R.J. Crewther, *Phys. Lett.* **70B** (1977) 349.
13. C. Rozenzweig, J. Schechter, and T. Trahern, *Phys. Rev.* **D21** (1980) 3388.
14. C. Vafa and E. Witten, *Nucl. Phys.* **B234** (1984) 173.

Wilson's Area Law and String Effective Actions for Confining Gauge Theories

D. Antonov* and D. Ebert[†]

*INFN-Sezione di Pisa, Università degli studi di Pisa,
Dipartimento di Fisica, Via Buonarroti, 2 - Ed. B - 56127 Pisa, Italy
and
Institute of Theoretical and Experimental Physics,
B. Cheremushkinskaya 25, RU-117 218 Moscow, Russia
[†] Theoretical Physics Division, CERN
CH - 1211 Geneva 23
and
Institut für Physik, Humboldt-Universität
10115 Berlin, Germany

Abstract. After reviewing the rôle of centre vortices and monopoles in Wilson's area law of confinement, we estimate the string tension for an ensemble of non-interacting centre vortices. Next, we discuss the string representations of dual superconductor type $SU(2)$- and $SU(3)$-gauge models and their application to the evaluation of field strength correlators. The large-distance asymptotic behaviours of the latter are shown to be in agreement with the stochastic vacuum model of QCD and existing lattice data.

INTRODUCTION

A very important question in quantum chromodynamics (QCD) is a deeper understanding of the mechanism of confinement, which should enable us to construct a realistic low-energy Lagrangian of hadron physics. Clearly, a necessary first step in this program is to find the dynamical degrees of freedom, which play a dominant role in confinement. As it turned out, the method of Abelian projection proposed by 't Hooft [1] is a powerful concept, which, by the use of so-called Abelian gauges, fixes the non-Abelian part of the gauge redundancy, thus reducing a given gauge group $SU(N)$ ($N = 2, 3, \ldots$) to its maximal Abelian subgroup $U(1)^{N-1}$ containing the centre $Z(N)$ as a subgroup. This procedure introduces magnetic excitations such as vortices (strings) and monopoles, which are generated from singular gauge transformations and eventually represent the IR degrees of freedom relevant to confinement.

A commonly accepted criterion of quark confinement is the area law of the Wilson loop [2], which follows from the non-trivial linking of magnetic centre vortices with the Wilson loop [3]– [5]. The vortices in such a "centre vortex picture" of confinement form closed magnetic flux lines (closed magnetic Dirac strings) in three space dimensions and two-dimensional world-sheets in $D = 4$ Euclidean space-time. In particular, the quantization of the magnetic flux leads to an expression for the Wilson loop in terms of centre elements $z \in Z(N)$ [5], [6]. Notice that, for N colours, $N - 1$ different kinds of vortices (strings) are obtained corresponding to the $N - 1$ non-trivial centre elements of $SU(N)$. In the first part of this talk, we illustrate the centre vortex picture of confinement for a simple model given by a gas of non-interacting centre vortices. To fix notations, we find it useful to first review the classification of magnetic vortices as well as magnetic monopoles in terms of the magnetic weight lattice and the magnetic root lattice [6].

Another very useful concept for explaining confinement is the dynamical scheme of a dual superconductor proposed twenty years ago by 't Hooft and Mandelstam [7]. This approach emphasizes, in particular, the role of magnetic monopoles. The condensation of Cooper pairs of magnetic monopoles squeezes the chromoelectric flux into (open) Abrikosov–Nielsen–Olesen (ANO) type vortices [8], which confine the quark and antiquark sitting at their ends. Recently, considerable evidence has been accumulated for an existing dual Meissner effect in numerical lattice calculations, demonstrating condensation of magnetic monopoles below a critical temperature T_c [9]. In the following we are mainly interested in analytical approaches towards confinement based on the derivation of a low-energy effective Lagrangian of the Ginzburg–Landau (GL) type. Indeed, the summation of monopole trajectories with inclusion of a suitable repulsion leads to a GL-theory with magnetic Higgs fields [10], [11]. The use of path integral duality transformations [12] then leads in a straightforward way, to a dual formulation, where monopole interactions take a simple form. In particular, on this basis it is possible to derive a string representation for $SU(2)$- and $SU(3)$-Abelian-projected gauge theories given in the form of effective GL-models [13–15]. This enables one to determine the string tension and to calculate other interesting quantities like field strength correlators.

In the second part, we review recent results of the dual superconductor approach to confinement for Abelian-projected $SU(2)$- and $SU(3)$-models [13], [14]. Here, it turns out that the estimated behaviour of field strength correlators is in agreement with the stochastic vacuum model (SVM) of QCD [16] and with lattice data [17].

WILSON'S AREA LAW: CENTRE VORTICES VERSUS MONOPOLES

An important electric order parameter of QCD characterizing the confinement of quarks is the average of the Wilson loop

$$< W(C) > = < \frac{1}{N} \operatorname{Tr} P \exp ig \oint_C A_\mu dx_\mu > , \qquad (1)$$

which is nothing else than an averaged amplitude of the process of creation, propagation and annihilation of a quark–antiquark pair. Here P denotes path ordering and, for generality, we consider the gauge group $SU(N)$. In the confinement phase, the Wilson loop average satisfies for large contours C the area law

$$\langle W(C) \rangle \sim e^{-\sigma|\Sigma_{\text{min.}}(C)|} \sim e^{-(\sigma R)T}, \tag{2}$$

Here, in the last relation the contour C has been chosen as a rectangular one in space-time, $C = R \times T$, and the area of the minimal surface $\Sigma_{\text{min.}}(C)$, bounded by this contour has been denoted as $|\Sigma_{\text{min.}}(C)|$. Note that the prefactor of the time T in the exponent of (2) defines the linearly increasing confinement potential $V(R) \sim \sigma R$ for a static quark–antiquark pair, σ being the string tension. The evaluation of this quantity is one of the central tasks in studying confining gauge theories. In particular, there arises an important question: what kind of field configurations (semiclassical "background" ones and/or quantum fluctuations) in the path-integral average $< \ldots >$ of eq. (1) is required to saturate Wilson's area law? In what follows we shall argue that the string tension might get contributions from the semiclassical background of centre vortices as well as from quantum correlations of monopoles.

Singular Field Configurations

A very useful concept for studying confinement is the method of Abelian projection proposed by 't Hooft [1]. In this approach it is convenient to decompose the gauge fields into Abelian (neutral) fields and non-Abelian (charged) parts. For example, in the case of $SU(3)$:

$$A_\mu(x) = (A_\mu^3(x)T^3 + A_\mu^8(x)T^8) + \sum_{a \neq 3,8} A_\mu^a(x)T^a, \tag{3}$$

where $T^{3,8}$ are diagonal generators living in the maximal Abelian subgroup $U(1)^2$ of $SU(3)$. The idea of Abelian projection is then to remove, by partially fixing the gauge, as many non-Abelian degrees of freedom as possible, leaving for $SU(N)$ the maximal Abelian subgroup $U(1)^{N-1}$ unbroken. This procedure allows, in particular, for singular gauge transformations, which define a suitable scheme for introducing magnetic monopoles and centre vortices:

$$A'_\mu = UA_\mu U^{-1} - \frac{i}{g}(\partial_\mu U)U^{-1} \equiv A_\mu^q + A_\mu^{M,V}. \tag{4}$$

Here the first non-singular term describes usual quantum fluctuations of the gauge field, whereas the monopole and vortex parts arise from singular (multivalued) gauge transformations. Notice, in particular, that centre vortices require gauge transformations from the N-fold, connected factor group $SU(N)/Z(N)$, i.e., $U \in SU(N)/Z(N)$, satisfying the condition $U(2\pi) = zU(0)$, $z \in Z(N)$, when going around the vortex core [18]. Here the centre Z_N is the discrete subgroup of $SU(N)$ whose elements commute with all other group elements,

$$Z_N = \left\{ e^{\frac{2\pi i}{N} k} \mathbb{1} \right\}, \qquad k = 0, 1, \ldots, N-1. \tag{5}$$

In the evaluation of (1) it is useful to employ the Stokes theorem for Abelian-projected (neutral) fields

$$\oint_C A_\mu(x) \, dx_\mu = \frac{1}{2} \int_{\Sigma(C)} d\sigma_{\mu\nu}(x(\xi)) F_{\mu\nu}(x(\xi)), \tag{6}$$

where $d\sigma_{\mu\nu}(x(\xi))$ is the surface element and $\xi \equiv (\xi_1, \xi_2)$ parametrizes the surface $\Sigma(C)$ bounded by the contour C. For singular gauge transformations (4) the singular field strength contributions describing monopoles and centre vortices take the form of surface integrals over open or closed Dirac sheets, respectively,

$$F'^{M,V}_{\mu\nu} = \frac{i}{g}(U[\partial_\mu, \partial_\nu]U^{-1})^a T^a \equiv \vec{m}_V^M \vec{T} \int_{\Sigma^{open}_{closed}} d\tilde{\sigma}_{\mu\nu}(x(\xi)) \, \delta^4(x - x(\xi)), \tag{7}$$

where $\vec{m}$ is a $(N-1)$-dimensional vector describing the magnetic charges of monopoles or centre vortices, and $d\tilde{\sigma}_{\mu\nu} = \frac{1}{2}\epsilon_{\mu\nu\rho\lambda}d\sigma_{\rho\lambda}$. Moreover, $\vec{T} = \{T^a\}$ denotes a vector whose components are given by the $(N-1)$ diagonal generators of $U(1)^{N-1}$. It is important to notice that in order to describe centre vortices or magnetic monopoles, a singular gauge transformation U should satisfy the following conditions [5], [6]

$$\text{monopoles}: \; P \, e^{\oint (\partial_\mu U) U^{-1}} = \mathbb{1}; \qquad \text{vortices}: \; P \, e^{\oint (\partial_\mu U) U^{-1}} \in Z_N. \tag{8}$$

The first requirement (8) is equivalent to the relation

$$e^{ig\vec{m}_M \cdot \vec{T}} = \mathbb{1}, \tag{9}$$

which leads to a quantization of the magnetic charge by generalized Dirac quantization conditions. The magnetic charges satisfying (9) take the form

$$\vec{m}_M = \frac{4\pi}{g} \sum_{a=1}^{N-1} n_a \vec{\epsilon}_a, \tag{10}$$

where n_a are integers, and $\vec{\epsilon}_a$ are the root vectors of the magnetic root lattice [6]. Analogously, the second condition (8) is equivalent to the quantization condition

$$e^{ig\vec{m}_V \cdot \vec{T}} = e^{\frac{2\pi i}{N} k} \mathbb{1} \equiv z(k) \in Z_N, \tag{11}$$

which can be solved by expressing the magnetic vortex charge in terms of weight vectors $\vec{\eta}_a \equiv (\vec{T})_{aa}$:

$$\vec{m}_V = \frac{4\pi}{g} \sum_{a=1}^{N-1} n_a \vec{\eta}_a .\tag{12}$$

For subsequent applications, it is convenient to rewrite the second condition (8) into a form relating the Wilson loop integral to the so-called linking number. Indeed, by inserting $F_{\mu\nu}^V$ (7) into (6), we get

$$\oint_C dx_\mu(\tau) A_\mu^V(x(\tau)) = \vec{m}_V \cdot \vec{T} \left[\frac{1}{2} \int_{\Sigma(C)} d\sigma_{\mu\nu}(x(\xi)) \int_{\Sigma_{\text{clos}}} d\tilde{\sigma}_{\mu\nu}(x(\xi')) \delta^4(x(\xi) - x(\xi')) \right]$$
$$\equiv \vec{m}_V \cdot \vec{T} \; I \; (\Sigma(C), \Sigma_{\text{clos}})$$
$$\equiv \vec{m}_V \cdot \vec{T} \; L \; (C, \Sigma_{\text{clos}}) .\tag{13}$$

Here I denotes the intersection number of the two two-dimensional surfaces $\Sigma(C)$ and Σ_{clos} which is equal to the linking number L of the contour C with the closed surface Σ_{clos}, given by [19]

$$L = \frac{1}{4\pi^2} \oint_{\Sigma_{\text{clos}}} d\sigma_{\mu\nu}(x) \oint_C dy_\beta \frac{1}{2} \epsilon_{\alpha\beta\mu\nu} \frac{(y-x)_\alpha}{|y-x|^4} .\tag{14}$$

Using (11) and (13) leads to the following expression

$$P \; e^{ig \oint A_\mu^V dx_\mu} = z^{L(C, \Sigma_{\text{clos}})} ,\tag{15}$$

where $z \in Z_N$. Equation (15) generalizes a similar expression obtained in Ref. [20] to the case of general centre vortices living in the magnetic weight lattice of $SU(N)$. Let us next apply (15) in order to determine the string tension for a simple model of non-interacting centre vortices.

String tension from centre vortices

For simplicity, let us consider the gauge group $SU(2)$, which has only one nontrivial centre element $z = e^{i\pi} = -1$. It is useful to embed the area $\Sigma(C)$ spanned by the contour C of the Wilson loop into a larger surface F, which at the end tends to "infinity". An ensemble of N centre vortices can then be decomposed into n_{in} vortices that pierce the minimal surface $\Sigma(C)_{\text{min.}}$ of the Wilson loop and n_{out} vortices that pierce the outer surface $F - \Sigma_{\text{min.}}(C)$, where

$$N = n_{in} + n_{out} .\tag{16}$$

Let the probability for a centre vortex to pierce "in" or "out" of $\Sigma_{\text{min.}}(C)$ be given by

$$p_{in} = B|\Sigma_{\text{min.}}(C)|e^{-S_0} ,$$
$$p_{out} = B|F - \Sigma_{\text{min.}}(C)|e^{-S_0} .\tag{17}$$

Here B is a characteristic quantity proportional to the inverse vortex area, and the weight factor e^{-S_0} is fixed by the vortex action $S_0 \sim const/g^2$ playing the role of a chemical potential. For a grand canonical ensemble of centre vortices, the Wilson loop average is then given by

$$\langle W(C) \rangle \propto \sum_{N=0}^{\infty} \sum_{n_{\rm in}, n_{\rm out}} \frac{1}{n_{\rm in}!} \left[B |\Sigma_{\rm min.}(C)| e^{-S_0} (-1)^{L(C, \Sigma_{\rm closed})} \right]^{n_{\rm in}}$$
$$\times \frac{1}{n_{\rm out}!} \left[B |F - \Sigma_{\rm min.}(C)| e^{-S_0} \right]^{n_{\rm out}} = \exp\{-\sigma_V |\Sigma_{\rm min.}(C)|\}, \qquad (18)$$

where the string tension σ_V, saturated by centre vortices, reads

$$\sigma_V = B\, e^{-S_0} \left[(-1)^{1+L} + 1 \right] \stackrel{L=1}{\to} 2 B e^{-S_0}. \qquad (19)$$

Clearly, the quantities B and S_0 have to be determined by a more quantitative analysis from the underlying gauge theory. The centre vortex picture of confinement with randomly distributed, uncorrelated ensembles of vortices has also recently been used for studying the nature of the deconfining phase transition in QCD [21].

What about monopole contributions to the area law? From (9), (7) and using Stokes theorem, one easily obtains

$$P e^{ig \oint A_\mu^M dx_\mu} = \mathbf{1}^I = \mathbf{1}. \qquad (20)$$

Thus, monopoles do not contribute in leading order to Wilson's area law. On the other hand, monopoles are important for topological reasons, e.g., for getting a non-vanishing Pontryagin index ν, which is required to describe the more general situation of a θ vacuum [22], [20]. Concerning confinement due to monopoles, it turns out that it is just the interaction of (condensed) monopoles with quantum fluctuations of the dual gauge field that leads to non-vanishing contributions to the string tension. Non-vanishing contributions to confinement are expected both from the cluster expansion of the SVM [16] and from dual superconductor models of the Ginzburg–Landau type [13], [14].

In the second part of this talk we shall consider the role of interacting monopoles for confinement and discuss the resulting string representations of gauge theories.

STRING REPRESENTATION OF DUAL SUPERCONDUCTOR TYPE MODELS

It is commonly argued that the Abelian-projected $SU(2)$-gluodynamics is just the Dual Abelian Higgs Model (DAHM). Its partition function reads [13]

$$Z = \int |\phi| D|\phi| D\theta D B_\mu \, \exp \left\{ -\int d^4x \left[\frac{1}{4}(F_{\mu\nu} - F_{\mu\nu}^E)^2 \right. \right.$$
$$\left. \left. + \frac{1}{2}|D_\mu \phi|^2 + \lambda(|\phi|^2 - \eta^2)^2 \right] \right\}, \qquad (21)$$

where $\phi(x) = |\phi(x)|e^{i\theta(x)}$ is the effective Higgs field of "Cooper pairs" of magnetic monopoles, B_μ and $F_{\mu\nu} = \partial_\mu B_\nu - \partial_\nu B_\mu$ are the dual gauge field and its field strength tensor, and $D_\mu = \partial_\mu - 2ig_m B_\mu$ is the covariant derivative with g_m standing for the magnetic coupling constant. Notice that $F^E_{\mu\nu}$ denotes the field strength tensor generated by external quarks,

$$F^E_{\mu\nu} = \frac{2\pi}{g_m} \int_{\Sigma(C)} d\tilde{\sigma}_{\mu\nu}(x(\xi))\delta^4(x - x(\xi)) \equiv \frac{2\pi}{g_m}\Sigma^E_{\mu\nu} . \qquad (22)$$

By comparing with (7), we see that in the dual theory external quarks are treated like monopoles. It is convenient to consider the London limit $\lambda \to \infty$, where the radial part of the monopole Higgs field becomes fixed to its v.e.v., $|\phi| \to \eta$. Next one performs a decomposition of the phase of the magnetic Higgs field, $\theta = \theta^{sing} + \theta^{reg}$, where the multivalued part $\theta^{sing}(x)$ describes a given electric string configuration, while $\theta^{reg}(x)$ stands for a single-valued fluctuation around such a configuration. Owing to the fact that the singularities of the phase of the magnetic Higgs field occur at the world-sheets of closed ANO-type strings [8], there exists a correspondence between θ^{sing} and closed-string world-sheets, given by the equation

$$\epsilon_{\mu\nu\lambda\rho}\partial_\lambda\partial_\rho\theta^{sing}(x) = 2\pi\Sigma_{\mu\nu}(x)$$
$$\equiv 2\pi \int_{\Sigma_{clos}} d\sigma_{\mu\nu}(x(\xi))\,\delta^4(x - x(\xi)) . \qquad (23)$$

Here $x(\xi) \equiv x_\mu(\xi_1, \xi_2)$ is a vector parametrizing the world-sheet Σ_{clos}. This correspondence then enables one to reformulate the integration over θ^{sing}'s as an integration over $x_\mu(\xi)$'s. Moreover, the integration over θ^{reg} leads to a δ-function constraint, which can be resolved by introducing a massive antisymmetric tensor field $h_{\mu\nu}(x)$ (Kalb–Ramond field). As a result of such "path integral duality transformations", the partition function (21) takes the form [13] [1]

$$Z = \int Dx_\mu(\xi) \int D\,h_{\mu\nu} \exp\left\{-\int d^4x \left[\frac{1}{12\eta^2}H^2_{\mu\nu\lambda}\right.\right.$$
$$\left.\left.+g_m^2\,h^2_{\mu\nu} + i\pi\,(2\Sigma^E_{\mu\nu} - \Sigma_{\mu\nu})h_{\mu\nu}\right]\right\} , \qquad (24)$$

where $H_{\mu\nu\lambda}$ denotes the kinetic field term, $H_{\mu\nu\lambda} = \partial_\mu h_{\nu\lambda} + \partial_\lambda h_{\mu\nu} + \partial_\nu h_{\lambda\mu}$ [2]. Finally, by performing the Gaussian integral over $h_{\mu\nu}$ and discarding closed surfaces $\Sigma_{\mu\nu}$, we obtain

$$Z = \exp\left[-\frac{\eta}{4g_m}\oint_C dx_\mu \oint_C dy_\mu \frac{K_1(m|x - y|)}{|x - u|}\right] \times$$
$$\times \int Dx_\mu(\xi) \exp\left[-\frac{g_m^2\eta^3}{2}\int d^4x \int d^4y\, \Sigma^E_{\mu\nu}(x)\frac{K_1(m|x - y|)}{|x - y|}\Sigma^E_{\mu\nu}(y)\right] , \qquad (25)$$

---

[1] We distinguish here between open and closed Dirac world-sheets.
[2] It is worth noticing that the string action (24) is a special case of a generic Julia–Toulouse mechanism [23] of confinement of $(h - 1)$ branes by condensation of $(D - h - 3)$ branes in a compact antisymmetric tensor field theory of rank h. In the above case we have $h = 1$.

where $m = 2g_m\eta$ is the mass of the dual gauge boson generated by the Higgs mechanism and K_1 is a modified Bessel function. Note that the first exponent on the right-hand side of (25) leads to the short-range Yukawa potential $V_{Yuk}(R) \propto \frac{1}{R} e^{-mR}$. The second exponent defines the string effective action $S_{Str}^{SU(2)}$. Obviously, it describes a bilocal interaction of surface elements on the world-sheet $\Sigma(C)$, spanned by the Wilson contour C of external quarks, via the exchange of massive (dual) gluons. Performing the expansion of the string effective action in powers of the derivatives w.r.t. ξ^a's, it can be shown that the first two terms of this expansion are the standard Nambu–Goto one and the so-called rigidity term. The coupling constant of the Nambu–Goto term (string tension) with logarithmic accuracy reads $\sigma \simeq \pi\eta^2 \ln \frac{\sqrt{\lambda}}{g_m}$, while the inverse coupling constant of the rigidity term has the form $1/\alpha_0 = -\pi/32g_m^2$. The Nambu–Goto action just yields the required linearly rising confinement potential. Notice also the negative sign of the coupling α_0, which reflects the stability of strings.

Another important issue investigated in Ref. [13] is the evaluation of the irreducible bilocal field strength correlator (cumulant) $\ll \tilde{F}_{\lambda\nu}(x)\tilde{F}_{\mu\rho}(0) \gg$ by a derivation of its string representation. Following the SVM [16], one can parametrize the bilocal correlator by two functions $D(x^2), D_1(x^2)$. For illustration, let us quote the IR asymptotic behaviour of the functions D, D_1 at large distances $|x| \gg 1/m$:

$$D \to \frac{m^4}{4\sqrt{2}\pi^{3/2}} \frac{e^{-m|x|}}{m|x|}, \quad D_1 \to \frac{m^4}{2\sqrt{2}\pi^{3/2}} \frac{e^{-m|x|}}{(m|x|)^{5/2}}. \quad (26)$$

This behaviour is very similar to the one observed in the lattice simulations of QCD in Ref. [17]. In particular, one can see that the rôle of the so-called correlation length of the vacuum T_g, at which the cumulant in SVM decreases, is played in DAHM by the inverse mass of the dual gauge boson, $T_g \sim 1/m$. The above considerations can easily be generalized to an effective Abelian-projected $SU(3)$ gauge theory with monopoles [11]. Discarding here external quarks and considering again the London limit, one arrives at the following string effective action [14]

$$S_{Str}^{SU(3)} = g_m\eta^3 \sqrt{\frac{3}{2}} \int d^4x \int d^4y \left[\Sigma_{\mu\nu}^1(x)\Sigma_{\mu\nu}^1(y) + \right.$$
$$\left. +\Sigma_{\mu\nu}^1(x)\Sigma_{\mu\nu}^2(y) + \Sigma_{\mu\nu}^2(x)\Sigma_{\mu\nu}^2(y)\right] \frac{K_1(m_B|x-y|)}{|x-y|}, \quad (27)$$

where $m_B = \sqrt{6}g_m\eta$ is the mass of the dual gauge fields $\vec{B} = (B_\mu^3, B_\mu^8)$. Thus, the most crucial difference of the $SU(3)$ string effective theory w.r.t. the $SU(2)$ case is the presence of two independent kinds of string world-sheets, which not only self-interact, but also interact with each other by the exchanges of the massive dual gauge bosons. Concerning cumulants of the field strength tensors $\tilde{F}_{\mu\nu}^{3,8}$, the vacuum again does exhibit a non-trivial correlation length $T_g = 1/m$. In particular, the IR asymptotics (26) of the functions D, D_1 remains valid for the cumulants with the replacement $m \to m_B$.

SUMMARY AND CONCLUSIONS

In this talk we have reviewed two approaches to confinement based on the hypothesis of Abelian dominance: firstly a "centre vortex picture" leading to Wilson's area law by using a gas of randomly distributed free vortices and, secondly, the "dual superconductor picture" with monopole condensation based on effective Lagrangians of the Ginzburg–Landau type. In the first kind of approach, a simple model of vortices and without monopoles has been shown to generate confinement. Clearly, a diluted gas of free vortices is only a rather crude model, where important dynamical quantities such as the characteristic inverse vortex area B and the vortex action have yet to be determined from the underlying gauge theory. Nevertheless, models of this type are close in spirit to the idea of "centre dominance" applied in recent lattice calculations, according to which centre (vortex) degrees of freedom rather than monopoles encode all long-distance physics. In particular, it has been shown that for vortex-free ensembles on the lattice, confinement is lost [24].

On the contrary, in the dual superconductor approach magnetic monopoles and dual Abelian gluons are considered as being the essential dynamical quantities in quark confinement. We emphasize that closed magnetic vortices play here a minor role and that it is just by the condensate of monopoles that the effective Lagrangian approach is able to produce electric confinement of quarks. Concerning this second approach, we have, in particular, discussed string representations of $SU(2)$- and $SU(3)$-gluodynamics by making use of the Abelian projection method and of so-called path integral duality transformations and then determined the string tension of the Nambu–Goto term and the coupling constant of the so-called rigidity term. Furthermore, the coefficient functions D, D_1 parametrizing the bilocal field strength correlator turn out to be in agreement with the SVM of QCD and existing lattice data. These results together with SVM evidently support the 't Hooft–Mandelstam conjecture of the dual superconductor nature of confinement.

ACKNOWLEDGEMENTS

One of the authors (D.E.) thanks Prof. J. da Providência and the Organizing Committee for the invitation to this conference and the kind hospitality. We are grateful to Profs. A. Di Giacomo and H. Reinhardt for fruitful discussions.

REFERENCES

1. 't Hooft, G., *Nucl.Phys.* **B190**, 455 (1981);
   Kronfeld, A.S., Schierholz, G., and Wiese, U.-J., *Nucl.Phys.* **B293**, 461 (1987).
2. Wilson, K.G., *Phys.Rev.* **D10**, 2445 (1974).
3. 't Hooft, G., *Nucl.Phys.* **B138**, 1 (1978).
4. Cornwall, M., *Nucl.Phys.* **B157**, 392 (1979).

5. Goddard, P., Nuyts, J. and Olive, D., *Nucl.Phys.* **B125**, 1 (1977);
   Englert, F. and Windey, P., *Phys.Rev.* **D14**, 2728 (1976).
6. Ezawa, Z.F. and Iwazaki, A., *Phys.Rev.* **D23**, 3036 (1981); *ibid* **D24**, 2264 (1981).
7. 't Hooft, G., in *High Energy Physics*, ed. Zichichi, A., Bologna: Editrice Compositori, 1976;
   Mandelstam, S., *Phys.Rep.* **C23**, 245 (1976).
8. Abrikosov, A.A., *Soviet Phys. J.E.T.P.* **5**, 1174 (1957);
   Nielsen, H.B. and Olesen, P., *Nucl.Phys.* **B61**, 45 (1973).
9. Di Giacomo, A. and Paffuti, G., *Phys.Rev.* **D56**, 6816 (1997).
10. Bardakci, K. and Samuel, S., *Phys.Rev.* **D18**, 2849 (1978);
    Samuel S., *Nucl.Phys.* **B154**, 62 (1979).
11. Maedan S. and Suzuki, T., *Progr.Theor.Phys.* **81**, 229 (1989).
12. Lee, K., *Phys.Rev.* **D48**, 2493 (1993);
    Orland, P., *Nucl.Phys.* **B428**, 221 (1994);
    Akmedov, E.T., Chernodub, M.N., Polikarpov, M.I. and Zubkov, M.A., *Phys.Rev.* **D53**, 2087 (1996).
13. Antonov, D. and Ebert, D., *Eur.Phys.J.* **C8**, 343 (1999).
14. Antonov, D. and Ebert, D., *Phys.Lett.* **B444**, 208 (1998).
15. Komarov, D.A. and Chernodub M.N., *JETP Lett.* **68**, 117 (1998)).
16. Dosch, H.G., *Phys.Lett.* **B190**, 177 (1987);
    Simonov, Yu.A., *Nucl.Phys.* **B307**, 512 (1988);
    Dosch, H.G., *Prog.Part.Nucl.Phys.* **33**, 121 (1994);
    Simonov, Yu.A., *Phys.Usp.* **39**, 313 (1996).
17. Di Giacomo, A. and Panagopoulos, H., *Phys.Lett.* **B285**, 133 (1992);
    Di Giacomo, A., Meggiolaro, E. and Panagopoulos, H., *Nucl.Phys.* **483**, 371 (1997).
18. Tze, H.C. and Ezawa, Z.F., *Phys.Rev.* **D14**, 1006 (1976); *ibid* **D15**, 1647 (1977).
19. Horowitz, G.T. and Srednicki, M., *Commun.Math.Phys.* **130**, 83 (1990).
20. Engelhardt, M. and Reinhardt, H., hep-th/9907139 (1999);
    See also: Reinhardt H., Talk at this Conference.
21. Engelhardt, M., Langfeld, K., Reinhardt, H. and Tennert, O., *Phys.Lett.* **B431**, 141 (1998); hep-ph/9908370 (1999).
22. Ezawa, Z.F. and Iwazaki, A., *Phys.Rev.* **D26**, 631 (1982).
23. Quevedo, F. and Trugenberger, C.A., *Nucl.Phys.* **B501**, 143 (1997) ;
    Julia, B. and Toulouse, G., *J.Physique-Lett.* **40**, 396 (1979).
24. De Forcrand, Ph. and D'Elia, M., *Phys.Rev.Lett.* **82**, 4582 (1999).

Magnetic Monopoles, Center Vortices, Confinement and Topology of Gauge Fields

H. Reinhardt, M. Engelhardt, K. Langfeld, M. Quandt, A. Schäfke

Institut für Theoretische Physik
Universität Tübingen
Auf der Morgenstelle 14
D-72076 Tübingen, Germany

Abstract. The vortex picture of confinement is studied. The deconfinement phase transition is explained as a transition from a phase in which vortices percolate to a phase of small vortices. Lattice results are presented in support of this scenario. Furthermore the topological properties of magnetic monopoles and center vortices arising, respectively, in Abelian and center gauges are studied in continuum Yang-Mills-theory. For this purpose the continuum analog of the maximum center gauge is constructed.

INTRODUCTION

Recent lattice calculations have given strong evidence for two confinement scenarios: 1. the dual Meissner effect [1], which is based on a condensate of magnetic monopoles in the QCD vacuum and 2. the center vortex picture [2], where the vacuum consists of a condensate of magnetic flux tubes which are closed due to the Bianchi identity. There are also lattice calculations which indicate that the spontaneous breaking of chiral symmetry, which can be related to the topology of gauge fields, is caused by these objects, i.e. by either magnetic monopoles [5] or center vortices [6]. In this talk we would like to discuss the confinement and topological properties of magnetic monopoles and center vortices. We will first discuss the two confinement scenarios based on magnetic monopoles and vortices, respectively, and subsequently investigate the topological properties of these objects. In particular, we will study the nature of the deconfinement phase transition in the center vortex picture. We will also show that in Polyakov gauge the magnetic monopoles completely account for the non-trivial topology of the gauge fields. Subsequently, we will extend the notion of center vortices to the continuum. We will present the continuum analog of the maximum center gauge fixing and the Pontryagin index of center vortices.

CONFINEMENT

The magnetic monopoles arise in Yang-Mills-Theories in the so called Abelian gauges [7]. Recent lattice calculations have shown that below a critical temperature T_C these monopoles are condensed [10] and give rise to the dual Meißner effect. In particular in the so called maximal Abelian gauge where all links are made as diagonal as possible, one observes Abelian and monopole dominance in the string tension [1]. However, very recent lattice calculations [11] also show that the Yang-Mills ground state does not look like a Coulombic monopole gas but rather indicate a collimation of magnetic flux, which is consistent with the center vortex picture of confinement, proposed in refs. [12], [13], [15], [14].

Center vortices represent closed magnetic flux lines in three space dimensions, describing closed two-dimensional world-sheets in four space-time dimensions. The magnetic flux represented by the vortices is furthermore quantized such that a Wilson loop linking vortex flux takes a value corresponding to a nontrivial center element of the gauge group. In the case of $SU(2)$ colour the only such element is (-1). For N colours, there are $N-1$ different possible vortex fluxes corresponding to the $N-1$ nontrivial center elements of $SU(N)$. Center vortices can be regarded as a fraction of a Dirac string: N superimposed center vortices form an unobservable Dirac string.

Consider an ensemble of center vortex configurations in which the vortices are distributed randomly, specifically such that intersection points of vortices with a given two-dimensional plane in space-time are found at random, uncorrelated locations. In such an ensemble, confinement results in a very simple manner. Let the universe be a cube of length L, and consider a two-dimensional slice of this universe of area L^2, with a Wilson loop embedded into it, circumscribing an area A. On this plane, distribute N vortex intersection points at random, cf. Fig. 1 (left). According to the specification above, each of these points contributes a factor (- 1) to the value of the Wilson loop if it falls within the area A spanned by the loop; the probability for this to occur for any given point is A/L^2.

The expectation value of the Wilson loop is readily evaluated in this simple model. The probability that n of the N vortex intersection points fall within the area A is binomial, and, since the Wilson loop takes the value $(-1)^n$ in the presence of n intersection points within the area A, its expectaton value is

$$<W> = \sum_{n=0}^{N}(-1)^n \binom{N}{n} \left(\frac{A}{L^2}\right)^n \left(1-\frac{A}{L^2}\right)^{N-n}$$
$$= \left(1-\frac{2\rho A}{N}\right)^N \stackrel{N\to\infty}{\longrightarrow} exp(-2\rho A), \qquad (1)$$

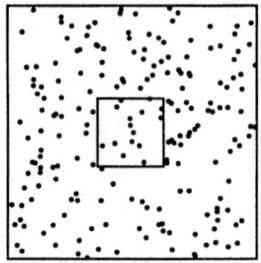 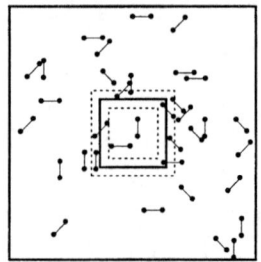

Figure 1. Two-dimensional plane of a random vortex gas with a planar Wilson loop. The dots represent the intersection points of the vortices with the plane considered. Left figure: confining vortex ensemble of uncorrelated intersection points. Right figure: deconfining vortex ensemble in which the intersection points occur pairwise within a distance d away from each other. (The intersection points of a pair are connected by a line.)

where in the last step, the size of the universe L has been sent to infinity while leaving the planar density $\rho = N/L^2$ of vortex intersection points constant. Thus, one obtains an area law for the Wilson loop, with the string tension $\sigma_{rvm} = 2\rho$.

In fact, in lattice calculations the vortex area density ρ has been shown to obey the proper scaling behaviour [3] as dictated by the renormalization group and thus represents a physical observable. Using a string tension of $\sigma \sim (440 MeV)^2$ as input one finds $\rho \approx 3.4 F^{-2}$ corresponding to a string tension $\sigma_{rvm} = (521 MeV)^2$ in the random model above which overestimates the input value. This overabundance of string tension can be easily understood by noticing that there are both dynamical [17] and kinematical correlations between the vortex intersection points, which have been discarded in the random vortex model considered above, which assumes that all intersection points are completely random, i.e. uncorrelated. This is, however, not true since the vortices are closed loops in $D = 3$ or closed surfaces in $D = 4$. Therefore the intersection points in the plane of the Wilson loops come in pairs. But a pair of intersection points does not (non-trivially) contribute to the Wilson loop. Only for large vortices exceeding the size of the Wilson loop the intersection points inside the Wilson loop are uncorrelated and can contribute (-1). On the other hand all vortices contribute to the area vortex density ρ measured on the lattice. This effect leads to a lower value of the string tension than the value $\sigma_{rvm} = 2\rho$ resulting from the random vortex model.

DECONFINEMENT

The above presented vortex picture of confinement naturally explains also the deconfinement transition as a transition from a phase of large vortices percolating throughout space-time to a phase of small vortices in a sense to be specified more precisely below. Indeed, assume that all vortices have a maximal size d. Then only the intersection points in a strip of width d along the perimeter of the Wilson loop can randomly contribute (-1), (while other intersection points come in pairs and hence do not contribute). The expectation value of the Wilson loop is then still given by eq. (1), however, with the full area A of the Wilson loop replaced by the area of the strip of width d along the perimeter P of the Wilson loop, $d \cdot P$, resulting in a perimeter law

$$< W > = exp(-2\rho dP) \qquad (2)$$

implying deconfinement. This picture of the deconfinement phase transition arising in the random vortex model as a transition from a phase of percolating vortices to a phase of small vortices is supported by the lattice calculations. Fig. 2 shows the vortex matter distribution as function of the vortex cluster extension at various temperatures for a 3-dimensional slice resulting from the 4-dim. lattice by omitting one spatial direction [18], see also ref. [17]. Far below the critical temperature T_C of the deconfinement phase transition a dominant portion of the vortex matter is contained in a big cluster extending over the whole lattice universe. As the temperature rises smaller clusters are more and more formed and well above the deconfinement phase transition large vortices have ceased to exist, the connectivity of the clusters is lost and all vortex matter being contained in small clusters.

If one analyzes the small vortex clusters dominating the deconfined phase in more detail, one finds that a large part of these vortices wind in the (Euclidean) temporal direction, i.e. the space-time direction whose extension is identified with the inverse temperature. Therefore, one finds that the typical configurations in the two phases can be characterized as displayed in Fig. 3 in a three-dimensional slice of space-time, where one space direction has been left away. Note that Fig. 3 also furnishes an explanation of the spatial string tension in the deconfined phase. A spatial Wilson loop embedded into Fig. 3 (right) can exhibit an area law since intersection points of winding vortices with the minimal area spanned by the loop can occur in an uncorrelated fashion despite those vortices having small extension. Note also the dual nature of this (magnetic) picture as compared with electric flux models [19]. In such models, electric flux percolates in the *deconfined* phase, while it does not percolate in the confining phase.

MAGNETIC MONOPOLES AND TOPOLOGY

Spontaneous breaking of chiral symmetry can be triggered by topologically non-trivial gauge fields, which give rise to zero modes of the quarks [21]. Magnetic monopoles and percolated vortices are long range fields and should hence be relevant for the global topological properties.

Topological properties of gauge configurations as measured by the Pontryagin index

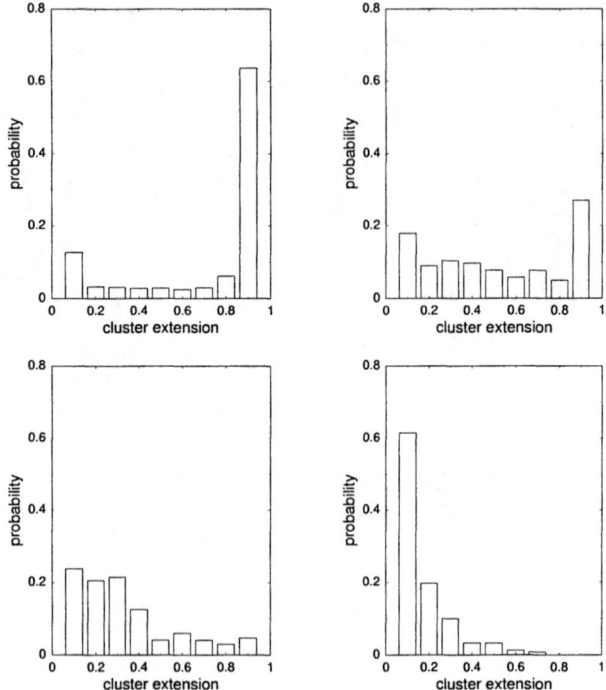
Figure 2. Vortex material distributions.

$$\nu = \frac{-1}{16\pi^2} \int Tr F_{\mu\nu}\tilde{F}_{\mu\nu} = \frac{1}{4\pi^2} \int d^3x \vec{E}(x)\vec{B}(x) \tag{3}$$

are preferably studied in the continuum theory. For the study of the topological properties of magnetic monopoles in the continuum theory the Polyakov gauge is particularly convenient. In this gauge one diagonalizes the Polyakov loop

$$\Omega(\vec{x}) = Pe^{-\int_0^T dx_0 A_0(x_0,\vec{x})} = V^\dagger \omega V \tag{4}$$

which fixes $V \in SU(2)/U(1)$ i.e. the coset part of the gauge group, which we assume, for simplicity, to be $SU(2)$. Magnetic monopoles arise as defects of the gauge fixing, which occur when at isolated points in space $\vec{x}_i$ the Polyakov loop becomes a center element

$$\Omega(\vec{x}_i) = (-1)^{n_i}, \quad n_i : \text{integer} \tag{5}$$

The field $A^V = VAV^\dagger + V\partial V^\dagger$ develops then magnetic monopoles located at these points. These monopoles have topologically quantized magnetic charge [8] given by the winding number

$$m[V] \in \Pi_2(SU(2)/U(1)) \tag{6}$$

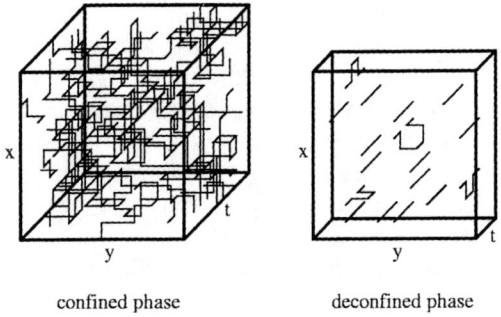

confined phase   deconfined phase

Figure 3. Typical vortex configurations in the confining (left) and the deconfined phase (right).

of the mapping $V(\vec{x})$ from a sphere S_2 around the magnetic monopole into the coset $SU(2)/U(1)$ of the gauge group.

In the Polyakov gauge the Pontryagin index can be exactly expressed in terms of magnetic charges [8], [22], [20]. If we assume a compact space-time manifold and that there are only point-like defects of gauge fixing, i.e magnetic monopoles are the only magnetically charged objects arising after gauge fixing, the Pontryagin index is given by [8]

$$\nu = \Sigma_i n_i m_i \tag{7}$$

The summation runs here over all magnetic monopoles with m_i being the magnetic charge of the monopole defined by equation (6) and the integer n_i is defined by the value of the Polyakov-loop at the monopole position (5). This relation shows that the magnetic monopoles completely account for the non-trivial topology of gauge fields, at least in the Polyakov gauge. Unfortunately, in other Abelian gauges like maximum Abelian gauge, such a simple relation between Pontryagin index and magnetic charges is not yet known and perhaps does not exist [20]. However, in the maximum Abelian gauge correlations between instantons and monopoles have been found, in both analytical and lattice studies [5].

CENTER VORTICES IN THE CONTINUUM

On the lattice center vortices are detected by going to the maximum center gauge and subsequently projecting the links onto center elements [2]. In the maximum center gauge

$$\sum_{x,\mu}(TrU_\mu(x))^2 \to max , \tag{8}$$

which is obviously insensitive to center gauge transformations, one exploits the gauge freedom to rotate a link variable as close as possible to a center element. Once the maximum center gauge has been implemented, center projection implies to replace all links by their closest center element. One obtains then a $Z(2)$ lattice which contains $D-1$ dimensional hypersurfaces Σ on which all links take a non-trivial center element, that is $U = -1$ in the case of $SU(2)$. The $D-2$ dimensional boundaries $\partial\Sigma$ of the hypersurfaces Σ represent the center vortices, which, when non-trivially linked to a Wilson loop, yield a center element for the latter. The notion of center vortices can be extended to the continuum theory by putting a given smooth gauge field $A_\mu(x)$ on a lattice in the standard fashion by introducing the link variables $U_\mu(x) = exp(-aA_\mu(x))$.

A careful analysis shows that the continuum analogies of the center vortices are defined by the gauge potential [9],

$$\mathcal{A}_\mu(x, \Sigma) = E \int_\Sigma d^{D-1}\tilde{\sigma}_\mu \delta^D(x - \bar{x}(\sigma)) \tag{9}$$

where $d^{D-1}\tilde{\sigma}_\mu$ is the dual of the $D-1$ dimensional volume element. Furthermore, the quantity $E = E_a H_a$ with H_a being the generators of the Cartan algebra represents (up to a factor of 2π) the so called co-weights which satisfy $exp(-E) = Z \in Z(N)$. Due to this fact the Wilson-loop calculated from the gauge potential (9) becomes,

$$W[\mathcal{A}](C) = \exp(-\oint_C \mathcal{A}) = Z^{I(C,\Sigma)} \tag{10}$$

where $I(C, \Sigma)$ is the intersection number between the Wilson-loop C and the hypersurface Σ. The representation, (9), is referred to as ideal center vortex. One should emphasize that the hypersurface Σ can be arbitrarily deformed by a center gauge transformation keeping, however, its boundary $\partial\Sigma$, i.e. the position of the center vortex, fixed. Thus for fixed $\partial\Sigma$ the dependence of the gauge potential (9) on the hypersurface itself is a gauge artifact.

The dependence on the hypersurface Σ can be removed by performing the gauge transformation

$$\varphi(x, \Sigma) = exp(-E\Omega(x, \Sigma)) \tag{11}$$

where $\Omega(x, \Sigma)$ is the solid angle subtended by the hypersurface Σ as seen from the point x. One finds then

$$\mathcal{A}_\mu(x, \Sigma) = \varphi(x, \Sigma)\partial_\mu \varphi^\dagger(x, \Sigma) + a_\mu(x, \partial\Sigma) \tag{12}$$

where

$$a_\mu(x, \partial\Sigma) = E \int_{\partial\Sigma} d^{D-2}\tilde{\sigma}_{\mu\nu} \partial_\nu D(x - \bar{x}(\sigma)) \tag{13}$$

depends only on the vortex position $\partial\Sigma$ and is referred to as "thin vortex". Here $D(x - \bar{x}(\sigma))$ represents the Green function of the D dimensional Laplacian. In fact, one can show [9] that the thin vortex represents the transversal part of the ideal vortex $a_\mu(x, \partial\Sigma) = P_{\mu\nu}\mathcal{A}_\nu(x, \Sigma)$ where $P_{\mu\nu} = \delta_{\mu\nu} - \frac{\partial_\mu \partial_\nu}{\partial^2}$ is the usual transversal projector. A careful and lengthy analysis [9] yields that the continuum analog of the maximum center gauge fixing is defined by

$$\min_{\partial \Sigma} \min_g \int \text{Tr}(A^g - a(\partial \Sigma))^2 \qquad (14)$$

where the minimization is performed with respect to all (continuum) gauge transformations $g \in SU(2)/Z(2)$ (which represent per se coset gauge transformations) and with respect to all vortex surfaces $\partial \Sigma$. For fixed thin center vortex field configuration $a(\partial \Sigma)$ minimization with respect to the continuum gauge transformation g yields the background gauge condition

$$[\partial_\mu + a_\mu(\partial \Sigma), A_\mu] = 0 \qquad (15)$$

where the thin vortex field $a_\mu(x, \partial \Sigma)$ figures as background gauge field. One should emphasize, however, that the background field has to be dynamically determined for each given gauge field $A_\mu(x)$ and thus depends on the latter. Obviously in the absence of a vortex structure in the considered gauge field $A_\mu(x)$ the background gauge condition reduces to the Lorentz gauge $\partial_\mu A_\mu = 0$.

TOPOLOGY OF CENTER VORTICES

Once the center vortex configurations in the continuum are at our disposal, it is straightforward to calculate their Pontryagin index. In the continuum formulation where center vortices live in the Abelian subgroup by construction the direction of the magnetic flux of the vortices is fully kept. The explicit calculation [9] shows that the Pontryagin index ν of the center vortices is given by

$$\nu = \frac{1}{4} I(\partial \Sigma, \partial \Sigma) \qquad (16)$$

where $I(\partial \Sigma, \partial \Sigma)$ represents the self-intersection number of the closed vortex sheet $\partial \Sigma$ defined by

$$I(\partial \Sigma_1, \partial \Sigma_2) = \frac{1}{2} \int_{\partial \Sigma_1} d\sigma_{\mu\nu} \int_{\partial \Sigma_2} d\tilde{\sigma}_{\mu\nu} \delta^4(\bar{x}(\sigma) - \bar{x}(\sigma')) . \qquad (17)$$

A careful analysis shows that for closed oriented surfaces the self intersection number vanishes. In order to have a non-trivial Pontryagin index the vortex surfaces have to be not globally oriented, i.e., they have to consist of oriented pieces. One can further show that at the border between oriented vortex patches magnetic monopole currents flow. It is these monopole currents which make the vortex sheet non-oriented since they change the orientation of the magnetic flux. Thus we obtain that even for the center vortices the non-trivial topology is generated by magnetic monopole currents flowing on the vortex sheets. This is consistent with our finding in the Polyakov gauge (see eq. (7)) where the Pontryagin index was exclusively expressed in terms of magnetic monopoles [8]. In fact, for a compact space-time manifold one can show that under certain mild assumptions the Pontryagin index can be expressed as

$$\nu = -\frac{1}{4}L(\partial\Sigma, C), \qquad (18)$$

where $L(\partial\Sigma, C)$ is the linking number between the vortex $\partial\Sigma$ and the monopole loop C on the vortex.

By implementing the maximum center gauge condition in the continuum one can derive, in an approximate fashion, an effective vortex theory [9], where the vortex action can be calculated in a gradient expansion. The leading order gives the Nambu-Goto action while in higher orders curvature terms appear. A model based on such an effective vortex action, in fact, reproduces the gross features of the center vortex picture found in numerical Yang-Mills lattice simulations.

ACKNOWLEDGMENT:

This work was supported in part by the DFG-Re 856/4-1 and DFG-En 415/1-1.

REFERENCES

1. G .S. Bali, C. Schlichter and K. Schilling, Prog. Theor. Phys. Suppl. **131** (1998) 645 (1998) and references therein.
2. L. Del Debbio, M. Faber, J. Greensite and S. Olejník, Phys. Rev. **D55** (1997) 2298;
   L. Del Debbio, M. Faber, J. Greensite, and Š. Olejník, hep-lat/9708023;
   L. Del Debbio, M. Faber, J. Giedt, J. Greensite, and Š. Olejník, Phys. Rev. **D 58**, 094501 (1998)
3. K. Langfeld, H. Reinhardt and O. Tennert, Phys. Lett. **B419** (1998) 317.
4. K. Langfeld, O. Tennert, M. Engelhardt and H. Reinhardt, Phys. Lett. **B452** (1999) 301.
5. S. Thurner, M. Feurstein, H. Markum and W. Sakuler, Phys. Rev. **D54** (1996) 3457;
   H. Suganuma, S. Sasaki, H. Ichie, F. Araki and O. Miyamura, Nucl. Phys. Proc. Suppl. **53** (1997) 528;
   S. Sasaki and O. Miyamura, Phys. Rev. **D59** (1999) 094507.
6. Ph. de Forcrand, M. D'Elia, Phys. Rev. Lett. **82** (1999) 4582.
7. G 't Hooft, Nucl. Phys. **B190** (1981) 455
8. H. Reinhardt, Nucl. Phys. **B503** (1997) 505;
   M. Quandt, H. Reinhardt and A. Schafke, Phys. Lett. **B446** (1999) 290.
9. M. Engelhardt, H. Reinhardt, hep-th/9907139
10. G. Di Cecio, G. Di Giacomo, G. Pafutti, M. Trigiante, Nucl. Phys. **B489** (1997) 739
    A. Di Giacomo, D. Martelli, G. Pafutti, hep-lat/9905007
11. J. Ambjørn, J. Giedt, J. Greensite, hep-lat/9907021

12. G. 't Hooft, Nucl. Phys. **B138**, 1 (1978).
13. J.M. Cornwall, Nucl. Phys. **B157**, 392 (1979).
14. H.B. Nielsen, and P. Olesen, Nucl. Phys. **B160**, 380 (1979);
    J. Ambjørn, and P. Olesen, Nucl. Phys. **B170** [FS1], 60, 265 (1980);
    P. Olesen, Nucl. Phys. **B200** [FS4]m 381 (1982).
15. G. Mack, and V.B. Petkova, Ann. Phys. (NY) **123**, 442 (1979)
    G. Mack, and V.B. Petkova, Ann. Phys. (NY) **125**, 117 (1980);
    G. Mack, Phys. Rev. Lett **45**, 1378 (1980).
16. E.T. Tomboulis, Phys. Rev. **D 23**, 2371 (1981);
    E.T. Tomboulis, Phys. Lett. **B303**, 103 (1993).
17. M. Engelhardt, K. Langfeld, H. Reinhardt, O. Tennert, Phys. Lett. **B 431** (1998) 141.
18. M. Engelhardt, K. Langfeld, H. Reinhardt, O. Tennert, hep-lat/9904004, Phys. Rev. D, in press.
19. A. Patel, Nucl. Phys. **B243**, 411 (1984).
20. O. Jahn, hep-th/9909004.
21. T. Banks, A. Casher, Nucl. Phys. **B169** (1980) 103
22. C. Ford, U. G. Mitreuter, T. Tok, A. Wipf and J. M. Pawlowski, Ann. Phys. **269** (1998) 26
    F. Lenz, O. Jahn, hep-th/9803177

How could quark polarization be measured

A.V. Efremov[1]

Joint Institute for Nuclear Research, Dubna, 141980 Russia

Abstract. The perspectives of two new nonstandard methods of transversal quark polarization measurement are considered: the jet handedness and the so-called "Collins effect" due to spin dependent T-odd fragmentation function responsible for the left-right asymmetry in fragmenting of transversally polarized quarks. Recent experimental indications in favor of these effects are observed:

1.The correlation of the T-odd one-particle fragmentation functions found by DELPHI in $Z \to$ 2-jet decay. Integrated over the fraction of longitudinal and transversal momenta, this correlation is of 1.6% order, which means order of 13% for the analyzing power.

2.A rather large ($\approx$ 10%) handedness transversal to the production plane observed in the diffractive production of ($\pi^-\pi^+\pi^-$) triples from nuclei by the $40\,GeV/c$ π^-–beam. It shows a clear dynamic origin and resembles the single spin asymmetry behavior.

All this makes us hope to use these effects in polarized DIS experiments for transversity measurement. The first estimation of transversity was done by using the azimuthal asymmetry in semi-inclusive DIS recently measured by HERMES and SMC.

I INTRODUCTION

My talk concerns recent progress in the possibility of transverse quark polarization measurement. This is interesting in many aspects, and one of the most important of them is quarks transversity distribution in proton $h_1(x)$ measurement. Let me recall that there are three most important (twist-2) parton distribution functions (PDF) in a nucleon: a non-polarized distribution function $f_1(x)$, longitudinal spin distribution $g_1(x)$ and transversal spin distribution $h_1(x)$. The first two have been more or less successfully measured experimentally in classical deep inelastic scattering (DIS) experiments, but the measurement of the last one is especially difficult since it belongs to the class of the so-called helicity odd structure functions and can not be seen there. To do this, one needs to know the transversal polarization of a quark scattered from a transversally polarized target.

There are several ways to do this:

---

[1] Supported by Russian Foundation for Basic Research under the Grant 98-02-16508.
E-mail: efremov@thsun1.jinr.ru

1. To measure the polarization of a self-analyzing hadron into which the quark fragmentizes in a semi-inclusive DIS, e.g. Λ-hyperon [1]. The drawback of this method, however, is a rather low rate of quark fragmentation into Λ-particle ($\approx 2\%$) and especially that it is mostly sensitive to s-quark polarization.

2. To measure a spin-dependent T-odd parton fragmentation function (PFF) [2–4] responsible for the left-right asymmetry in fragmentation of a transversally polarized quark with respect to quark momentum–spin plane. (The so-called "Collins asymmetry" [5].)

3. To measure the transversal handedness in multiparticle parton fragmentation [6], i.e. the correlation of quark spin 4-vector s_μ and particle momenta k_ν, $\epsilon_{\mu\nu\sigma\rho} s^\mu k_1^\nu k_2^\sigma k^\rho$ ($k = k_1 + k_2 + k_3 + \cdots$).

The last two methods are comparatively new, and only the last years some experimental indications the the T-odd PFF and the transversal handedness have appeared [7–9]. I am going to present the result of these experiments and their first applications for estimation of the proton transversity distribution.

II T-ODD QUARK FRAGMENTATION FUNCTION

The transfer of nucleon polarization to quarks is investigated in deep-inelastic polarized lepton – polarized nucleon scattering experiments [10]. The corresponding nucleon spin structure functions for the longitudinal spin distribution g_1 and transversal spin distribution h_1 for a proton are well known. The *inverse* process, the spin transfer from partons to a final hadron, is also of fundamental interest. Analogies of f_1, g_1 and h_1 are functions D_1, G_1 and H_1, which describe the fragmentation of a non-polarized quark into a non-polarized hadron and a longitudinally or transversely polarized quark into a longitudinally or transversely polarized hadron, respectively [2].

These fragmentation functions are integrated over the transverse momentum $\mathbf{k_T}$ of a quark with respect to a hadron. With $\mathbf{k_T}$ taken into account, new possibilities arise. Using the Lorentz- and P-invariance one can write, in the leading twist approximation, 8 independent spin structures [2,3]. Most spectacularly it is seen in the helicity basis where one can build 8 twist-2 combinations, linear in spin matrices of the quark and hadron $\boldsymbol{\sigma}$, $\mathbf{S}$ with momenta $\mathbf{k}$, $\mathbf{P}$. Especially interesting is a new T-odd and helicity even structure that describes a left–right asymmetry in the fragmentation of a transversely polarized quark: $H_1^\perp \boldsymbol{\sigma}(\mathbf{P} \times \mathbf{k_T})/P\langle k_T\rangle$, where the coefficient $H_1^\perp$ is a function of the longitudinal momentum fraction z, quark transversal momentum k_T^2 and $\langle k_T\rangle$ is an average transverse momentum.

In the case of fragmentation to a non-polarized or a zero spin hadron, not only D_1 but also the $H_1^\perp$ term will survive. Together with its analogies in distribution

---
[2] We use the notation of the work [2–4].

functions f_1 and $h_1^\perp$, this opens a unique chance of doing spin physics with non-polarized or zero spin hadrons! In particular, since the $H_1^\perp$ term is helicity-odd, it makes possible to measure the proton transversity distribution h_1 in semi-inclusive DIS from a transversely polarized target by measuring the left-right asymmetry of forward produced pions (see [4,11] and references therein).

The problem is that, first, this function is completely unknown both theoretically and experimentally and should be measured independently. Second, the function $H_1^\perp$ is the so-called T-odd fragmentation function: under the naive time reversal $\mathbf{P}$, $\mathbf{k_T}$, $\mathbf{S}$ and $\boldsymbol{\sigma}$ change sign, which demands a purely imaginary (or zero) $H_1^\perp$ in the contradiction with hermiticity. This, however, does not mean the break of T-invariance but rather the presence of an interference term of different channels in forming the final state with different phase shifts, like in the case of single spin asymmetry phenomena [12]. A simple model for this function could be found in [5]. It was also conjectured [13] that the final state phase shift can be averaged to zero for a single hadron fragmentation upon summing over unobserved states X. Thus, the situation here is far from being clear.

Meanwhile, the data collected by DELPHI (and other LEP experiments) give a unique possibility to measure the function $H_1^\perp$. The point is that despite the fact that the transverse polarization of a quark (an antiquark) in Z^0 decay is very small ($\mathcal{O}(\mathfrak{M}_\mathrm{u}/\mathcal{M}_Z)$), there is a nontrivial correlation between transverse polarizations of a quark and an antiquark in the Standard Model: $C_{TT}^{q\bar{q}} = (v_q^2 - a_q^2)/(v_q^2 + a_q^2)$, which reaches rather high values at Z^0 peak: $C_{TT}^{u,c} \approx -0.74$ and $C_{TT}^{d,s,b} \approx -0.35$. With the production cross section ratio $\sigma_u/\sigma_d = 0.78$ this gives the value $\overline{C_{TT}} \approx -0.5$ for the average over flavors.

The spin correlation results in a peculiar azimuthal angle dependence of produced hadrons (the so-called "one-particle Collins asymmetry"), if the T-odd fragmentation function $H_1^\perp$ does exist [5,14]. The first probe of it was done three years ago [15] by using a limited DELPHI statistics with the result $\left|H_1^\perp/D_1\right| \leq 0.3$, as averaged over quark flavors.

A simpler method has been proposed recently by an Amsterdam group [3]. They predict a specific azimuthal asymmetry of a hadron in a jet around the axis in direction of the second hadron in the opposite jet [3]:

$$\frac{d\sigma}{d\cos\theta_2 d\phi_1} \propto (1 + \cos^2\theta_2) \cdot \left(1 + \frac{6}{\pi}\left[\frac{H_1^{q\perp}}{D_1^q}\right]^2 C_{TT}^{q\bar{q}} \frac{\sin^2\theta_2}{1 + \cos^2\theta_2}\cos(2\phi_1)\right) \quad (1)$$

where θ_2 is the polar angle of the electron and the second hadron momenta $\mathbf{P_2}$, and ϕ_1 is the azimuthal angle counted off the ($\mathbf{P_2}$, $\mathbf{e^-}$)-plane. This looks simpler since there is no need to determine the $q\bar{q}$ direction.

This analysis [7] covered the DELPHI data collected from 1991 through 1995. All particles were generically assumed to be pions. Only charged particles were

---

[3]) We assume the factorized Gaussian form of k_T dependence for $H_1^{q\perp}$ and D_1^q integrated over $|k_T|$.

analyzed. About 3.5 millions of Z^0 hadronic decays were selected by using the standard selection criteria [16]. Jets were reconstructed by the JADE algorithm. Only 2-jet events were retained for the analysis. A leading particle in each jet was selected both positive and negative.

To study the detector response, a sample of Monte-Carlo events, generated with JETSET and passed through the same analysis chain as the data, was used. With these events, the correction factor

$$f_{\text{corr}} = \frac{N_{\text{generated}}(\theta_2, \phi_1)}{N_{\text{simulated}}(\theta_2, \phi_1)}$$

was built for each bin in the azimuthal angle of the first leading particle ϕ_1 and in the polar angle of the leading particle from the opposite jet θ_2 (see Expr. (1)).

The true distribution was defined as $N_{\text{true}} = f_{\text{corr}} N_{\text{raw}}$ and histograms in ϕ_1 for each bin of θ_2 were fitted by the expression [4] $P_0(1 + P_2 \cos 2\phi_1 + P_3 \cos \phi_1)$. The θ_2-dependence of P_2^{true} in the whole interval of θ_2 is presented in Fig.1 with the corresponding fit

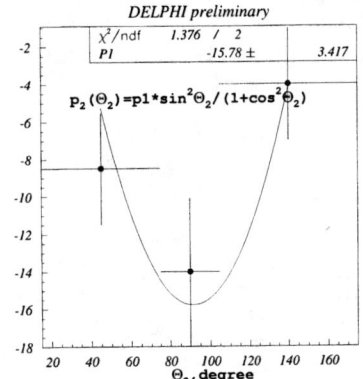

FIGURE 1. The θ_2-dependence of the P_2^{true}.

$$P_2^{\text{true}}(\theta_2) = -(15.8 \pm 3.4) \frac{\sin^2 \theta_2}{1 + \cos^2 \theta_2} \text{ppm}$$

The corresponding analyzing power, summed over z and averaged over quark flavors with $\overline{C_{TT}} \approx -0.5$ (assuming $H_1^\perp = \sum_H H_1^{\perp,q/H}$ is flavor-independent), according to Exp. (1) is

$$\left| \frac{<H_1^\perp>}{<D_1>} \right| = 12.9 \pm 1.4\% \ . \qquad (2)$$

The systematic errors, however, are by all means larger than the statistical ones and need further investigation.

III THE TRANSVERSE HANDEDNESS

The concept of jet handedness was introduced as a measure of polarization of parent partons (or hadrons) [6]. For a strong interaction process, parity conservation requires that at least three particles (either spinless or spin-averaged) in a final state or a pair of particles and jet direction were measured in order to build

---

[4] The term with $\cos \phi_1$ is due to the twist-3 contribution of usual one-particle fragmentation, proportional to the k_T/E.

a correlation of final momenta in the fragmentation (or decay) with initial polarization. Namely, from three particle momenta one can construct a pseudovector $n_\mu \propto \epsilon_{\mu\nu\sigma\rho} k_1^\nu k_2^\sigma k^\rho$ ($k = k_1 + k_2 + k_3 + \cdots$) which gives, when contracted with the initial polarization pseudovector, a scalar component in the strong process. Thus, measuring the handedness – the asymmetry in relative number of events N with respect to some projection of **n** to a direction **i** in the rest frame of the triple – can give an information on the initial polarization P_i in this direction (at least, for spin 1/2 and 1)

$$H_i = \frac{N(n_i > 0) - N(n_i < 0)}{N(n_i > 0) + N(n_i < 0)} = \alpha_i P_i , \qquad (3)$$

provided the analyzing power α_i is different from zero. The direction **i** could be chosen as longitudinal (L) with respect to the triple momentum **k** and as transversal ones ($T1$ or $T2$) perpendicular to **k** [5].

The transversal handedness was investigated for the reaction of diffractive production of pion triples [19]

$$\pi^- + A \to (\pi^-\pi^+\pi^-) + A, \qquad (4)$$

by π^- beam 40 GeV/c from a nucleus A.

For this reaction one can define the normal to the plane of production of a secondary pion triple $(\pi_f^- \pi^+ \pi_s^-)$

$$\mathbf{N} = (\mathbf{v}_{3\pi} \times \mathbf{v_b}) \qquad (5)$$

where $\mathbf{v_b} = \mathbf{k_b}/\epsilon_b$ and $\mathbf{v}_{3\pi} = \mathbf{k}_{3\pi}/\epsilon_{3\pi}$ are velocities of the initial π^- beam and the center of mass of the triple in Lab. r.f., and indices f and s label fast and slow π^-'s. The normal to the "decay plane" of the triple in its center of mass is defined as

$$\mathbf{n} = (\mathbf{v_f^-} - \mathbf{v^+}) \times (\mathbf{v_s^-} - \mathbf{v^+}) \qquad (6)$$

where $\mathbf{v_{f(s)}^-}$ or $\mathbf{v^+}$ are velocities of the fast (slow) π^- or π^+.

The transversal handedness according to (3) is

$$H_{T1} = \frac{N(\mathbf{Nn} > 0) - N(\mathbf{Nn} < 0)}{N(\mathbf{Nn} > 0) + N(\mathbf{Nn} < 0)}. \qquad (7)$$

Two other components of the handedness connected with $\mathbf{n} \cdot \mathbf{v}_{3\pi}$ and $\mathbf{n} \cdot (\mathbf{v}_{3\pi} \times \mathbf{N})$ are forbidden by the parity conservation in the strong interaction [6].

---

[5]) In fact, an idea similar to the handedness was earlier proposed in works [17]. Its application to certain heavy quark decays was studied in Ref. [18]. A similar technique was also studied in work [5].

[6]) It is easy to show that all these quantities are in fact Lorentz-invariant.

The transversal handedness (7) was measured for a wide sample of nuclear targets [9]. The dependence of H_{T1} on the atomic number A is presented in Fig.2. One can see that the handedness systematically decreases with increasing A, which resembles a depolarization effect in multiple scattering. An argument in this favor is the decrease of the effect as, approximately, inverse nuclei radius.

The value of the asymmetry (7) averaged over all nuclei is

$$H_{T1} = (5.96 \pm 0.21)\%. \tag{8}$$

Statistically, this is highly reliable verification of the existence of the correlation of the triple production and decay planes in process (4).

The values of two other asymmetries with respect to correlations $\mathbf{n} \cdot \mathbf{v}_{3\pi}$ and $\mathbf{n} \cdot (\mathbf{v}_{3\pi} \times \mathbf{N})$ were found to be comparable to zero from the same statistical material: $H_L = (0.25 \pm 0.21)\%$ and $H_{T2} = (0.43 \pm 0.21)\%$, respectively. This is by no means surprising, since they are forbidden by the parity conservation.

To understand the nature of the effect observed, the dependence of handedness (7) on the Feynmann variable X_F of the leading π^-, on the triple transversal momentum k_T and on the invariant mass of the triple $m_{3\pi}$ and its neutral subsystem $m_{\pi^+\pi^-}$ was studied. It was found that the handedness (7) increases with X_F and k_T, which resembles the behavior of the single spin asymmetry (e.g. the pion asymmetry or the Λ-polarization [20]).

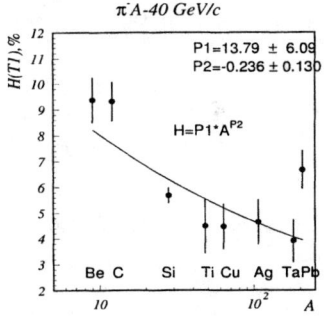

FIGURE 2. A-dependence of the handedness.

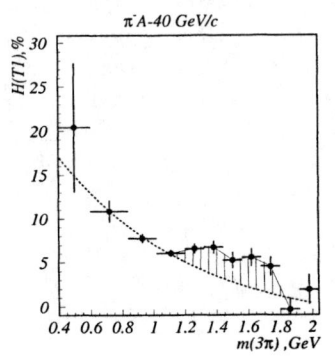
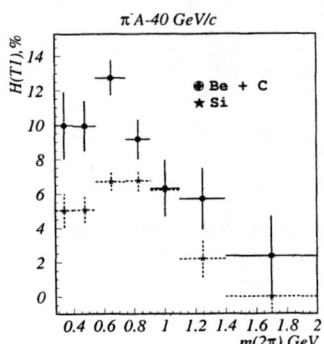

FIGURE 3. The handedness dependence on $m_{3\pi}$ (a) and $m_{\pi^+\pi_j^-}$ (b).

The dependence of H_{T1} on the triple invariant mass (Fig.3a) is especially interesting. It clearly indicates two different sources of H_{T1} with comparable contributions: a resonance and a nonresonance one. The resonance contribution is clearly seen in the mass of $a_1(1260)$ and $\pi_2(1670)$ region and by no means is due to a nonzero polarization of the resonances. The nonresonant background could also be polarized, provided that the 3π system is predominantly in a state with the total angular momentum $J \neq 0$, e.g. if a neutral pair $m_{\pi^+\pi^-}$ was predominantly produced from ρ-decay. Some indication of this can be seen from Fig.3b. In this context, the growth of H_{T1} in the region of small $m_{3\pi}$, i.e. in the region of small relative momenta of pions, looks quite intriguing.

All this demonstrates that the transverse handedness has a clear dynamic origin and in some features resembles the single spin asymmetry behavior. For a more detailed study of the phenomenon, a partial wave analysis of reaction (4) seems necessary.

IV PROTON TRANSVERSITY ESTIMATION

Recently azimuthal asymmetries in semi-inclusive hadron production on longitudinally (HERMES [21]) and transversally (SMC [22]) polarized targets where reported which together with DELPHI result (2) allows an estimation for transversity distribution.

The T-odd azimuthal asymmetry in semi inclusive DIS $ep \to e'\pi^+ X$ which HERMES try to measure consist of two sorts of terms (see [4] Eq. (115)): a twist-2 asymmetry $\sin 2\phi_h$ and a twist-3 asymmetry $\sin \phi_h$. The angle ϕ_h here is the azimuthal angle around z-axis in the direction of virtual γ momentum in the parton Breit frame counted from the electron scattering plane. The first asymmetry is proportional to the k_T-dependent transversal quark spin distribution in a longitudinally polarized proton, $h_{1L}^\perp$, while the second contains two parts: one term is again proportional to $h_{1L}^\perp$; and the second, to the twist-3 distribution function h_L.

The experimentally observed ϕ dependence in HERMES data shows no noticeable trace of $\sin 2\phi$ term. Thus, as a crude approximation one can assume a smallness of $h_{1L}^\perp \gg h_L$. For the same reason $h_L = h_1$ (see [4] Eq. (C15,C19)). This open a possibility to measure the proton transversity using the *longitudinally* polarized target.

The asymmetry measured by HERMES

$$A_{OL} = \frac{\int d\phi \sin\phi (d\sigma^+/d\phi)}{P_H^+ \int d\phi (d\sigma^+/d\phi)} - \frac{\int d\phi \sin\phi (d\sigma^-/d\phi)}{P_H^- \int d\phi (d\sigma^-/d\phi)}, \qquad (9)$$

where $P_H^\pm$ is the nucleon longitudinal polarization ($\pm$ sign means different spin directions) averaged over transversal momenta (assuming a Gaussian distribution) should reads as ([4] Eq. (115))

$$A_{OL} = \frac{2(2-y)\sqrt{1-y}(\frac{M}{Q})\sum_a e_a^2 x^2 h_1^a(x) H_1^{\perp a/\pi^+}(z)/z}{(1-y+y^2/2)\sum_a e_a^2 x f_1^a(x) D_1^{a/\pi^+}(z)} \cdot \frac{1}{\sqrt{1+<p_T^2>/<k_T^2>}} \quad (10)$$

where $<p_T^2>$ and $<k_T^2>$ are mean square of a transversal momenta of quark in the distribution and fragmentation functions.

Averaging separately the numerator and denominator over Q^2, y and z, and taking into account only the u-quark distribution in the proton ($f_1^u(x) \equiv u(x)$) which gives a dominant contribution for the π^+ production and assuming $<p_T^2>=<k_T^2>$, one can obtain for asymmetry (9)

$$A_{OL}(x) = \frac{2.1}{<z>\sqrt{2}} \cdot \left[\frac{h_1^u(x)}{u(x)}\right] \cdot \left[\frac{<H_1^{\perp u/\pi^+}(z)>}{<D_1^{u/\pi^+}(z)>}\right] \cdot x \quad (11)$$

Experimentally [21], $A_{OL}(x)$ for π^+ up to $x = 0.3$ looks like a linear function

$$A_{OL}(x) = (0.23 \pm 0.06)x \quad (12)$$

With $<z>= 0.41$ and DELPHI result (2) this gives an estimation for the ratio

$$\frac{h_1^u(x)}{u(x)} = const = 0.49 \pm 0.18 \quad (13)$$

Assuming the validity of (13) for valence parts in the whole interval of x, one could obtain an estimation for the u-quark contribution to the proton tensor charge

$$g_T^u \equiv \int dx \left(h_1^u(x) - h_1^{\bar{u}}(x)\right) = 0.96 \pm 0.36 \quad (14)$$

that is close to the result of the chiral quark-soliton model [23] $g_T^u \approx 1.12$ and to the limit followed from the density matrix positivity constraint at $Q^2 = 2.5\,GeV^2$ [24] $g_T^u \leq 1.09$.

Concerning the asymmetry observed by SMC [22] on transversely polarized target one can state that it agrees with the result of HERMES.

Really, SMC has observed the azimuthal asymmetry $d\sigma(\phi_c) \propto const \cdot (1 + a \sin \phi_c)$, where $\phi_c = \phi_h + \phi_S - \pi$ (ϕ_S is the azimuthal angle of the polarization vector) is the so-called Collins angle. The raw asymmetry $a = P_T \cdot f \cdot D_{NN} \cdot A_N$, where P_T, f, and $D_{NN} = 2(1-y)/[1+(1-y)^2]$ are the target polarization value, the dilution factor and the spin transfer coefficient.

The physical asymmetry A_N averaged over transverse momenta (assuming again a Gaussian form) is given by the expression (see [4] Eq. (116))

$$A_N = \frac{\sum_a e_a^2 x h_1^a(x) H_1^{\perp a/\pi^+}(z)/z}{\sum_a e_a^2 x f_1^a(x) D_1^{a/\pi^+}(z)} \cdot \frac{1}{\sqrt{1+<p_T^2>/<k_T^2>}} \quad (15)$$

Integration over x and z, and using again the approximation of the u-quark dominance and $<p_T^2> = <k_T^2>$ gives from the experimental value $A_N = 0.11 \pm 0.06$, $<z> = 0.45$ and DELPHI result (2)

$$\frac{<xh_1^u(x)>}{<xu(x)>} = 0.54 \pm 0.35 \qquad (16)$$

what in agreement with the HERMES ratio (13).

V CONCLUSIONS

In conclusion, I would like to stress that there are several ways allowing one to measure the transverse quark polarization among which the use of the T-odd PFF looks like the most perspective for future experiments in measuring of transversity, like COMPASS at CERN. I present the first experimental estimation for the absolute value of analyzing power of the method. Of course, a more accurate measurements of it is necessary. However, even now it allows the first crude estimation of the proton transversity from observed azimuthal asymmetry in semi-inclusive DIS. The most interesting discovery here is a good agreement of transversities obtained from transversally and longitudinally polarized targets due to small contribution of $h_{1L}^\perp(x)$. This allows measuring the transversity in the same experiments as for Δg.

I would like to thank K.Goeke, M.Polyakov and the Institute for Theoretical Physics II of Ruhr University Bochum, where part of this work was done, for discussions and warm hospitality.

REFERENCES

1. Augusten J.E. and Renard F.M., *Nucl. Phys.* **B162** 341 (1980)
2. Boer D. and Mulders P.J., *Phys. Rev.* **D57** 5780 (1998).
3. Boer D., Jakob R. and Mulders P.J., *Phys. Lett.* **B424** 143 (1998).
4. Mulders P.J. and Tangerman R., *Nucl. Phys.* **B461** 234 (1995); Boer D. and Tangerman R., *Phys. Lett.* **B381** 305 (1996).
5. Collins J., *Nucl.Phys.* **B396** 161 (1993); Artru X. and Collins J.C., *Z. Phys.* **C69** 277 (1996).
6. Efremov A., Mankiewicz L. and Törnqvist N., *Phys. Lett.* **B284** 394 (1992).
7. Efremov A.V., Smirnova O.G. and Tkatchev L.G., "Study of T-odd Quark Fragmentation Function in $Z^0 \to$ 2-jet Decay," in *Proc. of QCD98 conf., Nucl. Phys. Proc. Suppl.* **74** 49-52 (1999); hep-ph/9812522.
8. Efremov A.V., Ivanshin Yu.I., Tkatchev L.G. and Zulkarneev R.Ya., *JINR Rapid Commun.* No **3(83)** 5 (1997).
9. Efremov A.V., Ivanshin Yu.I., Tkatchev L.G. and Zulkarneev R.Ya., "Study of correlation of production and decay planes in $\pi \to 3\pi$ diffractive dissociation process on nuclei," Preprint JINR E1-98-371, (to be published in *Yadernaya Fizika* **63** No 3 2000); nucl-th/9901005.

10. Anselmino M., Efremov A. and Leader E., *Phys. Rep.* **265** 1 (1995).
11. Kotzinian A., *Nucl. Phys.* **B441** 236 (1995).
12. Gasiorowich S., *Elementary particle physics*, Wiley, New-York, 1966, p. 515.
13. Jaffe R.L., Jin X. and Tang J., *Phys. Rev.* **D57** 5920 (1998); *Phys. Rev. Lett.* **80** 1166 (1998).
14. Collins J. et al., *Nucl.Phys.* **B420** 565 (1994).
15. DELPHI-Collab., Bonivento W. et al. "A measurement of quark spin correlation in hadronic Z^0-decays," Int. Note DELPHI-95-81. *Contr. eps0549 to EPS-HEP conf., Brussels (1995).*
16. DELPHI Collab., Aarnio P. et al., *Phys. Lett.* **B240** 271 (1990).
17. Nachtmann O., *Nucl. Phys.* **B127** 314 (1977); Efremov A.V., *Sov. J. Nucl. Phys.* **28** 83 (1978).
18. Dalitz R.H., Goldstein G. and Marshall R., *Z. Phys.* **C42** 441 (1989).
19. Bellini G. et al., *Nucl.Phys.* **B199** 1 (1982); *Phys.Rev.Lett.* **48** 1697 (1982); *Let. Nuovo Cim.* **38** 433(1983) ; *Nuovo Cim.* **A79** 282 (1984).
20. See e.g. talks of Heller K. and Nurushev S.B. in *Proc. of 12th Int. Symp. on High Energy Spin Physics*, Amsterdam, 1996.
21. Avakian H., "Azimuthal single-spin asymmetries in semi inclusive DIS from HERMES," in *Proc. of DIS99 conf.*, Zeuthen, April 1999.
22. Bravar A., "Hadron azimuthal distributions and transverse spin asymmetries in DIS of leptons off transversely polarized targets from SMC," in *Proc. of DIS99 conf.*, Zeuthen, April 1999.
23. Kim H., Polyakov M.V. and Goeke K., *Phys. Lett.* **B387** 577 (1996); hep-ph/9604442.
24. Soffer J., "The h_1^q distributions and the nucleon tensor charge," in *Proc. of QCD97 conf., Nucl. Phys. Proc. Suppl.* **B64** 143-146 (1998)

Heavy Flavor Contributions to QCD Sum Rules and the Running Coupling Constant

W.L.van Neerven[1]

Instituut-Lorentz, Universiteit Leiden, P.O. Box 9506, 2300 RA Leiden, The Netherlands.

Abstract. We have calculated first and second order corrections to several sum rules measured in deep inelastic lepton-hadron scattering. These corrections, which are due to heavy flavors only, are compared with the existing perturbation series which is computed for massless quarks up to third order in the strong coupling constant α_s. A study of the perturbation series reveals that the large logarithms of the type $\ln^i Q^2/m^2$ dominate the perturbation series at much larger values than those given by the usual matching conditions imposed on the $\alpha_s(\mu)$. Therefore these matching conditions cannot be used to extrapolate the running coupling constant from small μ to very large scales like $\mu = M_Z$. An alternative description of the running coupling constant in the MOM-scheme is proposed.

INTRODUCTION

Structure functions, measured in deep inelastic lepton-hadron scattering (see Fig. 1)

$$l_1(k_1) + H(p) \to l_2(k_2) + 'X', \qquad (1)$$

where 'X'' denotes any inclusive final hadronic state, provide us with an excellent test of perturbative QCD. The in and outgoing leptons are represented by l_1 and l_2 respectively and the hadron is denoted by H. On the Born level the reaction proceeds via the exchange of one of the vector bosons V of the standard model which are given by γ, Z and $W^\pm$. The virtuality of the vector boson V and the C.M. energy are given by $q^2 = -Q^2 < 0$ and $S = (p+k_1)^2$ respectively. Further the scaling variables are defined by $x = Q^2/2p.q$ and $y = p.q/p.k_1$. The structure functions show up in the polarized and unpolarized cross sections. Starting with the spin-averaged cross section for $V = \gamma$ we obtain

---

[1] Work supported by the EC network 'QCD and Particle Structure' under contract No. FMRX-CT98-0194.

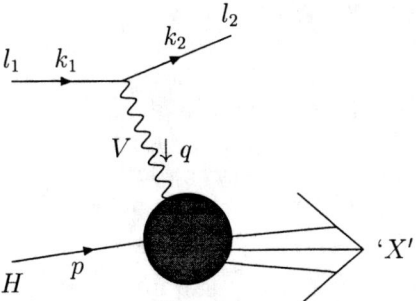

FIGURE 1. Kinematics of deep inelastic lepton-hadron scattering.

$$\frac{d^2\sigma}{dx\,dy} = \frac{4\pi\alpha^2}{(Q^2)^2} S \left[(1-y)F_2(x,Q^2) + x\,y^2 F_1(x,Q^2)\right]. \qquad (2)$$

For $V = W^\pm$ and $Q^2 \ll M_W^2$ the unpolarized cross section for the charged current process becomes

$$\frac{d^2\sigma^\pm}{dx\,dy} = \frac{G_F^2}{2\pi} S \left[(1-y)F_2(x,Q^2) + x\,y^2 F_1(x,Q^2) \mp x\,y\left(1-\frac{y}{2}\right)F_3(x,Q^2)\right]. \qquad (3)$$

Finally we are also interested in the spin structure functions which are measured in polarized scattering. In the case of $V = \gamma$ the cross section becomes

$$\frac{d^2\sigma^{H(\rightarrow)}(\rightarrow)}{dx\,dy} - \frac{d^2\sigma^{H(\leftarrow)}(\rightarrow)}{dx\,dy} = \frac{8\pi\alpha^2}{Q^2} \left[(2-y)g_1(x,Q^2)\right], \qquad (4)$$

from which one can extract the longitudinal spin structure function g_1. The transverse spin structure function g_2 is obtained from the cross section

$$\frac{d^3\sigma^{H(\uparrow)}(\rightarrow)}{dx\,dy\,d\phi} - \frac{d^3\sigma^{H(\downarrow)}(\rightarrow)}{dx\,dy\,d\phi}$$

$$= \cos\phi \frac{4\alpha^2}{Q^2} \left(\frac{4M^2 x(1-y)}{yS}\right)^{1/2} \left[y g_1(x,Q^2) + 2g_2(x,Q^2)\right], \qquad (5)$$

where ϕ is the angle between the spin of the hadron and the plane spanned by the lepton momenta $\vec{k}_i$. Further we have indicated between the brackets the spin of the incoming lepton l_1 and we have neglected in the expressions above all power corrections of the type M^2/S where M denotes the mass of the hadron H (see Fig. 1).

One of the most important predictions of QCD is the Q^2-evolution of the structure functions. However the x-dependence cannot be predicted yet. The latter follows from non-perturbative QCD which is at such a premature stage that the x-dependence cannot be determined. This problem can be avoided when one integrates over the latter variable which leads to the sum rules discussed below. Notice that this integration requires a full knowledge about the range $0 < x < 1$. Since there are no data available for $x < 10^{-4}$ and $x > 0.8$ one has to make extrapolations into this region. This will introduce an error which is very hard to estimate. In the past various techniques have been used to compute the sum rules. The most known among them are

1. Infinite momentum frame techniques used in current algebra.

2. Dispersion relations which are derived from the Regge behavior of the structure functions $F(x, Q^2)$ at small x.

3. Operator product expansion (OPE) techniques which in leading twist are equivalent to the parton model.

Using the last technique the derivation of the sum rules proceeds as follows. First let us define the hadronic structure tensor which appears in the derivation of the cross sections presented above. It is given by

$$W^{\mu\nu}(p,q,s) = \frac{1}{4\pi} \int d^4z\, e^{iq\cdot z} \langle p, s \mid [J^\mu(z), J^\nu(0)] \mid p, s\rangle$$

$$= W_S^{\mu\nu}(p,q) + iW_A^{\mu\nu}(p,q,s). \tag{6}$$

This tensor can be split up into a symmetric and a antisymmetric part i.e.

$$W_S^{\mu\nu}(p,q) = \left(-g^{\mu\nu} + \frac{q^\mu q^\nu}{q^2}\right) F_1(x, Q^2)$$

$$+ \left(p^\mu - \frac{p\,q}{q^2}q^\mu\right)\left(p^\nu - \frac{p\,q}{q^2}q^\nu\right)\frac{F_1(x, Q^2)}{p\,q}, \tag{7}$$

$$W_A^{\mu\nu}(p,q,s) = -\frac{M}{p\,q}\epsilon^{\mu\nu\alpha\beta}q_\alpha\left[s_\beta g_1(x, Q^2) + (s_\beta - \frac{s\,q}{p\,q}p_\beta)g_2(x, Q^2)\right]. \tag{8}$$

In the Bjorken limit ($Q^2 \to \infty$ and x is fixed) the integrand in Eq. (6) is dominated by the lightcone $z^2 = 0$. This allows us to make a lightcone expansion of the current commutator

$$[J(z), J(0)] \underset{z^2 \sim 0}{=} \sum_\tau \sum_N C^{N,\tau}(z^2\mu^2) O^{N,\tau}(\mu^2, 0), \tag{9}$$

where for convenience we have dropped the Lorentz indices of the currents. In the expression above τ and N denote the twist and spin of the operator $O^{N,\tau}$ respectively. The latter and the singular coefficient function $C^{N,\tau}$ are understood to be renormalized where μ represents the renormalization scale. After insertion of expression (11) into Eq. (6) one obtains the Nth moment of the structure function which equals

$$\int_0^1 dx\, x^{N-1} F(x,Q^2) = \sum_\tau \left(\frac{M^2}{Q^2}\right)^{\frac{\tau}{2}-1} A^{(N),\tau}(\mu^2)\, \mathcal{C}^{(N),\tau}\left(\frac{Q^2}{\mu^2}\right). \tag{10}$$

In momentum space the operator matrix element (OME) and the coefficient function are defined by

$$A^{(N),\tau}(\mu^2) = \langle p \mid O^{N,\tau}(\mu^2,0) \mid p \rangle, \tag{11}$$

and

$$\mathcal{C}^{(N),\tau}\left(\frac{Q^2}{\mu^2}\right) = \int d^4z\, e^{iq\cdot z} C^{N,\tau}(z^2\mu^2), \tag{12}$$

respectively. Limiting ourselves to twist *two* contributions only the non-singlet part (w.r.t flavor) of the structure functions is determined by the quark operator

$$O_k^N \equiv O_k^{\mu_1\mu_2\cdots\mu_N}(x) = \bar\psi(x)\gamma^{\mu_1}D^{\mu_2}\cdots D^{\mu_2}\lambda_k\psi(x), \tag{13}$$

where D^μ denotes the covariant derivative and λ_k are the generators of the flavor group $SU_F(n_f)$. In the case of $N=1$ the following operators are conserved

$$O_k^\mu(x) = \bar\psi(x)\gamma^\mu\lambda_k\psi(x) \quad \text{and,} \quad O_k^{\mu,5}(x) = \bar\psi(x)\gamma^\mu\gamma^5\lambda_k\psi(x), \tag{14}$$

which means that they do not have to renormalized. Hence the corresponding OME's and coefficient functions are independent of the scale μ i.e.

$$\int_0^1 dx F(x,Q^2) = A_k^{(1)} \mathcal{C}_k^{(1)}(Q^2), \quad \text{with} \quad \langle p \mid O_k^\mu(0) \mid p \rangle = A_k^{(1)} p^\mu, \tag{15}$$

where $A_k^{(1)}$ can be determined from $SU_F(n_f)$ and a low energy theorem. In general the coefficient function $\mathcal{C}^{(1)}$ can be expanded in a power series in $\alpha_s(\mu^2)$. However in two cases this function becomes trivial. The first example is the Adler sum rule [1]

$$\int_0^1 \frac{dx}{x}\left[F_2^{\bar\nu p}(x,Q^2) - F_2^{\nu p}(x,Q^2)\right] = K(n_f),$$

$$K(3) = 2 + 2\sin^2\theta_c \quad (SU_F(3)), \qquad K(4) = 2 \quad (SU_F(4)). \tag{16}$$

Here θ_c denotes the Cabibbo angle and for the constant $K(n_f)$ we have quoted the values given by the flavor group $SU_F(n_f)$ for $n_f = 3, 4$. The second example is the Burkhardt-Cottingham sum rule [2] given by

$$\int_0^1 dx\, g_2(x,Q^2) = 0. \tag{17}$$

The above sum rules above have the following properties

1. The values on the righthand side are independent of the method used for the derivation of the sum rules.
2. No power corrections of the type $(\Lambda^2/Q^2)^i$ (higher twist).
3. No mass corrections (e.g. due to heavy flavors).
4. No QCD corrections.

The coefficient function has a nontrivial form in the following cases provided the results follow from OPE. The first one is the polarized Bjorken sum rule [3]

$$\Delta g_1(Q^2) \equiv \int_0^1 dx \left[g_1^{ep}(x, Q^2) - g_1^{en}(x, Q^2) \right] = \frac{1}{6} \left| \frac{G_A}{G_V} \right| \mathcal{C}^{g_1}(n_f, Q^2). \tag{18}$$

The second one is represented by the unpolarized Bjorken sum rule [3]

$$\Delta F_1(Q^2) \equiv \int_0^1 dx \left[F_1^{\bar{\nu}p}(x, Q^2) - F_1^{\nu p}(x, Q^2) \right] = K(n_f)\, \mathcal{C}^{F_1}(n_f, Q^2),$$

$$K(3) = 1 + \sin^2\theta_c \quad (SU_F(3)), \qquad K(4) = 1 \quad (SU_F(4)), \tag{19}$$

The third one is the Gross-Llewellyn Smith sum rule [5]

$$\Delta F_3(Q^2) \equiv \int_0^1 dx \left[F_3^{\bar{\nu}p}(x, Q^2) + F_3^{\nu p}(x, Q^2) \right] = K(n_f)\, \mathcal{C}^{F_3}(n_f, Q^2),$$

$$K(3) = 6 - 2\sin^2\theta_c \quad (SU_F(3)), \qquad K(4) = 6 \quad (SU_F(4)). \tag{20}$$

The coefficient function can be expanded in α_s as

$$\mathcal{C}^r(n_f, Q^2) = \sum_{i=0}^{\infty} \left(\frac{\alpha_s(n_f, Q^2)}{\pi} \right)^i c_i^r(n_f), \quad r = g_1, F_1, F_3. \tag{21}$$

The sum rules in Eqs. (18)-(20) have the following properties

1. For $m_q = 0$ the c_i^r are known up to order α_s^3 for $i \leq 3$ (see [6], [7]).
2. Higher twist corrections are estimated in [8]
3. Heavy flavor corrections are computed up to order α_s^2 in [9]

In leading twist the coefficient functions can be computed in the QCD improved parton model. In this model the structure function can be written as

$$F(x, Q^2) = \sum_a e_a^2 \int_0^1 dz_1 \int_0^1 dz_2\, \delta(x - z_1 z_2) \hat{f}_a^H(z_1)\, \hat{\mathcal{F}}_a(z_2, \frac{Q^2}{m_q^2})$$

$$\equiv \sum_a e_a^2 \hat{f}_a^H \otimes \mathcal{F}_a(\frac{Q^2}{m_q^2}), \tag{22}$$

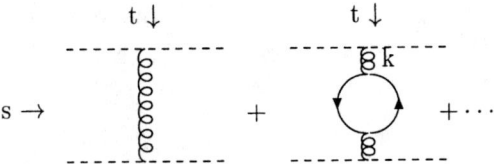

FIGURE 2. heavy flavor loop contribution to quark-quark scattering.

where $\hat{f}_a^H$ denotes the bare parton density and the parton structure function $\mathcal{F}_a$ represents the QCD corrections including the Born approximation. Notice that $\mathcal{F}_a$ contains mass singularities. They are regulated by giving the light quark a mass which is sufficient in the non-singlet case. Subsequently these singularities are removed via mass factorization which reads as follows

$$\mathcal{F}_a(\frac{Q^2}{m_q^2}) = A_{aa}(\frac{\mu^2}{m_q^2}) \otimes \mathcal{C}_a(\frac{Q^2}{\mu^2}). \tag{23}$$

Hence the hadronic structure function can be written as

$$F(Q^2) = \sum_a e_a^2 f_a^H(\mu^2) \otimes \mathcal{C}_a(\frac{Q^2}{\mu^2}), \quad \text{with} \quad f_a^H(\mu^2) = A_{aa}(\frac{\mu^2}{m_q^2}) \otimes \hat{f}_a^H, \tag{24}$$

where f_a^H denotes the renormalized parton density which like the coefficient function depends on the factorization scale μ. Performing the Mellin transformation one obtains

$$\int_0^1 dx\, x^{N-1} F(x, Q^2) = A^{(N)}(\mu^2) \mathcal{C}^{(N)}(\frac{Q^2}{\mu^2}), \tag{25}$$

so that one can make the following identifications

$$A^{(N)}(\mu^2) = \int_0^1 dz\, z^{N-1} \sum_a e_a^2 f_a^H(z, \mu^2), \quad \mathcal{C}^{(N)}(\frac{Q^2}{\mu^2}) = \int_0^1 dy\, y^{N-1} \mathcal{C}(y, \frac{Q^2}{\mu^2}), \tag{26}$$

which yields the coefficient function $\mathcal{C}^{(1)}$

HEAVY FLAVOR THRESHOLDS IN THE STRONG COUPLING CONSTANT

Coupling constant renormalization in the case of light flavors (massless quarks) conventionally proceeds in the $\overline{\text{MS}}$-scheme. Using n-dimensional regularization the coupling constant gets a dimension i.e. $g_s^u M^{4-n}$ where g_s^u is the bare (unrenormalized) coupling constant. The renormalization of the perturbation series is then carried out by the substitution

$$g_s^u \to g_s(\mu^2)\left[1 + \frac{\alpha_s(\mu^2)}{8\pi}\beta_0(n_f)\left\{\frac{2}{\varepsilon} + \gamma_E - \ln 4\pi + \ln\frac{\mu^2}{M^2}\right\} + \cdots\right], \qquad (27)$$

with $\alpha_s = g_s^2/4\pi$ and $\beta_0(n_f) = 11 - 2/3\, n_f$. The renormalized coupling constant α_s satisfies the differential equation

$$\mu^2 \frac{d\alpha_s(n_f,\mu^2)}{d\mu^2} = \beta(\alpha_s(n_f,\mu^2)), \quad \to \quad \alpha_s(n_f,\mu^2) = \frac{4\pi}{\beta_0(n_f)\ln\frac{\mu^2}{\Lambda^2}} + \cdots. \qquad (28)$$

The solution obtained from the equation above is quite appropriate since the scale μ is much larger than the light quark mass. However in the case of heavy quarks we encounter a problem which will be illustrated by the following example. Let us consider (light) quark-quark scattering as shown in Fig. 2. Including the heavy quark loop the cross section behaves like

$$\sigma \sim \alpha_s^2(n_f,\mu^2)\frac{1}{t} + 2\alpha_s^3(n_f,\mu^2)\frac{1}{t}\Pi_1\left(\frac{t}{m^2}\right) + \cdots, \qquad (29)$$

where $\Pi_1(x)$ denotes the one-loop vacuum polarization function which behaves asymptotically as

$$\Pi_1\left(\frac{t}{m^2}\right) \underset{-t \gg m^2}{\sim} -\beta_{0,Q}\ln\left(\frac{-t}{m^2}\right), \qquad \beta_{0,Q} = -\frac{2}{3}. \qquad (30)$$

This behavior leads to large corrections in the perturbation series which have to be resummed as follows. First we have to make the substitution

$$\alpha_s(n_f,\mu^2) \to \alpha_s(n_f+1,\mu^2)[1 + \alpha_s(n_f+1,\mu^2)\beta_{0,Q}\ln\frac{\mu^2}{m^2}\cdots], \qquad (31)$$

in Eq. (29) so that for $-t \gg m^2$ the cross section behaves like

$$\sigma \sim \alpha_s^2(n_f+1,\mu^2)\frac{1}{t} + 2\alpha_s^3(n_f+1,\mu^2)\frac{1}{t}\beta_{0,Q}\ln\frac{\mu^2}{-t}. \qquad (32)$$

Next $\alpha_s(n_f+1,\mu^2)$ satisfies the massless renormalization group equation in (28) but now for n_f+1 light flavors with the boundary condition [10]

$$\alpha_s(n_f,\Lambda_{n_f},\mu^2) = \alpha_s(n_f+1,\Lambda_{n_f+1},\mu^2), \qquad (33)$$

where the matching value is given by $\mu = m$. In this way one obtains a better behavior of the perturbation series than shown by Eq. (29) in particular for $\mu^2 = -t$. However the representation of the cross section in (32) at n_f+1 light flavors is only correct if $-t \gg m^2$. This follows from the vacuum polarization function $\Pi_1(-t/m^2)$ which gets close to the logarithm in Eq. (30) for $-t > 21\, m^2$ within 10%. For $-t \approx m^2$ one does not reach the asymptotic form of $\Pi_1(-t/m^2)$ so that in this case it is much better to use the representation for the cross section at n_f flavors in Eq. (29). It also shows that the matching scale μ in Eq. (31) has to chosen at a much larger value e.g. $\mu^2 = 21\, m^2$ instead of $\mu^2 = m^2$.

HEAVY FLAVOR CONTRIBUTIONS TO THE POLARIZED BJORKEN SUM RULE

The sum rule in Eq. (18) receives contributions from the following subprocesses (see Fig. 2 in [9])

$q + \gamma^* \to q + Q + \bar{Q}$,

$q + \gamma^* \to q + g$ one-loop gluon self energy correction,

$q + \gamma^* \to q$ two-loop vertex correction with one-loop gluon self energy. (34)

The correction to the sum rule up to order α_s^2 can be expressed as

$$C^{g_1}(n_f, Q^2) = 1 + \frac{\alpha_s(n_f, \mu^2)}{\pi} a_1 + \left(\frac{\alpha_s(n_f, \mu^2)}{\pi}\right)^2 \left[-a_1 \beta_0(n_f) \ln \frac{Q^2}{\mu^2} + a_2 + n_f b_2 \right.$$

$$\left. + H_Q\left(\frac{Q^2}{m^2}\right)\right], \qquad (35)$$

where H_Q denotes the heavy quark contribution. The two regions of interest are

$$Q^2 \ll m^2 \to H_Q\left(\frac{Q^2}{m^2}\right) \sim \frac{Q^2}{m^2} \ln \frac{Q^2}{m^2}, \qquad (36)$$

$$Q^2 \gg m^2 \to H_Q\left(\frac{Q^2}{m^2}\right) = a_1 \beta_{0,Q} \ln \frac{Q^2}{m^2} + b_2. \qquad (37)$$

The first limit shows that heavy quarks of mass m decouple from the perturbation series when their mass is much larger than the virtuality of the photon. In the region $Q^2 \gg m^2$ we observe the same logarithmic behavior as for Π_1 in Eq. (30). However here $C^{g_1}(n_f, Q^2)$ becomes asymptotic for $Q^2 > 40\, m^2$ within 15%. Hence the value of Q^2 is about twice as large as the one found for $-t$ in quark-quark scattering. This we have checked for charm production. Here we choose $n_f = 3$ and $m = m_c = 1.5$ GeV/c^2 in Eq. (35) and the results are shown in table 1 (see [9]). Further we observe that the heavy flavor corrections to the sum rule (21)

TABLE 1. QCD corrections to the polarized Bjorken sum rule.

Q^2 (GeV/c)^2	light flavors	charm contribution[a]	charm contribution[b]	charm contribution[c]
3	0.795	$-0.688 \cdot 10^{-3}$	$0.451 \cdot 10^{-2}$	$-0.265 \cdot 10^{-1}$
10	0.883	$-0.788 \cdot 10^{-3}$	$0.569 \cdot 10^{-3}$	$-0.586 \cdot 10^{-2}$
100	0.931	$-0.107 \cdot 10^{-2}$	$-0.0912 \cdot 10^{-2}$	$-0.121 \cdot 10^{-2}$

[a] Exact formula Eq. (35).
[b] Asymptotic formula according to Eq. (37).
[c] Estimated order α_s^2 correction according to [11].

are very small. They are even smaller than the estimated order α_s^4 corrections in [11] in spite of the fact that the heavy flavor contribution H_Q starts in order α_s^2. The reason that the logarithmic behavior in process (36) emerges at a much larger scale than the reaction in Fig. 2 can be explained as follows. In Fig. 2 the gluon momentum k entering the self energy is an external variable since $t = k^2$. However in the case of process (36) one has to integrate over k so that the latter is not an external variable anymore. Therefore the role of t is taken over by Q^2. At very large Q^2 the perturbation series in Eq. (35) can be improved by following the same procedure as outlined below Eq. (30). The result is that instead of Eq. (35) one obtains the improved perturbation series

$$C^{g_1}(n_f + 1, Q^2) = 1 + \frac{\alpha_s(n_f + 1, \mu^2)}{\pi} a_1 + \left(\frac{\alpha_s(n_f + 1, \mu^2)}{\pi}\right)^2 \left[-a_1 \beta_0(n_f + 1) \ln \frac{Q^2}{\mu^2}\right.$$

$$\left. + a2 + (n_f + 1)b_2\right], \tag{38}$$

which behaves much better for the choice $\mu^2 = Q^2$ provided $Q^2 \gg m^2$. However the expression above is not a correct representation when $Q^2 \approx m^2$ as we have already seen below Eq. (33). Hence $\mu^2 = m^2$ is not a good matching scale where the running coupling constant jumps from n_f to $n_f + 1$ flavors. Therefore this scale has to be chosen at a much larger value than what is usually done in the literature. This also means that extrapolations from a measurement of $\alpha_s(n_f, \mu^2)$ at small μ to large values like $\mu = M_Z$ have to be distrusted. A scheme (hereafter called MOM) which incorporates the effect of the heavy flavor thresholds in the running coupling constant in a much better way is proposed in [12]. Here one simply resums the vacuum polarization function as follows

$$\alpha_s^{\text{MOM}}(\mu^2) = \frac{\alpha_s(3, \mu^2)}{1 + \frac{\alpha_s(3,\mu^2)}{\pi} U_1 + \frac{\alpha_s(3,\mu^2)}{\pi}(U_2/U_1) \ln\left(1 + \frac{\alpha_s(3,\mu^2)}{\pi} U_1\right)}, \tag{39}$$

with

$$U_i = \sum_{f=c,b,t} \left[\Pi_i\left(\frac{\mu^2}{m_f^2}\right) - \Pi_i\left(\frac{\mu_0^2}{m_f^2}\right)\right], \qquad i = 1, 2 \cdots, \tag{40}$$

where the coupling constant $\alpha_s(3, \mu^2)$ is represented in the $\overline{\text{MS}}$-scheme. Expression (39) agrees rather well with the numerical solution of the renormalization group equation in Eq. (28) where the β-function depends on m. This in particular holds when the two-loop self energy contributions Π_2 are included. From the results discussed above we have seen that the scales where both Π_1 and H_Q behave logarithmically are relatively much closer to each other than to $\mu = m$. Hence it is better to put $n_f = 3$ in Eq. (35) and to make the replacement

$$\alpha_s(3, \mu^2) = \alpha_s^{\text{MOM}}(\mu^2)\left[1 + \frac{\alpha_s^{\text{MOM}}(\mu^2)}{\pi} U_1 + \cdots\right], \tag{41}$$

than to follow the procedure outlined below Eq. (30). In this way one gets a continuous transition through the flavor thresholds, also with respect to derivatives, of the perturbation series of C^{g_1} in Eq. (35). Similar results were also found for the sum rules in Eqs. (19) and (20) (see [9]). Finally we conclude

1. Heavy flavor corrections to the sum rules are very small and they do not exceed the estimated order α_s^4 contributions due to light quarks.

2. The matching scale in the literature (here $\mu = m$) where the running coupling constant jumps from n_f to $n_f + 1$ flavors is chosen to be too small.

3. The MOM scheme as chosen in Eq. (39) gives a better representation for the perturbation series than the usual $\overline{\text{MS}}$-scheme with the matching condition in Eq. (33).

REFERENCES

1. Adler S.L., *Phys. Rev.* **143**, 1144 (1966), ibid. *Nucl. Phys.* **B371**, 467 (1992).
2. Burkhardt H. and Cottingham W., *Ann. Phys.* **56**, 453 (1970).
3. Bjorken J.D., *Phys. Rev.* **148**, 1467 (1966), ibid. **D1**, 1376 (1970).
4. Bjorken J.D., *Phys. Rev.* **163**, 1767 (1967).
5. D.J. Gross and Llewellyn Smith C.H., *Nucl. Phys.* **B14**, 337 (1969).
6. Larin S.A., Tkachov F.V. and Vermaseren J.A.M., *Phys. Rev. Lett.* **66**, 862 (1991).
7. Larin S.A. and Vermaseren J.A.M., *Phys. Lett.* **B259**, 345 (1991).
8. Balitsky I.I., Braun V.M. and Kolesnichenko A.V., *Phys. Lett.* **B242**, 245 (1990), erratum: ibid. **B318**, 648 (1993).
9. Blümlein J. and van Neerven W.L., *Phys. Lett.* **450**, 417 (1999).
10. Marciano W.J., *Phys. Rev.* **D29**, 580 (1984).
11. Kataev A.L. and Starshenko V.V., *Mod. Phys. Lett.* **A10**, 235 (1995).
12. Shirkov D.V., *Teor. Mat. Fiz.* **93**, 466 (1992), ibid. *Nucl. Phys.* **B371**, 467 (1992).

Color-Flavor Transformations and QCD Low-Energy Effective Action

J. Budzcies and Ya. Shnir

University of Cologne, Institute for Theoretical Physics

Abstract. The color-flavor transformation is applied in the QCD strong coupling limit to construct the effective chiral Lagrangian. The calculations are performed by placing the fermions onto two different sub-lattices and by making use of the long-distance approximation and gradient expansion.

The saddle point approximation could be applied on such a double lattice. In this framework the effect of the spontaneous chiral symmetry breaking is described and the numerical values of the chiral condensate, pion decay constant and pion mass are estimated by using the only parameter of model, the lattice spacing a. The effective chiral Lagrangian of QCD is recovered up to the terms of order $O(p^4)$.

INTRODUCTION

There is very strong evidence that a hierarchy of scales of QCD is provided by the scales for chiral symmetry breaking Λ_χ and confinement Λ_{QCD}. The perturbative QCD governs the scales of momenta p up to $\Lambda_\chi \sim 1\ GeV$, where the coupling constant $g(p) \sim 1$. It blows up at the scale of about $\Lambda_{QCD} \sim 0.18\ GeV$.

In the past two decades a great deal was learned about non-perturbative structure of QCD vacuum at the scales between Λ_χ and Λ_{QCD}. The guiding idea is that it is much more useful to introduce an effective low-energy Lagrangian which encodes the symmetries of the underlying QCD.

The very powerful method to obtain such an effective theory is to integrate out the high energy degrees of freedom of QCD (quarks and gluons) to construct an low-energy action in terms of physically significant variables (akin meson fields). In this way one could recover (see for example recent reviews [1–3]) the QCD chiral Lagrangian, phenomenologically introduced by S. Weinberg [4]. Another way to calculate non-perturbative QCD variables is well-known lattice regularization scheme [5].

Very recently it was suggested that the separation into effective and non-effective degrees of freedom of the continuum theory can also be done in a lattice [6]. The idea was that the lattice effective Lagrangian could contain the collective fields responsible for the long distance behavior of the fundamental lattice QCD. Then the QCD chiral Lagrangian could be obtained by expanding the lattice effective theory in powers of the lattice spacing and external momenta and all the coefficients of the Gasser-Leutwyler continuum effective Lagrangian can be recovered in such a way [7].

The basic element of this scheme is the techniques of bosonization in the strong coupling limit and large N expansion derived in papers [8,9]. On the other hand there is an elegant mathematical formalism developed by M.R. Zirnbauer [10] to match two different formulations of a certain class of theories. This is so-called "color-flavor" transformation. In particular, the low-energy effective QCD on a lattice could be connected with an effective chiral Lagrangian by making use of such a transformation.

DESCRIPTION OF THE MODEL

As we noted above, at the scale of energy below $\Lambda_\chi \sim 1\ GeV$ the QCD coupling constant $g(p)$ becomes strong and the kinetic term of the gluon field $\frac{1}{4g^2}F^2_{\mu\nu}$ is negligible by comparison with the terms of quark-gluon interaction and quark mass:

$$L \approx \bar\psi(x)\gamma_\mu D_\mu(x)\psi(x) + m\bar\psi(x)\psi(x),$$

where D_μ is Dirac operator. On this scale one has to take into account non-perturbative effects, that could gives rise to long-range correlations of quarks wave functions. The local quark-gluon interaction does not fit this picture. For instance, there are fermions zero modes on the instanton background delocalization of which leads to spontaneous chiral symmetry breaking. On the other hand on the large distances approaching Λ_{QCD}^{-1} a dynamically generated mass scale must be taken into account.

The model which we consider is the Euclidean $U(N)$ gauge theory strongly coupled to fermions. We put the fermions (quarks) onto sites of d-dimensional hypercubical lattice. On the each link the action could be written in a standard, naive form

$$L = \frac{1}{2a}\left\{\bar\psi(x+a)\gamma_\mu U_\mu(x)\psi(x) - \bar\psi(x)U^\dagger_\mu(x)\gamma_\mu\psi(x+a)\right\} + im\bar\psi(x)\psi(x), \quad (1)$$

where $U_\mu(x) \equiv U(x, x+\hat{a}_\mu) = \exp\{iagA_\mu(x+\frac{\hat{a}_\mu}{2})\} \in U(N)$ is lattice gauge variable, a is lattice spacing, a parameter of the model, describing the effect of long-range correlations, and m is a quark bare mass. The quark fields are $2^{d/2}$-component spinors occur in n different flavors and we suppose for simplicity that all quarks are degenerate in mass.

It is convenient to decompose the 4-component Dirac bispinors $\psi(x)$ into left- and right-handed Weyl spinors as

$$\psi_L = \frac{1-\gamma_5}{2}\psi; \qquad \psi_R = \frac{1+\gamma_5}{2}\psi$$
$$\bar{\psi}_L = \bar{\psi}\frac{1+\gamma_5}{2} \qquad \bar{\psi}_R = \bar{\psi}\frac{1-\gamma_5}{2}. \tag{2}$$

For simplicity we consider here only one fermion flavor. A further generalization of this approach onto the case of N_f flavours is clear. Taking into account properties of γ_5 matrix,

$$\gamma_5\gamma_\mu = -\gamma_\mu\gamma_5, \quad \gamma_5^2 = 1, \quad \gamma_5 = \begin{pmatrix} 0 & 1 \\ 1 & 0 \end{pmatrix}, \tag{3}$$

the expression (1) could be re-written via left and right components in the form, obviously chiral-invariant in massless limit

$$L = \frac{1}{2a}\Big\{\bar{\psi}_R(x+a)\gamma_\mu U_\mu \psi_R(x) + \bar{\psi}_L(x+a)\gamma_\mu U_\mu \psi_L(x)$$
$$- \bar{\psi}_L(x)U_\mu^\dagger \gamma_\mu \psi_L(x+a) - \bar{\psi}_R(x)U_\mu^\dagger \gamma_\mu \psi_R(x+a)\Big\}$$
$$+ im\bar{\psi}(x)\psi(x). \tag{4}$$

The global chiral transformations of the fields are

$$\bar{\psi}_R(x) \to \bar{\psi}_R(x)U_R^{-1}; \qquad \psi_R(x) \to U_R \psi_R(x);$$
$$\bar{\psi}_L(x) \to \bar{\psi}_L(x)U_L^{-1}; \qquad \psi_L(x) \to U_L \psi_L(x). \tag{5}$$

Obviously the standard hypercubical lattice does not have the rotation symmetry of the continuum. But all tensors of rank up to two symmetric under the group of lattice transformations are also symmetric under the group of rotation of Euclidean space. That is why the standard terms of second order in momentum are recovered in the chiral effective Lagrangian on the hypercubical lattice [8,9]. To obtain the terms of 4th order in lattice momentum matching the continuum counterparts one has to enhance the symmetry of the lattice space, for example by putting QCD on the body centered hypercubical lattice [6].

In this note we would like to apply another way to form the lattice theory with such an enhanced symmetry of the tensors of rank four. As it was suggested by A.Altland [11], one can decomposed the space on such a way that the fermions on nearest neighbors sites belongs to different sub-lattices A and B (see fig. 1). Then the action of the model is given by the summation over all such elementary cells:

$$S = \frac{1}{2a}\sum_{\substack{i\in A \\ j\in B}}\Big\{\bar{\phi}_R(i)\gamma_\mu U_{ij}^\mu \psi_R(j) + \bar{\phi}_L(i)\gamma_\mu U_{ij}^\mu \psi_L(j) \tag{6}$$

$$- \bar{\psi}_L(j)U_{ji}^{\mu\dagger}\gamma_\mu \phi_L(i) - \bar{\psi}_R(j)U_{ji}^{\mu\dagger}\gamma_\mu \phi_R(i)\Big\} + im\sum_{i\in A}\bar{\psi}(i)\psi(i) + im\sum_{i\in B}\bar{\phi}(i)\phi(i)$$

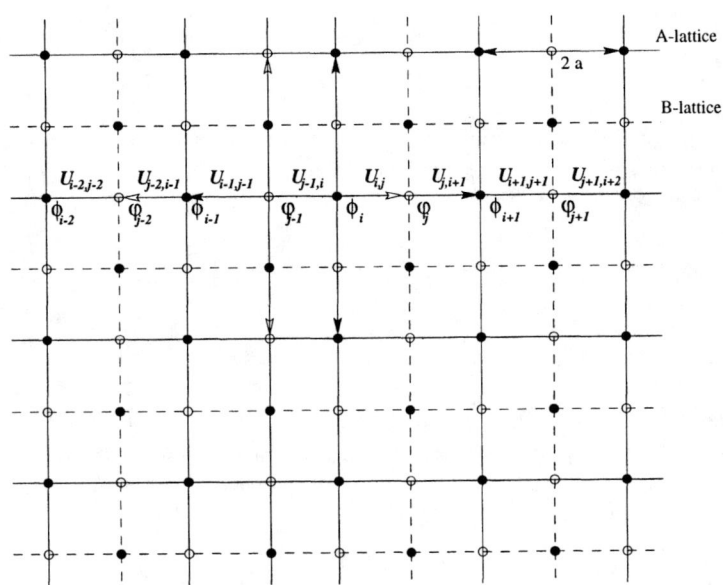

FIGURE 1. The model on two nested sub-lattices.

(here we introduced the notations $\psi(i) \equiv \psi(x_i)$ and $\phi(i) \equiv \psi(x_i+a)$ to sort out the fields onto different sub-lattices). We will see that such a double lattice provides a higher symmetry then a standard hypercubic lattice.

There are two chiral components of the spinor fields $\psi(i), \phi(j)$ coupled by group matrix U_{ij}. Denote two-component chiral fields as $\phi^a = (\phi_L, \phi_R)$, $\psi^a = (\psi_L, \psi_R)$. In terms of these fields the action (6) can be re-written in such a compact notation as

$$S = \frac{1}{2a} \sum_{\substack{i \in A \\ j \in B}} \sum_{a=1,2} \left\{ \bar{\phi}^a(i) \gamma_\mu U_{ij}^\mu \psi^a(j) + \bar{\psi}^a(j) U_{ji}^{\mu\dagger} \gamma_\mu \phi^a(i) \right\}$$
$$+ im \sum_{i \in A} \bar{\psi}(i)\psi(i) + im \sum_{i \in B} \bar{\phi}(i)\phi(i). \qquad (7)$$

COLOR-FLAVOR TRANSFORMATION AND SADDLE POINT CONFIGURATION

The basic idea of so-called "Color-Flavor transformation" [10] is an identity transforming a functional integral over the gauge group $U(N)$ into integral over matrix field Z:

$$\int_{U(N)} dU \exp\left(\bar{\phi}_\mu^i U^{ij}\psi_\mu^j + \bar{\psi}_\nu^j \widetilde{U}^{ij}\phi_\nu^i\right) = \int_{\mathbb{C}^{(n\times n)}} \mathcal{D}\mu(Z,\bar{Z}) \exp\left(\bar{\phi}_\mu^i Z_{\mu\nu}\phi_\nu^i - \bar{\psi}_\nu^j \bar{Z}_{\mu\nu}\psi_\mu^j\right). \quad (8)$$

Here the group (or "color") indices of the fermionic field are $i,j = 1,\ldots N$ and Z is (in general complex) $n \times n$ matrix parameterizing compact symmetric space $M = U(2n)/U(n) \times U(n)$ (or the "flavor" space). The integration measure is

$$\mathcal{D}\mu(Z,\bar{Z}) = N \times \det(1 + \bar{Z}_{\mu\nu}Z_{\mu\nu})^{-2n-N} \prod_{\mu,\nu=1}^{n} dZ_{\mu\nu}d\bar{Z}_{\mu\nu}.$$

The general proof and different generalizations of this remarkable formula both in the case of bosonic as well fermionic fields can be found in [10]. Such a transformation can also be applied in our case of the QCD partition function in strong coupling limit (7). The color-flavor transformation of the partition function turns out to be a color-chiral transformation: it re-arranges the fields by decoupling of the left and right components:

$$Z = \int \mathcal{D}Z\mathcal{D}Z^\dagger \mathcal{D}\phi \mathcal{D}\psi \exp\Big\{\frac{N}{2a} \sum_{a=1,2}\Big(\sum_{i\in A}\bar{\phi}^a(i)\gamma_\mu \sum_j Z^\mu_{ab}\phi^b(j)$$
$$+ \sum_{i\in B}\bar{\psi}^a(i)\gamma_\mu \sum_j Z^{\mu\dagger}_{ab}\psi^b(j)\Big) - (4+N)\sum_{\substack{i\in A\\ j\in B}} \ln\det\left(1 + Z^\mu{}_{ij}Z^{\mu\dagger}{}_{ij}\right)$$
$$+ iNm \sum_{i\in A}\bar{\psi}(i)\psi(i) + iNm \sum_{j\in B}\bar{\phi}(j)\phi(j)\Big\}.$$

Here the internal summation is given on the set of nearest neighbor sites of every i^{th} site. Integration over the fermion fields gives:

$$Z = \int \mathcal{D}Z\mathcal{D}Z^\dagger \exp\Big\{N\Big[\sum_{i\in A}\text{Tr}\ln\left(im + \frac{1}{2a}\sum_j Z^\nu_{ij}\gamma_\nu\right)\Big]$$
$$+ N\Big[\sum_{i\in B}\text{Tr}\ln\left(im + \frac{1}{2a}\sum_j Z^{\nu\dagger}_{ji}\gamma_\nu\right)\Big] + (4+N)\sum_{\substack{i\in A\\ j\in B}} \ln\det\left(1 + Z^\nu{}_{ij}Z^{\nu\dagger}{}_{ij}\right)\Big\}.$$

The saddle point approximation can be used to evaluate the partition function. One could suppose that the field Z is slowly variated at the distance of scale a. Thus an ansatz can be employed $Z^\mu_{ij} = z\gamma_\mu I_{ij}$ where I_{ij} is a unit matrix on the link $i \to j$ and z is a number. Substitution of this ansatz into the exponent in (9) gives:

$$S_{saddle} = -\frac{N}{2a^d}\text{Tr}\int d^4x \left\{\ln\left(im + \frac{dz}{a}\right) + \ln\left(im + \frac{d\bar{z}}{a}\right) - 2d\ln(1+z\bar{z})\right\}, \quad (9)$$

where we take into account that in the long-distance approximation the sum over links could be approximated by an integral as $\sum_{i\in A} \simeq \frac{1}{2a^d}\int d^dx$ and for the ansatz under consideration the nearest neighbor summation gives $\gamma_\mu \sum_i Z^\mu_{ji} = 2dz$, $\gamma_\mu \sum_i Z^{\mu\dagger}_{ji} = 2d\bar{z}$ and

$$\sum_{\substack{i \in A \\ j \in B}} \ln \det \left(1 + Z^\mu_{ij} Z^{\mu\dagger}_{ij}\right) \simeq \frac{2d}{2a^d} \text{Tr} \int d^d x \ln(1 + z\bar{z}).$$

Variation of (9) over the independent variables z and $\bar{z}$ gives the equations

$$\frac{1}{im + dz} = \frac{2\bar{z}}{1 + z\bar{z}}; \qquad \frac{1}{im + d\bar{z}} = \frac{2z}{1 + z\bar{z}}. \qquad (10)$$

Thus the saddle point is given by $z = \bar{z}$ and

$$z = \bar{z} = \frac{1}{2d - 1} \left\{ \sqrt{2d - 1 - (am)^2} \mp iam \right\} \xrightarrow{m \to 0} \frac{1}{\sqrt{2d - 1}}. \qquad (11)$$

Now we can consider small variations of the fields around of this configuration. Because the action (9) in massless limit depends only on the polynomial of the product $Z^\mu_{ij} Z^{\mu\dagger}_{ij}$, instead of I_{ij} one can plug into an ansatz for the Z-field a unitary matrix U_{ij}. But in $d = 4$ the most general form of the ansatz conforming to the chiral transformations of the quark fields in the massless limit could be employed:

$$Z^\mu_{ij} = z \left(\frac{1 - \gamma_5}{2} \gamma^\mu U_{ij} + \frac{1 + \gamma_5}{2} \gamma^\mu U^\dagger_{ij} \right) \equiv z \gamma^\mu U^{\gamma_5}_{ij} = Z^{\mu\dagger}_{ij}. \qquad (12)$$

Note that $Z^\mu_{ij} Z^{\mu\dagger}_{ij} = z^2 U_{ij} U^\dagger_{ij}$. Using the obvious relations

$$\bar{\phi}_L(i) \gamma_\mu Z^\mu \phi_L(j) = 0; \qquad \bar{\phi}_L(i) \gamma_\mu Z^\mu \phi_R(j) = z \bar{\phi}_L(i) U \phi_R(j)$$
$$\bar{\phi}_R(i) \gamma_\mu Z^\mu \phi_R(j) = 0; \qquad \bar{\phi}_R(i) \gamma_\mu Z^\mu \phi_L(j) = z \bar{\phi}_R(i) U^\dagger \phi_L(j)$$

etc for the chiral components of ψ,

one can see that the transformed action (9) on its saddle point configuration (12) can be represented as

$$Z = \int \mathcal{D}Z \mathcal{D}Z^\dagger \mathcal{D}\phi \mathcal{D}\psi \exp \Big\{ \frac{zN}{2a} \Big(\sum_{i \in A} \Big[\bar{\phi}_L(i) \sum_j U_{ij} \phi_R(j) + \bar{\phi}_R(i) \sum_j U^\dagger_{ij} \phi_L(j) \Big]$$
$$+ \sum_{i \in B} \Big[\bar{\psi}_L(i) \sum_n U_{ij} \psi_R(j) + \bar{\psi}_R(i) \sum_j U^\dagger_{ij} \psi_L(j) \Big] \Big) \qquad (13)$$
$$- (4 + N) \sum_{\substack{i \in A \\ j \in B}} \ln \det \left(1 + z^2 U_{ij} U^\dagger_{ij}\right) + Nm \sum_{i \in A} \bar{\psi}(i) \psi(i) + Nm \sum_{i \in B} \bar{\phi}(i) \phi(i) \Big\},$$

that is obviously not invariant under the chiral transformation in the massless limit. Indeed, an order parameter describing the spontaneous breaking of the chiral invariance is the quark condensate $<\bar{\psi}\psi>|_{Mink}= -i<\bar{\psi}\psi>|_{Eucl}$:

$$-i<\bar{\psi}\psi>|_{Mink}= -\frac{1}{V}\frac{\partial}{\partial m}S_{saddle}\Big|_{m=0} \simeq -\frac{N}{4za^3} \to <\bar{\psi}\psi>_{exp}\approx -(250\ MeV)^3.$$

That allows to estimate the scale

$$a \sim (200\ MeV)^{-1} \sim \Lambda_{QCD}.$$

RECOVERING OF THE LOW-ENERGY EFFECTIVE ACTION

It is known that the small quark bare mass m, due to effect of the spontaneous breaking of the chiral symmetry, gives mass to the pion. If we neglect the contribution of the Gaussian massive fluctuations around of the saddle point configuration[1] and take into account that in the large N limit the Z-field varies very slowly, the color-flavor transformed action in the long-distance approximation in $d=4$ takes the form

$$S = \frac{N}{2a^4}\operatorname{Tr}\int d^4x\left[\ln\left(im+\frac{4z}{a}U\right)+\ln\left(im+\frac{4z}{a}U^{-1}\right)-8\ln(1+z^2)\right]$$

$$= \frac{N}{2a^4}\operatorname{Tr}\int d^4x\left[\ln\left(1+\frac{iam}{4z}U^{-1}\right)+\ln\left(\frac{4z}{a}U\right)\right.$$

$$\left.+\ln\left(1+\frac{iam}{4z}U\right)+\ln\left(\frac{4z}{a}U^{-1}\right)-8\ln(1+z^2)\right]$$

$$= \frac{iNam}{8a^4z}\operatorname{Tr}\int d^4x\left([U+U^{-1}]+2\ln\left(\frac{4z}{a}\right)-8\ln(1+z^2)\right)$$

$$\simeq \frac{iNm\sqrt{7}}{8a^3}\operatorname{Tr}\int d^4x\left[U+U^{-1}\right].$$

In Minkowski space this has to be identified with the pion mass term in the standard QCD effective low-energy Lagrangian [1,4]

$$L_{m_\pi} = \frac{N}{4}F_\pi^2 m_\pi^2\operatorname{Tr}\int d^4x\left[U+U^{-1}\right]$$

where pion decay constant $F_\pi \sim 94\ MeV$. Thus the celebrated Gell-Mann–Oakes–Renner relation is recovered:

$$m<\bar{\psi}\psi>|_{Mink}= -4zm_\pi^2 F_\pi^2 \sim m_\pi^2 F_\pi^2.$$

---

[1] They are suppressed as $1/N$.

To put our model into complete correspondence to the low-energy chiral Lagrangian we consider the chiral limit of zero quark mass $m = 0$. Then the effective action in (9) could be written simply as

$$S_{kin} \simeq \frac{N}{2} \sum_{i \in A} \text{Tr} \ln \left(\sum_j U_{ij} \right) + \frac{N}{2} \sum_{i \in B} \text{Tr} \ln \left(\sum_j U_{ij}^{-1} \right). \quad (14)$$

Now using the expansion on a link

$$U_{ij} = U_i + \frac{a}{2} \partial_{i \to j} U_i + \frac{1}{2!} \frac{a^2}{4} \partial_{i \to j}^2 U_i + \frac{1}{3!} \frac{a^3}{8} \partial_{i \to j}^3 U_i + \frac{1}{4!} \frac{a^4}{16} \partial_{i \to j}^4 U_i + \ldots, \quad (15)$$

one can obtain

$$S_{kin} = \frac{N}{2} \sum_{i \in A} \text{Tr} \ln U_i - \frac{N}{2} \sum_{j \in B} \text{Tr} \ln U_j$$

$$+ \frac{N}{2} \sum_{i \in A} \text{Tr} \ln \left(1 + \frac{a^2}{8} U_i^{-1} \sum_j \partial_{i \to j}^2 U_i + \frac{1}{4!} \frac{a^4}{16} U_i^{-1} \sum_j \partial_{i \to j}^4 U_i \right)$$

$$+ \frac{N}{2} \sum_{j \in B} \text{Tr} \ln \left(1 + \frac{a^2}{8} U_j^{-1} \sum_i \partial_{j \to i}^2 U_j + \frac{1}{4!} \frac{a^4}{16} U_j^{-1} \sum_i \partial_{j \to i}^4 U_j \right), \quad (16)$$

where we take into account that the sum over all links $\sum_i \partial_{j \to i} U_j = 0$, $\sum_i \partial_{j \to i}^3 U_j = 0$. One can rewrite this expression by using Hermitian matrices $L_{i \to j} \equiv U_i^{-1} \partial_{i \to j} U_i$. To put our model into correspondence to the Gasser-Leutwyler chiral perturbation theory we make use of the long-distance approximation. Then, up to the four-derivative terms we find:

$$S_{kin} \approx \frac{Nz}{8a^2} \text{Tr} \int d^4x \, L_\mu^2 - \frac{Nz}{386} \text{Tr} \int d^4x \left[2(\partial_\mu L_\mu)^2 + L_\nu L_\mu L_\nu L_\mu \right], \quad (17)$$

that, taking into account above obtained estimation for $a \sim (200 \; MeV)^{-1}$ is in a rather good agreement with the corresponding terms of the derivative expansion of the effective chiral Lagrangian (see for example [1–3]):

$$\text{Re} \, S_{eff} = \frac{F_\pi^2}{4} \text{Tr} \int d^4x \, L_\mu^2 - \frac{N}{192\pi^2} \text{Tr} \int d^4x \left[2(\partial_\mu L_\mu)^2 + L_\nu L_\mu L_\nu L_\mu \right]. \quad (18)$$

Note that the expansion of the Lagrangian originated from the lattice model allows to recover the standard terms of the gradient expansion up to the terms of forth order in momentum. This is a direct consequence of the way we introduce the double lattice. Similar approach, based on the theory formulated on the body-centered hypercubical lattice was suggested in [6].

Finally, let us show that, using the color-flavor transformed action on its saddle point value (7), one can recover the Lagrangian of the Nambu-Jona-Lasinio model.

Indeed, in the long-distance approximation the corresponding partition function can be written as

$$Z \sim \int \mathcal{D}Z\mathcal{D}Z^\dagger \mathcal{D}\phi \mathcal{D}\psi \prod_{a,b} \exp\Big\{\frac{N}{2a^4}\Big[\underbrace{\int d^4x\, \bar\phi^a \gamma_\mu Z_{ab}^\mu \phi^b}_{A} + \underbrace{\int d^4x\, \bar\psi^a \gamma_\mu Z_{ab}^{\mu\,\dagger} \psi^b}_{B}$$

$$- 8\underbrace{\int d^4x\, \mathrm{Tr}\ln\left(1 + Z_{ab}^\mu Z_{ab}^{\mu\,\dagger}\right)}_{AB}\Big]\Big\},$$

where the indices $a,b = 1,2$ correspond to the chiral components of the fermion field (2).

Note that the factor $z^2 = 1/(2d-1)$ in four dimensions is rather small and on the saddle point we can write $\mathrm{Tr}\ln(1+z^2) \approx \mathrm{Tr}\, Z_{ab}^\mu Z_{ab}^{\mu\,\dagger}$. Thus the integral over Z turns to be a Gaussian one:

$$\int \mathcal{D}Z\mathcal{D}Z^\dagger \prod_{a,b}\exp\Big\{u_{ab}Z_{ab} + v_{ab}Z_{ab}^\dagger - Z_{ab}Z_{ab}^{\mu\,\dagger}\Big\} = \exp\Big\{\sum_{a,b} u_{ab}v_{ab}\Big\},$$

and we obtain from (19)

$$Z \sim \mathcal{D}\psi\, \exp\Big\{\frac{N}{8a^4}\Big[\int d^4x \sum_{a,b}(\bar\psi^a\gamma_\mu\psi^b)(\bar\psi^a\gamma_\mu\psi^b)\Big]\Big\}$$

$$= \mathcal{D}\psi\, \exp\Big\{\frac{N}{16a^4}\Big[\int d^4x\,(\bar\psi\gamma_\mu\psi)^2 + (\bar\psi\gamma_\mu\gamma_5\psi)^2\Big]\Big\},$$

where ψ is a standard Dirac bispinor. This establish a link between the QCD in the strong coupling limit and the phenomenological description provided by the Nambu-Jona-Lasinio model.

Conclusions

On the basis of the works [12] sketched above, we have derived a new way to link QCD in a strong coupling limit and low-energy effective meson theory. We find also that by placing the quarks onto two sub-lattices and expanding the color-flavor transformed action in powers of lattice spacing, the chiral Lagrangian is recovered up to the terms of order $O(p^4)$. To sum up, the color-flavor transformation gives useful insights into the non-perturbative QCD and provides a new point of view which could be developed further. Problems which need further study within this approach to QCD include the following:

- a possibility to apply the color-flavor transformation to the weak coupling perturbative QCD;
- an effect of massive fluctuations around the saddle point configuration;
- and the possible application of the sketched approach to the conventional (continuum) QCD.

Acknowledgments

This talk is a short version of the work in collaboration with A.Altland and M.R. Zirnbauer [12]. We are most indebted, for fruitful discussions, to M.R. Zirnbauer, to whom belongs the very idea to apply color-flavor transformation to construct QCD low-energy effective action and A.Altland, who developed the technique of the gradient expansion and suggested to use the set of nested sub-lattices. Ya.S. also would like to thank V.Petrov, P.Pobylitsa and M.Polyakov for useful discussions.

REFERENCES

1. Meißner U.G., *Rep. Prog. Phys.* **56** 903 (1993).
2. Schafer T. and Shuryak E.V., *Rev. Mod. Phys.* **70** 323 (1998).
3. Diakonov D., *Chiral Quark-Soliton Model*, Lectures given at Advanced Summer School on Nonperturbative Quantum Field Physics, Peniscola, Spain, June 1997; hep-ph/9802298.
4. Weinberg S., *Phys. Rev. Lett.* **18** 188 (1967); *Phys. Rev.* **166** 1568 (1968).
5. Wilson K., *Phys. Rev.* **D 10** 2445 (1974).
6. Myint S. and Rebbi C., *Nucl. Phys.* **B 421** 241 (1994).
7. Levi A.R., Lubicz V. and Rebbi C., *Phys. Rev.* **D 56** 1101 (1997).
8. Kluberg-Stern H., et al, *Nucl. Phys.* **B 190** 504 (1981).
9. Kawamoto N. and Smit J., *Nucl. Phys.* **B 192** 100 (1981).
10. Zirnbauer M.R., *J. Phys.* **A 29** 7113 (1996).
11. Altland A. and Simons B.D., *to be published in Nucl. Phys.*; cond-mat/9909152.
12. Altland A., Budzcies J., Shnir Ya. and Zirnbauer M.R., *to be published*.

CHIRAL PERTURBATION THEORY

Chiral QCD dynamics: Recent results

Ulf-G. Meißner

Forschungszentrum Jülich, IKP(Th), D-52425 Jülich, Germany

Abstract. I review recent chiral perturbation theory results for the electromagnetic hyperon and the strange nucleon form factors. I also present a precise chiral nucleon–nucleon potential based on a modified Weinberg power counting and discuss the pertinent results and implications.

INTRODUCTION

Effective field theory (EFT) is a powerful tool to study hadron structure and hadron dynamics at energies below the scale of chiral symmetry breaking. For the meson and meson–baryon sectors, this EFT is chiral perturbation theory (CHPT). Its formulation is based on the fact that the pseudo–Goldstone bosons of QCD interact weakly with themselves and with matter fields at low energies due to the Goldstone theorem. A detailed review is e.g. ref. [1]. Here, I will present some recent results from the three flavor sector, pertaining to the electromagnetic structure of the hyperons and the matrix–elements of the strange vector current in the nucleon. These results show that three flavor baryon chiral perturbation theory is indeed *effective*. I then consider the two–nucleon system. Here, an added complication arises due to the appearance of shallow bound states (or large scattering lengths). This has to be dealt with by a non–perturbative resummation. I will show that Weinberg's original proposal [2] can lead (after suitable modifications) to a chiral two–nucleon potential, which is as precise as semi–phenomenological models, but has the advantage of being controlled and systematic. I show some first promising results for phase shifts and deuteron properties and discuss some implications.

HYPERON FORM FACTORS

To third order in the chiral expansion, i.e. to leading one–loop order, the electromagnetic form factors (ffs) of the nucleon have been studied in refs. [3–5]. At that order, one has to deal with two counterterms in the electric and two in the magnetic ffs. Using e.g. the proton and neutron electric radii and magnetic moments as input, the ffs are fully determined to that order. In particular, no counterterms appear in the momentum expansion of the magnetic ffs. To this order in the chiral expansion, the ffs are precisely described for momentum transfer squared up to $Q^2 \simeq 0.2\,\text{GeV}^2$. It appears therefore natural to extend such an investigation to the three flavor case. Surprisingly, that has never been attempted until recently [6] despite a huge amount of studies in three flavor chiral perturbation theory. This investigation was triggered by the recent results on the Σ^- radius reported by the WA89 collaboration at CERN and by the SELEX collaboration at FNAL (note that the SELEX results are still preliminary), $\langle r_{\Sigma^-}^2 \rangle_{\text{exp}} = 0.92 \pm 0.32 \pm 0.40\,\text{fm}^2$ [7], $\langle r_{\Sigma^-}^2 \rangle_{\text{exp}} = 0.60 \pm 0.08 \pm 0.08\,\text{fm}^2$ [8], obtained by scattering a highly boosted hyperon beam off the electronic cloud of a heavy atom (elastic hadron–electron scattering). The pattern of the charge radii embodies information on SU(3) breaking and the structure of the groundstate octet. In a CHPT calculation of the corresponding ffs, the baryon structure is to some part given by the meson (pion and kaon) cloud and in part by short distance physics parametrized in terms of local contact interactions. In the general case, such a splitting depends on the regulator scheme and scale one chooses. Here, we work in standard dimensional regularization and set $\lambda = 1\,\text{GeV}$ throughout (since this is the natural hadronic scale). If one performs the SU(3) calculation to third order, one has no new counterterms as compared to the SU(2) case. Therefore, fixing the low–energy constants (LECs) from proton and neutron properties allows one to make parameter–free predictions for the hyperons. As an added bonus, kaon loops induce a momentum dependence in the isoscalar magnetic form factor of the nucleon, as first pointed out in ref. [9], whereas in the pure SU(2) calculation, $G_M^{I=0}(Q^2)$ is simply constant. This allows one to study the contribution of kaon loops (strangeness) to the em ffs of the nucleon (not to be confused with the strange ffs to be discussed below).

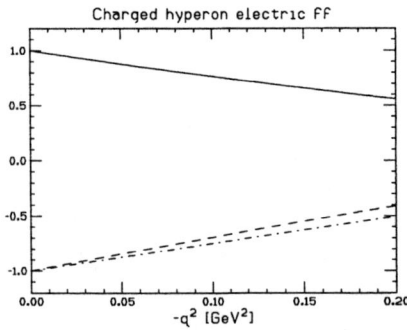

Figure 1. The electric form factors of the charged hyperons calculated in three flavor baryon CHPT. Solid, dashed, dot–dashed line: Σ^+, Σ^-, Ξ^-, in order.

Consider first the hyperons. The electric ffs of the charged hyperons are given in fig.1. The corresponding radii are (a more detailed discussion also of the neutral particles and magnetic radii is given in [6]) $\langle r_{\Sigma^+}^2 \rangle = 0.64\ldots 0.66\,\text{fm}^2$, $\langle r_{\Sigma^-}^2 \rangle = 0.77\ldots 0.80\,\text{fm}^2$, $\langle r_{\Xi^-}^2 \rangle = 0.61\ldots 0.65\,\text{fm}^2$. The given uncertainty does not reflect the contribution from higher orders, which should be calculated. The prediction for the Σ^- is in fair agreement with the recent measurements. The result for the Σ radii is at variance with quenched lattice QCD calculations, which give $0.56(5)\,\text{fm}^2$ and $0.72(6)\,\text{fm}^2$ for the negative and positive Σ, respectively [10]. However, quenched lattice calculations should be taken with a grain of salt (the true error due to the quenching is only known for very few quantities, certainly not for the radii). In the CHPT approach, the difference of the radii is due to some short distance physics encoded in the LEC d_{102}^0 [11] and to the Foldy term. The loop contributions are almost equal, but the difference due to the counterterm and the Foldy term for the Σ hyperons is $\langle r_{\Sigma^+}^2 \rangle - \langle r_{\Sigma^-}^2 \rangle = -\frac{8d_{102}}{(4\pi F_\phi)^2} + \frac{b_D}{m^2} = -0.10\ldots-0.15\,\text{fm}^2$, depending on how one fixes the electric LEC d_{102} and the magnetic LEC b_D. Here, $F_\phi = 100$ MeV is the average pseudoscalar decay constant. A more detailed discussion of the parameter dependence is given in [6]. All the numbers given here are based on a third order calculation. Clearly, a fourth order calculation is called for to further quantify these results. Finally, I remark that the chiral description of the neutron charge ff is clearly improved in SU(3) as compared to the two flavor case. Obviously, this sizeable kaon cloud effect will be reduced at next order since the effect of recoil only starts to show up at fourth order. The inclusion of such recoil effects is expected to improve already the SU(2) calculation, leaving less room for the kaon cloud effects. It is also worth pointing out that this effect from kaon loops is opposite to what one expects from a ϕ-coupling [12] and thus some cancellations are expected to take place.

STRANGE VECTOR FORM FACTORS

Recently, the first results from parity–violating electron scattering experiments, which allow to pin down the so–called strange form factors of the nucleon, have become available. These strange ffs parametrize the matrix elements of the strange vector current, $\langle N| \bar{s}\,\gamma_\mu\,s\,|N\rangle = \langle N|\,\bar{q}\,\gamma_\mu\,(\lambda^0/3 - \lambda^8/\sqrt{3})\,q\,|N\rangle$, with $q = (u,d,s)$ denoting the triplet of the light quark fields and $\lambda^0 = I$ (λ^a) the unit (the $a = 8$ Gell–Mann) SU(3) matrix. The singlet and octet currents are parametrized in terms of electric and magnetic ffs, which give the strange ffs via $G_{E/M}^{(s)}(Q^2) = G_{E/M}^{(s)}(0) + \langle r_{E/M,s}^2 \rangle Q^2/6 + \mathcal{O}(Q^4)$. The SAMPLE collaboration has reported the first measurement of the strange magnetic moment of the proton [13]. To be precise, they give the strange magnetic form factor in units of nuclear magnetons at a small momentum transfer $Q_S^2 = 0.1$ GeV^2, $G_M^{(s)}(Q_S^2) = +0.23 \pm 0.37 \pm 0.15 \pm 0.19$. The rather sizeable error bars document the difficulty of such type of experiment. The HAPPEX collaboration has chosen a different kinematics which is more sensitive to the strange electric form factor [14]. Their measurement implies $G_E^{(s)}(Q_H^2) + 0.39\,G_M^{(s)}(Q_H^2) = 0.023 \pm 0.034 \pm 0.022 \pm 0.026$, at $Q_H^2 = 0.48$ GeV^2. Of course, this momentum transfer might be too high for the CHPT analysis at third order to hold, but in the absence of data at lower Q^2, let us assume that we can still use the HAPPEX result. This loophole should be kept in mind. There have been many theoretical speculations about the size of the strange form factors, some of them clearly in conflict with the data. These data have been analyzed in the framework of chiral perturbation theory [15], extending previous work [16]. It was shown in [9] that one can make a parameter–free prediction for the momentum dependence of the nucleons' strange magnetic (Sachs) form factor based on the chiral symmetry of QCD solely. The value of the strange magnetic moment, which contains an unknown low–energy constant (b_0), can be deduced from the SAMPLE experiment using the momentum–dependence derived in [9]. Furthermore, the SU(3) analysis of the octet electromagnetic form factors performed in [6] allows one to pin down the octet component of the strange vector current. Thus, to leading one–loop order, there is only one new singlet counterterm (d^0_{102}), the strength of which can be determined from the value found by HAPPEX. This allows to give a band for the strange electric form factor and make a prediction for the MAMI A4 experiment, which intends to measure $G_E^{(s)}(Q_M^2)+0.22\, G_M^{(s)}(Q_M^2)$ with a four–momentum transfer (squared) $Q_M^2 = 0.23$ GeV^2 of approximately half the HAPPEX value. Under the assumptions mentioned, one can determine the LECs b_0 and d^0_{102} with sizeable uncertainties reflecting the experimental input. The central values are of natural size and the corresponding results for the strange electric ff is shown in fig.2 by the solid line. The dashed lines reflect the theoretical uncertainty based on a very conservative analysis. The corresponding strange radii and the strange magnetic moment are [9,15]: $\langle r^2_{E,s}\rangle = (0.05 \pm 0.09)\,\text{fm}^2$, $\langle r^2_{M,s}\rangle = -0.14\,\text{fm}^2$ and $\mu_s = (0.18 \pm 0.44)$ n.m. , where the uncertainty in the strange radius stems mostly from the uncertainty in the singlet LEC d^0_{102}, whereas the prediction for the magnetic radius at this order is parameter–free. The uncertainty in μ_s is completely given by the error of the SAMPLE analysis.

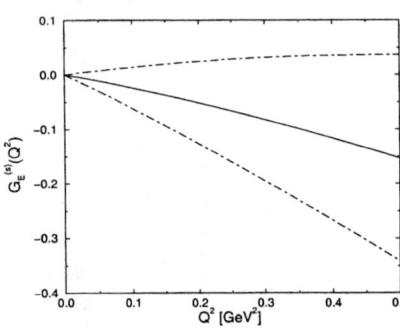

Figure 2. The strange electric form factor of the nucleon from heavy baryon chiral perturbation theory (to third order in the chiral expansion).

A few more remarks on the strange electric ff are in order. The radius is fairly small and *positive*, and even given the sizeable uncertainty, it is on the lower side of the predictions based on dispersive approaches including maximal OZI violation [17,18]. It is more compatible with models that include $\pi\rho$ [19] contributions in the isoscalar spectral functions besides the vector meson poles ($\omega, \phi, \ldots$) or dispersive analysis of the $\bar{K}K$ [20] continuum. Note also that from the octet current the strange electric radius inherits the chiral singularity $\sim \ln(M_K)$. It is also worth pointing out that the momentum dependence of the strange electric form factor is rather different from the one of the neutron charge form factor, which also vanishes at zero momentum transfer. We also note that using the central values for the LECs, the prediction for the MAMI A4 experiment, which attempts to measure $G_E^{(s)}(Q_M^2)$ + 0.22 $G_M^{(s)}(Q_M^2)$ at a four-momentum transfer (squared) of $Q_M^2 = 0.23$ GeV^2, is very small but afflicted with a large uncertainty. A more detailed discussion is given in ref. [15]. It is also important to stress that a dispersive analysis of the $K\bar{K}$ continuum leads to much smaller values for the strange radii [21]. That approach is based on an analytic continuation of the empirical KN scattering amplitudes and uses unitarity bounds. In principle, CHPT calculations and dispersion relations can be mapped one–to–one as has been shown, see e.g. ref. [22]. Both calculations need to be improved. On the CHPT side, the next order has to be investigated in order to check the convergence. The dispersion relations need better data, since so far the analytic continuation cannot be made stable without extra assumptions. More direct data from Jefferson Lab and MAMI should help to clarify the situation.

THE NN INTERACTION FROM EFT

In what follows, I will be concerned with the effective potential between nucleons as defined (and somewhat modified) in ref. [2] and studied quantitatively in [23]. In contrast to the meson–baryon sector discussed before, the appearance of shallow bound states requires a non–perturbative resummation. This can be done in different ways. I follow here the EFT approach suggested by Weinberg, where one has to deal with two different types of interactions. First, there is one–, two–pion exchange (OPE,TPE) and so on, to describe the long and medium range physics. Second, there are four–nucleon contact interactions to describe the short (and to some extent the medium) range physics. So we have a scale separation, the dividing line being somewhere above twice the pion mass and below the typical scale of chiral symmetry breaking, $\Lambda_\chi \simeq 1$ GeV. The problem at hand can be treated exactly by integrating out the pionic degrees of freedom from the Fock space using the projection formalism of Okubo, Fukuda, Sawada and Taketani [25]. Based on this approach, we have set up the following scheme. First, one constructs the chiral NN potential based on a power counting in harmony with the projection formalism. This is outlined in detail in ref. [26]. To third order in small momenta, this potential is given by the following contributions (LO = leading order, NLO = next-to-leading order a.s.o.): **LO** OPE with lowest order insertions and two 4N contact interactions without derivatives. **NLO** Vertex and self–energy corrections to the LO interactions, TPE with lowest order insertions and seven 4N contact interactions with two derivatives. **NNLO** Vertex and self–energy corrections to OPE as well as TPE with exactly one insertion from the dimension two πN Lagrangian. These terms encode non–trivial information about the pion–nucleon interaction beyond leading order and are thus sensitive to the chiral structure of QCD. This potential is divergent. All divergences can be dealt with by subtracting divergent loop integrals, which leads to an overall renormalization of the axial–vector coupling g_A and seven of the nine coupling constants related to the 4N interactions. The precise procedure is discussed in detail in ref. [27]. The renormalized potential still has a bad high energy behaviour. Some of the contact interactions (NNLO TPE contributions) grow quadratically (cubically) with increasing momenta. Even the momentum–independent contact interactions necessitate regularization, which is performed on the level of the Lippmann–Schwinger equation. That is done in the following way: $V(\vec{p},\vec{p}\,') \to f_R(\vec{p})\,V(\vec{p},\vec{p}\,')\,f_R(\vec{p}\,')$, where $f_R(\vec{p})$ is a regulator function chosen in harmony with the underlying symmetries. In ref. [27], two different regulator functions are used, the sharp cutoff $f_R^{\text{sharp}}(\vec{p}) = \theta(\Lambda^2 - p^2)$, and an exponential form, $f_R^{\text{expon}}(\vec{p}) = \exp(-p^{2n}/\Lambda^{2n})$, with $n = 2, 3, \ldots$. The latter form is more suitable for the calculation of some observables. Bound and scattering states can then be obtained by solving the Lippmann–Schwinger equation with the regularized potential. Next, one has to pin down the parameters. Those related to the pion–nucleon interaction beyond leading order can be fixed by a fit of the chiral perturbation theory pion–nucleon amplitude [28] to the dispersion–theoretical one inside the Mandelstam triangle [29]. In addition, we have nine coupling constants related to four–nucleon contact interactions. These can be uniquely determined by a fit to the S- and P-waves together with the mixing parameter ϵ_1. While both S-waves contain two parameters, the P-waves and ϵ_1 depend on one (more precisely, one can form linear combinations of the LECs which appear as these parameters in the considered partial waves). So we can perform global fits to the S- and P-waves for nucleon lab energies from 0 to 100 MeV. For example, such a global fit with $\Lambda = 500$ MeV at NLO and 875 MeV at NNLO leads to a deuteron binding energy of $E_d = -2.17$ and -2.21 MeV at NLO and NNLO, respectively. Therefore, without any fine tuning, we can reproduce the empirical value within 2 percent and 1 permille at NLO and NNLO, in order. The increase of the cut–off value when going from NLO to NNLO is related to the fact that at NNLO, the chiral TPEP includes mass scales above the two–pion mass, which is the scale related to the uncorrelated TPEP appearing at NLO. It is also worth mentioning that the quality of the fits increases visibly as one goes from LO to NLO to NNLO (for details, see ref. [27]). This is, of course, expected from the underlying power counting. With that, one can predict these partial waves for energies above 100 MeV. All other partial waves with angular momentum ≥ 2 and the deuteron properties are *predictions*.

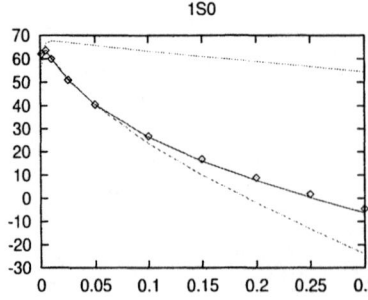

Figure 3. Predictions for the $^1S_0$ partial wave (in degrees) at LO (upper dashed curve), NLO (lower dashed curve) and NNLO (solid curve) in comparison to the Nijmegen PSA (diamonds) for nucleon laboratory energies up to 0.3 GeV.

The resulting S–waves are shown in figs.3,4 in comparison to the Nijmegen phase shift analysis (PSA) [30]. The improvement when going from LO to NLO to NNLO is clearly visible.

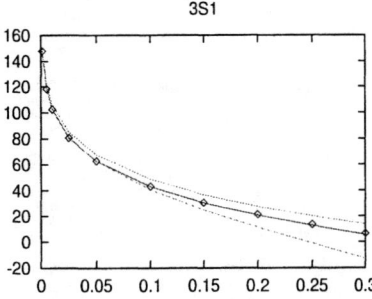

Figure 4. Predictions for the $^3S_1$ partial wave (in degrees) at LO (upper dashed curve), NLO (lower dashed curve) and NNLO (solid curve) in comparison to the Nijmegen PSA (diamonds) for nucleon laboratory energies up to 0.3 GeV.

I would also like to discuss briefly the so–called effective range expansion. For any partial wave, one can write $p \cot \delta = -1/a + r\, p^2/2 + v_2\, p^4 + v_3\, p^6 + \mathcal{O}(p^8)$, with δ the phase shift, p the nucleon cm momentum, a the scattering length and r the effective range. It has been stressed in ref. [31] that the shape parameters v_i are a good testing ground for the range of applicability of the underlying EFT. At NNLO, we find e.g. $a = 5.424\,(5.420)\,\mathrm{fm}$, $r = 1.741\,(1.753)\,\mathrm{fm}$, $v_2 = 0.046\,(0.040)\,\mathrm{fm}^3$, and $v_3 = 0.67\,(0.67)\,\mathrm{fm}^5$ for $^3S_1$. Similarly, for $^1S_0$, we have $a = -23.72\,(-23.74)\,\mathrm{fm}$, $r = 2.68\,(2.67)\,\mathrm{fm}$, $v_2 = -0.61\,(-0.48)\,\mathrm{fm}^3$, and $v_3 = 5.1\,(4.0)\,\mathrm{fm}^5$. The numbers in the brackets refer to the np system from the Nijmegen II potential. Note that one can also perform the fit in such a way that the scattering lengths and effective ranges are exactly reproduced. This leads only to modest changes in the values of $v_{2,3}$, e.g. for such a fit in $^1S_0$ one has $v_2 = -0.53\,\mathrm{fm}^3$ and $v_3 = 5.0\,\mathrm{fm}^5$. This rather good agreement illustrates again that the long range physics associated with pion exchanges is incorporated correctly and it demonstrates the predictive power of such an EFT approach. The P–waves are mostly well described, although the NNLO TPEP is a bit too strong in $^3P_1$ and $^3P_2$. Of particular interest is ϵ_1 since it has also been calculated at NLO [24] and NNLO [32] in the approach of Kaplan, Savage and Wise (KSW), in which only the leading 4N contact interactions are resummed. Our results are shown in comparison to the ones of refs. [24,32] in fig.5 as a function of nucleon cm momentum up to 350 MeV. For energies below 150 MeV, the KSW results are comparable to ours, but it is obvious from that figure that their approach is tailored to work at low energies. If one wants to go to momenta above 100 MeV, it appears that pion exchange should be treated non–perturbatively. Consider now the D– and F–waves. These are free of parameters and most problematic since the NNLO TPEP can be very strong. In some potential models, TPEP is simply cut at distances of (approximately) less than 1 fermi. Nevertheless, we find a rather satisfactory description of these partial waves. Of particular interest is $^3D_1$ since it is related to the deuteron channel. Also, $^1D_2$ is supposed to be very sensitive to contributions from the Δ–resonance, which in our approach is subsumed in some of the LECs related to the dimension two πN interaction. Both these partial waves are well described. Even the small $^3D_3$ partial wave is very well reproduced up to the opening of the pion production threshold (in the Bonn potential, this partial wave is dominated by correlated TPE). Note, however, that the D-waves are most sensitive to the choice of the regulator cut-off.

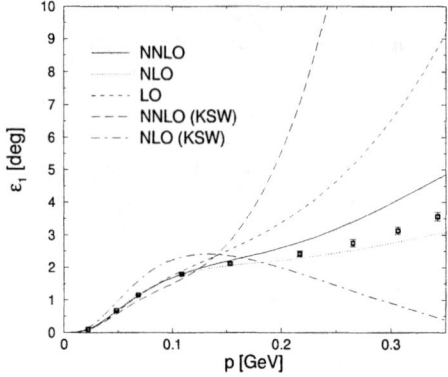

Figure 5. The $^3S_1-^3D_1$ mixing parameter ϵ_1 for our approach and the KSW scheme in comparison to the Nijmegen PSA as a function of the nucleon cms momentum.

Furthermore, the so-called peripheral waves ($l \geq 4$) have already been considered by the Munich group [33] (see also ref. [34]). Their calculation is based on Feynman graphs using dimensional regularization. The potential was constructed perturbatively by proper partial wave projection of the NNLO OPE and TPE. While in most of the peripheral waves OPE is dominant, there are a few exceptions where chiral NNLO TPEP is needed to bring the predictions in agreement with the data or PSA results. Our calculation, which is based on a completely different regularization scheme and treats the potential non-perturbatively, leads to the same results. This is rather gratifying. Finally, I consider the bound state properties. At NNLO (NLO), we use an exponential regulator with $\Lambda = 1.05\,(0.60)\,\text{GeV}$, which reproduces the deuteron binding energy within an accuracy of about one third of a permille (2.5 percent). No attempt is made to reproduce this number with better precision. In table 1 the deuteron properties are collected and compared to the data and two realistic potential model predictions. We note that the deviation of our prediction for the quadrupole moment compared to the empirical value is slightly larger than for the realistic potentials. Still, it remains to be checked whether this problem persists when one includes the meson-exchange currents (compare also the discussion in ref. [35]). The asymptotic D/S ratio, called η, and the strength of the asymptotic wave function, A_S, are well described. The D-state probability, which is not an observable, is most sensitive to variations in the cut-off. At NLO, it is comparable and at NNLO somewhat larger than that obtained in the CD-Bonn or the Nijmegen-93 potential. This increased value of P_D is related to the strong NNLO TPEP. At N^{3}LO, I expect this to be compensated by dimension four counterterms. It is also worth mentioning that at NNLO, we have two additional very deeply bound states. These have, however, no influence on the low-energy physics and can be projected out. Furthermore, these states are an artefact of

TABLE 1. Deuteron properties derived from our chiral potential compared to two "realistic" potentials (Nijmegen–93 and CD–Bonn) and the data. Here, r_d is the root–mean–square matter radius. An exponential regulator with $\Lambda = 600\,\text{MeV}$ and $\Lambda = 1.05\,\text{GeV}$ at NLO and NNLO, in order, is used.

	NLO	NNLO	Nijm93	CD-Bonn	Exp.
E_d [MeV]	−2.1650	−2.2238	−2.224575	−2.224575	−2.224575(9)
Q_d [fm^2]	0.266	0.262	0.271	0.270	0.2859(3)
η	0.0248	0.0245	0.0252	0.0255	0.0256(4)
r_d [fm]	1.975	1.967	1.968	1.966	1.9671(6)
A_S [fm$^{-1/2}$]	0.8662	0.8845	0.8845	0.8845	0.8846(16)
P_D [%]	3.62	6.11	5.67	4.83	−

the too strong potential and will most probably vanish at N^{3}LO. Altogether, the description of the deuteron as compared to ref. [23] is clearly improved. In the EFT approach, it is simple and unambiguous to couple external probes to the two–nucleon system. Therefore, electromagnetic properties of the deuteron can and will be studied based on a systematic power counting. Furthermore, the extension of these ideas to systems with more than two nucleons is underway.

CONCLUDING REMARKS

I have first shown that three flavor baryon chiral perturbation theory can be used to gain insight into the structure of the hyperons. In particular, hyperon radii can be predicted *parameter-free* because the pertinent low–energy constants can be pinned down from proton and neutron properties alone. These predictions agree with recent measurements performed at CERN and FNAL scattering high momentum hyperon beams in the electronic cloud of heavy atoms. Furthermore, a leading one–loop analysis sheds light on the nucleon matrix elements of the strangeness vector current. The presently available data from BATES and JLAB point towards a small strange electric form factor and more pronounced strange magnetic ff. Clearly, more data and more accurate (fourth order) calculations are called for. Despite lots of mumbling and talking, SU(3) baryon chiral perturbation theory is effective. In the two–nucleon system, we now have an EFT approach which can compete in accuracy with semi–phenomenological boson–exchange models. Even better, it is based on a power counting scheme and thus corrections can be calculated systematically. I believe that calculations including external electroweak probes and more than two nucleons will lead to a deeper understanding of QCD dynamics in the few–nucleon system. After quite a bit of time and work, we are now in the position to make *model-independent* statements in a sector of hadron physics which was long believed to be a playground for model–builders only. These times are changing.

ACKNOWLEDGEMENTS

I thank Véronique Bernard, Paul Büttiker, Evgeny Epelbaum, Nadia Fettes, Walter Glöckle, Thomas Hemmert, Bastian Kubis, Guido Müller and Sven Steininger for enjoyable collaborations and the organizers for their hospitality.

REFERENCES

1. V. Bernard, N. Kaiser and Ulf-G. Meißner, Int. J. Mod. Phys. **E4**, 183 (1995).
2. S. Weinberg, Phys. Lett. **B251**, 288 (1990); Nucl. Phys. **B363**, 3 (1991).
3. J. Gasser, M.E. Sainio and A. Švarc, Nucl. Phys. **B307**, 779 (1988).
4. V. Bernard et al., Nucl. Phys. **B388**, 315 (1992).
5. V. Bernard et al., Nucl. Phys. **A635**, 121 (1998).
6. B. Kubis, T.R. Hemmert and Ulf-G. Meißner, Phys. Lett. **B456**, 240 (1999).
7. M.I. Adamovich et al. (The WA89 Collaboration), Eur. Phys. J. **C8**, 55 (1999).
8. I. Eschrich (SELEX collaboration), hep-ex/9811003.
9. T.R. Hemmert, Ulf-G. Meißner and S. Steininger, Phys. Lett. **B437**, 184 (1998).
10. D.B. Leinweber et al., Phys. Rev. **D43**, 1659 (1991).
11. G. Müller and Ulf-G. Meißner, Nucl. Phys. **B492**, 379 (1997).
12. M.F. Gari and W. Krümpelmann, Phys. Lett. **B274**, 279 (1992).
13. B. Mueller et al. (SAMPLE collaboration), Phys. Rev. Lett. **78**, 3824 (1997).
14. K.A. Aniol et al. (HAPPEX collaboration), Phys. Rev. Lett. **82**, 1096 (1999).
15. T.R. Hemmert, B. Kubis and Ulf-G. Meißner, Phys. Rev. **C60**, 045501 (1999).
16. M.J. Ramsey-Musolf and H. Ito, Phys. Rev. **C55**, 3066 (1997).
17. R.L. Jaffe, Phys. Lett. **B229**, 275 (1989).
18. H.-W. Hammer, Ulf-G. Meißner and D. Drechsel, Phys. Lett. **B367**, 323 (1996).
19. Ulf-G. Meißner et al., Phys. Lett. **B408**, 381 (1997).
20. M.J. Ramsey–Musolf and H.-W. Hammer, Phys. Rev. Lett. **80**, 2539 (1998); H.-W. Hammer and M.J. Ramsey–Musolf, hep-ph/9903367.
21. H.-W. Hammer and M.J. Ramsey–Musolf, hep-ph/9812261.
22. J. Gasser and Ulf-G. Meißner, Nucl. Phys. **B357**, 90 (1991).
23. C. Ordóñez, L. Ray and U. van Kolck, Phys. Rev. **C53**, 2086 (1996).
24. D.B. Kaplan, M.J. Savage and M.B. Wise, Nucl. Phys. **B534**, 329 (1998).
25. S. Okubo, Prog. Theor. Phys. **12**, 603 (1954); N. Fukuda, K. Sawada and M. Taketani, Prog. Theor. Phys. **12**, 156 (1954).
26. E. Epelbaoum, W. Glöckle and Ulf-G. Meißner, Nucl. Phys. **A637**, 107 (1998).
27. E. Epelbaum, W. Glöckle and Ulf-G. Meißner, preprint FZJ-IKP(TH)-1999-19.
28. N. Fettes, Ulf-G. Meißner and S. Steininger, Nucl. Phys. **A640**, 199 (1998).
29. P. Büttiker and Ulf-G. Meißner, hep-ph/9908247.
30. V.G.J. Stoks et al., Phys. Rev. **C48**, 792 (1993).
31. T.D. Cohen and J.M. Hansen, Phys. Rev. **C59**, 13 (1999).
32. S. Fleming, T. Mehen and I.W. Stewart, nucl-th/9906056.
33. N. Kaiser, R. Brockmann and W. Weise, Nucl. Phys. **A625**, 758 (1997).
34. M.R. Robilotta and C.A. da Rocha, Nucl. Phys. **A615**, 391 (1997).
35. R.B. Wiringa et al., Phys. Rev. **C51**, 38 (1995).

Up to two-loop calculations in HBCPT

Judith A. McGovern, Michael C. Birse and K. B. Vijaya Kumar

*Theoretical Physics Group, Department of Physics and Astronomy,
University of Manchester, Manchester M13 9PL, UK*

Abstract. We start with a brief introduction to HBCPT, and discuss spin-dependent forward Compton scattering both as a pedagogic example and to present our recent results for the fourth-order contribution. We then present two recent two-loop calculations in HBCPT; the fifth-order contribution to the nucleon mass and the lowest-order contribution of the Wess-Zumino-Witten anomalous Lagrangian to forward spin-dependent Compton scattering. In both cases there are checks on the results which confirm the consistency of HBCPT at two-loop level. While the smallness of the mass correction is encouraging for the convergence of the theory, the fourth-order contribution to the spin polarisability turns out to be very large.

INTRODUCTION

Chiral perturbation theory is establishing itself as the principal tool for determining the consistency of data in disparate low energy processes involving pions, nucleons and photons. The audience at this conference can confidently be assumed to have a good knowledge of chiral symmetry, but perhaps to be less familiar with chiral perturbation theory *per se*, so I am including a brief introduction here.

A crucial result of old-fashioned current algebra, embodied in the Weinberg-Tomozawa relations among others, is that in the chiral limit pions cease to interact with other hadrons, including other pions, as their energy goes to zero. This raises the possibility of constructing an effective theory of pionic interactions in which amplitudes can be expanded as a power series in term of the pion mass (M_π) and momentum (q), with some typical hadronic energy scale such as m_ρ in the denominator. The first realisation of this, initially by Weinberg [1] and elaborated by Gasser and Leutwyler [2], was in the purely pionic sector. It was shown that it was possible to write down a Lagrangian expanded order by order in the number of derivatives and powers of the pion mass, and use it as a full field theory including loops. The crucial point, which allows the theory to be systematic, is that a loop diagram with vertices from the nth order Lagrangian give contributions which are at least two orders higher. These loops of course generate divergences, which are cancelled by the higher-order counterterms. Thus although the theory is not renormalisable in the conventional sense, it can be renormalised order by order. It is axiomatic in the effective field theory approach that, at a given order, *all* possible terms consistent with chiral symmetry and gauge and Lorentz invariance should be included. The finite parts of the coefficients of these terms are *a priori* unknown low energy constants (LEC's), which must be fit by comparison with data. Though the number of these constants rises sharply with the order, only a subset contribute to any given process, and the hope, largely realised to date, is that predictive power will be retained. The purely mesonic theory is now on a very firm footing, and two-loop calculations are becoming commonplace [3].

Initial attempts to construct an effective theory of pions and nucleons, however, quickly ran into trouble [4]. The reason is that a new dimensioned parameter, the nucleon mass (m_N), is introduced, and it is certainly not small. When nucleon lines appear within loops, powers of m_N can appear, lowering the chiral dimension. Thus loops with vertices from higher order terms can give contributions at lowest order, and the power counting is destroyed.

The solution turns the problem on its head. Instead of worrying that m_N is not small, we treat it as large, and carry out a simultaneous expansion in powers of $1/m_N$ [5]. Technically, this is done by integrating out the lower components of the nucleon wavefunction, generating in the process extra terms in the Lagrangian for the remaining "non-relativistic" nucleon. These terms have fixed coefficients, with one or more power of m_N in the denominator. An example of such a term is the magnetic moment. The process of integrating out the lower components generates a term in the second-order Lagrangian which gives the nucleon a magnetic moment of $eQ/2m_N$. Another term is already present in the second-order relativistic Lagrangian, with a free coefficient, which after the non-relativistic reduction gives an anomalous magnetic moment. This coefficient has to be obtained from experiment; at lowest order it coincides with the physical value, but at higher order the anomalous magnetic moment will have an additional contributions, giving a deviation between the bare and physical values. (It should be pointed out that, because of the way it enters the covariant derivative, and also because of a numerical similarity between $\sqrt{\alpha}_{\rm em}$ and M_π/m_N, e is treated on the same footing as the pion mass or pion and photon momenta. Thus the charge coupling of the photon to the proton is a first order vertex, but the coupling to the magnetic moment, which vanishes for soft photons, is second order.)

This heavy baryon chiral perturbation theory (HBCPT) has been extensively tested at one-loop order, proving consistent with—and indeed providing another method of demonstrating—all low-energy theorems (LET's) based on such considerations as Lorentz and gauge invariance and chiral symmetry. In this paper we will present results for the next-to-leading-order (but still one-loop) forward spin-dependent Compton scattering, giving a demonstration of the realisation of the LET of Low, Gell-Mann and Goldberger [6], and results for the spin polarisability.

Until now calculations in HBCPT have been almost exclusively one-loop, for the excellent reason that two-loop diagrams enter only at fifth order, while the fourth-order Lagrangian is still in the process of been worked out [7]. There are however some processes where the leading contribution is two-loop; an example is the imaginary part of the nucleon electromagnetic form factors, which involves the anomalous $\gamma \to 3\pi$ vertex, with all the pions coupling to the nucleon. This process was considered by Bernard et al. in ref. [8]. However since the imaginary part comes from the kinematical regime where the pions are on-shell, this is not a two-loop calculation in the sense of having two internal momenta to integrate over. Here we present two calculations which are two-loop in the full sense.

The two calculations have one thing in common; for different reasons, they can have no contribution from counterterms at the order to which we are working. Thus they must both be finite, and the results are particularly clean. In fact neither requires knowledge of $\mathcal{L}_{\pi N}$ beyond third order, and in fact the final results are entirely free of LEC's beyond lowest order. The first calculation is the fifth-order piece of the chiral expansion of the nucleon mass, in which the only non-vanishing contribution turns out to come from the expansion of the relativistic one-loop graph in powers of $1/m_N$. The other two-loop calculation is the leading (seventh-order) contribution of the anomalous Wess-Zumino-Witten Lagrangian [9] to forward spin-dependent Compton scattering in the limit that the photon energy goes to zero. To satisfy the LET this should vanish. We find that it does so, quite non-trivially, and the result enhances our confidence in the consistency of HBCPT, as well as showing the compatibility of the WZW Lagrangian and HBCPT, and testing the structure of the former in a novel way.

COMPTON SCATTERING: NLO CALCULATION

Compton scattering from the nucleon has recently been the subject of much work, both experimental and theoretical. For the case of unpolarised protons the experimental amplitude is well determined, and in good agreement with the results of heavy-baryon chiral perturbation theory (HBCPT). However the situation with regard to scattering from polarised targets is less satisfactory, not least because until very recently no direct measurements of polarised Compton scattering had been attempted.

The usual notation for spin-dependent pieces of the forward scattering amplitude for real photons of energy ω and momentum q is

$$\epsilon_2^\mu \Theta_{\mu\nu} \epsilon_1^\nu = ie^2 \omega W^{(1)}(\omega) \boldsymbol{\sigma} \cdot (\boldsymbol{\epsilon}_2 \times \boldsymbol{\epsilon}_1) + \ldots \quad (1)$$

From a theoretical perspective there is particular interest in the low-energy limit of the amplitude: $e^2 W^{(1)}(\omega) = 4\pi(f_2(0) + \omega^2 \gamma) + \ldots$, where γ is the forward spin-polarisability. The LET of Low, Gell-Mann and Goldberger [6] states that $f_2(0) = -\alpha_{em}\kappa^2/2m_N^2$.

While direct measurements of f_2 at zero energy are not currently feasible, it can be related to the photon absorption cross-section at energies above the pion production threshold by the Gerasimov-Drell-Hearn sum rule [10]; the cross-sections in turn can be estimated from pion electroproduction data. All analyses so far have shown a significant discrepancy between the sum rule and the LET [11–13].

FIGURE 1. Classes of diagrams which contribute to spin-dependent Compton scattering in the $\epsilon \cdot v = 0$ gauge at LO. The open circles are vertices from $\mathcal{L}^{(2)}$ and the solid dot is a vertex from $\mathcal{L}^{(3)}$. Crossed graphs and graphs with other orderings of vertices are not shown.

Electroproduction data have also been used to extract the polarisability; Sandorfi et al. [12] find $\gamma_p = -1.3 \times 10^{-4}$ fm^4 and $\gamma_n = -0.4 \times 10^{-4}$ fm^4, while the more recent analysis of Drechsel et al. [14] gives a rather smaller value of $\gamma_p = -0.6 \times 10^{-4}$ fm^4. (We shall use units of 10^{-4} fm^4 for polarisabilities from now on.)

Spin-dependent forward Compton scattering was calculated at leading (third) order in the early days of HBCPT [15], but the NLO calculation has been done only very recently [16,17]. At each order the LET is satisfied. However the value obtained for the spin polarisability does not seem to be converging so far, as we shall see.

At third order the diagrams which contribute in HBCPT are shown in Fig. 1. Let us consider first the Born terms. In a relativistic calculation, only the first diagram and the corresponding crossed graph exist (there is no seagull at this order), and they give the LET. The crossed Born term for relativistic nucleons however includes the nucleon "z-graph" in which the intermediate line is an antinucleon. In HBCPT, this contribution is absent from the crossed Born graph. However in integrating out the antinucleons, a fixed-coefficient seagull term term is generated, which has the same effect. Thus in HBCPT the LET is satisfied by the Born plus seagull graphs, which give contributions proportional to $-(Q+\kappa)^2$ and $Q(Q+2\kappa)$ respectively, leaving a final contribution proportional to the square of the bare anomalous magnetic moment as required.

At third order, however, loop graphs also enter. The net contribution of all loop graphs to the LET is zero, as required, but they do give a polarisability of $\gamma = \alpha_{em} g_A^2 / (24\pi^2 f_\pi^2 M_\pi^2) = 4.54$ for both the proton and the nucleon. As this is much larger than, and opposite in sign to, the experimental estimates, the NLO contribution is of interest.

The graphs which contribute at NLO (fourth order) are shown in Fig. 2. There are no seagull contributions at this order, either fixed or with LEC's: crossing symmetry requires the amplitude to contain only odd powers of the photon energy, ω; odd powers of M_π are therefore needed to give a contribution of even chiral order. However odd powers of M_π (corresponding to fractional powers of the quark mass) can only come from loops; they cannot be present in the Lagrangian. Thus only the vertices from $\mathcal{L}_{\pi N}^{(2)}$ are required for the calculation.

This time the loop contributions to $W^{(1)}$ do not vanish at $\omega = 0$. This is not in contradiction to the LET, however, because the isovector magnetic moment has

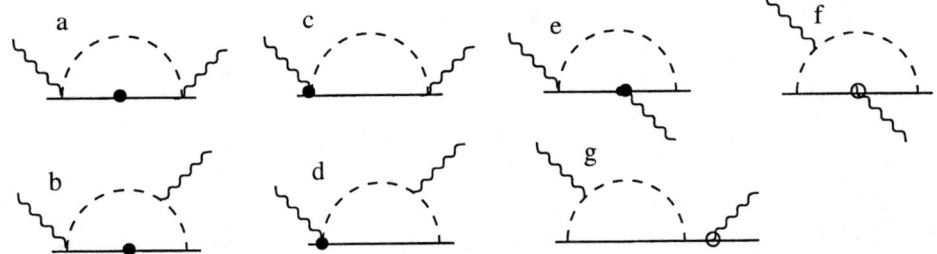

FIGURE 2. Diagrams which contribute to spin-dependent forward Compton scattering in the $\epsilon \cdot v = 0$ gauge at NLO. The dots and circles are vertices from $\mathcal{L}^{(2)}$, the solid and open dots at the photon-nucleon vertices representing the couplings proportional to the charge current and magnetic moment respectively.

a loop correction at third order. At fourth order, a correction is required to bring the bare and dressed values of κ^2 into line, which is exactly what we obtain. The polarisability to NLO is [16,17]

$$\gamma = \frac{\alpha_{em} g_A^2}{24\pi^2 f_\pi^2 M_\pi^2} \left[1 - \frac{\pi M_\pi}{8 m_N}(15 + 3\kappa_v + (6 + \kappa_s)\tau_3) \right]. \quad (2)$$

Although this has a factor of M_π/m_N compared with the leading piece, the numerical coefficient is large. Using the physical values of the masses and anomalous magnetic moments gives $\gamma = 4.5 - (6.9 + 1.6\tau_3)$. The NLO contributions are disappointingly large, and call the convergence of the expansion into question. The polarisability is a quantity where the contribution from intermediate Δ's is expected to be large. In the absence of explicit Δ fields, their contribution shows up in counterterms of fifth order. Since all estimates of these contributions have found them to be negative [15,18], the full result is likely to be even further from the "experimental" value.

THE NUCLEON MASS AT TWO LOOPS

Just about the simplest HBCPT calculation at any order is the nucleon mass shift. Here the chiral expansion is just an expansion in powers of M_π. The even orders can receive counterterm contributions, but the odd orders, corresponding to fractional powers of the quark mass, can only come from loops; the lowest one-loop contribution is of order M_π^3. So far the counterterms at second and fourth order have not been well determined, so it is of interest to consider the convergence of the odd terms in the series separately.

The heavy-baryon propagator is given by

$$S^{-1} = \omega - \Sigma(\omega, \mathbf{k}), \quad (3)$$

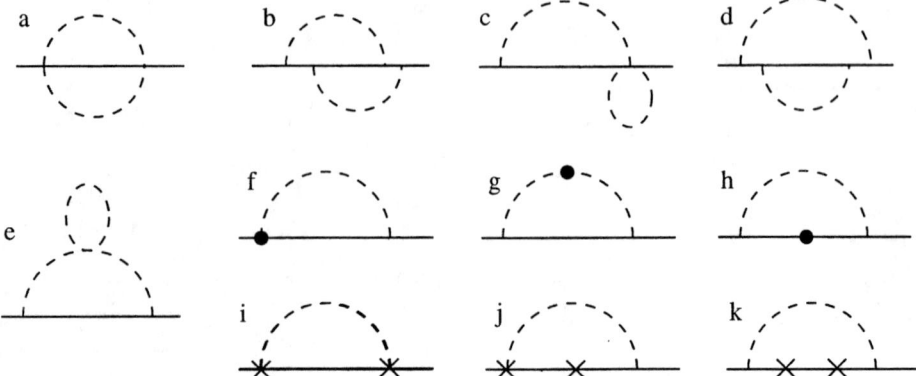

FIGURE 3. Contributions to $\Sigma^{(5)}$. Solid dots represent insertions from $\mathcal{L}_{\pi N}^{(3)}$ and $\mathcal{L}_{\pi\pi}^{(4)}$, and crosses from $\mathcal{L}_{\pi N}^{(2)}$ (both include fixed terms from the expansion in $1/m_N$).

where the nucleon momentum is written as $p = mv + k$, m is the bare mass, and $\omega = v \cdot k$. The mass shift $\delta m = m_N - m$ is the value of ω for which the propagator has a pole at zero three-momentum:

$$\delta m - \Sigma(\delta m, 0) = 0. \tag{4}$$

To order M_π^3, we have $\Sigma(\delta m) = \Sigma(0)$, and so [19]

$$\delta m^{(2)} + \delta m^{(3)} = -4c_1 M_\pi^2 - \frac{3g_A^2 M_\pi^3}{32\pi F_\pi^2}, \tag{5}$$

where the second term comes from the one-loop self-energy diagram. Writing as $\Sigma^{(n)}$ the expression for the $O(q^n)$ part of Σ, which can in turn be expanded in powers of ω/M_π, we obtain

$$\delta m^{(5)} = \Sigma^{(5)}(0) + \delta m^{(2)} \Sigma^{(4)\prime}(0) + \delta m^{(3)} \Sigma^{(3)\prime}(0) + \tfrac{1}{2}(\delta m^{(2)})^2 \Sigma^{(3)\prime\prime}(0) \tag{6}$$

where primes indicate derivatives with respect to ω.

The set of graphs which contribute to $\Sigma^{(5)}$ are shown in Fig. 3. Those which contribute to the first derivative of $\Sigma^{(4)}$ are like Figs 3f and h, but with second-order vertices. The two-loop diagrams 3a, b and d all contain the same single irreducible two-loop integral

$$I = \int \frac{d^d l \, d^d k}{(2\pi)^{2d}} \frac{1}{v \cdot k \, (M^2 - l^2)(M^2 - (k-l)^2)}. \tag{7}$$

This integral can be done by using Feynman parameters, yielding

$$I = -\frac{M^{2d-5}\pi^{\frac{1}{2}}\Gamma(\frac{5}{2} - d)}{(4\pi)^d} \int_0^1 (x - x^2)^{(1-d)/2} dx$$
$$= -\frac{M^{2d-5} 2^{d-2}\pi\Gamma(\frac{5}{2} - d)\Gamma(\frac{3-d}{2})}{(4\pi)^d \Gamma(2 - \frac{d}{2})}. \tag{8}$$

which tends to zero as $d \to 4$. Thus only pieces which are the product of two one-loop integrals are left. These contain divergences, which are cancelled by graphs with counterterm insertions, Figs. 3f-h. The full contribution from Eq. 6 is as follows:

$$\delta m^{(5)} = \frac{3g^2 M^5}{32\pi F^2} \left(\frac{2\bar{l}_4 - 3\bar{l}_3}{F^2} - \frac{4(2\bar{d}_{16} - \bar{d}_{18})}{g} + \frac{g^2}{8\pi^2 F^2} + \frac{1}{8m^2} \right). \tag{9}$$

The LEC's $\bar{l}_i$ are from the fourth-order mesonic Lagrangian, and the $\bar{d}_i$ are from the third-order pion-nucleon Lagrangian as given in ref. [20]. They are defined with a renormalisation scale of M_π, hence the absence of chiral logarithms.

This is not however the final result for the fifth-order mass term. It is customary to express results in CPT not in terms of the bare parameters that appear in the Lagrangian, for instance M, g, F and m, but in terms of the physical values M_π, g_A, F_π and m_N; this ensures that the values of lower order results do not change when a higher-order calculation is done. Since the bare and physical parameters are equal to lowest order, we can simply make the substitution in Eq. 9; the error introduced is of seventh order. However, when we do the same for $\delta m^{(3)}$, we generate corrections of fifth order, which must also be included. In this case, since the one-loop graph involves two pion-nucleon couplings, we choose to use the physical pion-nucleon coupling constant $g_{\pi NN}$, rather than g_A/f_π. The fifth order corrections so generated then cancel almost completely with the explicit fifth order terms, leaving as our final result

$$\delta m^{(3)} + \delta m^{(5)} = -\frac{3g_{\pi NN}^2 M_\pi^3}{32\pi m_N^2} \left(1 - \frac{M_\pi^2}{8m_N^2} \right). \tag{10}$$

Further details are given in ref. [21]. Thus in the heavy-baryon limit the contribution of order M_π^5 vanishes, with all corrections being absorbed in the physical pion mass and pion-nucleon coupling constant in the M_π^3 contribution. For finite nucleon mass, the correction is just that obtained if the relativistic one-loop contribution is expanded in powers of M_π/m_N [4]. Since the $1/m_N$ terms in the HBCPT Lagrangian are constructed to respect Lorentz invariance, this agreement is reassuring but certainly not surprising. It is also worth noting that the fifth-order piece is only 0.3% of the third-order piece, which is more encouraging for the convergence of the theory.

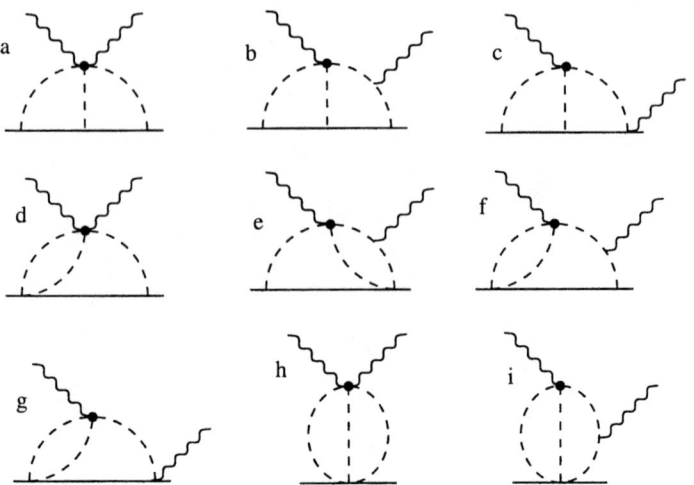

FIGURE 4. Diagrams with anomalous vertices (solid dots) which contribute to Compton scattering in the $\epsilon \cdot v = 0$ gauge.

COMPTON SCATTERING REVISITED

Returning to spin-dependent Compton scattering, we look at a subset of two-loop diagrams, namely those which involve an anomalous photon-pion coupling, as shown in Fig. 4. Since these have a factor of N_c in the amplitude, they can be cleanly distinguished from all the other diagrams which would contribute at the same order. This time, to satisfy the LET, the total contribution to $f_2(0)$ has to be zero.

We have calculated the contributions from all of the diagrams of Fig. 4. The first three are proportional to g_A^3, the rest to g_A. It is relatively straightforward to show that the second set cancel and make no net contribution. The first set do also cancel, but to prove it integration-by-parts identities have to be used to find relationships between various two-loop integrals. The details are given in ref. [22]. This provides a highly non-trivial test of the consistency of HBCPT at the two loop level, and also of the consistency of the WZW and HBCPT Lagrangians.

REFERENCES

1. S. Weinberg, Physica **96A** 327 (1979).
2. J. Gasser and H. Leutwyler, Ann. Phys. (N.Y.) **158** 142 (1984); Nucl. Phys. **B250** 465 (1985).
3. For a review see J. Bijnens, hep-ph/9710341.
4. J. Gasser, M. E. Sainio and A. Švarc, Nucl. Phys. **B307** 779 (1988).

5. E. Jenkins and A. V. Manohar, Phys. Lett. B **255** 558 (1991).
6. F. Low, Phys. Rev. **96** 1428 (1954); M. Gell-Mann and M. Goldberger, Phys. Rev. **96** 1433 (1954).
7. U-G. Meißner, G. Müller and S. Steininger, hep-ph/9809446.
8. V. Bernard, N. Kaiser and U-G. Meißner, Nucl. Phys. **A611** 429 (1996).
9. E. Witten, Nucl. Phys. **B223** 422 (1983).
10. S. B. Gerasimov, Sov. J. Nucl. Phys. **2** 430 (1966); S. D. Drell and A. C. Hearn, Phys. Rev. Lett. **16** 908 (1966).
11. I. Karliner, Phys. Rev. **D 7** 2717 (1973);
12. A. M. Sandorfi, C. S. Whisnant and M. Khandaker, Phys. Rev. **D 50** R6681 (1994).
13. D. Drechsel and G. Krein, Phys. Rev. **D 58** 116009 (1998)
14. D. Drechsel, G. Krein and O. Hanstein, Phys. Lett. **B 420** 248 (1998).
15. V. Bernard, N. Kaiser, J. Kambor and U-G. Meißner, Nucl. Phys. **B 388** 315 (1992).
16. X. Ji, C-W. Kao and J. Osborne, hep-ph/9908526.
17. K. B. V. Kumar, J. A. McGovern and M. C. Birse, hep-ph/9909442.
18. T. R. Hemmert, B. R. Holstein J. Kambor and G. Knöchlein, Phys. Rev. **D 57** 5746 (1998).
19. V. Bernard, N. Kaiser and U-G. Meißner, Int. J. Mod. Phys. E **4** 193 (1995).
20. N. Fettes, U-G. Meißner and S. Steininger, Nucl. Phys. **A640** 199 (1998).
21. J. A. McGovern and M. C. Birse, Phys. Lett. B **446** 300 (1999).
22. J. A. McGovern and M. C. Birse, hep-ph/9908249.

Nonequilibrium chiral perturbation theory and Disoriented Chiral Condensates

Angel Gómez Nicola

Departamento de Física Teórica
Universidad Complutense, 28040, Madrid, Spain

Abstract. We analyse the extension of Chiral Perturbation Theory to describe a meson gas out of thermal equilibrium. For that purpose, we let the pion decay constant be a time-dependent function and work within the Schwinger-Keldysh contour technique. A useful connection with curved space-time QFT allows to consistently renormalise the model, introducing two new low-energy constants in the chiral limit. We discuss the applicability of our approach within a Relativistic Heavy-Ion Collision environment. In particular, we investigate the formation of Disoriented Chiral Condensate domains in this model, via the parametric resonance mechanism.

INTRODUCTION AND MOTIVATION

The forthcoming experiments on Relativistic Heavy Ion Collision (RHIC) at BNL and CERN will be able to test accurately the dynamics of the QCD plasma. After the collision, the plasma formed in the central rapidity region cools down via hydrodynamic expansion, and nonequilibrium effects become important in that regime. Among them, one of the most interesting suggestions has been the formation of the so called Disoriented Chiral Condensates (DCC). The DCC were proposed originally in [1] as misaligned vacuum regions, where the chiral field points out in directions in isospin space different from that where the vacuum expectation value of the pion field vanishes. If such regions were formed, one could observe large clusters of pions emitted coherently from the plasma as the pion field relaxes to the normal vacuum. This kind of behaviour is indeed observed in Centauro and anti-Centauro events in cosmic ray experiments [2]. However, one should point out that a clear signal for DCC formation has not been observed yet in the RHIC at Fermilab [3], although it could well happen that the DCC's are too small to be directly detected and one has to think of alternative observables (see below).

On the other hand, after the hadronisation time, a proper description of the microscopic meson dynamics makes it compulsory to use an effective low-energy theory for QCD. In this context, the chiral symmetry plays a fundamental role. The effective theory must incorporate all the QCD symmetries and the chiral spontaneous symmetry breaking (SSB) pattern, so that the Nambu-Goldstone bosons (NGB) are the lightest mesons (π, K, η). The light quark masses are then introduced perturbatively. One possible choice is simply an $O(N)$ model, with the standard classical SSB potential. Its fundamental fields are the $N-1$ pions and the σ, which acquires a nonzero vacuum expectation value v. This is the Linear Sigma Model (LSM) description. However, one should bear in mind that the LSM becomes nonperturbative in the coupling constant at low energies, so that alternative perturbative expansions have to be used, such as large N. Besides, the LSM only shares the QCD chiral symmetry breaking pattern for $N=4$. An alternative approach is to construct an effective theory as an infinite sum of terms with increasing number of derivatives, only for the NGB fields. This is the description based in the Nonlinear Sigma Model (NLSM), which is the lowest order action one can write down in this expansion. Higher order corrections come both from NGB loops and higher order lagrangians and can be renormalised order by order in energies, yielding finite predictions for the meson observables. The unknown coefficients, which encode all the information on the underlying theory, absorb the loop infinities and can be fitted to experiment. This framework constitutes the so called chiral perturbation theory (ChPT) [4,5]. The perturbative expansion is carried out in terms of the ratio of the $\mathcal{O}(p)$ meson energy scales of the theory (masses, external momenta, temperature and so on) and the chiral scale $\Lambda_\chi \simeq 1$ GeV (see [6–9] for a review).

Nonequilibrium effects such as the DCC's have been investigated in the literature using $O(N)$ models with initial thermal equilibrium conditions $\sigma(t=0)=0$ and $\pi^a(t=0)=0$. In this context, two different scenarios for DCC formation have been proposed: the first one takes place in the early stages of the plasma evolution. Roughly speaking, as the field rolls down along the potential, long wavelength modes grow exponentially and enhance the formation of DCC's. There have been several approaches in the literature to implement this idea in the $O(N)$ model, like classical simulations [10], large N [11], or analysis based on reasonable assumptions on the kinematics [12,13]. Typical DCC sizes within this approach are of the order of 2-3 fm, containing $n_\pi \simeq 0.2$ fm^3 pions, whereas the plasma cools down in a proper time of about $\tau \simeq 5$-10 fm/c. As commented above, these numbers yield too small DCC's to be observed directly. A second suggestion, which has been recently proposed [14–16] is based on the parametric resonance mechanism and inherits the idea from inflationary reheating [17]. In this approach, the σ field is very close to the bottom of the potential but it is still *oscillating* around it (it clearly overshoots the vacuum if the initial conditions are imposed on the top) in the late stage of the plasma evolution. Those oscillations transfer energy to the pion modes, giving rise to exponentially growing pion solutions for certain bands in momentum space, via parametric resonance. Recent work within the LSM in this approach predicts rather larger DCC, of sizes up to 5 fm [15]. More details about this mechanism will be given below.

In the present work, we will explore the applicability of ChPT to describe the meson plasma out of thermal equilibrium. So far, this formalism has been applied only in equilibrium, to study the low T ($T = \mathcal{O}(p)$) meson gas and the chiral phase transition [18,19]. The key idea of our approach is to make use of the derivative expansion consistently defined in ChPT in order to study the system not far from equilibrium. It is therefore best suited for the late stage evolution and has the additional advantages typical of standard ChPT, i.e, it deals only with NGB fields and is equally applicable to three flavours. We will show that a systematic power counting can be established in this case and, furthermore, that the renormalisation program can be consistently implemented. Details can be found in [20]. In addition, in the last section we will explore the possibility of describing DCC formation within ChPT, via parametric resonance.

THE MODEL AND CHIRAL POWER COUNTING

Our starting point is the nonlinear sigma model (NLSM) where we let the pion decay constant– the only relevant parameter to the lowest order in derivatives– be time dependent. In the context of a RHIC, such time dependence can be thought of as proper time evolution within the so called Bjorken initial conditions [21], where observables depend only on proper time and not on rapidity. This picture is consistent with the experimental observations. We take the initial time $t = 0$, having in mind that it would correspond to a proper time $\tau_0 \simeq$ 1-2 fm/c, a typical hadronisation time. Thus, we will consider the following NLSM action

$$S[U] = \int_C dt \int d^3\vec{x}\, \frac{f^2(t)}{4}\, \text{tr}\, \partial_\mu U^\dagger(\vec{x},t)\partial^\mu U(\vec{x},t) \qquad (1)$$

Here, C is the Schwinger-Keldysh contour (see [20] for details), which parametrises the nonequilibrium path integral where we are considering thermal equilibrium for $t \leq 0$ at a temperature $T_i = \beta_i^{-1}$, as the initial condition. Note that the action (1) is chiral invariant ($U \to LUR^\dagger$) by construction, which will play an important role in what follows. As a first approximation, we will be interested only in the strict chiral limit for two light flavours, i.e, massless pions. Therefore, we are not including any explicit symmetry-breaking term in the action. Thus, $f(t \leq 0) = f \simeq 93$ MeV to leading order ($f \neq f_\pi$ to higher orders) and for $t > 0$ the system departs from equilibrium. Note that, since we choose that departure to be instantaneous, $f(t)$ cannot be analytical at $t = 0$. This is just an artifact of the approximation and should not have any effect on the long-time behaviour. Finally, as customary, $U(x)$ is parametrised in terms of pion fields π^a as:

$$U(\vec{x},t) = \frac{1}{f(t)}\left\{\left[f^2(t) - \pi^2\right]^{1/2} I + i\tau_a \pi^a\right\} \qquad (2)$$

and $\pi^a(t_i - i\beta_i) = \pi^a(t_i)$ is the equilibrium boundary condition, with $t_i < 0$.

The new ingredient we need to incorporate in the power counting in order to be consistent with ChPT is then

$$\frac{\dot{f}(t)}{f^2(t)} \simeq \mathcal{O}\left(\frac{p}{\Lambda_\chi}\right), \qquad \frac{\ddot{f}(t)}{f^3(t)}, \frac{[\dot{f}(t)]^2}{f^4(t)} \simeq \mathcal{O}\left(\frac{p^2}{\Lambda_\chi^2}\right), \qquad (3)$$

and so on. Obviously, our results will depend upon the choice of $f(t)$. One can think of $f(t)$ as an external source, to which we wish to obtain the nonequilibrium response of the system. Alternatively, this model can be thought of to lowest order as the LSM with the time-dependent constraint $\sigma^2 + \pi^2 = f^2(t)$. We shall discuss below a reasonable assumption for $f(t)$ in connection with DCC formation. Meanwhile, we shall keep $f(t)$ arbitrary.

To lowest order in the pion fields, the above NLSM action can be written as

$$S_0[\pi] = -\frac{1}{2}\int_C d^4x\, \pi^a(\vec{x},t)\left[\Box + m^2(t)\right]\pi^a(\vec{x},t) \qquad (4)$$

where $\int_C d^4x = \int_C dt \int d^3\vec{x}$ and $m^2(t) = -\ddot{f}(t)/f(t)$. That is, the model accommodates a time-dependent pion mass term, without breaking explicitly the chiral symmetry. This effect is the same as switching on an external curved space-time background, as we will see in the next section.

RENORMALISATION AND CURVED SPACE-TIME

Once we have defined our nonequilibrium power counting, we can apply ChPT to calculate the time evolution of the observables. In doing so, we must pay special attention to renormalisation. The fact that there is a time-dependent mass term indicates that there can be new time dependent infinities in the chiral loops. However, we are in the chiral limit, so we are not allowed to introduce the usual $\mathcal{O}(p^4)$ mass and wave function counterterms breaking the chiral symmetry [5]. In other words, we should be able to construct the most general fourth order action, which in particular should include new terms (and hence new low-energy constants) to cancel those extra divergences, preserving exactly the chiral symmetry.

In order to find this $\mathcal{O}(p^4)$ lagrangian, we will make use of a very fruitful analogy: the action (1) is equivalent to formulate the NLSM on a curved space-time background corresponding to a spatially flat Robertson-Walker metric, with scale factor $a(t) = f(t)/f(0^+)$ (see [20] for details). Note that in this language, $m^2(t)$ in (4) represents the minimal coupling with the RW metric preserving chiral invariance.

Therefore, we can construct the $\mathcal{O}(p^4)$ action as:

$$S_4[U, g, R] = \int_C d^4x \sqrt{-g}\left[\mathcal{L}_4[U,g] + (L_{11}Rg^{\mu\nu} + L_{12}R^{\mu\nu})\mathrm{tr}\partial_\mu U^\dagger \partial_\nu U\right] \qquad (5)$$

where g is the metric determinant, $\mathcal{L}_4[U,g]$ stands for the standard (equilibrium) lagrangian [5] with indices raised and lowered with the $g^{\mu\nu}$ metric and the rest are new $\mathcal{O}(p^4)$ invariant couplings with the scalar curvature $R(x)$ and the Ricci tensor $R_{\mu\nu}(x)$ in the chiral limit. These are the new terms we need, where L_{11} and L_{12} are the new coupling constants. In fact, this problem has been already considered in [22] in order to study the energy-momentum tensor of QCD at low energies. In that work it has been found that L_{11} is renormalised in dimensional regularisation, whereas L_{12} is already finite. Their numerical values can be obtained from the experimental information on the QCD energy-momentum form factors. They yield $L_{12} \simeq -2.7 \times 10^{-3}$ and $L_{11}^r(\mu = 1 GeV) \simeq 1.4 \times 10^{-3}$ where μ is the renormalisation scale. In our case, with our RW metric we get to $\mathcal{O}(\pi^2)$,

$$S_4[\pi, g] = -\frac{1}{2} \int_C d^4 x \pi^a \left[f_1(t) \partial_t^2 - f_2(t) \nabla^2 + m_1^2(t) \right] \pi^a + \mathcal{O}(\pi^4) \tag{6}$$

with

$$f_1(t) = 12 \left[(2L_{11} + L_{12}) \frac{\ddot{f}(t)}{f^3(t)} - L_{12} \frac{[\dot{f}(t)]^2}{f^4(t)} \right]$$

$$f_2(t) = 4 \left[(6L_{11} + L_{12}) \frac{\ddot{f}(t)}{f^3(t)} + L_{12} \frac{[\dot{f}(t)]^2}{f^4(t)} \right]$$

$$m_1^2(t) = -\left[\frac{f_1(t)\ddot{f}(t) + \dot{f}_1(t)\dot{f}(t)}{f(t)} + \frac{1}{2} \ddot{f}_1(t) \right] \tag{7}$$

The above lagrangian should take care of the nonequilibrium infinities we might find in the pion two-point function. We will see below that this is indeed the case.

THE PION DECAY FUNCTIONS $f_\pi(T)$

The first observable one might think of calculating in ChPT is the pion decay constant to one loop. In the nonequilibrium model, it will become a time-dependent function $f_\pi(t)$. One should point out that the definition of f_π is subtle even in thermal equilibrium [19,23]. In addition, one has in general $f_\pi^s(T) \neq f_\pi^t(T)$ corresponding to the axial current spatial and temporal components and due to the loss of Lorentz covariance in the thermal bath [24]. We refer to [20] for details on how to define properly $f_\pi(t)$ out of equilibrium. Once this has been done, one has to consider the one loop diagrams for the pion two-point function coming from (1) plus the tree level ones from (6). The final result up to $\mathcal{O}(p^4)$ reads [20]

$$[f_\pi^s(t)]^2 = f^2(t) \left[1 + 2f_2(t) - f_1(t) \right] - 2iG_0(t) \tag{8}$$

$$[f_\pi^t(t)]^2 = f^2(t) \left[1 + f_2(t) \right] - 2iG_0(t) \tag{9}$$

for $t > 0$, with $f_{1,2}(t)$ in (7) and $G_0(t)$ is nothing but the equal-time pion two-point function $G_0(t) = G_0(x,x)$ with $G_0(x,y)$ the solution of the differential equation

$$\left\{ \Box_x + m^2(x^0) \right\} G_0(x,y) = -\delta_C(x^0 - y^0)\delta^{(3)}(\vec{x} - \vec{y}) \tag{10}$$

with KMS equilibrium conditions $G_0^>(\vec{x}, t_i - i\beta_i; y) = G_0^<(\vec{x}, t_i; y)$, $G_0^>$ and $G_0^<$ being advanced and retarded correlation functions. Clearly, this equation cannot be solved analytically for an arbitrary $f(t)$, but it can be managed numerically. Therefore, one must remember that $G_0(t)$ depends implicitly on $f(t)$ through (10).

As a consistency check, the results (8)-(9) reproduce the equilibrium result [25] when we switch off the time derivatives of $f(t)$:

$$[f_\pi^s(T)]^2 = \left[f_\pi^t(T)\right]^2 = f^2\left(1 - \frac{T^2}{6f_\pi^2}\right) \qquad (11)$$

An interesting consequence of our result is that $f_\pi^s(t) \neq f_\pi^t(t)$ to one-loop, unlike the equilibrium case. In addition, from (8)-(9) and (7) we see that the difference $[f_\pi^s(t)]^2 - [f_\pi^t(t)]^2$ is finite, so that $f_\pi^s(t)$ and $f_\pi^t(t)$ can be renormalised at the same time, which is another consistency check. We remark that $G_0(t)$ contains in general UV divergences, to be absorbed by $f_1(t)$ and $f_2(t)$ in the renormalisation of L_{11}. An explicit check of this renormalisation procedure will follow in the next section.

DISORIENTED CHIRAL CONDENSATES IN CHPT

In this section we will consider a particular choice of $f(t)$ and apply our previous results. Our motivation is the possibility of generating DCC-like structures in this context. We shall sketch some of our preliminary results here, while details of the calculation and further work will be postponed to a forthcoming paper.

As we have discussed above, our approach is meant to be useful in a stage of the plasma evolution where the departure from equilibrium is of the same order as the meson energies. Hence, we should be able to obtain similar results as the analysis performed in the LSM in the parametric resonance regime [14–16], where the rolling down of the σ field is in its late oscillatory period. This is the same behaviour of the inflaton field in reheating [17]. One then allows for a time-dependent classical background $\sigma(t)$ in the LSM, splitting the field as

$$\sigma(\vec{x}, t) = \sigma(t) + \delta\sigma(\vec{x}, t) \qquad (12)$$

where $\delta\sigma$ is the quantum fluctuation. As a first approximation, one can neglect the pion fluctuations $\langle \pi^2 \rangle \ll v^2$ [14,16] and solve the equation of motion, which yields just $\sigma(t) = \sigma_0 \cos m_\sigma t$. Here, σ_0 is the initial field amplitude, which in this approximation is a small quantity. Even though, one can still produce exponentially growing pion fields (DCC) which in the end will be responsible for the damping of the oscillations as the field relaxes to equilibrium. One should bear in mind that neglecting $\langle \pi^2 \rangle$ to lowest order is a rather crude approximation, as pointed out in [15], which is clearly not valid for large times when the pion correlator grows significantly. Nonetheless, we will carry on with this simple case, just to understand qualitatively how ChPT can also account for the description of DCC's. A better approximation would be to solve the coupled equations for the σ and π fields, which yields the solution for $\sigma(t)$ in terms of elliptic functions [15]. Therefore, in this simple picture, we take our $f(t)$ of the same form as the lowest order $\sigma(t)$ in the LSM, i.e,

$$f(t) = f\left[1 - \frac{q}{2}(\cos Mt - 1)\right] \tag{13}$$

Here, q is a small parameter, playing the role of σ_0 in the LSM. Notice that our nonequilibrium chiral power counting demands $qM^2 = \mathcal{O}(p^2)$ and so on. Thus, for definiteness, we will take $q = \mathcal{O}(p^2/\Lambda_\chi^2)$, so that the $\mathcal{O}(p^4)$ corrections remain under control (see below), and M arbitrary. In the end, we will discuss how the results are affected by T_i, q and M. Therefore, we have $m^2(t) = -(qM^2/2)\cos Mt(1 + \mathcal{O}(q))$, so that the differential equation (10) becomes to leading order

$$\left[\frac{d^2}{dt^2} + \frac{4k^2}{M^2} - 2q\cos Mt\right] G_0^>(k,t,t') = 0 \tag{14}$$

where we have Fourier transformed in the spatial coordinates only ($k^2 = |\vec{k}|^2$). The above equation is nothing but the Mathieu equation, which has several well-known interesting properties [26,27]. Among them, it admits unstable solutions exponentially growing in time, for certain values of $4k^2/M^2$. This is the simplest version of the parametric resonance mechanism. In particular, the instabilities develop in bands in k, centered at $k_n = nM/2$, of width $\Delta k_n = \mathcal{O}(q^n)$. Hence, in the approximation we are working, only the first band is relevant, i.e, unstable solutions only exist for $M/2 - \Delta k_1 < k < M/2 + \Delta k_1$. This is known in the Cosmology literature as the narrow resonance approximation [17]. A typical unstable solution

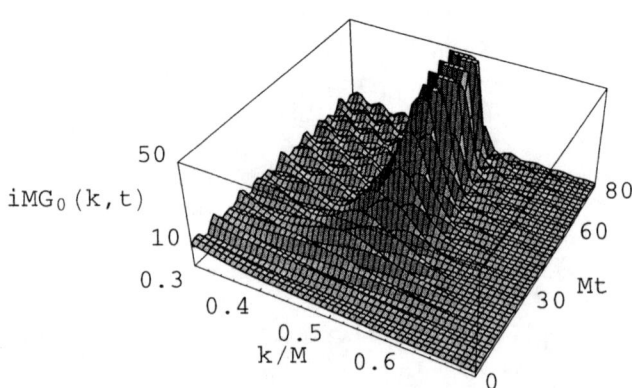

FIGURE 1. Profile of $G_0(k,t)$ for $T_i = M$ and $q = 0.1$. Both spatial momentum and time are measured in units of M. The instability band for this case lies roughly between $0.4M < k < 0.6M$

$G_0(k,t)$ has been plotted in Figure 1 for a particular choice of the parameters in the first band. The solutions typically oscillate with an exponentially growing amplitude inside the unstable region. Therefore, we see that our ChPT approach allows for DCC-type configurations.

Next, we will apply our results for $f_\pi(t)$ to this particular case. The equal-time correlation function $G_0(t) = \int d^3k G_0(k,t,t)$ turns out to be UV divergent, as expected. After standard manipulations in dimensional regularisation ($d = 4 - \epsilon$) one can cast the divergent part for $t > 0$ as

$$iG_0^{div}(t) = -\frac{qM^2}{16\pi^2}\cos Mt \left(\frac{1}{\epsilon} + \frac{1}{2}\log\frac{\mu^2}{M^2}\right) \quad (15)$$

which is an example of the new time-dependent divergences we were talking about in previous sections. In fact, we see that it has exactly the same form as $\ddot{f}(t)/f(t)$. Furthermore, replacing in (8)-(9), we find that the result is rendered *finite* and *scale independent* with the same renormalisation of L_{11} derived in [22].

The final results for $f_\pi^s(t)$ are plotted in Figure 2 for different choices of the parameters. We clearly observe the damping effect on the amplitude due to the unstable solutions at long times. In other words, the DCC's accelerate thermalisation. We also observe that this mechanism becomes less efficient for smaller q and M. Typically, the unstable corrections to the amplitude of $f_\pi(t)$ are proportional to $(qM^2/4\pi f_\pi^2)\exp(qMt)$. On the other hand, this effect seems to be rather insensitive to the initial temperature and thus we expect to catch all the important qualitative behaviour concerning the DCC's, regardless of the initial conditions. It should be pointed out that the curves have been cut off at the times where the one-loop contribution becomes of the same size as the tree level one. From that point onwards, the exponentially growing correlator dominates, yielding unphysical results. As commented above, we do not expect our simple cosine shape for $f(t)$ to be valid for all times, since it is derived neglecting the pion correlator. This final time t_f roughly defines the applicability range of our results. We expect that this range is enough to account for all the plasma time evolution of a realistic RHIC. For instance, for $M=1$ GeV and $q = 0.1$, we get $t_f \simeq 35M^{-1} \simeq 8$ fm/c. This is exactly the same as extrapolating the equilibrium result (11) to predict the critical temperature at $T = T_c \simeq 6f_\pi^2$, where all the higher order corrections become of the same order. Nonetheless, that formula predicts the right behaviour of $f_\pi(T)$ as it approaches the transition. In the same way, our results reproduce the expected qualitative behaviour as we extrapolate them up to times $t \simeq t_f$. Therefore, $f_\pi(t)$ can be regarded as an alternative observable (it is the residue of an axial-axial correlator and it can be measured in semileptonic decays) to test the size of DCC-like configurations in the late stage of the plasma expansion.

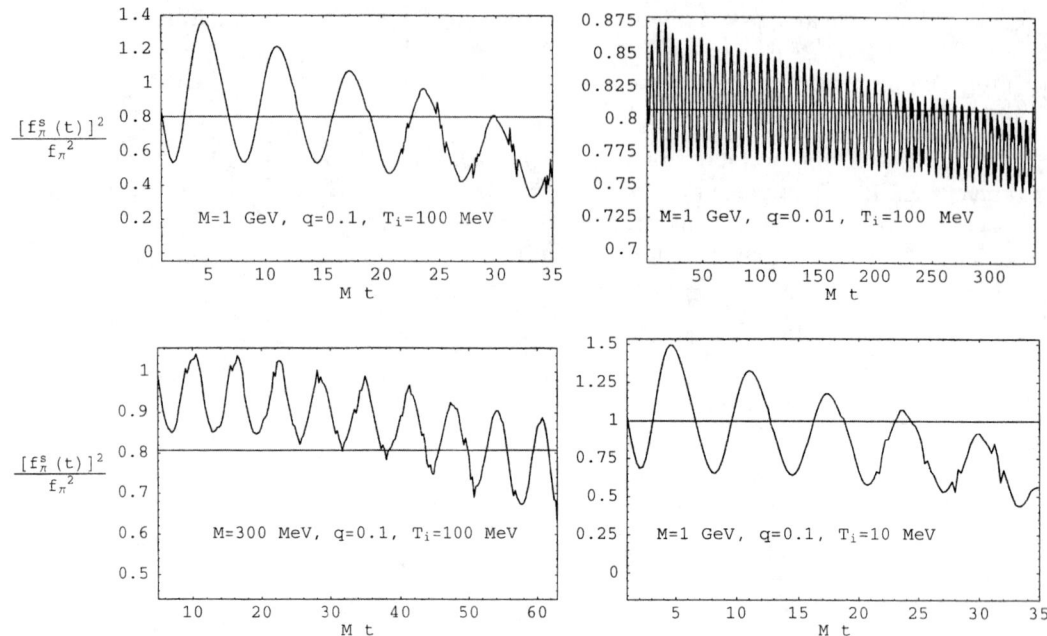

FIGURE 2. Results for $f_\pi^s(t)$ in the parametric resonance regime. The horizontal line is the value of f_π at the initial temperature T_i.

CONCLUSIONS AND OUTLOOK

We have reviewed recent work on the extension of ChPT to a nonequilibrium situation. The NLSM with a time dependent pion decay constant provides a nonequilibrium effective model with a well-defined perturbative expansion and power counting near equilibrium. The analogy of this model with curved space-time QFT allows to consistently construct higher order lagrangians and implement renormalisation. As a first application, we have obtained the renormalised one-loop $f_\pi(t)$.

We have also shown how this model can be applied to describe DCC-like structures in the late stage of the expansion of a hot plasma formed after a RHIC. Work in progress includes a more realistic study of the parametric resonances, including consistently the pion correlations in $f(t)$ and calculating the correlation length and the number of pions. One can also think of including pion masses, extending the results to three flavours, using large N methods and cosmological applications as other interesting aspects to be investigated in this context.

ACKNOWLEDGMENTS

I wish to thank the organisers of the "Hadron Physics" conference and the Theory group in Coimbra for their kind help and hospitality. Financial support from CICYT, Spain, project AEN97-1693, is also acknowledged.

REFERENCES

1. A.Anselm, *Phys. Lett.* **B217** (1989) 169; A.Anselm and M.Ryskin, *Phys. Lett.* **B226** (1991) 482 J.D.Bjorken, *Int. J. Mod. Phys.* **A7** (1992) 4189; J.P.Blaizot and A.Krzywicki, *Phys. Rev.* **D46** (1992) 246.
2. L.T.Baradzei et al, *Nucl. Phys.* **B370** (1992) 365.
3. T.Brooks et al, *Phys. Rev.* **D55** (1997) 5667.
4. S.Weinberg, *Physica* **A96** (1979) 327.
5. J.Gasser and H.Leutwyler, *Ann. Phys.* (N.Y.) **158** (1984) 142, *Nucl. Phys.* **B250** (1985) 465.
6. J.F.Donoghue, E.Golowich and B.R.Holstein, *Dynamics of the Standard Model*, Cambridge University Press 1992.
7. U-G.Meißner, *Rep.Prog.Phys.* **56**, 903-996, 1993.
8. A.Pich, *Rep.Prog.Phys.* **58**, 563-610, 1995.
9. A.Dobado, A.Gómez Nicola, A.López-Maroto and J.R.Peláez, *Effective lagrangians for the Standard Model*, Springer 1997.
10. K.Rajagopal and F.Wilczek, *Nucl. Phys.* **B404** (1993) 577.
11. D.Boyanowsky, H.J. de Vega and R.Holman, *Phys. Rev.* **D51** (1995) 734.
12. F.Cooper, Y.Kluger, E.Mottola and J.P.Paz, *Phys. Rev.* **D51** (1995) 2377.
13. M.A.Lampert, J.F.Dawson and F.Cooper, *Phys. Rev.* **D54** (1996) 2213.
14. S.Gavin and B.Müller, *Phys. Lett.* **B329** (1994) 486; S.Mrowczynski and B.Müller, *Phys. Lett.* **B363** (1995) 1.
15. D.Kaiser, *Phys. Rev.* **D56** (1997) 706; *Phys. Rev.* **D59**, 117901
16. H.Hiro-Oka and H.Minakata, *Phys. Lett.* **B425** (1998) 129; B434 (1998) 461-462 (E).
17. L.Kofman, A.Linde and A.Starobinsky, *Phys. Rev.* **D56** (1997) 3258.
18. P.Gerber and H.Leutwyler, *Nucl. Phys.* **B321** (1989) 387.
19. A.Bochkarev and J.Kapusta, *Phys. Rev.* **D54** (1996) 4066.
20. A.Gómez Nicola and V.Galán-González, *Phys. Lett.* **B449** (1999) 288-298.
21. J.D.Bjorken, *Phys. Rev.* **D27** (1983) 140.
22. J.F.Donoghue and H.Leutwyler, *Z. Phys.* **C52** (1991) 343.
23. J.I.Kapusta and E.V.Shuryak, *Phys. Rev.* **D49** (1994) 4694.
24. R.D.Pisarski and M.Tytgat, *Phys. Rev.* **D54** (1996) R2989.
25. J.Gasser and H.Leutwyler, *Phys. Lett.* **B184** (1987) 83.
26. N.W.Mac Lachlan, *Theory and Application of Mathieu Functions*, Dover (New York), 1961.
27. M.Abramowitz and I.A.Stegun, *Handbook of Mathematical Functions*, Dover (New York), 1970.

EFFECTS OF HOT AND DENSE MATTER

The Sigma Meson and Chiral Restoration in the Hot and/or Dense Nuclear Matter

T. Kunihiro

Faculty of Science and Technology, Ryukoku University, Seta, Ohtsu, 520-2194, Japan

Abstract. We first indicated that if the collective modes with the mass 500-700 MeV exists in the $I = J = 0$ channel, various empirical facts in hadron physics can be naturally accounted for, which otherwise would remain mysterious.

We proposed several experiments to produce and detect the σ in nuclei using nuclear and electro-magnetic projectiles. An emphasice was put on the fact that the proper quantity is the spectral function which describes the nuclear medium having a hadron. The recent CHAOS data which show a spectral enhancement near the $2\,m_\pi$ threshold in the σ channel from the reactions $A(\pi, 2\pi)A'$ where A and A' denotes nuclei is interpreted as a possible evidence of a partial restoration of chiral symmetry in nuclei. Comments on conventional manybody-theoretical approaches on the experiments were also given. Experiments using nuclear targets were proposed to explore the underlying mechanism of the enhancement seen in the CHAOS experiment.

The major part of this talk is based on the following references: [1], [2], [3] and [4]. If you are interested in this talk, please see the above references, especially the last one, with which the present talk is almost overlapped.

REFERENCES

1. T. Hatsuda and T. Kunihiro, Phys. Rep. **247**, 221 (1994).
2. T. Kunihiro, Prog. Theor. Phys. Suppl. **120**, 75 (1995).
3. T. Hatsuda, T. Kunihiro and H. Shimizu, Phys. Rev. Lett. **82**, 2840 (1999).
4. T. Kunihiro, hep-ph/9905262.

Matter-induced hadronic processes [1,2]

W. Broniowski*, W. Florkowski*, and B. Hiller[†]

*H. Niewodniczański Institute of Nuclear Physics
ul. Radzikowskiego 152, 31342 Kraków, Poland
[†]Center for Theoretical Physics, University of Coimbra
P-3004 516 Coimbra, Portugal

Abstract. Two examples of "exotic" phenomena which become possible and important in the presence of nuclear matter are discussed: $\omega \to \pi\pi$ decay, and $\rho - \omega$ mixing. Significance of these processes for the low-mass dilepton production in relativistic heavy-ion collisions is indicated.

INTRODUCTION

An interesting factor brought in by the presence of the medium is that processes which are forbidden in the vacuum by symmetry principles now become possible. The constraints of Lorentz-invariance, G-parity, or isospin invariance, are no longer effective. Below we discuss two examples of such processes: $\omega \to \pi\pi$ decay, and $\rho - \omega$ mixing. Strictly speaking, ρ and ω mix in the vacuum but this is a negligible effect caused by the small explicit breaking of the isospin symmetry. Similarly, the partial width for the decay $\omega \to \pi\pi$ is only ~ 0.2MeV in the vacuum, which is again a tiny isospin-violation effect.

In this paper we summarize the results of Refs. [1–3] which extend the work presented in Refs. [4] and [5]. We emphasize that the *matter-induced width* for the $\omega \to \pi\pi$ decay is large: for ω moving with respect to the medium with a momentum above ~ 200MeV the corresponding width, at the nuclear saturation density, is of the order of 100MeV. We also show that even a moderate excess of neutrons over protons in nuclear matter, such as in ^{208}Pb, can lead to large ρ-ω mixing.

The in-medium broadening of the ω meson, as well as the shifts of the positions of the resonances (due to their mixing) are examples of the so-called in-medium modifications of hadron properties, which are predicted in a variety of theoretical calculations [6–19]. The recent interest in studying such modifications (for a review see [20]) has been trigerred by the experimental observation of the enhanced

---

[1)] Talk presented by W. Florkowski.
[2)] Research supported by PRAXIS grants XXI/BCC/429/94 and PRAXIS/P/FIS/12247/1998, and by the Polish State Committee for Scientific Research grant 2P03B-080-12.

production of low-mass dileptons in relativistic heavy-ion collisions [21,22]. The data are most easily described by the assumption that either the masses of vector mesons decrease in medium or their widths become larger.

The non-vanishing amplitude for the decay $\omega \to \pi\pi$ indicates that the processes of pion annihilation into dilepton pairs in the ω channel are also possible, as first pointed in Ref. [4]. However, due to the smallness of the $\omega\gamma$ coupling they cannot compete with the annihilation occuring in the ρ channel [2,4]. Nevertheless, the large width of the ω mesons should cause a depletion in their population. In our opinion such an effect should be included in simulations of heavy-ion collisions. In fact, the results of some recent transport calculations [28,29] show an excess in the dilepton yield at $q^2 = m_\omega^2$, attributed to the direct $\omega \to e^+e^-$ decay. With an increased hadronic width of the ω a better agreement with the data may be achieved.

The effects of the isospin asymmetry in nuclear matter for the $\rho - \omega$ mixing were studied in Ref. [5] in the framework of the Walecka model. The results of Ref. [5] indicate that at asymmetries such as in ^{208}Pb and at nuclear saturation density, the ρ and ω mix with an angle of about $\sim 2\%$. In our approach, performed on a broader footing, we show that the matter-induced $\rho - \omega$ mixing can be in fact much larger. We expect that it may show up, among other medium-induced effects, in future high-accuracy relativistic heavy-ion collisions.

$\omega \to \pi\pi$ DECAY IN NUCLEAR MATTER

Our calculation of the $\omega \to \pi\pi$ width is done in the framework of an effective hadronic theory. Mesons (ω, σ, π) interact with nucleons and $\Delta(1232)$. We work to the leading order in nuclear density, hence only the diagrams shown in Fig. 1 are taken into account. The "bubble" diagram (a) was studied by Wolf, Friman, and Soyeur [4], who pointed out the significance of the $\omega - \sigma$ mixing for the in-medium $\omega \to \pi\pi$ decay. The "triangle" diagrams (equally important in any formal scheme) were taken into consideration in Ref. [1]. The complete set of diagrams (a-d) was included in Ref. [2].

The solid line in Fig. 1 denotes the in-medium nucleon propagator [23]

$$G(k) = (\not{k} + M)\left[\frac{1}{k^2 - M^2 + i\varepsilon} + \frac{i\pi}{E_k}\delta(k_0 - E_k)\theta(k_F - |\mathbf{k}|)\right], \quad (1)$$

where k is the nucleon four-momentum, M denotes the nucleon mass, $E_k = \sqrt{M^2 + \mathbf{k}^2}$, and k_F is the Fermi momentum. Diagram (a) involves the intermediate σ-meson propagator, which we take in the form

$$G_\sigma(k) = \frac{1}{k^2 - m_\sigma^2 + im_\sigma\Gamma_\sigma - \frac{1}{4}\Gamma_\sigma^2}. \quad (2)$$

Here the mass and the width of the σ meson are chosen in such a way that they reproduce effectively the experimental $\pi\pi$ scattering length at $q^2 = m_\omega^2 = (780\text{MeV})^2$, which is the relevant kinematic point for the process at hand. From this fit we find $m_\sigma = 789 \text{MeV}$ and $\Gamma_\sigma = 237 \text{MeV}$. Note that m_ω and m_σ are very close to each other, which enhances the amplitude obtained from diagram (a) [4].

The double line in diagrams (c-d) denotes the Δ propagator

$$G_\Delta^{\alpha\beta}(k) = \frac{\slashed{k} + M_\Delta}{k^2 - M_\Delta^2 + i M_\Delta \Gamma_\Delta - \frac{1}{4}\Gamma_\Delta^2} \left[-g^{\alpha\beta} + \frac{1}{3}\gamma^\alpha\gamma^\beta + \frac{2k^\alpha k^\beta}{3M_\Delta^2} + \frac{\gamma^\alpha k^\beta - \gamma^\beta k^\alpha}{3M_\Delta} \right]. \tag{3}$$

This formula corresponds to the usual Rarita-Schwinger definition [24,25] with the denominator modified in order to account for the finite width of the Δ resonance, $\Gamma_\Delta = 120 \text{MeV}$.

We assume that the ωNN and $\omega\Delta\Delta$ vertices have the form which follows from the minimum-substitution prescription and vector-meson dominance applied to the nucleon and the Rarita-Schwinger [24] Lagrangians:

$$V_{\omega NN}^\mu = g_\omega \gamma^\mu, \tag{4}$$

$$V_{\omega\Delta\Delta}^{\mu\alpha\beta} = g_\omega \left[-\gamma^\mu g^{\alpha\beta} + g^{\alpha\mu}\gamma^\beta + g^{\beta\mu}\gamma^\alpha + \gamma^\alpha\gamma^\mu\gamma^\beta \right]. \tag{5}$$

The results presented below do not qualitatively depend on the form of the coupling, as long as it remains strong. The coupling constant g_ω can be estimated from

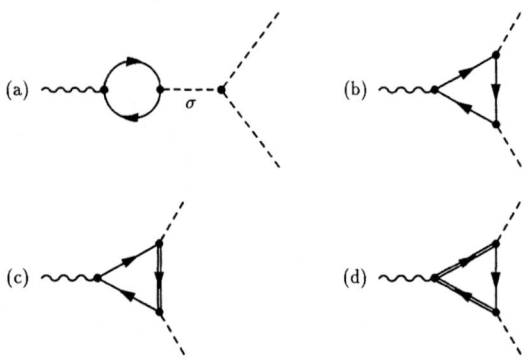

FIGURE 1. Diagrams contributing to the $\omega \to \pi\pi$ amplitude in nuclear medium. The incoming ω has momentum q and polarization ϵ. The outgoing pions have momenta p and $q-p$. Diagrams (b-d) have corresponding crossed diagrams, not displayed.

the vector dominance model. We use $g_\omega = 9$. For the πNN vertex we use the pseudoscalar coupling, with the coupling constant $g_{\pi NN} = 12.7$. The same value is used for $g_{\sigma NN}$. The $\sigma\pi\pi$ coupling constant is taken to be equal to $g_{\sigma\pi\pi} = 12.8\,m_\pi$, where $m_\pi = 139.6\text{MeV}$ is the physical pion mass (this value follows from the fit done to $\pi\pi$ scattering phase shifts done in Ref. [4]). The $\pi N\Delta$ vertex has the form $V^\mu_{\pi N\Delta} = (f_{\pi N\Delta}/m_\pi)p^\mu \vec{T}$, where p^μ is the pion momentum, $\vec{T}$ is the $\frac{1}{2} \to \frac{3}{2}$ isospin transition matrix, and the coupling constant $f_{\pi N\Delta} = 2.1$ [26].

The amplitude, evaluated according to the diagrams depicted in Fig. 1 (a-d) can be uniquely decomposed in the following Lorentz-invariant way:

$$\mathcal{M} = \epsilon^\mu(Ap_\mu + Bu_\mu + Cq_\mu), \qquad (6)$$

where p is the four-momentum of one of the pions, q is the four-momentum of the ω meson, u is the four-velocity of nuclear matter, and ϵ specifies the polarization of ω. Our calculation is performed in the rest frame of nuclear matter, where $u = (1,0,0,0)$. In this reference frame the amplitude $\mathcal{M}$ vanishes for vanishing 3-momentum $\mathbf{q}$, as requested by rotational invariance. Hence, the process $\omega \to \pi\pi$ occurs only when the ω moves with respect to the medium.

In Fig. 2 we present our numerical results at the nuclear saturation density, $\rho_B = \rho_0 = 0.17\text{fm}^{-3}$. We show the width of longitudinally polarized ω mesons, Γ^L, as a function of $|\mathbf{q}|$. In our calculation, we reduce the value of the in-medium nucleon mass to 70% of its vacuum value, $M^* = 0.7M$, which is a typical number at the nuclear saturation density. We also reduce by the same factor the mass of the Δ, i.e. $M^*_\Delta = 0.7M_\Delta$, since it is expected to behave similarly to the nucleon.

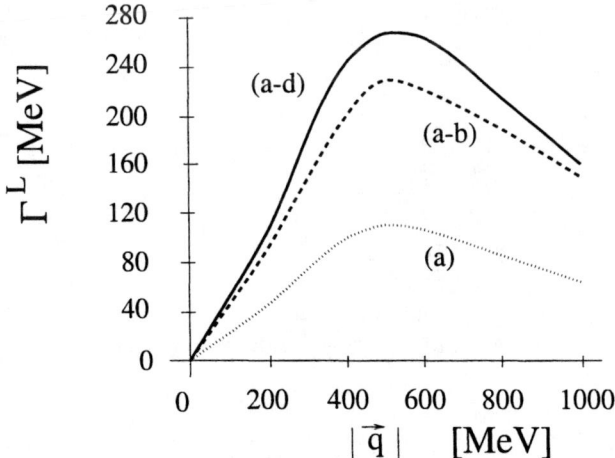

FIGURE 2. The in-medium width pf the ω meson plotted as a function of its 3-momentum $|\mathbf{q}|$. The labels (a), (a-b) and (a-d) refer to Fig.1. They indicate the diagrams included in the calculation.

The labels indicate which diagrams of Fig. 1 have been included. The complete result corresponds to the case (a-d). The case (a) reproduces the result of Ref. [4]. The width of the transverse modes (~ 1 MeV) is negligible and we do not show it in the plot.

In Fig. 1 the mass of ω is kept at the vacuum value. For $m_\omega^* = 0.70 m_\omega$ our results decrease by a factor of 2.5. Still, the widths remain substantial and the effects discussed above are important.

$\rho - \omega$ MIXING IN ASYMMETRIC NUCLEAR MATTER

For vanishing 3-momentum $\mathbf{q}$, the vector-meson correlator in the coupled ρ^0 and ω channels has the following structure

$$\Pi^{\alpha\beta}(\nu, \mathbf{q} = 0) = \begin{pmatrix} \Pi_\rho^{\alpha\beta}(\nu) & \Pi_{\rho\omega}^{\alpha\beta}(\nu) \\ \Pi_{\rho\omega}^{\alpha\beta}(\nu) & \Pi_\omega^{\alpha\beta}(\nu) \end{pmatrix}, \tag{7}$$

where ν is the energy variable. For the diagonal parts of (7) we choose a simple form which can mimic the results of various calculations of in-medium vector mesons:

$$\Pi_v^{\alpha\beta}(\nu) = Z_v^{*-1}\left((\nu - i\Gamma_v^*/2)^2 - m_v^{*2}\right)T^{\alpha\beta}, \quad v = \rho, \omega. \tag{8}$$

The asterisk denotes here the in-medium values of the resonance position, m_v, width, Γ_v, and the wave-function renormalization, Z_v. In the case $\mathbf{q} = 0$ the tensor $T^{\alpha\beta}$ has a simple form, $T^{\alpha\beta} = \text{diag}(0, 1, 1, 1)$ [9,23]. Our parameterization incorporates basic features of mesons propagating in nuclear medium, such as the shift of the resonance position, broadening, and wave-function renormalization.

Applying the same formalism [23] as in the previous Section, we find that the off-diagonal matrix element in Eq. (7), describing the mixing of ρ^0 and ω (at $\mathbf{q} = 0$) is given by expression

$$\Pi_{\rho\omega}^{\alpha\beta}(\nu, \mathbf{q} = 0) = -i \int \frac{d^4k}{(2\pi)^4} \left\{ \text{Tr}[V_\rho^\alpha(\nu) G_D^p(k^0 + \nu, \mathbf{k}) V_\omega^\beta(-\nu) G_F^p(k)] - \right.$$
$$\left. \text{Tr}[V_\rho^\alpha(\nu) G_D^n(k^0 + \nu, \mathbf{k}) V_\omega^\beta(-\nu) G_F^n(k)] \right\} + (F \leftrightarrow D)$$
$$\equiv T^{\alpha\beta} \Pi_{\rho\omega}(\nu), \tag{9}$$

where $G_D^{p,n}(k)$ and $G_F^{p,n}(k)$ denote the density part and the free part of the Dirac propagator for the proton and neutron (compare our notation in Eq. (1)). The quantity

$$V_{v,\alpha} = g_v\left(\gamma_\alpha - \frac{\kappa_v}{2M}\sigma_{\alpha\beta}\partial^\beta\right) \tag{10}$$

is the vector-mesons nucleon vertex which includes the tensor coupling κ. Following Ref. [12] we use two parameter sets:

$$\text{I}: \quad g_\rho = 2.63, \quad \kappa_\rho = 6.0, \quad g_\omega = 10.1, \quad \kappa_\omega = 0.12,$$
$$\text{II}: \quad g_\rho = 2.72, \quad \kappa_\rho = 3.7, \quad g_\omega = 10.1, \quad \kappa_\omega = 0.12.$$

This parameterization follows from the vector meson dominance model [30]. The basic difference between the two sets is the value of κ_ρ [31].

Explicit evaluation gives

$$\Pi_{\rho\omega}(\nu) = \frac{2}{3} g_\rho g_\omega \int \frac{d^3k}{(2\pi)^3 E_k^*} \frac{\theta(k_n - |\mathbf{k}|) - \theta(k_p - |\mathbf{k}|)}{\nu^2 - 4(E_k^*)^2} \times$$
$$\left[8(E_k^*)^2 + 4M^{*2} + 3(\kappa_\rho + \kappa_\omega) \frac{M^*}{M} \nu^2 + \kappa_\rho \kappa_\omega \frac{(E_k^*)^2 + 2M^{*2}}{M^2} \nu^2 \right], \qquad (11)$$

where k_p and k_n are the proton and neutron Fermi momenta and M^* is the nucleon mass in medium. In symmetric matter, where $k_p = k_n = k_F$, the proton and neutron contribution to Eqs. (9) and (11) cancel, and $\Pi_{\rho\omega}(\nu)$ vanishes. In asymmetric matter $k_n > k_p$, and we get a net contribution to $\Pi_{\rho\omega}(\nu)$. We note that the proton and neutron densities are equal to $\rho_{p,n} = k_{p,n}^3/(3\pi^2)$, and the baryon density ρ_B and the isospin asymmetry x are equal to $\rho_B = \rho_p + \rho_n$ and $x = (\rho_n - \rho_p)/\rho_B$. At low x it can be easily shown that $\Pi_{\rho\omega}(\nu)$ is linear in x. It remains linear for asymmetries accessible in heavy-ion collisions. If in addition we expand Eq. (11) at small ρ_B, we notice that $\Pi_{\rho\omega}(\nu) \sim x\rho_B = \rho_n - \rho_p$, in agreement with the low-density theorem for the scattering amplitude.

Finding the eigenvalues of the matrix (7) is equivalent to solving the following equation

$$\text{Det} \begin{pmatrix} (\nu - i\Gamma_\rho^*/2)^2 - m_\rho^{*2} & \sqrt{Z_\rho^* Z_\omega^*} \Pi_{\rho\omega}(\nu) \\ \sqrt{Z_\rho^* Z_\omega^*} \Pi_{\rho\omega}(\nu) & (\nu - i\Gamma_\omega^*/2)^2 - m_\omega^{*2} \end{pmatrix} = 0. \qquad (12)$$

Equation (12) yields eigenvalues ν_1 and ν_2, and the corresponding eigenstates $|1\rangle$ and $|2\rangle$. Our convention is that in the absence of mixing, i.e. for $x = 0$, we have $|1\rangle = |\rho\rangle$ and $|2\rangle = |\omega\rangle$. A commonly used measure of mixing of states is the mixing angle. Since the problem (12) is not hermitian, the eigenstates $|1\rangle$ and $|2\rangle$ are not orthogonal and we cannot define a single mixing angle. We find it useful to introduce two mixing angles, θ_1 and θ_2, through the relations

$$|1\rangle = \cos\theta_1 |\rho\rangle + \sin\theta_1 |\omega\rangle, \quad |2\rangle = -\sin\theta_2 |\rho\rangle + \cos\theta_2 |\omega\rangle. \qquad (13)$$

Since the matrix in (12) is complex, the mixing angles are also complex.

Our results are shown in Table I, which contains 8 representative cases for $\rho_B = 2\rho_0$. We assume that at this density $M^*/M = 0.5$. The table should be read from top to bottom. The first row labels the case. Five input rows contain m_ρ^*, m_ω^*, Γ_ρ^*, Γ_ω^* and $\sqrt{Z_\rho^* Z_\omega^*}$ for symmetric matter of density ρ_B.

To summarize, we observe that the mixing effects are sizable for all sensible cases, with mixing angles of the order of $10°$, or larger. As a consequence, the

Input:	1	2	3	4	5	6	7	8
m_ρ^* (MeV)	500	500	500	550	550	650	550	550
m_ω^* (MeV)	500	500	500	450	450	450	450	450
Γ_ρ^* (MeV)	0	200	200	300	300	300	300	300
Γ_ω^* (MeV)	0	50	50	50	200	200	50	200
$\sqrt{Z_\rho^* Z_\omega^*}$	0.7	0.7	0.35	0.7	0.7	0.7	0.7	0.7
Output:	Set I						Set II	
$\mathrm{Re}(\nu_1)$ (MeV)	535	509	502	557	559	656	555	557
$\mathrm{Re}(\nu_2)$ (MeV)	469	494	499	445	443	446	446	444
$2\mathrm{Im}(\nu_1)$ (MeV)	0	170	193	298	306	309	297	302
$2\mathrm{Im}(\nu_2)$ (MeV)	0	83	57	56	198	198	55	200
$\mathrm{Re}(\theta_1)$ (deg)	45	12	3	11	17	11	9	14
$\mathrm{Re}(\theta_2)$ (deg)	45	11	2	8	13	7	7	12
$\mathrm{Im}(\theta_1)$ (deg)	0	-31	-13	-7	-1	2	-6	-2
$\mathrm{Im}(\theta_2)$ (deg)	0	-32	-13	-6	-2	0	-6	-2

TABLE 1. ρ-ω mixing in asymmetric matter, $x = x_{\mathrm{Pb}}$, $\rho_B = 2\rho_0$, $M^*/M = 0.5$.

resonance positions and widths of the vector mesons are shifted significantly. Our analysis shows also (see [3] for more details) that the mixing effect will continue to be important at moderate temperatures. Therefore we expect that our results of large ρ-ω mixing may show up, among other possible medium-induced effects, in future high-accuracy relativistic heavy-ion experiments. In particular, the results to be obtained with the HADES spectrometer at the SIS accelerator at GSI, whose anticipated mass resolution in the discussed region will reach 1% [32], should be influenced by the phenomenon of ρ-ω mixing.

REFERENCES

1. Broniowski, W., Florkowski, W., and Hiller, B., *Acta Phys. Pol.* **B30**, 1079 (1999).
2. Broniowski, W., Florkowski, W., and Hiller, B., *Preprint INP1822/PH*, nucl-th/9905040.
3. Broniowski, W., and Florkowski, W., *Phys. Lett.* **B440**, 7 (1998).
4. Wolf, G., Friman, B., and Soyeur, M., *Nucl. Phys.* **A640**, 129 (1998).
5. Dutt-Mazumder, A. K., Dutta-Roy, B., and Kundu, A., *Phys. Lett.* **B399**, 196 (1997).
6. Brown, G. E., and Rho, M., *Phys. Rev. Lett.* **66**, 2720 (1991).
7. Celenza, L. S., Pantziris, A., Shakin, C. M., and Sun, W.-D., *Phys. Rev.* **C45**, 2015 (1992).
8. Hatsuda, T., and Lee, S. H., *Phys. Rev.* **C46**, R34 (1993).
9. Jean, H.-C., Piekarewicz, J., and Williams, A. G., *Phys. Rev.* **C49**, (1994) 1981.
10. Cassing, W., Ehehalt, W., and Ko, C. M., *Phys. Lett.* **B363**, 35 (1995).
11. Li, G. Q., Ko, C. M., and Brown, G. E., *Nucl. Phys.* **A606**, 568 (1996).

12. Hatsuda, T., Shiomi, H., and Kuwabara, H. , *Prog. Theor. Phys.* **95**, 1009 (1996).
13. Rapp, R., Chanfray, G., and Wambach, J., *Nucl. Phys.* **A617**, 472 (1997).
14. Friman, B., and Pirner, H. J., *Nucl. Phys.* **A617**, 496 (1997).
15. Klingl, F., Kaiser, N., and Weise, W., *Nucl. Phys.* **A624**, 527 (1997).
16. Leupold, S., Peters, W. , and Mosel, U., *Nucl. Phys.* **A628**, 311 (1998).
17. Eletsky, V. L., Ioffe, B. L., and Kapusta, J. I., *Eur. J. Phys.* **A3**, 381 (1998).
18. Friman, B., *Acta Phys. Pol.* **B29**, 3195 (1998).
19. Bratkovskaya, E. L., and Ko, C. M., *Phys. Lett.* **B445**, 265 (1999).
20. *Quark Matter 97, Proc. 13th Int. Conf. on Ultra-Relativistic Nucleus-Nucleus Collisions*, Tsukuba, Japan, 1997, *Nucl. Phys.* **A638**, and references therein.
21. CERES Collab., Agakichiev, G., et al., *Phys. Rev. Lett.* **75**, 1272 (1995).
22. HELIOS/3 Collab., Masera, M., et al., *Nucl. Phys.* **A590**, 3c (1995).
23. Chin, S. A., *Ann. Phys. (NY)* **108**, 301 (1977).
24. Rarita, W., and Schwinger, J., *Phys. Rev.* **60**, 61 (1941).
25. Benmerrouche, M., Davidson, R., and Mukhopadhyay, N. C., *Phys. Rev.* **C39**, 2339 (1989).
26. Durso, J. W., Jackson, A., and VerWest, B., *Nucl. Phys.* **A345**, 471 (1980).
27. Koch, P., *Z. Phys.* **C57**, 283 (1993).
28. Koch, V., *Proc. of the XXXVII International Winter Meeting on Nuclear Physics*, Bormio, Italy, January 1999, nucl-th/9903008.
29. Koch, V., *Acta Phys. Pol.* **29**, 3233 (1998).
30. Sakurai, J. J., *Currents and Mesons*, Chicago: Univ. of Chicago Press, 1969.
31. Machleidt, R., in *Advances in Nuclear Physics*, edited by Negele, J. W., and Vogt, E., New York: Plenum, 1989, Vol. 19.
32. contribution of Stroth, J., et al. in *Hadrons in Nuclear Matter*, edited by Feldmeier, H., and Nörenberg, W., GSI, Darmstadt, 1995, *Proc. Int. Workshop XXIII on Gross Properties of Nuclei and Nuclear Excitations*, Hirschegg, Austria, 1995.

The deconfinement phase transition, hadronization and the NJL model

Sibaji Raha*

*Physics Department, Bose Institute
93/1, Acharya Prafulla Chandra Road
Calcutta 700 009, INDIA
Electronic Mail : sibaji@bosemain.boseinst.ernet.in

Abstract. One of the confident predictions of QCD is that at sufficiently high temperature and/or density, hadronic matter should undergo a thermodynamic phase transition to a colour deconfined state of matter - popularly called the Quark-Gluon Plasma (QGP). In low energy effective theories of Quantum Chromodynamics (QCD), one usually talks of the chiral transition for which a well defined order parameter exists. We investigate the dissociation of pions and kaons in a medium of hot quark matter decsribed by the Nambu - Jona Lasinio (NJL) model. The decay widths of pion and kaon are found to be large but finite at temperature much higher than the critical temperature for the chiral (or deconfinement) transition, the kaon decay width being much larger. Thus pions and even kaons (with a lower density compared to pions) may coexist with quarks and gluons at such high temperatures. On the basis of such premises, we investigate the process of hadronization in quark-gluon plasma with special emphasis on whether such processes shed any light on acceptable low energy effective theories of QCD.

Quantum Chromodynamics (QCD) is believed to be *the underlying theory* of strong interactions. Although enormously successful, inasmuch as it is part of the standard model of physical interactions, the applicability of QCD is studying nuclear interactions or hadronic processes at low energies is still limited. The difficulties are largely technical, associated with the non-perturbative features of the theory and hence, a large amount of effort is being devoted to finding effective lagrangians which contain features compatible with the low energy limit of QCD. This is the thrust of the present workshop. The philosophy of my talk is somewhat complementary; I am going to discuss QCD phenomena which occur in very energetic nuclear collisions, where one hopes perturbative QCD does play an important role, at least during the initial time period.

A confident prediction of QCD is that at very high temperature and/or density, the bulk properties of strongly interacting matter would be governed by the coloured

QCD degrees of freedom - quarks and gluons- rather than the usual hadrons. Such a phase is called quark gluon plasma (QGP) [1] in the literature. Conditions conducive to the formation of QGP may have existed in the early universe during the first few microseconds after the Big Bang. Also, such conditions may be transiently created in highly energetic collsions of large nuclei and the search for such a novel phase of matter constitutes a major area of current research in the field of high energy physics.

Whether QGP is separated from the hadronic world by an actual thermodynamic phase transition is an open question. It has been widely postulated that such a phase transition may indeed occur, where the quarks and gluons convert into colourless hadrons. Recent results, showing the lack of thermodynamic equilibrium [2] in the quark-gluon phase in ultrareletivistic heavy ion collisions, indicate however that such an ideal situation is unlikely. It should also be noted that although the persistence of non-perturbative effects till very high temperatures was suggested in the literature quite early on [3], it is only recently that the lattice results have confirmed that non-perturbative hadron like excitations could survive at temperatures far above the chiral phase transition temperature [4]. The lattice result for pion screening mass has been studied in ref. [5]. The analysis of [5] has been contradicted by Boyd et al. in ref. [6]. The conclusion of these authors [6] is consistent with the existence of free quarks at high temperatures. On the other hand, Shuryak [7] argued in a subsequent work that the non-perturbative modes, especially pion- like excitations, could indeed survive till temperatures above T_c. Furthermore, similar results for pion screening masses are obtained in σ- model as well [6]. It is thus imperative to understand the behaviour of such hadronic resonances, their formation, stability and so on, in a quark gluon medium at high temperature. We confine our attention to the case of pions and kaons only; these, being lighter than other hadrons, account for the bulk of the multiplicity.

Formation of light mesons like pions and kaons, a bound state of light relativistic quarks, is an extremely difficult problem to handle in QCD, where all the troublesome features of non-perturbative QCD appear. We therefore employ the usual practice of looking at the pion and kaon as Goldstone bosons arising from the spontaneous breaking of the chiral symmetry, a feature most suitably addressed in the Nambu Jona-Lasinio (NJL) model [8].

The formulation of NJL model in flavour SU(3) was first introduced by Hatsuda et al. [9] and Bernard et al. [10]. The three flavour NJL model Lagrangian is written in terms of u, d and s quarks, the interaction between them being constrained by the $SU(3)_L \otimes SU(3)_R$ chiral symmetry, explicit symmetry breaking due to the current quark masses and the $U(1)_A$ breaking due to the axial anamoly [9]. The full Lagrangian with KMT (Kobayashi- Maskawa -'t-Hooft) term is given below [9].

$$\mathcal{L} = \bar{q}(i\gamma \cdot \partial - \mathbf{m})q + \frac{1}{2}g_s\sum_{a=0}^{8}[(\bar{q}\lambda_a q)^2 + (\bar{q}i\lambda_a\gamma_5 q)^2]$$
$$+ g_D[det\bar{q}_i(1-\gamma_5)q_j + h.c.] \qquad (1)$$

where the quark fields q_i has three colours ($N_c = 3$) and three flavours ($N_f = 3$)

and λ_a ($a = 1, 8$) are the Gell-Mann matrices. The quark mass matrix is given by $\mathbf{m} = diag(m_u, m_d, m_s)$.

In the mean field approximation, the quark condensates at finite temperature are given by,

$$<< \bar{q}_i q_i >> = -2N_c \int \frac{d^3 p}{2\pi^3} \frac{M_i}{E_{ip}} f(E_{ip}) \tag{2}$$

where E_{ip} is the quark single particle energy for the i-th specie and $f(E_{ip}) = 1 - n_{ip} - \bar{n}_{ip}$, n_{ip} and $\bar{n}_{ip}$ being the Fermi-Dirac distributions for quarks and anti-quarks, respectively. If quark chemical potential is zero, then $n_{ip} = \bar{n}_{ip} = [exp(E_{ip}/T) + 1]^{-1}$.

The temperature dependent constituent quark masses M_i are obtained from the expressions below,

$$M_u = m_u - 2g_s\alpha - 2g_D\beta\gamma$$
$$M_d = m_d - 2g_s\beta - 2g_D\alpha\gamma$$
$$M_s = m_s - 2g_s\gamma - 2g_D\alpha\beta \tag{3}$$

where

$$<< \bar{u}u >> \equiv \alpha, \quad << \bar{d}d >> \equiv \beta, \quad << \bar{s}s >> \equiv \gamma \tag{4}$$

We now want to investigate the decay of pionic and kaonic excitations, the properties of which we assume to be given by the NJL model. It should be mentioned here that at temperatures above the critical temperature, these mesonic excitations are more like resonances with large effective masses [4,5]. In the following, we study [11] the decay width of such pseudoscalar excitations in the hot quark medium as a function of temperature, starting with the Lagrangian given above in equation (1).

The quark mass M_i appearing in eq. (2) and in eq. (3) is a very important ingredient in our calculation. In the absence of any medium and/or dynamic effect, M_i is the current quark mass. On the other hand, we know that due to the spontaneous breakdown of the chiral symmetry, quarks attain the value of the constituent quark mass [12].

In the present calculation we have used the parametrisation of ref. [10] ($\Lambda = 631.4$, $g_s\Lambda^2 = 3.67$, $g_D\Lambda^5 = -9.29$ and current mass $m_{u,d}(m_s) = 5.5$ (135.7) MeV) to calculate the quark and meson masses. The constituent quark masses are calculated using the gap equations(eq. 3). These quark masses are then put into the dispersion equation for mesons to get dynamical masses of mesons (π and K, here).

$$1 + 2G_\phi \Pi_{ij}(\omega, \vec{q} \to 0) = 0 \tag{5}$$

where Π_{ij} is the one loop polarization due to u and d quark for pions and u or d and s quark for kaons. G_ϕ is the coupling constant with ϕ coerresponding to π or K. The general expression for polarization function is

$$\Pi(q_0, \vec{q}) = \frac{N_c}{(2\pi)^3} \int_0^\Lambda \frac{d^3p}{E_p E_k} \Bigg[(n_k - n_p) \left\{ \frac{1}{E_p - E_k + q_0 + i\epsilon} + \frac{1}{E_p - E_k - q_0 - i\epsilon} \right\}$$
$$\times (-E_p E_k + \vec{p}.\vec{k} + M_1 M_2)$$
$$+ (n_k + n_p - 1) \left\{ \frac{1}{E_p + E_k + q_0 + i\epsilon} + \frac{1}{E_p + E_k - q_0 - i\epsilon} \right\}$$
$$\times (E_p E_k + \vec{p}.\vec{k} + M_1 M_2) \Bigg] \quad (6)$$

where N_c is the number of colours and $\vec{k} = \vec{p} + \vec{q}$. The energies $E_p = \sqrt{p^2 + M_1^2}$ and $E_k = \sqrt{(\vec{p} + \vec{q})^2 + M_2^2}$. For pion, $M_1 = M_2 = M_u$. For kaon, $M_1 = M_{u(d)}$ and $M_2 = M_s$. n_k and n_p are the Fermi-Dirac distribution functions defined earlier. The pseudoscalar couplings are,

$$G_\pi = g_s + g_D \gamma$$
$$G_{K^\pm} = g_s + g_D \beta$$
$$G_{K^0} = g_s + g_D \alpha \quad (7)$$

where α, β and γ are defined in eq. (4).

The decay width is evaluated using the imaginary part of the eq.(6) as given below,

$$\Gamma_\phi = -\frac{G_{\phi q}^2 Im\Pi(\omega, \vec{q} \to 0)}{\omega} \quad (8)$$

where $G_{\phi q}$ is the empirical meson-quark coupling as obtained in NJL. Here we have used $G_{\pi q} = 3.5$ and $G_{Kq} = 3.6$ [10].

The variation of quark and meson masses is shown in figure 1. The u or d quark masses starting from 135 MeV drops to the current quark mass value just after a temperature of 200 MeV. On the other hand, the drop in the strange quark mass is much smaller around that temperature, showing the effect of explicitly broken chiral symmetry, by a larger amount, in the SU(3) sector. Pion and kaon both show a similar qualitative behaviour. The masses of pion and kaon remain constant at their free masses upto a temperature of 200 MeV but increases sharply after that, the pion mass rising to 900 MeV and kaon mass to 1000 MeV around a temperature of 450 MeV, thus giving a slower increment for kaons compared to pions. The difference in the behaviour of kaon and pion can be attributed to the difference in u and s quark masses.

Figure 2 shows that the decay width, for both pion and kaon, is very high at high temperature and decreases with decreasing temperature, going to zero at around $T = 0.2$ GeV. It is worth noticing that at around the same temperature, the effective pion mass attains the value of the free pion mass (figure 1). The decay width of kaon is around 3 GeV where as that of pion is around 1.4 GeV at T = 500 MeV. This is very significant for two reasons. Firstly, our results show that though there will be pions and kaons along with the quarks at high temperature phase, the numbers of mesons will be very small due to their large decay width. Moreover, the number of kaons will be much less compared to pions at high temperature phase, though both the mesons will become stable around the same temperature (below 200 MeV).

We thus find that even without any consideration of the detailed evolution and dynamics of the system, the mesonic modes in a hot quark medium are found to become important around a temperature of 200 MeV. Though the question whether this is a signature of a phase transition cannot be addressed within the present framework, the fact that most of the pions and kaons decay into quarks, owing to a large decay width at temperatures higher than $T = 200$ MeV, is a remarkable finding. Moreover, both the pionic as well as kaonic modes start becoming important at about the same temperature, thus providing a hint of some kind of a transition temperature. It is therefore tempting to attempt a microscopic investigation of the process of hadronization within the NJL model.

Recently there have been some attempts to study the formation of hadrons in quark matter using different semi-microscopic approaches. These studies can be

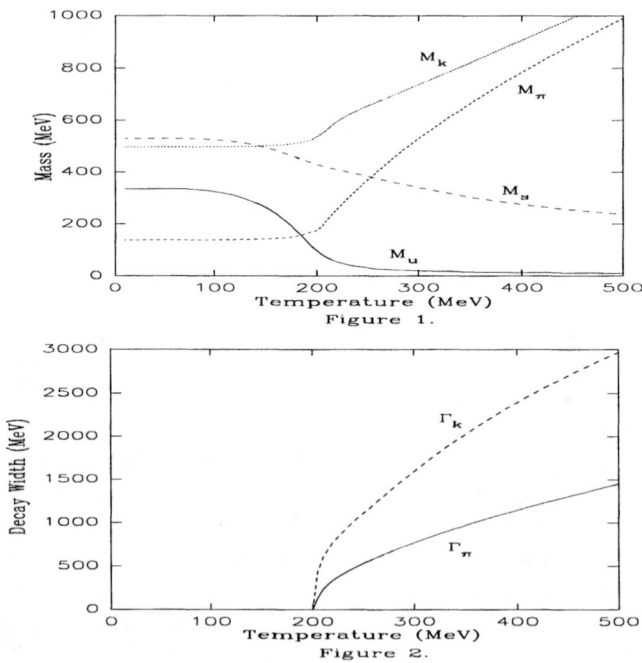

Figure 1 : Variation of quark and meson masses with temperature.
Figure 2 : Variation of meson decay widths with temperature.

characterized either as model dependent calculations [13], or the computer codes based on the string phenomenology [14], or other phenomenological description of hadronization [15]. None of these approaches account for the essential lack of equilibrium in the quark-gluon phase. Some efforts have also been made to estimate hadronization within the parton cascade model by introducing a cut-off to mimic the non-perturbative effects [16]. To the best of our knowledge, the first study aimed at investigating the dynamical process of hadronization in a non-equilibrated quark-gluon system from a physically transparent approach was in [17].

We have earlier shown [2] that perturbative estimates of the gluon-gluon, quark-gluon and quark-quark cross sections allow us to study the evolution of the quark-gluon matter formed in ultrarelativistic heavy ion collisions by visualizing the quarks as Brownian particles in a hot gluonic thermal bath. Let us start with such a premise, which, in our opinion, describes the non-equilibrium aspects of the evolution in a physically transparent manner.

Hadronization in such a system can be studied in the light of Smoluchowski's theory of coagulation in colloids, which was elaborated further by Chandrasekhar [18]. This theory suggests that coagulation results as a consequence of each colloidal particle being surrounded by a sphere of influence of a certain radius R such that the Brownian motion of a particle proceeds unaffected only so long as no other particle comes within its sphere of influence. When another Brownian particle does come within a distance R of the test particle, they form a two body cluster. This cluster also describes a Brownian motion but at a reduced rate due to its increased size/mass. The process continues further till a single cluster of all the particles is formed.

In order for this stochastic scenario of cluster formation to apply to a system of Brownian quarks in the hot gluon bath, it is essential that each quark has an appropriate sphere of influence of radius r. Obviously, this radius r will depend on the spin-isospin combination of final cluster (whether the final cluster is scalar, pseudoscalar, vector or axial vector meson or even a baryon). Mesons are formed when one quark and one antiquark with proper quantum numbers come within the spheres of influence of each other. (Clusters with greater numbers of quarks and antiquarks (e.g. baryons) can also be formed by imposing the conditions of colour neutrality and charge balance properly.) This implies that the radius of the sphere of influence corresponds to the correlation length between the quarks in the proper hadronic channel. In other words, this is the screening length of the corresponding hadrons in the hot quark-gluon matter. (One can immediately see, without further ado, that the rate of pion formation in the hot quark matter should be rather small at high temperatures and increase with falling temperature as the pion screening length (inverse of the screening mass) decreases with increasing temperature in all realistic pictures.) The pions formed at very high temperatures are most likely to decay back into quarks and antiquarks, because of the large decay width at high temperatures. We adopt, for the sake of consistency, the NJL model estimates for the pion screening mass [19] in the present work.

The rate of pion formation from Brownian quarks as a stochastic process, as is evident from the preceding discussion, depends on the number of quarks (antiquarks) falling into the sphere of influence of another antiquark (quark), the number of pions decaying back to quarks and antiquarks and also the change in the pion density due to the expansion of the system. Thus the rate equations can be written as,

$$\frac{dn_{\pi^a}}{dt} = n_{q_i} n_{\bar{q}_j} 4\pi r^2 <\vec{v}\cdot\hat{r}> - \Gamma^{total}_{\pi^a \to q_i \bar{q}_j} - \frac{n_{\pi^a}}{t} \quad (9)$$

$$\frac{dn_{\pi^0}}{dt} = \frac{1}{2}(n_u n_{\bar{u}} + n_d n_{\bar{d}})4\pi r^2 <\vec{v}\cdot\hat{r}> - \Gamma^{total}_{\pi^0 \to u\bar{u}(d\bar{d})} - \frac{n_{\pi^0}}{t} \quad (10)$$

$$\frac{dn_{q_i}}{dt} = \Gamma^{total}_{g \to q_i \bar{q}_i} + \Gamma^{total}_{gg \to q_i \bar{q}_i} - n_{q_i} n_{\bar{q}_j} 4\pi r^2 <\vec{v}\cdot\hat{r}> + \Gamma^{total}_{\pi \to q_i \bar{q}_j} - \frac{n_{q_i}}{t} \quad (11)$$

In eqs.(9,10), the first term is the rate of pion formation ($a \equiv +$ or $-$); the second term is the rate of pions decaying back to quarks and the third term is due to Bjorken (longitudinal) expansion of the system. $i(j)$ stands for u or d (we ignore s and other heavier flavours). $<\vec{v}\cdot\hat{r}>$ (the average relative velocity in the radial direction) is calculated using the Jüttner distribution,

$$f(x,p) = e^{-\beta p \cdot u(x)} \quad (12)$$

There would also be a corresponding rate equation for the antiquarks, which looks exactly like eq. (11) and hence not explicitly written. In eq. (11) the $\Gamma^{total}_{g \to q\bar{q}}$ as well as $\Gamma^{total}_{gg \to q\bar{q}}$ stand for the corresponding net quantities.

As mentioned earlier, we are considering a non-equilibrated quark matter and hence the pions formed will also be out of equilibrium. This is taken into account by multiplying the relevant distribution functions with the ratios $r_q = n_q/n_{eq}$ and $r_\pi = n_\pi/n_{e\pi}$ where n_q and n_{eq} are non-equilibrium and equilibrium densities of quarks and n_π and $n_{e\pi}$ are non-equilibrium and equilibrium densities for pions.

In all these expressions, the appropriate masses are the effective masses including the current as well as thermal contribution, , whose importance in determining the dynamics of the hot quark matter has been well established. For quarks (antiquarks), this is

$$m_{eff} = \sqrt{m_q(curr)^2 + m_q(thermal)^2}$$

where [20,21]

$$m_q^2(thermal) = (1 + \frac{r_q}{2})(\frac{g_s T}{3})^2 \quad (13)$$

and $m_q(curr)$ is taken to be 10 MeV. For gluons the thermal mass is,

$$m_g(thermal) = \frac{2}{3} g_s T \quad (14)$$

The running coupling constant α_s as a function of temperature is given by [22]

$$\alpha_s = \frac{12\pi}{(33-2n_f)ln\left[\frac{\bar{Q}^2}{\Lambda^2}\right]} \qquad (15)$$

with $\bar{Q}^2 = m_{eff}^2(T) + 9T^2$.

Simultaneously, we must take account of energy momentum conservation which, for a Bjorken flow, corresponds to the following equation

$$\frac{\partial \epsilon}{\partial t} = -\frac{\epsilon + P}{t} \qquad (16)$$

where $\epsilon \equiv \epsilon_{total} = \epsilon_g + \epsilon_q + \epsilon_\pi$. We also include the one loop correction to ϵ_g [23]. ϵ and P are related through the velocity of sound, as in [2]. For a complete description of the system, eqs. (9), (10), (11) and (16) must be solved self-consistently. The initial conditions are taken from [2] for RHIC energies. The initial time (t_g) is the time when gluons thermalise (=0.3 fm), where $r_q = 0.15$, $r_g=1$ and r_π is taken to be 0. The temperature at this time is 500 MeV. Note that we are working at $y = 0$ so that t and τ are the same and the baryon chemical potential is zero.

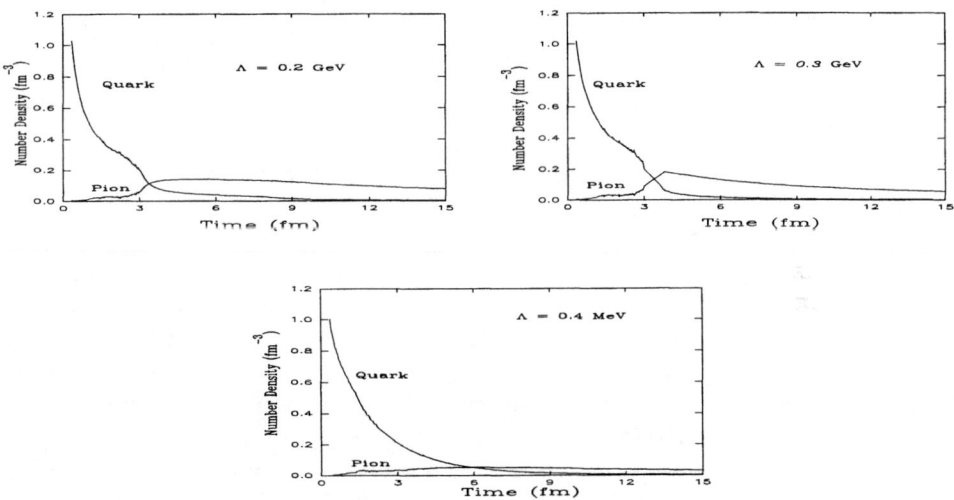

Figure 3 : Time evolution of quark and pion number densities for various values of QCD parameter Λ.

Figure 3 shows the variation of pion and quark number densities with time, for various values of the QCD parameter Λ. In all three cases, we find the same qualitative feature that pions start appearing in the system quite early on but they become appreciable in number only after some time. At late times the system is dominated by pions. This cross over occurs at $t \geq 4$ fm for $\Lambda = 0.2$ or 0.3 GeV while for $\Lambda= 0.4$ GeV this happens at $t \sim 6$ fm.

Figure 4 shows the variation of temperature with time. Obviously, there is a dramatic effect of the QCD parameter Λ. In all the cases, there is a change at $T \sim 215$ MeV, corresponding to $t \sim 3.5$ fm; the variation of temperature with time becomes slower, as is expected in the mixed phase of a first order phase transition. At $\Lambda=0.2$ GeV, this occurs for a very short period of time, before the system starts cooling again. The duration of the constant temperature configuration increases with Λ, and for $\Lambda=0.4$ GeV, it persist upto 9 fm before the temperature of the system starts falling again.

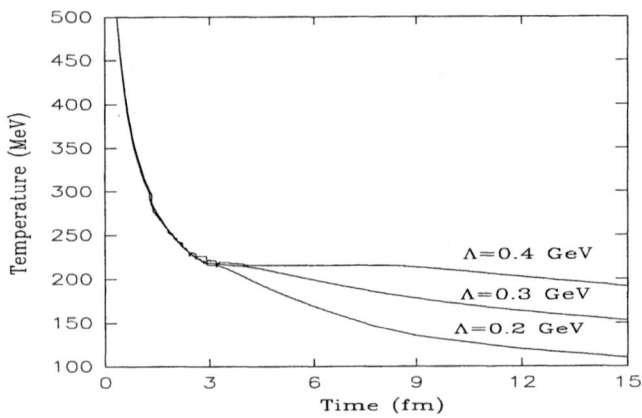

Figure 4 : Time evolution of the Temperature for various values of QCD parameter Λ.

Obviously, this is a clear indication of an *apparent* first order transition. Microscopically, the appearance of the mixed phase at a temperature of ~ 215 MeV can be understood from the fact that the pion decay width goes to zero at such a temperature. All the pions that were formed earlier in the system tended to decay back to quarks and antiquarks on a fast time scale. Only after the pion decay width becomes small would the formed pions become stable.

The apparently desirable features of the dynamical deconfining transition seem to arise quite naturally in this scenario. It is therefore important to test if all the conservation laws are obeyed during the evolution. To this end, let us check if entropy increases steadily during the entire evolution.

The decrease in entropy occurs precisely where the gluons drop out of the system. This, in hindsight, was to be expected, since the sudden decrease in the number of degrees of freedom in the system should lead to a decrease in entropy. The NJL model, in the present form, does not account for the gluonic degrees of freedom. It thus appears to us that in order to obtain an effective low energy lagrangian from QCD capable of describing the dynamical evolution of the quark-gluon system, the role of the gluons must be taken into account. Incorporation of the dilaton field to account for the QCD trace anomaly in the NJL model [24] is indeed a promising

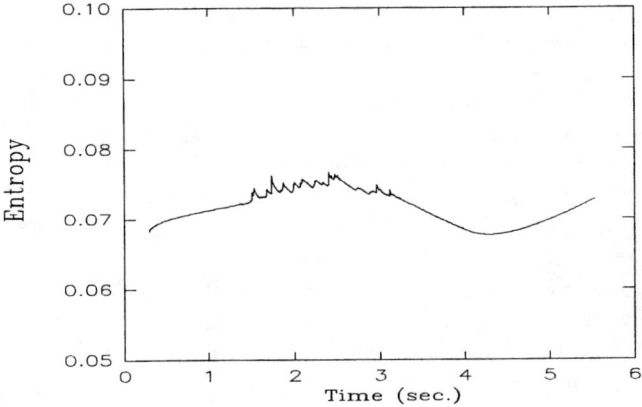

Figure 5 : Time evolution of entropy with time ($\Lambda = 0.3$ GeV).

step in this respect, but the entropy carried by the gluons at high temperature should also be accommodated. Such a study is on our current agenda.

It is a great pleasure to thank the organisers of the workshop, and especially Prof. J. da Providencia, for the kind invitation and their warm hospitality. This talk is based on a continuing collaboration the author has had with Jan-e Alam, Abhijit Bhattacharyya, Sanjay Kumar Ghosh and Bikash Sinha over the past several years. I take this opportunity to thank them all.

REFERENCES

1. Shuryak, E. V., *Phys. Rep.* **61**, 71 (1980).
2. Shuryak, E.V., *Phys. Rev. Lett.* **68**, 3270 (1992); Chakraborty, S., Raha, S. and Sinha, B., *Mod. Phys. Lett.* **A7**, 927 (1992); Alam, J., Raha, S. and Sinha, B., *Phys. Rev. Lett.* **73**, 1895 (1994); Biro, T.S., Müller, B., Thoma, M. H. and Wang, X.N., *Nucl. Phys.* **A566**, 543c (1994); Alam, J. Roy, P., Sarkar, S., Raha, S. and Sinha, B., *Int. J. Mod. Phys.* **A12**, 5151 (1997).
3. Plümer, M., Raha, S., and Weiner, R.M., *Phys. Lett.* **B139**, 198 (1984); *Nucl. Phys.* **A418**, 549c (1984).
4. Born, K. et al., *Phys. Rev. Lett* **67**, 302 (1991).
5. A. Gocksh, *Phys. Rev. Lett.* **67**, 1701 (1991).
6. Boyd, G., Gupta, S., Karsch, F. and Laermann, E., *Z. Phys.* **C64**, 331 (1994).
7. Shuryak, E.V. and Schaefer, T., *Nucl. Phys.* **B53**, 472 (1997).
8. Nambu, Y and Jona-Lasinio, G., *Phys. Rev.* **122**, 345 (1961).
9. Kunihiro, T. and Hatsuda, T., *Phys. Lett* **B206**, 385 (1988); *Phys. Rep.* **247**, 221 (1994).
10. Bernard, V., Jaffe, R. L. and Meissner, U - G., *Nucl. Phys.* **B308**, 753 (1988).
11. Bhattacharyya, A., Ghosh, S.K. and Raha, S., *Mod. Phys. Lett.* **A14**, 621 (1999).

12. Frederico, T. and Miller, G. A., *Phys. Rev.* **D45**, 4207 (1992).
13. Rehberg, P., Klevansky, S. P. and Hüfner, J., *Phys. Rev.* **C53**, 410 (1996).
14. Werner, K., *Phys. Rep.* **232**, 87 (1993).
15. Barz, H.W., Friman, B. L., Knoll, J. and Schulz, H., *Nucl. Phys.* **A519**, 831 (1990); Biró, T. S., Lévai, P. and Zimányi, J., *Phys. Lett.* **B347**, 6 (1995).
16. Geiger, K. and Ellis, J., *Nucl. Phys.* **A590**, 609c (1995)
17. Ghosh, S. K., Talk at 1997 Yukawa International Seminar on Non-perturbative QCD, Kyoto, Japan, December 1997.
18. Chandrasekhar, S., *Rev. Mod. Phys.* **15**, 1 (1943).
19. Florkowski, W. and Friman, B. L., *Acta Phys. Pol.* **25**, 49 (1994).
20. Altherr, T. and Seibert, D., *Phys. Rev.* **C49**, 1684 (1994).
21. Traxler, C. T. and Thoma, M. H., *Phys. Rev.* **C53**, 1348 (1996).
22. Geiger, K.,*Phys. Rev.* **D46**, 4965 (1992).
23. Plümer, M., Raha, S. and Weiner, R. M., *Phys. Lett.* **B139**, 198 (1984); *Nucl. Phys.* **A418**, 549c (1984).
24. Jaminon, M., *This volume*.

Interplay between Kaons and Kaon Like Excitations of the Medium

Maria C. Ruivo

Departamento de Fsica da Universidade de Coimbra[1]
P - 3004 - 516 Coimbra, Portugal

Abstract. The behavior of kaons in non strange quark matter, at zero and finite temperature, is investigated within two Nambu-Jona-Lasinio type models: one without vector pseudo vector interaction (I) and the other including this interaction (II). At zero temperature, the phase transition in model I is first order and a low energy collective excitation of the Fermi sea with kaon quantum numbers occurs; the vector pseudo vector interaction makes the phase transition a crossover one and inhibits the appearance of the low energy mode. It is found that the net effect of a hot and dense medium on the kaonic behavior is a slight reduction of the usual splitting between K^+ and K^- masses, which remains nevertheless meaningful. The low energy mode in model I occurs only at very low temperatures.

INTRODUCTION

It is believed that at high density and /or temperature the QCD vacuum is described by a weakly interacting gas of quarks and gluons, the quark-gluon plasma. The system is therefore supposed to undergo a phase transition at critical values of the density, ρ_c, and / or temperature, T_c. Restoration of chiral symmetry and deconfinement are then expected to occur. Lattice simulations [1] and different model calculations support this believe. The nature of the phase transition and the values of the critical parameters are current issues nowadays. Besides the relevance of these studies for astrophysics, they are as well an important direction of research in nuclear and particle physics. The question of how modifications of hadron properties might occur in hot and dense media is attracting a lot of attention, since, among other information, these modifications may carry a signal for the proximity of the phase transition. In particular, the study of possible modifications of pseudoscalar mesons, such as kaons, is extremely interesting, having in mind its role on the strong interaction and the issue of kaon condensation. In the study of the effects

---

[1] Work supported in part by projects PRAXIS/PCEX/C/13/96, PRAXIS/2/2.1/FIS/451/94 and PRAXIS/P/FIS/12247/98, FCT, Portugal

of the medium on hadronic behavior, one should have in mind that the medium is a complex system, where a great variety of medium particle-hole excitations occur, some of them with the same quantum numbers of the hadrons under study; the interplay of all these excitations might play a significant role in the modifications of hadron properties.

Most theoretical approaches dealing with kaons in flavor asymmetric media, since the pioneering work of Kaplan and Nelson [2], predict a slight raising of the K^+ mass and a more pronounced lowering of the K^- mass [3–7], a conclusion which is supported by the analysis of data on kaonic atoms [8]. Experimental results at GSI seem to be compatible with this scenario [9–11]. In fact, K^+ and K^- yields are comparable at the same available energy, in spite of the K^- production being suppressed in pp collisions by an order of magnitude [10]. Recent data on kaon production show that the K^-/K^+ rate is larger in Ni+Ni collisions than in nucleon-nucleon collisions, which indicate an enhancement of K^- production in the medium; in C+C collisions, the spectral slopes of K^- were found to be steeper than those of K^+ [11]. It is clear that kaons and antikaons behave differently in the medium and a decrease of the antikaon mass is compatible with the data.

From the theoretical point of view, the driving mechanism for the mass splitting between kaons and anti-kaons is attributed mainly to the selective effects of the Pauli principle, the interaction of K^- with the $\Lambda(1405)$ resonance playing an important role as well. In this talk I will report a work carried out in our group on the effects of flavor asymmetric media on the behavior of pseudoscalar mesons, specially the kaons, within Nambu-Jona-Lasinio type models [12,13,5]. NJL type models have already been used to study the behavior of kaons in media, at zero chemical potential [14] or zero temperature and asymmetric matter [3]. I will talk on the behavior of kaons in quark matter simulating symmetric nuclear matter: $\rho_u = \rho_d$, $\rho_s = 0$, at zero and finite temperature. The work addresses the following points: nature of the phase transition, role played by kaon-like excitations of the Fermi sea and influence of vector-pseudovector interaction on the behavior of the kaonic excitations. The combined effect of density and temperature will finally be discussed.

MODEL AND FORMALISM

We work in a flavor $SU(3)$ NJL type model with scalar-pseudoscalar and vector-isovector pieces, and a determinantal term, the 't Hooft interaction, which breaks the $U_A(1)$ symmetry. We use the following Lagrangian:

$$\begin{aligned}\mathcal{L} &= \bar{q}(i\gamma^\mu \partial_\mu - \hat{m})q + \tfrac{1}{2}g_S \sum_{a=0}^{8}[(\bar{q}\lambda^a q)^2 + (\bar{q}i\gamma_5 \lambda^a q)^2] \\ &\quad - \tfrac{1}{2}g_V \sum_{a=0}^{8}[(\bar{q}\gamma_\mu \lambda^a q)^2 + (\bar{q}\gamma_\mu \gamma_5 \lambda^a q)^2] \\ &\quad + g_D \{\det[\bar{q}(1+\gamma_5)q] + \det[\bar{q}(1-\gamma_5)q]\}\end{aligned} \qquad (1)$$

In order to discuss the effects of the vector-pseudovector interaction we consider the cases: $g_V = 0$ (model I) and $g_V \neq 0$ (model II). The model parameters, which are the bare quark masses $m_d = m_u, m_s$, the coupling constants g_S, g_V and g_D and the cutoff in three-momentum space, Λ, were fitted to the values in vacuum for the pion mass, $m_\pi = 138$ MeV, the pion decay constant $f_\pi = 93$ MeV, the kaon mass, $m_K = 493.5$ MeV, the quark condensates, $<\bar{u}u> = <\bar{d}d> = -(246.8\text{MeV})^3$, $<\bar{s}s>/<\bar{u}u> = 0.80$. The resulting parameter set becomes: model I: $\Lambda = 631.4$ MeV, $g\Lambda^2 = 2.2$, $m_u = m_d = 5.5$ MeV and $m_s = 132.9$ MeV; model II: $\Lambda = 800$ MeV, $g\Lambda^2 = 2.015$, $m_u = m_d = 4$ MeV and $m_s = 105.3$ MeV.

The model calculations will be presented within the formalism of the generating functional; the bosonization performed leads to results identical to those obtained within the mean field approach based on a Hamiltonian formalism that we used in previous works.

The 't Hooft term is a six quark interaction and some work has to be done in order to write in a form suitable to use the bosonization procedure. It can be shown the this term, $\mathcal{L}_D = g_D\, det\,\{[\bar{q}_i\,(1+\gamma_5)\,q_j] + det\,[\bar{q}_i\,(1-\gamma_5)\,q_j]\}$ can be written in the form (see [15–17]):

$$\mathcal{L}_D = \frac{1}{6}g_D\, D_{abc}\,(\bar{q}\lambda^c q)\,[(\bar{q}\lambda^a q)(\bar{q}\lambda^b q) - 3(\bar{q}i\gamma_5\lambda^a q)(\bar{q}i\gamma_5\lambda^b q)] \qquad (2)$$

with: $D_{abc} = d_{abc}, a,b,c \in \{1,2,..8\}$, (structure constants of SU(3)), $D_{000} = \sqrt{\frac{2}{3}}$, $D_{0ab} = -\sqrt{\frac{1}{6}}\delta_{ab}$

The usual procedure to obtain a four quark effective interaction from this six quark interaction is to contract one bilinear $(\bar{q}\lambda_a q)$. Then, from the two previous equations, an effective Lagrangian is obtained:

$$\begin{aligned} L_{eff} =\; & \bar{q}(i\gamma^\mu \partial_\mu - \hat{m})q \\ & + S_{ab}[(\bar{q}\lambda^a q)(\bar{q}\lambda^b q)] + P_{ab}[(\bar{q}i\gamma_5\lambda^a q)(\bar{q}i\gamma_5\lambda^b q)] \\ & - \frac{1}{2}g_V \sum_{a=0}^{8}[(\bar{q}\gamma_\mu\lambda^a q)^2 + (\bar{q}\gamma_\mu\gamma_5\lambda^a q)^2] \end{aligned} \qquad (3)$$

where:

$$\begin{aligned} S_{ab} &= g_S\,\delta_{ab} + g_D D_{abc} <\bar{q}\lambda^c q> \\ P_{ab} &= g_S\,\delta_{ab} - g_D D_{abc} <\bar{q}\lambda^c q> \end{aligned} \qquad (4)$$

By using the usual methods of bosonization one gets the following effective action:

$$\begin{aligned} I_{eff} =\; & -iTr\,\ln(i\partial_\mu\gamma_\mu - \hat{m} + \sigma_a\lambda^a + i\gamma_5\phi_a\lambda^a + \gamma^\mu V_\mu + \gamma_5\gamma^\mu A_\mu) \\ & - \frac{1}{2}(\sigma_a S_{ab}^{-1}\sigma_b + \phi_a P_{ab}^{-1}\phi_b) \\ & + \frac{1}{2G_V}(V_\mu^{a\,2} + A_\mu^{a\,2}), \end{aligned} \qquad (5)$$

from which we obtain the gap equations and meson propagators.

In order to introduce density and temperature effects, we use the thermal Green functions, which for a system of quarks q_i, at temperature T and chemical potential μ_i, read:

$$S(p,T,\mu_i) = (\gamma_\mu p_\mu + M_i) [\frac{1}{p^2 - M_i^2 + i\varepsilon} + 2\pi i \delta(p^2 - M_i^2)[\theta(p^0) n_i^+ + \theta(-p^0) n_i^-]] \qquad (6)$$

where:

$$n_i^{(\mp)} = [1 + \exp(\beta(E_i \mp \Delta E_i \pm \mu_i))]^{-1}$$

are the Fermi distribution functions, with $\beta = 1/T$, M_i the constituent quark mass, $E_i = (p^2 + M_i^2)^{1/2}$ and ΔE_i the energy gap introduced by the vector-pseudovector interaction [22]. The following gap equations are obtained:

$$M_i = m_i - 2 g_S < \bar{q}_i q_i > - 2 g_D < \bar{q}_j q_j >< \bar{q}_k q_k > \qquad (7)$$

$$\Delta E_i = 2 g_V < q_i^+ q_i > \qquad (8)$$

with i,j,k cyclic and $< \bar{q}_i q_i >$, $< q_i^+ q_i >$ are respectively the quark condensates and the quark densities.

The condition for the existence the poles in the propagators of kaons leads to the following dispersion relation:

$$(1 - K_P J_{PP})(1 - K_A J_{AA}) - K_P K_A J^2_{PA} = 0 \qquad (9)$$

with:

$$\omega J_{PA} = (M_u + M_s) J_{PP} + 2 < \bar{u}u + \bar{s}s > .$$

$$\omega J_{AA} = (M_u + M_s) J_{PA} + 2 < u^+u - s^+s > .$$

$$J_{PP} = 2 N_c \int \frac{d^3p}{(2\pi)^3} \left\{ \frac{M_u(M_s - M_u) - \omega E_u}{(E_s^2 - (\omega + E_u)^2)E_u} \tanh\frac{\beta(E_u + \bar{\mu}_u)}{2} + \right.$$

$$\left. \frac{M_u(M_s - M_u) + \omega E_u}{(E_s^2 - (\omega - E_u)^2)E_u} \tanh\frac{\beta(E_u - \bar{\mu}_u)}{2} + s \to u, \omega \to -\omega \right\} \qquad (10)$$

with $K_P = g_S + g_D < \bar{d}d >$ and $K_A = -g_V$, $\bar{\mu}_u = \mu_u - \Delta E_u$, $\omega = \pm m_{K^\pm} - (\Delta E_u - \Delta E_s)$ for $K^\pm$

Similar expressions are obtained for pions in asymmetric matter, by replacing $s \leftrightarrow d$ [13].

DISCUSSION OF THE RESULTS

A Phase transitions with chemical potential and temperature

The nature of the phase transition in NJL models has been discussed by several authors [19,22,20,21]. We reanalyze this problem here in order to establish a connection between the vacuum state and its excitations. We calculate therefore the pressure $P(\rho,T), = -[\Omega(\rho,T) - (\Omega(\rho=0,T)]$ were $\Omega(\rho,T)$ is the thermodynamical potential. We conclude (see Fig. 1) that model I exhibits a first order phase transition for $T \leq T_c \simeq 50$ MeV. The problem with this version of NJL model is that the vacuum becomes unstable at very low densities; at $T = 0$ there is range of densities $(0.1\rho_0 \leq \rho \leq 2\rho_0)$ were the system is in a mixed phase. The quark masses and excitation energies of this vacuum do not vary continously with density and within that region one can only talk about average values. Work in order to overcome this problem is in progress; meanwhile, we think of interest to discuss the behavior of kaonic excitations in this model. Model II does not have this problem because, we have a crossover phase transitions and the vacuum is stable.

Comparison of kaonic behavior in model I and II at $T = 0$

Let us first analyze the results obtained at zero temperature (see Figs. 2 and 3.). In both models we observe a splitting between kaons and antikaons as expected,

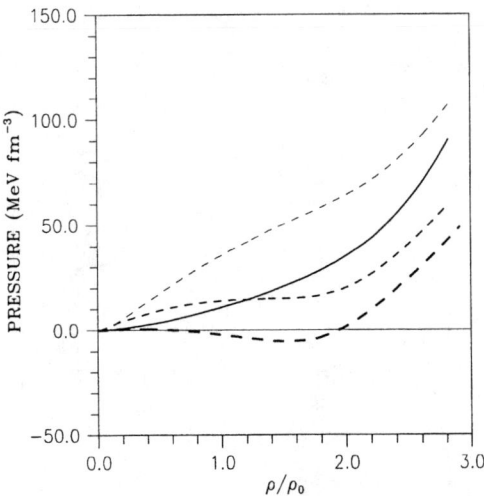

FIGURE 1. Pressure as a function of baryonic density: for model I (dashed curves) at $T = 0, 50, 100$MeV (thick, medium thick and thin curves); for model II at $T = 0$MeV (full curve).

the driving mechanism of this behavior being the Pauli blocking; however, the behavior of K^- is different in the two models. In model I we observe that the K^- mass decreases moderately up to around the normal nuclear density and afterwards increases until enters into the continuum. Moreover, a new low-lying branch with K^- quantum numbers occurs around a density $0.6\rho_0$. This excitation has a different origin from the K^+ and the upper K^-, which are excitations of the Dirac vacuum; the low energy mode is a particle-hole excitation of the Fermi sea. It is a typical effect in a many-body system that the Fermi sea continuum of excitations induces attraction below its boundaries and repulsion above, which explains the increase of the K^- mass above a certain density and the occurence of the low energy mode A similar effect is found for pions in neutron matter were a soft low energy π^+ -like mode occurs [13,23].

The sum rules are a very important tool to analyze the collectivity and relative importance of the modes [12,13,23,5]. One can derive a generalization of the PCAC relation in the medium from the Energy Weighted Sum Rule (EWSR), well known from Many Body Theories. For the mesonic state $|r>$ with energy ω_r associated with the transition operator Γ the strength function $F_r = \omega_r |<r|\Gamma|0>|^2$ satisfies the EWSR which reads

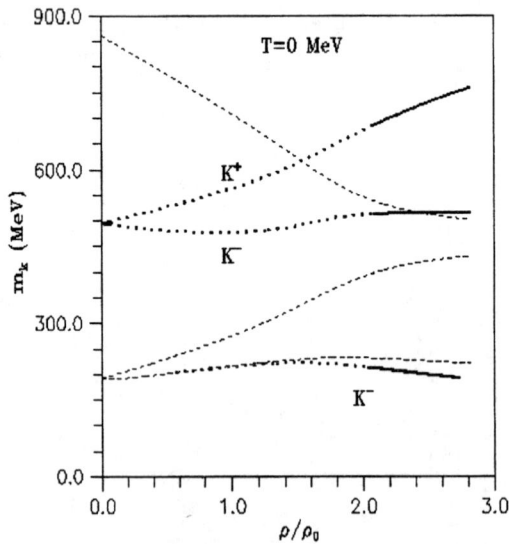

FIGURE 2. Kaon masses as a function of the baryonic density in model I. The lower curve refers to the $s\bar{u}$ low-lying mode starting at $\simeq 0.6\rho_0$. The dotted region coincides with the mixed phase. The two lower dashed curves represent the boundaries of the Fermi sea continuum and the upper dashed curve represents the threshold $M_u + M_s$.

$$m_1 = \sum_r \omega_r |<r|\Gamma|0>|^2 = \frac{1}{2} <\Phi_0|[\Gamma,[H,\Gamma]]|\Phi_0>, \qquad (11)$$

the transition operator being defined in the present case by $\Gamma = \Gamma_+ + \Gamma_-$, with $\Gamma_\pm = \gamma_5(\lambda_4 \pm i\,\lambda_5)/\sqrt{2}$. We obtain therfore a relation that can be written in terms of the decay constants of kaons:

$$\sum_\alpha m_{K,\alpha}^2 f_{K,\alpha}^2 \simeq -(m_u + m_s)[<\overline{\Psi}_u\Psi_u> + <\overline{\Psi}_s\Psi_s>]. \qquad (12)$$

As the hadronic quantities are calculated at finite temperature and density this equation provides a generalization of the PCAC relation, and allows to determine the percentage of the total strength located in the kaonic modes.

The sum rule would be exactly fulfilled if all the modes, including the continuum modes, were taken into account, as expressed by the above relation. However, the pseudoscalar bound state solutions are highly collective and this sum rule is almost exactly satisfied in the vacuum, neglecting the continuum modes. We verified that in the medium the degree of satisfaction of the sum rule is good, provided, naturally, that all the bound state solutions are considered. The strength associated to the low energy mode can not be neglected as the density increases; it becames of the same order of magnitude as the upper K^- around twice the normal nuclear matter density [12,13].

Other authors found modes with a similar origin [18,4], which were interpreted as hyperon effects. It is well known that the hyperons located below the $\overline{K} - N$ threshold are of particular importance for the $\overline{K} - N$ dynamics. Since NJL type models do not have bound nucleons and hyperons, an extrapolation from a quark

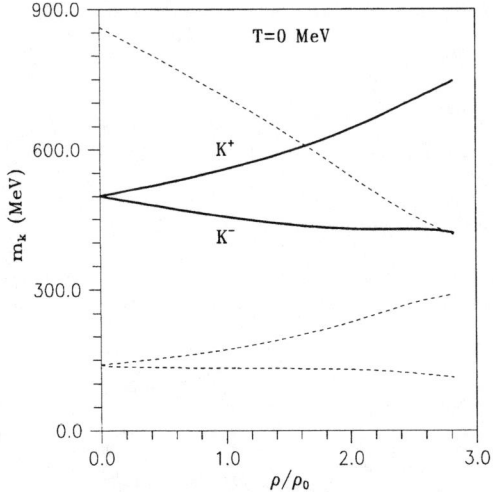

FIGURE 3. Kaon masses as a function of baryonic density at $T = 0$ MeV for model II; The dashed lines have a similar meaning of those of Fig. 2.

system to a baryon system should be made. From this point of view this low-lying mode (an s-particle-u-hole excitation) may be interpreted as as simulating the effects of a $\Lambda(1116)$-particle-proton-hole state. This interpretation is plausible having in mind parity analysis and the excitation energy of the mode ($\simeq 200$ MeV at $\rho \simeq 0.6\rho_0$). However, the hyperon $\Lambda(1405)$, that is supposed to play a very important role in the K^- behavior, was not found in the present approach.

The results for model II plotted in Fig. 3 show that the vector-pseudovector interaction term partially cancels some Fermi sea effects, leading to a moderate decrease of the mass of the K^- and inhibiting the occurrence of the low-lying s-particle-u-hole mode. In fact, the vector-pseudovector interaction narrows the Fermi sea continuum of $s\bar{u}$ excitations, therefore reducing the repulsive effect above its boundaries and the attractive effect below [5]. So, there is not enough energy gain to decouple the low energy mode from the continuum. As already pointed out, within this model the phase transition is crossover. There is probably a connection between the appearance of the Fermi sea excitations and the order of the phase transition.

Combined effect of density and temperature

One first conclusion about the combined effect of density and temperature is that the low energy mode disappears in model I above very low temperatures. As the temperature increase, the behavior of kaons and antikaons in both models is qualitatively similar, so we plot only the results for model II in Fig.4. The dominant

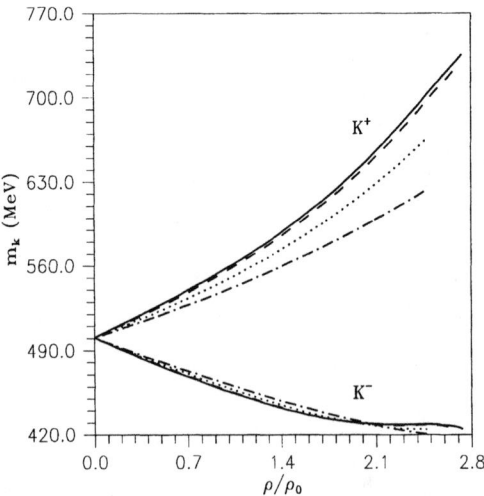

FIGURE 4. Kaon masses as a function of the baryonic density in model II: $T = 0$MeV (full curves), $T = 20$MeV (dashed curves), $T = 50$MeV (dotted curves) and $T = 100$MeV (dashed-dotted curves)

effect is the reduction of the splitting between K^+, K^- masses (see fig 2.) however, at $T = 100$ MeV, $\rho \simeq 2.5\rho_0$ an appreciable splitting, around $200 MeV$, is still found and the K^- has a more pronounced decrease with density. This means that, under the combined effect of temperature and density, the K^- are still easier to be produced in medium, which is compatible with experimental results that indicate an enhancement of the K^- production in the medium.

REFERENCES

1. Kanaya, K., *Prog. Theor. Phys. Sup.* **129**, 197 1997.
2. Kaplan, D. B. and Nelson, A. E., *Phys. Lett.* **B 175**, 57 (1986). ; **B 192**, 193(1987).
3. Lutz, M., Steiner, A. and Weise, W., *Nucl. Phys.* **A 574**. 755 (1994).
4. Waas, T., Rho, M. and Weise,W., *Nucl. Phys.*, **A 617**, 449 (1997); Lutz, M.*Phys. Lett.* **B 426**, 12 (1998).
5. Ruivo, M. C., Sousa C. A. and Providência C.,*Nucl. Phys.* **A651**, 59 1999.
6. Cassing, W. and Bratkovskaya, E. L., *Phys. Rep.* **308**, 65 1999.
7. Schaffner,J. et al *Phys. Lett.* **B 334**, 268 (1994).
8. Friedmann, E., Gal A. and Batty, C. J., *Nucl. Phys.* **A 579**, 518 (1994).
9. Schröter,A. H. et al, *Z. Phys.* **A 350**, 101 (1994).
10. Herrman, H. (FOPI Collaboration), *Nucl. Phys.* **A 610**, 49c (1996).
11. Barth, R. et al (KaoS Collaboration), *Phys. Rev. Lett.* **78**, 4007 1997; **82**, 1640, 1999.
12. Ruivo, M. C. and Sousa C. A., *Phys. Lett.***B385**, 39 1996.
13. Sousa C. A. and Ruivo M. C.,*Nucl. Phys.***A625**, 713 (1997).
14. Hatsuda, T. and Kunihiro, T., *Phys. Rep.* **247**, 221 (1994).
15. Vogl, H. and Weise W., *Prog. Part. Nucl. Phys*, **27**, 195 (1991).
16. Ripka, G. *Quarks Bound by Chiral Fields*, Oxford: Clarendon Press, 1997 pp. 14-17.
17. Volkov, M. K. and Yudichev, V. L., *hep-ph/9906371* (1999).
18. Yabu, H., Nakamura, S.Myhrer F. and Kubodera, K. *Phys. Lett.* **B 315**, 17 (1993).
19. Asakawa, M. and Yazaki, K., *Nucl. Phys.* **A 504**, 668 (1989).
20. Klevansky, S. P., *Rev. Mod. Phys.* **64**,649 (1992).
21. Cugnon, J., Jaminon, M. and Van den Bossche, B., *Nucl. Phys.* **A 598**,515 (1996).
22. Klimt, S. Lutz, M. and Weise W., *Phys. Lett.* **B 249**, 386 (1990).
23. Broniowski, W and Hiller, B., *Nucl. Phys.*, **A 643**, 161 (1998).

ρ-Mass Modification in Dense Medium Some Recent Studies in Finite Nucleus

Abhijit Bhattacharyya

Variable Energy Cyclotron Centre,
1/AF, Bidhannagar, Calcutta 700 064, INDIA

Abstract. The modification of the ρ-meson mass inside a finite nucleus will be discussed from an effective model of nuclear matter. Recent experiments have shown that there is a substantial decrease in the ρ-meson mass inside the He^3 nucleus. Some authors have claimed that this is a signature of the partial restoration of chiral symmetry. I will discuss that even in the absence of chiral symmetry effective mean field nuclear matter models can explain these findings.

INTRODUCTION

The study of in-medium properties of hadrons is a field of high current interest for a long time. It is expected that in ultra-relativistic collisions of heavy ions a hot and dense hadronic and/or quark matter may be formed. In such an environment of hot and dense matter the modification of hadronic properties can indeed have important physical consequences. It has recently been observed that there is an enhancement of dilepton production in the low invariant mass domain in heavy ion collider experiments [1-3]. This interesting observation has triggered the speculation [4] that the effective ρ-meson mass in the nuclear medium is decreased. On the other hand, theoretical studies based on chiral symmetry restoration have led to the expectation that even at finite densities (at or above nuclear density) there may be a partial restoration of chiral symmetry, leading to the decrease of vector meson masses from their free values [5]. All these questions assume great importance also in the context of quark-gluon plasma searches in heavy ion collisions. In such a scenario the hadronic matter constitute the background to the sought-for signals and thus must be controlled to a high degree of accuracy before conclusive evidence for the quark-gluon plasma can be extracted from the experimental data [6].

The evidence of the ρ mass modification in dense medium, as concluded from the data obtained from the ultra-relativistic collisions of heavy ions, is indirect. Very recently, however, a direct measurement of the invariant ρ mass in photoproduction of ρ^0 on He^3 has been reported in the literature [7,8]. This experiment has found a substantial decrease in the ρ mass ($\delta m_{\rho_0} \sim 280 \pm 40$ MeV). It has been argued [7] that such large decrease cannot be explained by the mean field picture of nuclear matter [9]. These authors suggest that this should be taken as a signature of the (partial) restoration of chiral symmetry [10] in ground state nuclei. In this talk I will mainly discuss that such a conclusion that the dropping ρ-mass is a signature of partial restoration of chiral symmetry is premature. If one includes the interaction of ρ-meson with nucleons in a proper manner one can obtain a substantial drop in the ρ-mass. There is no need to invoke any discussion of the partial restoration of chiral symmetry for this purpose.

MODEL CALCULATION OF ρ-MASS IN DENSE MATTER

It is established now that QCD is *the theory* of the strongly interacting matter where quarks and gluons are the fundamental particles. However, one can not use it to describe hadronic matter. The non-perturbative features of QCD, which is not very well understood, prevents us from doing so. As a result one has to look for the effective models. There have been many models proposed so far to look at the in-medium properties of hadrons. One of the most popular mean field models for nuclear matter is the Walecka model [11], which was proposed first in 1974. The model then has been greatly modified over the years by a number of authors. (For a recent comparative study among the various versions and also with other models, see [12]). As far as the discussion of the nuclear matter goes the Walecka model gives very satisfactory results. For the purpose of the present work we concentrate our attention only to this model. The Lagrangian density for the Walecka model is given by [13]:

$$\mathcal{L} = \bar{\psi}(i\gamma_\mu\partial^\mu - m_n)\psi - \left[g_\omega\bar{\psi}\gamma_\mu\psi\omega^\mu - \frac{1}{4}G_{\mu\nu}G^{\mu\nu} + \frac{1}{2}m_\omega^2\omega_\mu\omega^\mu\right]$$
$$+ \frac{1}{2}(\partial_\mu\sigma\partial^\mu\sigma - m_\sigma^2\sigma^2) + g_\sigma\bar{\psi}\sigma\psi - \frac{1}{4}F_{\mu\nu}F^{\mu\nu}$$
$$+ \frac{1}{2}m_\rho^2\rho^2 + g_\rho\bar{\psi}\gamma^\mu\tau.\psi\rho_\mu + f_\rho\bar{\psi}\sigma^{\mu\nu}\tau.\psi\frac{\partial_\mu}{2m}\rho_\nu \quad (1)$$

In the above equation ψ, σ, ω^μ and ρ^μ are, respectively, the nucleon, the σ, the ω and the ρ meson fields; m_n, m_σ, m_ω and m_ρ are the corresponding masses; g_σ and g_ω are the couplings of the nucleon to σ and ω mesons, respectively; g_ρ and f_ρ are the vector and tensor couplings of rho meson; $G_{\mu\nu} = \partial_\mu\omega_\nu - \partial_\nu\omega_\mu$ and $F_{\mu\nu} = \partial_\mu\rho_\nu - \partial_\nu\rho_\mu + ig_\rho[\rho_\mu, \rho_\nu]$.

The procedure adopted to calculate the properties of the medium and heavy nuclei is as follows. One first writes down the field equations of all the mesons and baryons at the mean filed level. This results in a set of coupled differential equations. These coupled differential equations are then solved self consistently with some input from experimentally known quantities. As a result one gets different fields as a function of r and also the density profile. However, for a light nucleus, like He^3, the mean field approximation is not very reliable. In this work we have used a simple approach, a la Saito et al. [9], to calculate the effective ρ mass in helium. The density profile that we have used is a simple Gaussian form for the density distribution of He^3, in which the width parameter β_3 is fitted to reproduce the rms charge radius of He^3 i.e. 1.88 fm [9]. The density profile is given in figure 1. So once we know the density distribution of He^3, one can easily calculate the effective ρ mass, m_ρ^*, as a function of radius, all the fields being known as a function of baryon density.

It should me mentioned at this stage that the meson masses do not change at the men field level. One has to go beyond MF (one loop) level to look at the meson mass change. The effective mass, which is known as the pole mass, is calculated from the pole of the full propagator. The full propagator D is defined by the Dyson-Schwinger equation

$$D = D_0 + D_0 \Pi D \tag{2}$$

where D_0 is the free propagator and Π is the polarisation function. The solution of the above equation, for the zero three momentum, gives the equation for effective mass of ρ-meson. This is given by

$$m_\rho^{*2} = m_\rho^2 + \Pi_{vac} + \Pi_{med} \tag{3}$$

where m_ρ^* is the effective mass of the ρ-meson, m_ρ is the bare mass, Π_{med} is the medium part of the polarisation function and Π_{vac} is the zero density part which comes from the renormalisation of hte vacuum part. Now the medium and the vacuum part of the polarisation function are given by

$$\Pi_{med} = \sum_{B=n,p} \frac{8g_\rho^2}{\pi^2} \int_0^{k_{FB}} \frac{p^2 dp}{E_p \left(m_\rho^{*2} - 4E_p^2\right)}$$
$$\times \left[\frac{2}{3} \left(2p^2 + 3m_n^{*2}\right) + m_\rho^{*2} \left\{2m_n^* \left(\frac{c_\rho}{2m_n}\right)\right.\right.$$
$$\left.\left. - \frac{2}{3} \left(\frac{c_\rho}{2m_n}\right)^2 \left(p^2 + 3m_n^{*2}\right)\right\}\right] \tag{4}$$

$$\Pi_{vac} = \frac{g_\rho^2}{\pi^2} m_\rho^{*2} \left[I_1 + m_n^* \left(\frac{c_\rho}{2m_n}\right) I_2 \right.$$
$$\left. + \frac{1}{2} \left(\frac{c_\rho}{2m_n}\right)^2 \left(m_\rho^{*2} I_1 + m_n^{*2} I_2\right)\right] \tag{5}$$

$$I_1 = \int_0^1 dx\, x(1-x) \ln\left[\frac{m_n^{*2} - m_\rho^{*2} x(1-x)}{m_n^2 - m_\rho^2 x(1-x)}\right] \tag{6}$$

model	g_ρ	c_ρ
Bonn Potential [16]	2.63	6.1
QCD Sum rule [17]	2.5 ± 0.2	8.0 ± 2.0
Walecka Model	8.912	6.1

TABLE 1. Parameter values for different models.

$$I_2 = \int_0^1 ln\left[\frac{m_n^{*2} - m_\rho^{*2}x(1-x)}{m_n^2 - m_\rho^2 x(1-x)}\right] \quad (7)$$

In the above set of equations $c_\rho \equiv f_\rho/g_\rho$, k_{FB} is the Fermi momentum of baryon B (neutron or proton), m_n^* is the in-medium nucleon mass and m_n is the corresponding bare mass.

There are two coupling constants involved in the above set of equations. One is the vector coupling of the ρ-meson g_ρ and the other is the tensor coupling f_ρ (or equivalently c_ρ) [14]. This is where the difference between our approach and the earlier works arises; previous authors [9] neglected the tensor coupling of the ρ to the nucleon. We will come to the discussion of the effect of this tensor coupling later. We have used three sets of coupling constants, shown in table 1. The density dependence of the ρ-meson mass, for these three sets of parameters, has been shown in figure 2.

In order to compare the results for the effective ρ mass with the experimental values, we calculate the average mass of the ρ-meson in the He^3 nucleus. The average mass is defined as

$$\langle m_\rho^* \rangle = \frac{\int d^3 r m_\rho^*(r) \rho_B(r)}{\int d^3 r \rho_B(r)} \quad (8)$$

In table 2, we show the average ρ-mass for the different sets of parameters.

RESULTS

We have calculated the density variation of the ρ-meson mass inside He^3 nucleus using an effective model of hadrons. recently there have been two papers on the density variation of ρ-mass inside the He^3 nucleus. Both of them are the results from the ρ^0 photoproduction experiment of He^3. The first one is in the energy range $E_\gamma = 800 - 1120 MeV$ and the second for $E_\gamma = 380 - 700 MeV$. The first paper finds a drop in the ρ-mass of $160 \pm 35 MeV$, i.e. m_ρ^* is in the range $575 - 645 MeV$. The other study finds an effective ρ-mass in the range $450 - 530 MeV$.

Model	Average Mass (MeV)
Bonn Potential [16]	536
QCD Sum rule [17]	449 − 565
Walecka Model	304

TABLE 2. Average mass of ρ meson for different models.

In almost all the previous studies of the ρ-meson inside a light nucleus from the mean field approach, the tensor coupling of the ρ-meson to the nucleon [15,9] was not included, as already mentioned. As a result, the variation of ρ-mass was rather soft in all the previous cases. Here, the incorporation of the tensor coupling leads to a change in the ρ-meson mass which is much larger and we get results which are very close to the experimental findings. For example, the Bonn potential parameter set [16] yields $\langle m_\rho^* \rangle = 536 MeV$. On the other hand, for the QCD sum rule case [17], we get $\langle m_\rho^* \rangle = 449 - 565 MeV$. For the Walecka model parameter set, the value of $\langle m_\rho^* \rangle$ is somewhat lower.

On the basis of above observations we argue that the reduction of the effective mass of ρ- meson in He^3 need not be an unambiguous signal for the restoration of chiral symmetry, as suggested by the authors of ref. [8]. In particular, even the mean field model of nuclear matter is capable of accommodating such substantial changes in the effective ρ mass, if all the interactions are properly taken into account.

We would like to mention here that the present calculation is really an estimate of the behaviour of ρ meson mass inside a nucleus. To get a quantitative estimate and compare with the experiments mentioned above, one should do a full calculation of photoproduction processes. Such a non-trivial calculation is presently under study.

This work has been done in collaboration with Sanjay K. Ghosh and Sibaji Raha.

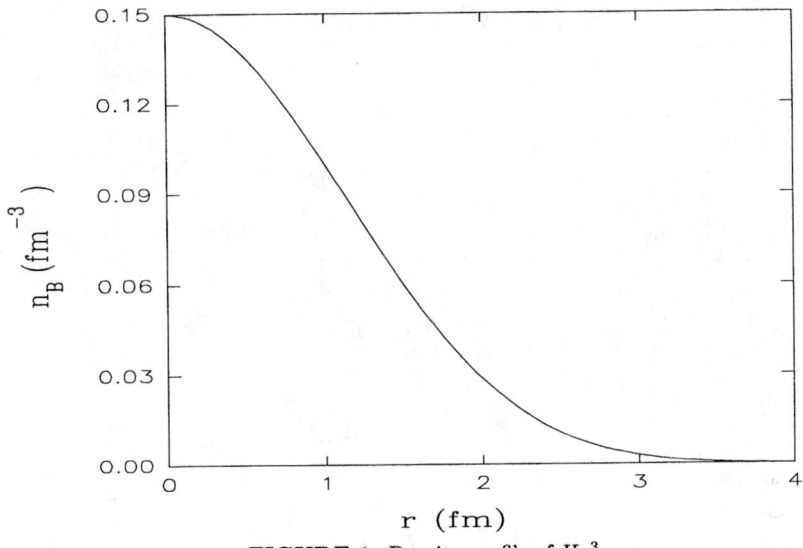

FIGURE 1. Density profile of He^3

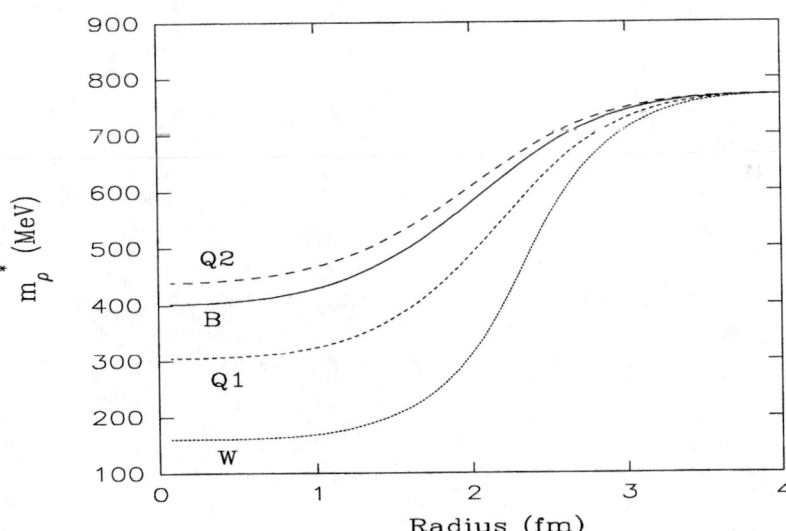

FIGURE 2. Density dependence of rho mass, Q1 and Q2 are the upper and lower mass limits for the QCD sum rule parameter set, B for Bonn potential parameters and W is for Walecka model parameters.

REFERENCES

1. CERES collaboration, Th. Ulrich et al., Nucl. Phys. **A610**, 313c (1996).
2. HELIOS Collaboration, M. Masera et al., Nucl. Phys. **A590**, 93c (1995).
3. NA50 collaboration, E. Scomparin et al., Nucl. Phys. **A610**, 331c (1996).
4. G. Q. Li, C. M. Ko and G. E. Brown, Phys. Rev. Lett. **75**, 4007 (1995).
5. G. E. Brown et al., Nucl. Phys. **A343**, 295 (1995).
6. J. Alam, S. Raha and B. Sinha, Phys. Rep. **273**, 243 (1996) and references therein.
7. G. J. Lolos et. al., Phys. Rev. Lett. **80**, 241 (1998).
8. G. M. Huber et. al., Phys. Rev. Lett. **80**, 5285 (1998).
9. K. Saito, A. W. Thomas and K. Tsushima, Phys. Rev. **C56**, 566 (1997).
10. G. E. Brown, M. Buballa and M. Rho, Nucl. Phys. **A609**, 519 (1996).
11. J.D.Walecka, Ann. Phys. (N.Y.) **83**, 491 (1974).
12. A. Bhattacharyya and S. K. Ghosh, Int. J. Mod. Phys. **7**, 495 (1998).
13. J. Piekarewicz and A. G. Williams, Phys. Rev. **C47**, 2462 (1993).
14. H. Shiomi and T. Hatsuda, Phys. Lett. **B334**, 281 (19994).
15. K. Saito, K. Tsushima and A. W. Thomas, Phys. Rev. **C55**, 2637 (1997).
16. R. Machleidt, K. Holinde and Ch. Elster, Phys. Rep. **149**, 1 (1987).
17. S-L. Zhu, Phys. Rev. **C59**, 435 (1999).

Magnetic Oscillations in the Nambu – Jona-Lasinio model

D. Ebert* and K.G. Klimenko[†]

*Theoretical Physics Division, CERN
CH - 1211 Geneva 23, Switzerland
and
Institut für Physik, Humboldt-Universität, D-10115 Berlin, Germany
[†]Institute for High Energy Physics, 142284 Protvino, Moscow Region, Russia

Abstract. The phase structure of a simple Nambu–Jona-Lasinio model has been investigated at non-zero values of μ and H, where H is an external magnetic field and μ is the chemical potential. On this basis magnetic oscillations effects were considered. It was shown that there are standard (periodic) van Alphen–de Haas magnetic oscillations of some thermodynamical quantities, including magnetization, pressure and particle density in the NJL system. Besides, we have found non-standard, i.e. non-periodic, magnetic oscillations, since the frequency of oscillations is a H-dependent quantity. Finally, there arises an oscillating behaviour not only for thermodynamical quantities, but also for a dynamical quantity like the quark mass.

Magnetic oscillations effects are well-known phenomena in condensed matter physics. In particular, the oscillation effect of the magnetization, which is called now the van Alphen–de Haas effect, was for the first time predicted by Landau and then experimentally observed in some non-relativistic systems (in metals) more than sixty years ago [1,2]. At present, a lot of the attention of researchers dealing with magnetic oscillations is focused on relativistic condensed matter systems (mainly on QED at nonzero values of the chemical potential μ and external magnetic field H), since the results of these studies may be applied to cosmology, astrophysics and high energy physics [3,4].

It was shown in the framework of QED that the thermodynamical potential $\Omega(\mu, H)$ of the system has in 1-loop approximation the following form $\Omega(\mu, H) = \Omega_{mon}(\mu, H) + \Omega_{osc}(\mu, H)$, where $\Omega_{mon}(\mu, H)$ is the monotonic part of $\Omega(\mu, H)$, and all magnetic oscillations are contained in the so-called oscillating part

$$\Omega_{osc}(\mu, H) = \sum_{k=1}^{\infty} [A_k(H)\cos(2\pi k\omega) + B_k(H)\sin(2\pi k\omega)], \qquad (1)$$

where $\omega = (\mu^2 - m^2)/(2eH)$ (e, m are electric charge and mass of fermions, respectively), and $A_k(H), B_k(H)$ are smoothly varying functions. Due to the presence of trigonometric functions, expression (1) obviously oscillates over the variable $(2eH)^{-1}$ with the frequency $(\mu^2 - m^2)$, which is not an H-dependent quantity. In condensed matter physics such kind of oscillations are usually called periodic ones.

In the present talk magnetic oscillation effects are considered in the framework of quantum field theory with four-fermion interactions

$$L = \sum_{k=1}^{N} \bar{q}_k i\hat{\partial} q_k + \frac{G}{2N} [(\sum_{k=1}^{N} \bar{q}_k q_k)^2 + (\sum_{k=1}^{N} \bar{q}_k i\gamma_5 q_k)^2], \quad (2)$$

which is the N-fermionic extension of the simplest Nambu – Jona-Lasinio model (NJL) [5].[1] Obviously, the model (2) is invariant under (global) $SU(N)$ and $U(1)_V$ transformations as well as continuous $U(1)_A$ chiral transformations: $q_k \to e^{i\theta\gamma_5} q_k$; $(k = 1, ..., N)$.

We shall find the thermodynamic potential $\Omega(\mu, H)$, which is related to the corresponding effective potential $V_{H\mu}(\Sigma)$ of the NJL system (2) by

$$\Omega(\mu, H) = V_{\mu H}(\Sigma) \big|_{\Sigma = \Sigma_{min}} \quad (3)$$

and contains all the information about thermodynamical quantities such as magnetization, particle density, etc. In the relation (3), one should first of all calculate the effective potential $V_{H\mu}(\Sigma)$. So, before considering the magnetic oscillations, we can study the vacuum properties of the NJL model.

Notice that special attention has been paid to the analysis of the vacuum structure of NJL-type models at non-zero temperature and chemical potential [6,7], in the presence of external (chromo-)magnetic fields [8–10], with allowance for curvature and non-trivial space-time topology [11,12]. The combined influence of external electromagnetic and gravitational fields on the dynamical chiral symmetry breaking (DCSB) effect in four-fermion field theories was investigated in [13,14]. However, the influence of both an external magnetic field H and chemical potential μ on the phase structure of the NJL model was not considered up to now.

Phase structure of the model. The necessary information about the phase structure of a given field theoretical model is contained in the global minimum point of the corresponding effective potential. In the presence of μ, H the effective potential $V_{H\mu}(\Sigma)$ of the NJL model has in leading order of large N the following form

$$V_{H\mu}(\Sigma) = V_H(\Sigma) - \frac{eH}{4\pi^2} \sum_{k=0}^{\infty} \alpha_k \theta(\mu - s_k) \left\{ \mu\sqrt{\mu^2 - s_k^2} - s_k^2 \ln\left[\frac{\mu + \sqrt{\mu^2 - s_k^2}}{s_k}\right] \right\}, \quad (4)$$

where $\alpha_k = 2 - \delta_{k0}$, $s_k = \sqrt{\Sigma^2 + 2eHk}$. $V_H(\Sigma)$ is the effective potential at $\mu = 0$, $H \neq 0$

$$V_H(\Sigma) = \frac{H^2}{2} + V_0(\Sigma) - \frac{(eH)^2}{2\pi^2}\left\{\zeta'(-1, x) - \frac{1}{2}[x^2 - x]\ln x + \frac{x^2}{4}\right\}, \quad (5)$$

---

[1] For simplicity, we consider in the following fermions ("quarks") of equal electric charge.

where $x = \Sigma^2/(2eH)$, $\zeta(\nu,x)$ is the generalized Riemann zeta-function, $\zeta'(-1,x) = d\zeta(\nu,x)/d\nu|_{\nu=-1}$, and

$$V_0(\Sigma) = \frac{\Sigma^2}{2G} - \frac{1}{16\pi^2}\left\{\Lambda^4 \ln\left(1+\frac{\Sigma^2}{\Lambda^2}\right) + \Lambda^2\Sigma^2 - \Sigma^4 \ln\left(1+\frac{\Lambda^2}{\Sigma^2}\right)\right\} \quad (6)$$

is the effective potential at $H, \mu = 0$. In (6) Λ is the ultraviolet cut off parameter. Finally, let us remark that Σ is an auxiliary scalar field, which, at the tree level, is proportional to $\bar{q}q$ by the equations of motion. The global minimum point of the potential (4) defines the vacuum expectation value of Σ and is equal to the dynamical quark mass.

At $\mu, H = 0$ and $G < G_c = 4\pi^2/\Lambda^2$ the global minimum point of $V_0(\Sigma)$ equals to the value $\Sigma = 0$. Hence, in this case quarks are massless and chiral symmetry remains intact. If $G > G_c$, the effective potential (6) has a nontrivial global minimum point, which we shall denote as M. (Evidently, M depends on the values of G and Λ [7].)

At $\mu = 0$, $H \neq 0$ the chiral symmetry of the model is spontaneously broken for arbitrary values of the bare coupling constant G. This is due to the fact, that the global minimum point $\Sigma_0(H)$ of the potential $V_H(\Sigma)$ is unequal to zero [8,10].

In order to study the properties of the NJL model vacuum for the general case, when both μ and H are nonzero, one should find all solutions of the stationarity equation

$$\frac{\partial}{\partial \Sigma}V_{H\mu}(\Sigma) = \frac{\partial}{\partial \Sigma}V_H(\Sigma) + \frac{2eH\Sigma}{4\pi^2}\sum_{k=0}^{\infty}\alpha_k\theta(\mu-s_k)\ln\left[\frac{\mu+\sqrt{\mu^2-s_k^2}}{s_k}\right] = 0 \quad (7)$$

and select that one, at which the potential $V_{H\mu}(\Sigma)$ takes its smallest value. This is the global minimum point for the function (4). The properties of this point as a function of μ and H give us a lot of information about the ground state. We omit here the detailed consideration of this procedure and present directly the phase structure description of the model (Figure 1).

In this figure, in the plane $(\mu, \sqrt{eH})$ the phase portrait of the model is qualitatively represented for the case $G_c < G < (1.225...)G_c$, where M is the quark mass at $\mu = H = 0$, $M_1 = (\Lambda^2/2 - 2\pi^2/G)^{1/2}$. Here one can see infinite sets of symmetric massless $A_0, A_1, ...$ phases, as well as massive phases $C_0, C_1, ...$ with DCSB. In addition, there is another massive phase B. Dashed and solid lines in Figure 1 are critical curves of first- and second-order phase transitions, respectively. One can also see on this phase portrait infinitely many tricritical points t_k, s_k ($k = 0, 1, 2, ...$) which lie on the boundary between massless and massive phases (chiral boundary). (A point of the phase diagram is called a tricritical one if, in an arbitrarily small vicinity of it, there are first- as well as second-order phase transitions.) Numerical investigation gives the following values of the external magnetic field corresponding to tricritical points t_0 and s_0 at different values of the bare coupling constant G: $eH_{t_0}/\Lambda^2 = 0.01...$; $0.08...$; $0.13...$ as well as $eH_{s_0}/\Lambda^2 = 0.006...$; $0.056...$; $0.103...$ for

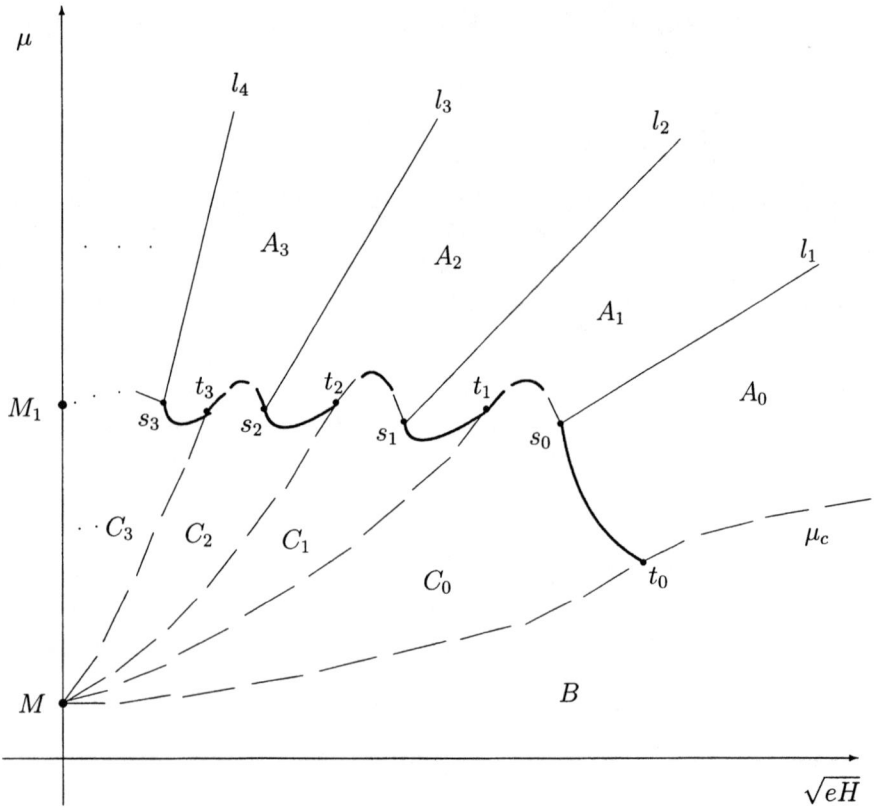

FIGURE 1. Phase portrait of the NJL model. (Detailed description is given in the text.)

G/G_c=1.01; 1.1; 1.2, respectively. We should also remark that the part $\widehat{t_0\mu_c(H)}$ of the chiral boundary is described by the equation $V_{H\mu}(0) = V_{H\mu}(\Sigma_0(H))$.

Points (μ, H) of the phase diagram, lying above the chiral boundary, correspond to the chirally symmetric ground state of the NJL model. One-fermion excitations of this vacuum have zero masses. At first sight, it might seem that the properties of this symmetric vacuum are slightly varied, when parameters μ and H are changed. However, this is not the case, and in this region, as was mentioned above, we have infinitely many massless symmetric phases of the theory corresponding to infinitely many Landau levels, as well as a variety of critical curves of second-order phase transitions. Let us next show this.

It is well-known that the state of thermodynamic equilibrium (the ground state) of an arbitrary quantum system is described by the thermodynamic potential

(TDP) Ω, which is just the value of the effective potential at its global minimum point (see (3)). In the case under consideration, the TDP $\Omega(\mu, H)$ at $\mu > M_1$ (see Figure 1) has the form

$$\Omega(\mu, H) \equiv V_{H\mu}(0) = V_H(0) -$$
$$- \frac{eH}{4\pi^2} \sum_{k=0}^{\infty} \alpha_k \theta(\mu - \epsilon_k) \left\{ \mu \sqrt{\mu^2 - \epsilon_k^2} - \epsilon_k^2 \ln\left[\left(\sqrt{\mu^2 - \epsilon_k^2} + \mu\right)/\epsilon_k\right]\right\}, \quad (8)$$

where $\epsilon_k = \sqrt{2eHk}$. We shall use the following criterion of phase transitions: if at least one first (second) partial derivative of $\Omega(\mu, H)$ is a discontinuous function at some point, then this is a point of a first- (second-) order phase transition.

Using this criterion, let us show that lines $l_k = \{(\mu, H) : \mu = \sqrt{2eHk}\}$ ($k = 1, 2, ...$), are critical lines of second-order phase transitions.

Indeed, from (8) one easily finds

$$\left.\frac{\partial \Omega}{\partial \mu}\right|_{(\mu,H) \to l_{k+}} - \left.\frac{\partial \Omega}{\partial \mu}\right|_{(\mu,H) \to l_{k-}} = 0, \quad (9)$$

as well as:

$$\left.\frac{\partial^2 \Omega}{(\partial \mu)^2}\right|_{(\mu,H) \to l_{k+}} - \left.\frac{\partial^2 \Omega}{(\partial \mu)^2}\right|_{(\mu,H) \to l_{k-}} = -\left.\frac{eH\mu}{2\pi^2 \sqrt{\mu^2 - \epsilon_k^2}}\right|_{\mu \to \epsilon_{k+}} \to -\infty. \quad (10)$$

Equation (9) means that the first derivative $\partial \Omega/\partial \mu$ is a continuous function on all lines l_k. However, the second derivative $\partial^2 \Omega/(\partial \mu)^2$ has an infinite jump on each line l_k (see (10)), so these lines are critical curves of second-order phase transitions. (Similarly, we can prove the discontinuity of $\partial^2 \Omega/(\partial H)^2$ and $\partial^2 \Omega/\partial \mu \partial H$ on all lines l_n.)

The presence of an infinite set of massive phases C_k on the phase portrait is conditioned by a special structure of the stationarity equation (7). Analytical and numerical considerations of it show that below the chiral boundary the effective potential global minimum point $\Sigma(\mu, H)$, which is identical to the quark mass, has μ and H dependences. The function $\Sigma(\mu, H)$ is a continuous one inside each of regions C_k. However, it is a discontinuous one on each of the curves $\widetilde{Mt_k}$, where the quark mass changes its value by a jump. That is why boundaries between C_k-regions are the first order phase transition lines. In contrast, in the phase B, the global minimum point is equal to $\Sigma_0(H)$ ($\equiv$ quark mass in the case $\mu = 0, H \neq 0$), which is a μ-independent quantity. This means that the particle density $n \equiv -\partial \Omega/\partial \mu$ in the ground state of the phase B is identically equal to zero, whereas in each phase C_k this quantity differs from zero.

Magnetic oscillations. Now we want to show that there arise, from the presence of infinite sets of massless A_k phases as well as of massive C_k ones, magnetic oscillations (the so-called van Alphen–de Haas-type effect) of some physical parameters in the NJL model gauged by an external magnetic field.

Let the chemical potential be fixed, i.e. $\mu = \text{const} > M_1$ (see Figure 1). Then on the plane $(\mu, \sqrt{eH})$ (see Figure 1) we have a line that crosses the critical lines $l_1, l_2, ...$ at points corresponding to some values $H_1, H_2, ...$ of the external magnetic field. The particle density n and the magnetization m of any thermodynamic system are defined by the TDP in the following way: $n = -\partial\Omega/\partial\mu$, $m = -\partial\Omega/\partial H$. At $\mu = \text{const}$ these quantities are continuous functions of the external magnetic field only, i.e. $n \equiv n(H)$, $m \equiv m(H)$. We know that all the second derivatives of $\Omega(\mu, H)$ are discontinuous on every critical line l_n (see (10)). The functions $n(H)$ and $m(H)$, being continuous in the interval $H \in (0, \infty)$, therefore have first derivatives that are discontinuous on an infinite set of points $H_1, ..., H_k, ...$ Such a behaviour manifests itself a phenomenon usually called oscillations.

Analogously to QED and condensed-matter physics [1,2], let us again separate the expression for a physical quantity with oscillations into two parts: the first monotonic one does not contain any oscillations, whereas the second part, which is of particular interest here, contains all the oscillations. Following this rule, we can write down, say, the TDP (8) of the NJL model in the form $\Omega(\mu, H) = \Omega_{mon}(\mu, H) + \Omega_{osc}(\mu, H)$. In order to present the oscillating part $\Omega_{osc}(\mu, H)$ in an analytical form, we shall use the technique elaborated in [4], where manifestly analytical expressions for this quantity were found in the case of a perfect relativistic electron–positron gas. This technique can be used without any difficulties in our case, too. So, applying in (8) the Poisson summation formula [1]

$$\sum_{n=0}^{\infty} \alpha_n \Phi(n) = 2 \sum_{k=0}^{\infty} \alpha_k \int_0^{\infty} \Phi(x) \cos(2\pi k x) dx, \quad (11)$$

where $\alpha_n = 2 - \delta_{n0}$, one can get for $\Omega_{osc}(\mu, H)$ the following expression

$$\Omega_{osc} = \frac{\mu}{4\pi^{3/2}} \sum_{k=1}^{\infty} \left(\frac{eH}{\pi k}\right)^{3/2} [Q(\pi k \nu) \cos(\pi k \nu + \pi/4) + P(\pi k \nu) \cos(\pi k \nu - \pi/4)], \quad (12)$$

where $\nu = \mu^2/(eH)$. The functions $P(x)$ and $Q(x)$ in (12) are connected with the Fresnel integrals $C(x)$ and $S(x)$ [16]: $C(x) = \frac{1}{2} + \sqrt{\frac{x}{2\pi}}[P(x)\sin x + Q(x)\cos x]$, $S(x) = \frac{1}{2} - \sqrt{\frac{x}{2\pi}}[P(x)\cos x - Q(x)\sin x]$. They have, at $x \to \infty$, the following asymptotics [16]: $P(x) = x^{-1} - 3x^{-3}/4 + ...$, $Q(x) = -x^{-2}/2 + 15x^{-4}/8 + ...$ Formula (12) presents, in a manifestly analytical form, the oscillating part of the TDP (8) for the NJL model at $\mu > M_1$. In the case under consideration, since the TDP is proportional to the pressure of the system, one can conclude that the pressure in the NJL model oscillates when $H \to 0$, too. It follows from (12) that the frequency of oscillations over the parameter $(eH)^{-1}$ equals $\mu^2/2$ and does not depend on H. So, in this case we have periodic magnetic oscillations. Then, starting from (12), one can easily find the corresponding expressions for the oscillating parts of $n(H)$ and $m(H)$. These quantities oscillate at $H \to 0$ with the same frequency $\mu^2/2$ and have a rather involved form, so we do not present them here.

Finally, we should note that the character of magnetic oscillations in the NJL model at $\mu > M_1$ resembles the magnetic oscillations in massless quantum electrodynamics [3,4]. Indeed, in this case in both models one can find periodic magnetic oscillations of some thermodynamic parameters.

Now let us show that at a fixed value of the chemical potential and $M < \mu < M_1$ the character of magnetic oscillations is changed. In this case on the plane $(\mu, \sqrt{eH})$ we have a line drawn through an infinite set of the C_k-phases. Hence, the thermodynamic potential of the NJL system has the following form: $\Omega(\mu, H) = V_{H\mu}(\Sigma(\mu, H))$, where $\Sigma(\mu, H)$ is the global minimum point of the potential $V_{H\mu}(\Sigma)$. Applying in (4) again the formula (11), one can find the following expression for the oscillating part of TDP

$$\Omega_{osc} \sim \sum_{k=1}^{\infty} \left(\frac{eH}{\pi k}\right)^{3/2} [Q(\pi k\nu)\cos(2\pi k\omega + \pi/4) + P(\pi k\nu)\cos(2\pi k\omega - \pi/4)], \quad (13)$$

where $\nu = \mu^2/(eH)$, $\omega = (\mu^2 - \Sigma^2(\mu, H))/(2eH)$. From (13) one can see that the TDP $\Omega(\mu, H)$ oscillates with frequency $(\mu^2 - \Sigma^2(\mu, H))/2$ if the variable $(eH)^{-1}$ tends to infinity. Since $\Omega(\mu, H)$ is, up to a sign, equal to the pressure in the ground state of the system, also in the present case the pressure in the NJL model is an oscillating quantity. Moreover, other thermodynamic quantities such as particle density $n = -\partial\Omega/\partial\mu$ and magnetization $m = -\partial\Omega/\partial H$ oscillate with the same frequency.

Here we should do an important remark. In the NJL model at $M < \mu < M_1$, in contrast to QED, the magnetic oscillation frequency is a H-dependent quantity. (Since the quark mass $\Sigma(\mu, H)$ has H-dependency.) So, strictly speaking, in the NJL model magnetic oscillations are not periodic ones. Recently, similar peculiarities of magnetic oscillations are observed in some ferromagnetic semiconductive materials such as $HgCr_2Se_4$ [17], where non-periodic magnetic oscillations over the variable $(eH)^{-1}$ were found to exist for electric conductivity as well as magnetization.

Finally, we should remark that in the NJL model not only thermodynamic quantities oscillate, but some dynamical parameters of the system do as well. This concerns, in particular, oscillations of the dynamical quark mass. In fact, by applying the Poisson summation formula (11) to the stationarity equation (7) and searching for the solution $\Sigma(\mu, H)$ of this equation in the form $\Sigma(\mu, H) = \Sigma_{mon} + \Sigma_{osc}$, one can easily find the following expressions for $H \to 0$:

$$\Sigma_{osc}(\mu, H) \sim \frac{(eH)^{3/2}}{\mu \tilde{M}} \sum_{k=1}^{\infty} \frac{\sin(2\pi k\tilde{\omega} - \pi/4)}{k^{3/2}}, \quad (14)$$

where $\tilde{\omega} = (\mu^2 - \tilde{M}^2)/(2eH)$, and $\tilde{M} \equiv M(\mu)$ is the quark mass at $H = 0$, $\mu \neq 0$.

Conclusions: Let us point out once more that for strongly correlated fermionic systems there is a possibility to observe nonperiodic magnetic oscillations. Moreover, in such systems in the presence of an external magnetic field some dynamical quantities (for example, fermion masses) should oscillate as well. Our results may be applicable in astrophysics, in the physics of neutron stars etc, where one should take into account the relativistic character of different phenomena.

Note that the strength of the surface magnetic field of a neutron star is about 10^{12} G and in the interior it is probably 10^{18} G [18]. Our numerical estimates of the H_{s_0} values using $\Lambda = 700$ Mev show that the magnetic field corresponding to the tricritical point s_0 varies in the interval 10^{17} G $\div$ 10^{18} G, when $1.01 < G/G_c < 1.2$. Hence, the typical neutron star magnetic field strengths are much smaller, than the value of H_{s_0}, and are located in the oscillation region of the NJL model (see Figure 1). So, the H-dependency of different physical parameters (such as particle density, magnetization, quark mass, etc) inside neutron stars possibly has a nonperiodic oscillating character.

Despite the relativistic character of our investigations, we believe that qualitatively the presented results are valid for nonrelativistic electronic systems, and may be applicable in condensed matter physics, too.

More complete information about phase structure as well as magnetic oscillations in several NJL-type models one can find in our recent paper [19].

ACKNOWLEDGMENTS

We would like to thank David Blaschke for useful discussions. One of the authors (D.E.) gratefully acknowledges the kind support and warm hospitality of the colleagues of the Theory Division at CERN. This work was supported in part by the Russian Fund for Fundamental Research, project 98-02-16690, and by DFG-project 436 RUS 113/477.

REFERENCES

1. Landau, L.D., *Collections of Works*, Moscow: Nauka, 1969, vol.1 [in Russian]; L.D. Landau, L.D., and Lifshitz, E.M., *Statistical Physics*, Moscow: Nauka, 1976, vol.1 [in Russian]; Lifshitz, E.M., and Pitaevski, L.P., *Statistical Physics*, Oxford: Pergamon Press, 1980.
2. De Haas, W.J., and Van Alphen, P.M., *Proc. Amsterdam Acad.* **33**, 1106 (1936); Shoenberg, D., *Magnetic Oscillations in Metals*, Cambridge: Cambridge University Press, 1984; Lifshitz, I.M., *Selected Works. Electronic Theory of Metals, Physics of Polymers and Bipolymers*, Moscow: Nauka, 1994 [in Russian].
3. Elmfors, P., Persson, D., and Skagerstam, B.-S., *Phys. Rev. Lett.* **71**, 480 (1993); *Astropart. Phys.* **2**, 299 (1994); Persson, D., and Zeitlin, V., *Phys. Rev.* **D 51**, 2026 (1995); Andersen, J.O., and Haugset, T., *Phys. Rev.* **D 51**, 3073 (1995); Zhukovsky, V.Ch., Shoniya, T.L., and Eminov, P.A., *J. Exp. Theor. Phys.* **80**, 158 (1995);

Zhukovsky, V.Ch., Vshivtsev, A.S., and Eminov, P.A., *Phys. Atom. Nucl.* **58**, 1195 (1995); Khalilov, V.R., *Phys. Atom. Nucl.* **61**, 1520 (1998).
4. Vshivtsev, A.S., and Klimenko, K.G., *J. Exp. Theor. Phys.* **82**, 514 (1996).
5. Nambu, Y., and Jona-Lasinio, G., *Phys. Rev.* **122**, 345 (1961); **124**, 246 (1961).
6. Kawati, S., and Miyata, H., *Phys. Rev.* **D 23**, 3010 (1981); Bernard, V., Meissner, U.-G., and Zahed, I., *Phys. Rev.* **D 36**, 819 (1987); Christov, Chr.V., and Goeke, K., *Acta Phys. Pol.* **B 22**, 187 (1991); Ebert, D., Kalinovsky, Yu.L., Münchow, L., and Volkov, M.K., *Int. J. Mod. Phys.* **A 8**, 1295 (1993).
7. Vshivtsev, A.S., Zhukovsky, V.Ch., and Klimenko, K.G., *J. Exp. Theor. Phys.* **84**, 1047 (1997).
8. Klevansky, S.P., and Lemmer, R.H., *Phys. Rev.* **D 39**, 3478 (1989).
9. Ebert, D., and Volkov, M.K., *Phys. Lett.* **B 272**, 86 (1991); Shovkovy, I.A., and Turkowski, V.M., *Phys. Lett.* **B 367**, 213 (1995); Ebert, D., and Zhukovsky, V.Ch., *Mod. Phys. Lett.* **A 12**, 2567 (1997); Babansky, A.Yu., Gorbar, E.V., and Shchepanyuk, G.V., *Phys. Lett.* **B 419**, 272 (1998).
10. Gusynin, V.P., Miransky, V.A., and Shovkovy, I.A., *Phys. Lett.* **B 349**, 477 (1995).
11. Inagaki, T., Muta, T., and Odintsov, S.D., *Mod. Phys. Lett.* **A 8**, 2117 (1993); Kim, D.K., and Koh, I.G., *Phys. Rev.* **D 51**, 4573 (1995); Ferrer, E.J., Gusynin, V.P., and de la Incera, V., *Phys. Lett.* **B 455**, 217 (1999).
12. Vshivtsev, A.S., Vdovichenko, M.A., and Klimenko, K.G., *J. Exp. Theor. Phys.* **87**, 229 (1998).
13. Inagaki, T., Muta, T., and Odintsov, S.D., *Progr. Theor. Phys. Suppl.* **127**, 93 (1997).
14. Geyer, B., Granda, L.N., and Odintsov, S.D., *Mod. Phys. Lett.* **A 11**, 2053 (1996); Elizalde, E., Shil'nov, Yu.I., and Chitov, V.V., *Class. Quant. Grav.* **15**, 735 (1998).
15. Prudnikov, A.P., Brychkov, Yu.A., and Marichev, O.I., *Integrals and Series*, New York: Gordon and Breach, 1986.
16. Bateman, H., and Erdeyi, A., *Higher Transcendental Functions*, New York: McGrawHill, 1953.
17. Balaev, A.D., et al., *Zh. Eksp. Teor. Fiz.* **113**, 1877 (1998) [in Russian].
18. Shapiro, S.L., and Teukolsky, S.A., *Black Holes, White Dwarfs and Neutron Stars, The Physics of Compact Objects*, New York: Wiley, 1983.
19. Ebert, D., Klimenko, K.G., Vdovichenko, M.A., and Vshivtsev, A.S., preprint CERN-TH/99-113; hep-ph/9905253 (to be published in Phys. Rev. D).

MESONS

On Valence Gluons in Heavy Vector Quarkonia: Implications for Spectra and Decays

S.B.Gerasimov

Bogoliubov Laboratory of Theoretical Physics, Joint Institute for Nuclear Research, 141980 Dubna, Moscow reg., Russia.

Abstract. We consider, within an adiabatic approach applied to a constituent gluon, bound to a pair of heavy quarks, the properties of the lowest vector hybrid $\bar{Q}Qg$-mesons and estimate the effects of their mixing with low-lying $\bar{c}c$-charmonia on masses, leptonic widths, and interaction with hadrons constructed of light quarks.

INTRODUCTION

The spectroscopy of systems constructed of a heavy quark Q and a heavy antiquark $\bar{Q}$ is known to be well described by the phenomenological potential models [1], where free parameters in the potentials are determined by fitting the calculated observables to the data. The gluon degrees of freedom are assumed to be integrated out in these models. The validity of this procedure is however questionable due to the presence of slow, large-scale nonperturbative vacuum fluctuations of the gluon field, as it follows from the lattice QCD approaches.

On the other hand, recent progress in understanding the production and decay processes of heavy quarkonia [2] is related to the idea of the valence gluon admixture in heavy quarkonium state vectors and the presence of the colour-octet $\bar{Q}Q$ - subsystem inside the colour-singlet bound state

$$|heavy\ meson\rangle = a_0|Q\bar{Q}\rangle + a_1|Q\bar{Q}g\rangle + ... \qquad (1)$$

Therefore, those quarkonium states which are outside the potential regime should not be used to fix parameters in the potential models, while for the states containing the gluon admixture the application of the potential models alone should leave a sufficient "room" for improvement assigned to subsequent inclusion of the gluon degrees of freedom.

The key idea of our approach is nonperturbative mechanism of higher Fock - components (i.e. the state vectors with the constituent gluons, Eq.(1)) generation in heavy quarkonia via the mixing with low-lying hybrid states.

THE SCHEME OF LOWEST VECTOR HYBRID MASS CALCULATION

In the hybrid states, gluons are confined in a bound state by the confining interaction with the colour-octet quark core. Following [3], and further references to earlier works therein, we asume that this interaction is represented by an effective potential acting in three-body system - two heavy quarks and massless vector "particle" - the gluon. Turning to this particular picture of the heavy hybrid composition, we notice existence of slow ($\bar{Q}Q$) and fast (g) sub-systems. Therefore, it is natural to proceed [3] in the spirit of the Born-Oppenheimer, or adiabatic approximation, *i.e.* to solve first a relativistic wave equation for the gluon moving in a (presumably, confining) field of fixed center, and then to make use of the found gluon energy ϵ_g as a part of the potential entering into the Schrödinger equation for the slow $\bar{Q}Q$ - sub-system

$$(\frac{\vec{p}^2}{m_Q} + \frac{1}{6}\frac{\alpha_s}{R_Q} + V_{\bar{Q}Q}(\varepsilon_g) - E_Q)\Psi(\vec{R}_Q) = 0 \qquad (2)$$

We note that the "Coulomb" potential is repulsive here because the $\bar{Q}Q$-pair is in the colour octet state. Our further main assumptions are as follows. We take the lowest magnetic (M1)- and electric (E1)- modes for the spin-orbital wave function of gluons

$$\vec{Y}^{M1}_{j,l,m} = \vec{Y}_{j,j,m}(\theta,\varphi)\mid_{j=1} \qquad (3)$$

$$\vec{Y}^{E1}_{j,l,m} = [\sqrt{\frac{2j}{2j+1}}\vec{Y}_{j,j-1,m}(\theta,\varphi) + \sqrt{\frac{j}{2j+1}}\vec{Y}_{j,j+1,m}(\theta,\varphi)]_{j=1} \qquad (4)$$

where $\vec{Y}_{j,l,m}(\theta,\varphi)$ are the vector spherical harmonics. This means that for the $J^{PC} = 1^{--}$ - hybrid mesons, we are going to consider, the orbital momentum l of a $\bar{Q}Q$-system should be $l = 0(1)$ for the $M1(E1)$ - gluon modes. The radial part of the gluon wave function is assumed to obey the Klein-Gordon-Fock (KGF) equation with an external field including the (strong) Coulomb potential and the squared form of the linear confinement potential, properly scaled against analogous potentials for colourless $\bar{Q}Q$ -states (the scaling factor being the ratio of the corresponding Casimir operators equal to 9/4). For the assumed interaction between the two colour-octet, point-like particles, the KGF-equation would be of the form

$$(\vec{p}_g^2 + V_s^{g2}(r_g) + 2\varepsilon_g V_v^g(r_g) - V_v^{g2}(r_g) - \varepsilon_g^2)\psi(\vec{r}_g) = 0 \qquad (5)$$

As far as the colour "charge" of the $\bar{Q}Q$ - sub-system is spatially distributed, we define the "form-factor-modified" potentials through the folding integral

$$V_{s(v)}(\vec{r}_g) = \int V_{s(v)}(\mid \vec{r}_g - \vec{r}' \mid)\rho(\vec{r}')d^3r' \qquad (6)$$

where the density function $\rho(\vec{r})$ is related to an (unknown) wave function of heavy quarks.

Finally, our calculation scheme acquires the variation form. We take, as a trial wave function of quarks, simple expressions of the exponential form (with the pre-exponential centrifugal or nodal factors) which contain one variable parameter. This parameter propagates to the gluon energy, and then it appears again in the equation of motion for quarks. The last step is the minimization of the Schrödinger energy functional leading to the numerical value of this variable parameter and all energies, hence, to the hybrid meson mass. Summing up, with the trial wave functions of exponential form and on the basis of the adiabatic approximation, the masses of the vector hybrid states were estimated to be 4.02 (or 4.21) GeV for charmed quarks with mass m_c=1.4 GeV and a valence gluon of the M1(or E1)-type, while for the bottom quarks with mass m_b =4.8 GeV the corresponding masses are 10.65 (or 10.75) GeV. The mean values of r_g and $R_{\bar{Q}Q}$ characterizing spatial extension of hybrid wave functions are equal to .45 fm and .4 fm for the charmed states, and .47 fm and .3 fm for the b-flavored hybrids. The approximate independence of the characteristics of light particle (i.e. the gluon) on masses of quarks is familiar manifestation of the heavy quark symmetry. The values of the "Coulomb" constant and slope of the linear potential have been taken equal to $\kappa=\frac{4}{3}\alpha_s = .49$ and $a = .16$ GeV^2. The obtained values are rather close to estimates from different models [4]. In particular, they are very close to the string model elaborated in [5] where the gluon-quark interaction has acquired the form following from our formula (6) if

$$\rho(\vec{r}) = 1/2(\delta(\vec{r} - \vec{r}_Q) + \delta(\vec{r} - \vec{r}_{\bar{Q}}))$$

CHARMONIA-HYBRID MIXING PHENOMENA: A MODEL ESTIMATION

It is quite natural to expect that proper estimation of the hybrid meson(s) mixing with nearby quarkonia will be important to understand some peculiarities of the charmonium spectra and decays slightly over $4 GeV$ [6].

As a first step, we confine ourselves to consideration of the four-level mixing in the charmonium spectrum choosing the ground state $J/\Psi(1S)$, $\Psi(2S)$, $\Psi(3S)$ and the presumed hybrid state H_c with the calculated mass around $4 GeV$ as mixing states. The nondiagonal elements m_{nH} in the 4x4 - mass-matrix

$$m_{nH} = \langle H_c; Q\bar{Q}g | \mathcal{H}_{int} | nS; Q\bar{Q} \rangle \quad (7)$$

$$\mathcal{H}_{int} = g_s \sum_i \frac{1}{2}\vec{\lambda}(i)(i\epsilon_g)(\vec{A}^{E1}(r,\Theta,\phi) \cdot \vec{r}(i))\delta(\vec{r} - \vec{r}(i)) +$$

$$+ \frac{g_s}{2m_Q} \sum_i \frac{1}{2}\vec{\lambda}(i)(\sigma(i) \cdot \vec{A}^{M1}(r,\Theta,\phi))\delta(\vec{r} - \vec{r}(i))$$

are calculated with the explicit radial wave functions

$$R_{3S}(r) = N_3 \cdot (1 - a_3(\gamma r)^m + b_3(\gamma r)^{2m}) \exp\left(1/2(\gamma r)^m\right), m = \frac{4}{3} \tag{8}$$

$$R_{1S} = R_{3S}("3" \to "1"; a = b = 0), \tag{9}$$
$$R_{2S} = R_{3S}("3" \to "2"; b = 0), \tag{10}$$

parametrized to reproduce approximately the spatial dimensions ($i.e. < r^2 >$),the location of the radial function nodes and the values of the wave functions of the $1S$-$3S$ -quarkonia states at "zero" distance, which correspond to the QCD-motivated potentials (e.g. [7] and references therein).
As representative sets of $[\gamma_n; a_n; b_n]$ for the (nS)-states of charmonia we take $[\gamma_1 = .883; a_1 = b_1 = 0]$, $[\gamma_2 = .715; a_2 = .57; b_2 = 0]$ $[\gamma_3 = .628; a_3 = 1.101; b_3 = .197]$, where all γ's are in units of GeV. The lowest vector hybrid state $h_c(g_{M1}Q\bar{Q})$ with the $M1$-type gluon mode, i.e., with $l_g = 1$ and $L_{Q\bar{Q}} = 0$ also called "the gluon-excited state", should presumably be rather narrow due to the dynamical selection rule discussed in a number of earlier papers [8,9], which prevents the decays of this state into the ground-state charmed mesons. This selection rule is not acting for hybrids with the $E1$-type gluon mode ($l_g = 0, L_{Q\bar{Q}} = 1$), or "the quark-excited state", which should therefore have very large width [10]. Hence, in what follows, we consider the mixing of the low-lying vector charmonia with the gluon-excited, $M1$-type vector hybrid state. The radial wave function of this hybrid meson is taken in the factorized form in accord with the adopted adiabatic approximation

$$R_{gQ\bar{Q}}(r_g, R_Q) = N_g N_Q r_g \exp\left(1/2(\alpha_g r)^m + 1/2(\beta_Q R_Q)^m\right), m = \frac{4}{3} \tag{11}$$

with the numerical values $\alpha_g = 1.235$ GeV, $\beta_Q = .973$ GeV. The nondiagonal elements of the symmetric 4×4- mass-matrix have been calculated as matrix elements of the interaction hamiltonian (7) over the nS-charmonia states ($n = 1, 2, 3$), and the ("fourth") hybrid state and their values are: $m_{n4} = .25, .074, .044$ GeV for $n = 1, 2, 3$, respectively, with all other nondiagonal elements equal to zero.
To obtain "physical" eigenvalues of the diagonalized matrix near to the known masses of the Ψ-family, we take the "bare" masses, which stand along the main diagonal having values $m_{nn} = 3.153; 3.695; 4.05$ GeV for $n = 1, 2, 3$ and $m_{44}(h_c) = 4.07 GeV$. The diagonalization procedure leads to the "physical" masses: $m(J/\Psi) = 3.089[3.097], m(\Psi(2S)) = 3.685[3.686], m(\Psi(3S)) = 4.03[4.04], m(H_c) = 4.17[4.16]$ where masses of the known Ψ-mesons in GeV are indicated in parentheses. The corresponding eigenfunctions reveal the following quark-gluon configuration mixing

$$J/\Psi = .968\psi(1S) + .0302\psi(2S) + .0113\psi(3S) - .247\psi(h_c), \tag{12}$$
$$\Psi(2S) = -.0628\psi(1S) + .989\psi(2S) + .0161\psi(3S) - .134\psi(h_c), \tag{13}$$
$$\Psi(3S) = -.0908\psi(1S) - .0697\psi(2S) + .940\psi(3S) - .321\psi(h_c), \tag{14}$$
$$H_c = .223\psi(1S) + .141\psi(2S) + .332\psi(3S) + .906\psi(h_c). \tag{15}$$

Leptonic Width Ratios

Assuming the dynamical dominance of the quarkonia-components in the mentioned states while computing the leptonic decay widths of the corresponding vector mesons, we can compare the model and phenomenological ratios of the meson wave functions "at the zero relative $Q - \bar{Q}$-distance":

$$R^2_{J/\Psi}(0) : R^2_{2S}(0) : R^2_{3S}(0) : R^2_{H_c}(0) = 1 : .65[.64] : .34[.27 \pm .05] : .34[.29 \pm .09], \tag{16}$$

where the values in parentheses are calculated using the proportionality between $R^2_V(0)$ and $m^2(V)\Gamma(V \to l^+l^-)$. The puzzling equality of the leptonic widths of the $\Psi(4.04)$ and $\Psi(4.16)$ states is explained in our approach by a coherent sum of the admixture (separately, looking small) amplitudes of the $(1S) - (3S)$ quarkonia states in the dominantly hybrid $\Psi(4.16)$-resonance.

Colour Polarizability of the Pion and Low-Energy Pion-Charmonium Interaction

We shall take for granted, that there is no clear-cut evidence for the long-range power-behaved interaction between hadrons, and that the most long-range part thereof is due to either one-pion- or two-pion-exchange diagrams. This focus attention on the pion interaction with heavy quarkonia, while we consider the interaction of isoscalar heavy quarkonia with light hadrons composed mainly of the u-, and d quarks. We estimate here one of possible non-perturbative mechanisms of this interaction connected with the earlier estimated presense of the bound valence gluons in the state-vector of a heavy quarkonium, namely, via the hybrid-state admixture in the quarkonium wave function. Following analogy with electromagnetic interactions, we describe the blok of the pion-gluon interaction via a new structure constants - the gluon, or colour polarizabilities of the pion. Owing to the chromoelectric and chromomagnetic polarizability of pions, the incoming pion will interact or scatter on the standing wave of the chromoelectric and chromomagnetic wave generated by the gluon distributions in quarkonia through the hybrid configuration admixtures in the quarkonia state vectors.

We base our numerical estimation of these new structure parameters of pions on the following observations. Earlier, in the course of evaluation of light meson structure parameters [11], the numerical vale of the electric polarizability of charged pions was obtained [12]

$$\alpha^{\pi^\pm}_\gamma = \frac{e^2}{8\pi^2 m_\pi F_\pi^2} \tag{17}$$

via the combination of the quark-loop evaluation of the axial-structure constant in the charged pion radiative decay with the help of the DMO sum rule [14], and the

Teren'tev relation [15], based on the current algebra and PCAC, between this constant and the pion electric polarizability. The same result has later been obtaned through the direct calculation of the one-loop, quark-meson diagrams applied to the low energy photon-meson scattering [16,17].

Now, we apply the chiral lagrangian of the linear σ-model [17], including pions and quarks, to calculate the coefficients of colour polarizability of pions. Following closely the work of Lvov [17], where the needed one-loop integrals have been listed, we obtain for the chromoelectric, (α_g^π), and chromomagnetic, (β_g^π), polarizability of the pion

$$\alpha_g^\pi = -\frac{2}{3}\frac{m_\pi^2}{m_q^2}d \tag{18}$$

$$\beta_g^\pi = \frac{7}{9}\frac{m_\pi^2}{m_q^2}d, \, d = \frac{\alpha_s}{\pi m_\pi F_\pi^2} \tag{19}$$

where $\alpha_s = g_s/4\pi$, g_s is the quark-gluon coupling constant to be taken at a characteristic scale of the mean virtuality of involved gluons, $F_\pi \simeq 93 MeV$, and we put $m_\sigma = 2m_q$, $m_{\sigma(q)}$ being mass of the σ-meson or constituent quark. We have singled out the factor of the pion and quark mass ratio to emphasize the difference with the electromagnetic polarizabilities which have non-zero values in the leading order in the pion-to-quark mass ratio. Another qualitqtive difference consists in that the usual electric $(\alpha_\gamma^{\pi^\pm})$ and magnetic $(\beta_\gamma^{\pi^\pm})$ polarizabilities have, respectively, positive and negative values, while their sum is equal to zero in the leading approximation. The colour polarizabilities tell that with respect to applied chromoelectric and chromomagnetic fields the pion responds as the "dia-electric" and "para-magnetic", respectively. The extreme "ideal dia-electric" property is usually attributed to nonperturbative QCD vacuum. The found qualitative similarity of the pion response to applied chromoelectric field can signify on the influence of the pion "medium" surrounding the quark core of light hadrons, e.g. the nucleon, on the distribution of gluon field inside a given hadron, hence on the dynamics of quark confinement.

The effective lagrangian for a low-energy gluon-pion interaction can now be written down in full analogy with the electromagnetic case, e.g., [18]. To transfer now to the pion-quarkonium interaction, one should take into account that the chromoelectric and chromomagnetic fields entering the lagrangian will correspond to the distribution of a given multipole-type solution of the Klein-Gordon-Fock equation with the confining interaction, introduced in the previous sections. To illustrate the simplest situation, we write down the $\pi - J/\Psi$ - scattering cross section at "zero-energy", that is in the scattering length approximation. It shows the very small value

$$\sigma_{\pi J/\Psi}(|\vec{q}_\pi| \to 0) = .0019 \text{mb}. \tag{20}$$

This value corresponds to the presence of the $M1$-type "valence" gluon in the J/Ψ state vector due to mixing with the lowest mass hybrid charmonium, presumably the Ψ4.16-resonance, according to our abovementioned estimates. The found value is in accord with the value calculated in Ref. [19] at the non-zero pion momentum, while extrapolation of their cross section to zeroth value demonstrate the further rapid fall. This feature demonstrates the difference of dynamics with the " moderately-hard" bound gluons and with the soft " gluon stuff " which was used, to our understanding, in [19].

The heavy quarkonia interaction with the nuclear medium is under discussion for a long time in view of importance of this topic for identification of the quark-gluon plasma in relativistic heavy-ion collisions via the "quenching" of the produced quarkonium cross section. What could be the role of the valence "moderately-hard" gluons in these processes needs more detailed elaboration and discussion. We hope to return to these and related questions of the "cascade" hadronic quarkonia decays, e.g., $\Psi' \to J/\Psi + 2\pi$, in the first place, elsewhere.

CONCLUDING REMARKS

Our estimation of the mixing parameters of lowest H_c-state and low-lying charmonia states gives, quantitatively, much larger values, than the values of Ref. [10] and, at variance with [10], it is in accord with earlier advanced hypothesis that approximately equal leptonic widths of the $\Psi(4.04)$ - and $\Psi(4.16)$-resonances is due to the hybrid-quarkonia mixing phenomenon. However, in our picture the $\Psi(4.16)$-resonance is predominantly hybrid state, while $\Psi(4.04)$ is dominantly the $3S$-radially excited quarkonium. Whether such an interpretation can be experimentally distinguished from the Ono-Close-Page scenario [6,20] (the approximately equal partition of the $(3S)$-state and the H_c-state between the $\Psi(4.04)$ and $\Psi(4.16)$ charmonium state vectors) remains to be considered. The involvement problem of the broad $(g_{E1}Q\bar{Q})$-type hybrid state(s) also needs to be clarified.

ACKNOLEDGMENTS

The author would like to express his gratitude to Organizing Committee of the "Hadron-99" Workshop, Coimbra, Portugal, for invitation to present a talk and the support.

REFERENCES

1. Lucha, W., Schöberl, F. F., and Gromes, D., *Phys. Rep.* **200**, 127 (1991).
2. Bodwin, G., Braaten, E., and Lepage, G. P., *Phys. Rev.* **D46**, R1914 (1992); *ibid.* **D51**, 1125 (1995).
3. Gerasimov, S. B., " On valence gluons in heavy quarkonia", in *Proceedings of the 14th Int. Conference on Problems in Quantum Field Theory*, Joint Institute for Nuclear Research, Dubna, 1999, pp. 487-491; hep-ph/9812509.

4. Close, F. E., *Rep. Prog. Phys.* **51**, 833 (1988).
5. Kalashnikova, Yu. S., and Yufryakov, Yu. B., *Yad. Fiz.* **60**, 374 (1997).
6. Close, F. E., and Page, P. R., *Phys. Lett.* **B366**, 323 (1996).
7. Eichten, E., and Quigg, C., *Phys. Rev.* **D52**, 1726 (1995).
8. Tanimoto, M., *Phys. Lett.* **B116** 186, (1982).
9. Close, F. E., and Page, P. R., *Nucl. Phys.* **B443**, (1995) 233.
10. Iddir, F., Safir, S., and Pene, O., *Phys. Lett.* **B433**, 125 (1998).
11. Gerasimov, S. B., *Sov. J. Nucl. Phys.* **29**, 513 (1979).
12. Gerasimov, S. B., (unpulished); reproduced in Ref. [13].
13. Petrun'kin, V. A., *Sov. J. Part. Nucl.* **12**, 278 (1981).
14. Das, T., Mathur, V. S., and Okubo, S., *Phys. Rev. Lett.* **19**, 859 (1967).
15. Teren'tev, M. V., *Sov. J. Nucl. Phys.* **16**, 162 (1972).
16. Volkov, M. K., and Ebert, D., *Sov. J. Nucl. Phys.* **34**, 182 (1981).
17. Lvov, A. I., *Sov. J. Nucl. Phys.* **34**, 289 (1981).
18. Lvov, A. I., *Int. J. Mod. Phys.* **A8**, 5267 (1993).
19. Fujii, H., and Kharzeev, D., hep-ph/9903495.
20. Ono, S., *Z.Phys.* **C26**, 307 (1984).

Brief Review of some modified Versions of the Nambu-Jona-Lasinio Model

M. Jaminon

*Université de Liège, Institut de Physique B5,
Sart Tilman, B-4000 Liège 1, Belgium*

Abstract. Two modified versions of the NJL model are investigated. The first one has the SU(2) chiral symmetry and describes the spin 1 ρ meson starting from a 4-point tensor interaction. The second one has the SU(3) symmetry and implements the trace anomaly of QCD.

INTRODUCTION

The Nambu-Jona-Lasinio model [1], reinterpreted as a schematic quark model, has the merit to be much simpler than QCD while reproducing most of its symmetries. In particular, chiral symmetry breaking leading to a Goldstone pion has been considered up to quite recently [2] as correctly described in this model. The axial anomaly can also be introduced quite easily : in the SU(2) version, this is done keeping only the scalar isoscalar and the pseudoscalar isovector 4-point interactions; in its SU(3) version, it can be introduced via a t'Hooft determinant or a mass term for the η' [3]. In the present paper we present two modified versions of the NJL model which preserve its original symmetries. One consists in describing the spin 1 ρ meson using a 4-point tensor interaction rather than the usual vector interaction. The second one consists in implementing the trace anomaly of QCD at the level of the classical Lagrangian. Their respective merits are stressed in the corresponding sections.

NJL WITH A 4-POINT TENSOR INTERACTION

It is well known that when the ρ vector meson is described with a vector 4-point interaction in the NJL model, chiral symmetry requires the introduction of an additional pseudovector interaction:

$$\mathcal{L}_V = -\frac{1}{G_V}[(\bar{q}\vec{\tau}\gamma_\mu q)^2 - (\bar{q}\vec{\tau}\gamma_5\gamma_\mu q)^2]. \tag{1}$$

This entails a mixing between the pseudoscalar and pseudovector fields which imposes a redefinition of the physical fields. This feature has undesirable consequences. Firstly, the $\rho\pi\pi$ form factor acquires a strong momentum dependence yielding to a much too low value for the width of the $\rho \to 2\pi$ decay, even when the 2π channel is included in the self-energy of the ρ [4]. Secondly, the off-shell momentum dependence of the pion wave function is far to be in better agreement with lattice QCD studies than in the case $G_V = 0$ [5]. Finally, from a technical point of view, this $A - \pi$ mixing complicates drastically the computation of physical pionic observables as scattering lengths and radii, even if their precise values do not dramatically depend on the choice $G_V = 0$ or $G_V \neq 0$. The values of a_0^0 and a_0^2 are around 0.20 and -0.045 respectively- to be compared with $a_0^{0,exp} = 0.26 \pm 0.05$ and $a_0^{2,exp} = -0.028 \pm 0.012$.

In connection with these problems, we consider a very simple model which corresponds to a SU(2) NJL model in which the spin 1 ρ meson is generated by a tensor 4-quark interaction instead of a vector one [6]- [8]:

$$\mathcal{L}_T = -\frac{1}{G_T}[(\bar{q}\vec{\tau}\sigma_{\mu\nu}q)^2 + (\bar{q}i\gamma_5\sigma_{\mu\nu}q)^2]$$
$$= -\frac{1}{G_T}[(\bar{q}\vec{\tau}\sigma_{\mu\nu}q)^2 - (\bar{q}\sigma_{\mu\nu}q)^2] \quad (2)$$

yielding the following bosonized effective action:

$$I_{eff} = -i\, Tr\, \ln[i\partial_\mu\gamma_\mu - m + \sigma + i\gamma_5\vec{\pi}.\vec{\tau} + \sigma^{\mu\nu}(\vec{T}_{\mu\nu}.\vec{\tau} + T^0_{\mu\nu})]$$
$$- \int d^4x \frac{1}{2G_S}(\sigma^2 + \vec{\pi}^2) + \int d^4x \frac{1}{2G_T}(\vec{T}_{\mu\nu}.\vec{T}^{\mu\nu} - T^0_{\mu\nu}T^{0\mu\nu}). \quad (3)$$

The quantity m denotes the current quark mass, $\vec{\pi}$ the pion field, σ the field of its chiral partner, $\vec{T}_{\mu\nu}$ is associated with ρ and $T^0_{\mu\nu}$ with ω. G_S is the strength of the scalar interaction. The interpolating field is considered either as the quadridivergence of a tensor current [6] either as a vector current [7]. At the pole, the two descriptions are equivalent. The main advantage of this model is that there is no $A - \pi$ mixing. The mathematical treatment of $\pi - \pi$ scattering is then greatly simplified with regard to similar calculation with vector interaction : only one box diagram has to be computed instead of 16! In addition to this box diagram, the Weinberg theorem only requires the exchange of the chiral partner of the pion (σ): the exchange of the ρ does not contribute to the low energy scattering lengths at variance with the case of a vector interaction. Besides its simplicity, the model has the merit to give rather good agreement for the π and ρ observables (see Table 1). It increases a_0^0 closer to its central value while a_0^2 keeps its Weinberg value. It reproduces $g_{\rho\pi\pi}$ and then the total width of the ρ. The former is defined from $[q^2 = (p - p')^2, P_\mu = p_\mu + p'_\mu]$:

$$\langle \pi^b(p)|j_\mu^a|\pi^c(p')\rangle = g_{\rho\pi\pi}(q^2)\epsilon^{abc}P_\mu, \quad a,b,c = 1,2,3, \quad (4)$$

TABLE 1. π and ρ observables compared with experimental values

	The model	Exper.
M (MeV)	387.5	
m_π (MeV)	138.5	139
f_π (MeV)	91.6	93
m_ρ (MeV)	770.0	770
$g_{\rho\pi\pi}$	6.02	
$\Gamma_{\rho\to 2\pi}$ (MeV)	150.4	
a_0^0	0.24	0.026± 0.05
a_0^2	-0.042	-0.028±0.012
$\langle \bar{q}q \rangle^{1/3}$ (MeV)	-208.0	
$g_{\rho q\bar{q}}$	1.03	
$\langle r_\pi^2 \rangle^{1/2}$ (fm) (a)	0.52	
$\langle r_\pi^2 \rangle^{1/2}$ (fm) (a)+(b)	0.63	0.678±0.012

where $j_\mu^a \sim \partial_\nu j^{a,\mu\nu}$ is the source of the ρ field $\rho_\mu^a \sim \partial_\nu T^{a,\mu\nu}$. One gets:

$$g_{\rho\pi\pi} \equiv g_{\rho\pi\pi}(q^2 = m_\rho^2) \; ; \; g_{\rho\pi\pi}(q^2) = 8M g_{\rho\bar{q}q} g_{\pi\bar{q}q}^2 m_\rho I_3(p,p') \quad (5)$$

with

$$I_3(p,p') = -iN_c \int \frac{d^4k}{(2\pi)^4} \Delta(k)\Delta(k-p)\Delta(k-p') \quad (6)$$

$$\Delta(k) = \frac{1}{k^2 - M^2}. \quad (7)$$

In Eq. (5), M is the constituent quark mass which satisfies the usual gap equation; m_ρ is the mass of the meson ρ, $g_{\rho\bar{q}q}$ its coupling constant to the quarks. $g_{\pi\bar{q}q}$ is the coupling constant of the π to the quarks.

The pion charge radius is also in rather good agreement with its experimental value. It is calculated from its electromagnetic form factor:

$$\langle r_\pi^2 \rangle = -6 \frac{dF_\pi(t)}{dt}|_{t=0}, \quad (8)$$

where $F_\pi(t), t=q^2$ receives contribution from the two graphs of Fig.1:

$$F_\pi(t) = F_\pi^{(a)}(t) + F_\pi^{(b)}(t). \quad (9)$$

The exact expressions for $F_\pi^{(a)}(t)$ and $F_\pi^{(b)}(t)$ can be found in ref. [6]. They both contribute to the pion form factor (see Fig.2) while the charge radius is mainly due to graph (a), which takes into account the quark structure of the pion. Gauge invariance is preserved since $F_\pi^{(a)}(0) = 1$ and $F_\pi^{(b)}(0) = 0$ due to its explicit dependence in t. The values of Table 1 are obtained for the set of parameters:

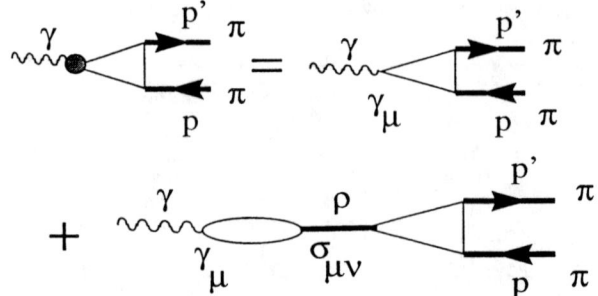

FIGURE 1. Photon vertex: (a) bare vertex, (b) ρ exchange.

$$G_S = 21.0 \ GeV^{-2} \quad G_T = 5.8 \ GeV^{-2} \quad m = 8.9 \ MeV \quad \Lambda = 0.730 \ GeV. \qquad (10)$$

The cutoff Λ is only introduced in the diverging integrals and G_T is fixed by fitting the ρ mass. The latter corresponds to the pole of the propagator:

$$P^{\mu\nu,\rho\sigma} = P_{T1}(q^2)\left(\hat{q}^\mu \hat{q}^\sigma g^{\nu\rho} - \hat{q}^\mu \hat{q}^\rho g^{\nu\sigma} - \hat{q}^\nu \hat{q}^\sigma g^{\mu\rho} + \hat{q}^\nu \hat{q}^\rho g^{\mu\sigma}\right) \\ + P_{T2}(q^2)\left(g^{\mu\rho}g^{\nu\sigma} - g^{\mu\sigma}g^{\nu\rho}\right) \qquad (11)$$

where

$$P_{T1}(q^2) = -\frac{G_T^2 4q^2 S(q^2, M, \Lambda)}{A_-(q^2, M, G_T, \Lambda) A_+(q^2, M, G_T, \Lambda)} \qquad (12)$$

$$P_{T2}(q^2) = \frac{G_T}{2A_-(q^2, M, G_T, \Lambda)}. \qquad (13)$$

The expressions for $A_\pm(q^2, M, G_T, \Lambda)$ and $S(q^2, M, \Lambda)$ are given in [6]. The ρ mass satisfies:

$$A_+(m_\rho^2, M, G_T, \Lambda) = 0. \qquad (14)$$

The absence of $A - \pi$ mixing entails that the correlation function in the π channel is in better agreement with QCD lattice calculations (see upper part of Fig.3). This correlation function is calculated from:

$$P_\pi(x) \equiv i\langle 0 \mid T\{(\overline{q}(x)i\gamma_5\tau^a q(x))(\overline{q}(0)i\gamma_5\tau^a q(0))\} \mid 0\rangle \\ = \int \frac{d^4q}{(2\pi)^4} exp(iq.x)\Pi_\pi(q). \qquad (15)$$

The momentum space correlation function $\Pi_\pi(q)$ is computed using dispersion relation. One then has:

$$P_\pi(x) = \text{Res } \Pi_\pi(m_\pi^2)D(m_\pi, x) + \frac{1}{\pi}\int_{4M^2}^\infty \text{Im } \Pi_\pi(s)_{cont} D(\sqrt{s}, x) ds \qquad (16)$$

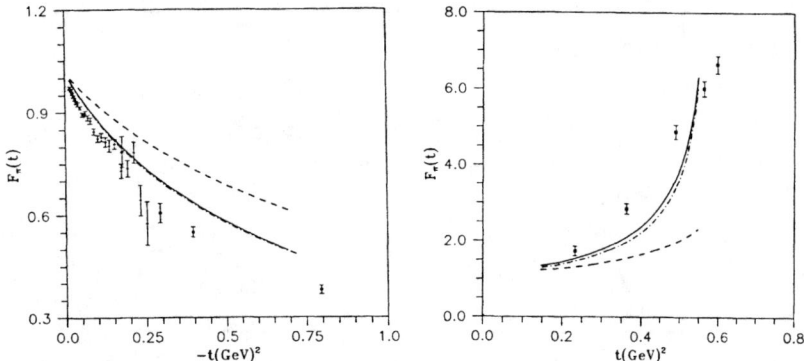

FIGURE 2. Pion charge form factor for t < 0 and t > 0: Graphs (a)+(b) of Fig.1 (full curve) and graph (a) (dashes) with parameters (10). For the dashed-dotted curve, the cutoff has been included in all the integrals over the momentum of the quark. The experimental points are from ref.[9].

where $D(m_\pi, x)$ is the free massive bosonic propagator. In the ρ channel, one has a similar expression to Eq.(16) with m_π replaced by m_ρ and $P_\pi(x)$ by $P_\rho(x) \equiv P_{\rho,\mu}^\mu(x)$. The cases $G_V = 0, G_T \neq 0$ and $G_V \neq 0, G_T = 0$ nearly yield the same results while $G_V = 0, G_T = 0$ gives too low values for $P_\rho(x)$ at large x. This is due to the fact that, at large x, the contribution to the correlator is mainly ascribed to the pole (first term of the RHS of Eq.(16)) which is missing in the case $G_V = 0, G_T = 0$. In QCD, $P_\rho(x)$ is dominated, at low x, by the contribution of the continuum: due to the unrenormalizabilty of the model, none of the three cases has the suitable contribution.

In spite of its merits, the model presents some drawbacks that we briefly review now. Firstly, it presents the same problems as the usual NJL with vector interaction: it does not confine the quarks and it is not renormalizable. In addition, it is a model whose extension to SU(3) seems difficult. Indeed, the simplest generalization yields a trivial 4-point interaction since $(\bar{q}\lambda_a\sigma_{\mu\nu}q)^2 + (\bar{q}\lambda_a i\gamma_5\sigma_{\mu\nu}q)^2 = 0$ due to properties of Dirac gamma's.

Let us conclude this section making some remarks. Firstly, the ρ meson so described does not account for the vector meson dominance: it does not contribute to the low energy scattering lengths and the charge radius of the pion is mainly ascribed to the bare photon quark coupling. Even if the shape of the electromagnetic form factor is well reproduced, the strong increase near the ρ pole does not account for the $\rho \to \pi\pi$ decay but only reflects the quark structure of the ρ. Secondly, the good agreement with experiment has to be taken with cautious since the open channel $\rho \to \pi\pi$ has not be included in the calculation of the self-energy of the ρ. It is known [4] that with a vector interaction, such a channel increases the latter of around 65 MeV. It also increases the width of the ρ, roughly doubling its value.

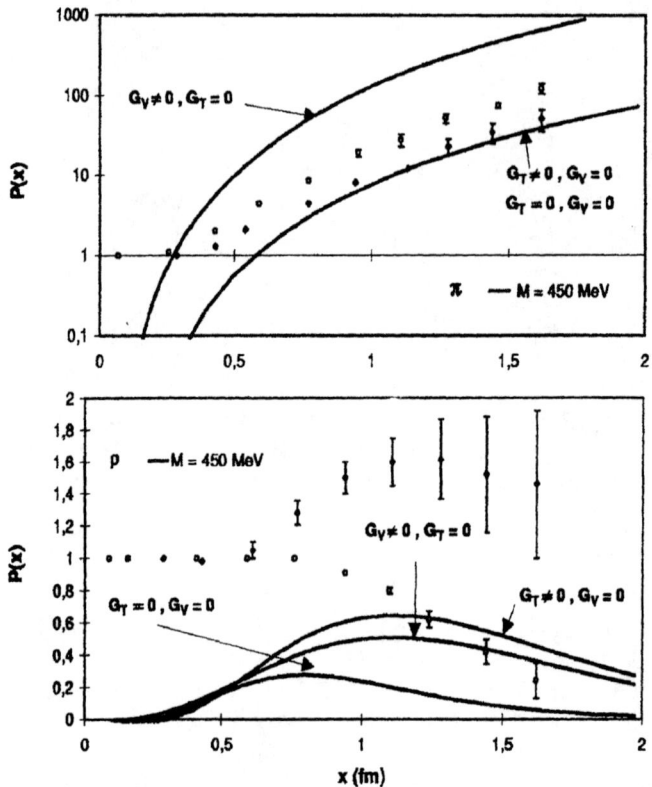

FIGURE 3. Euclidean point to point correlation functions relative to the massless quarks ones, in the π (upper part) and ρ (lower part) channels. Results are shown for $M = 450$ MeV.

In our approach, the chiral partner of the ρ meson is the $b_1(1235)$ $J^{PC} = 1^{+-}$ and not the $a_1(1260)$ $J^{PC} = 1^{++}$. However it has no bound state $[A_-(q^2, M, G_T, \Lambda) \neq 0$ for $G_T > 0]$. In the same way, there is no bound state for the ω, whose propagator should be given by Eqs. (11-13) with G_T replaced by $-G_T$. One can solve this problem generating the ω by a mixing between a tensor and a vector 4-point interactions. This will lead to a different behavior with density for the mass of the ρ and of the ω [10]. Such a model does not leave place for an interpretation of the nucleons in terms of droplets of massless quarks surrounded by the vacuum. Indeed, to generate the ω, the vector coupling constant has to be still larger than in the case without tensor interaction [10], yielding to a repulsive contribution to the Fermi pressure which prevents this interpretation in terms of droplets [11]. This feature is exhibited in Fig.4.

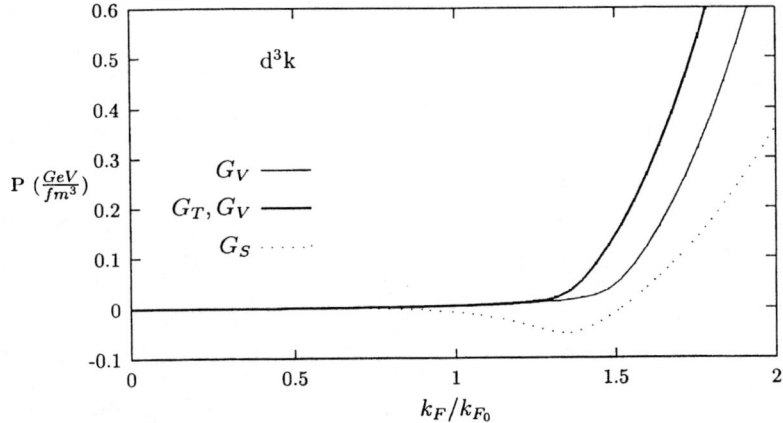

FIGURE 4. Pressure with respect to the vacuum versus the quark Fermi momentum.

SCALED NJL MODEL

We now turn to another modified version of the NJL model which implements the trace anomaly of QCD. Introduced in [12] in the SU(2) case, it has been extended to three flavors in [3]. Except in [13], only the scalar and pseudoscalar sectors have been considered. In a first step the original NJL action is made scale invariant multiplying (i) the cutoff by one single point-like scalar dilaton field χ whose mean value χ_0 is identified with the vacuum gluon condensate (ii) the $(\varphi_a \varphi_a)$ term by the squared dilaton field χ^2. In a second step, the trace anomaly is implemented by the introduction of a dilaton potential. The total effective action written in the

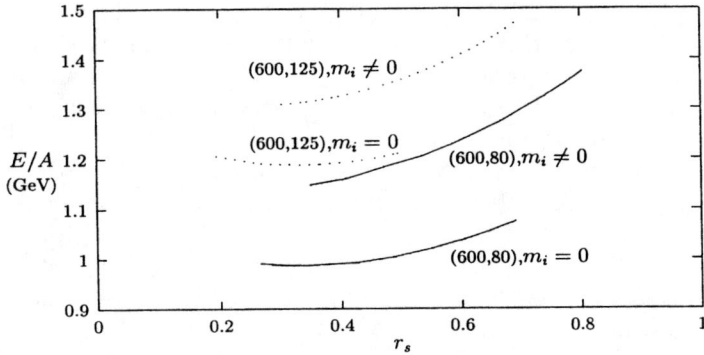

FIGURE 5. Binding energy per baryon number versus r_s.

SU(3) symmetry is[1]:

$$I_{eff} = - Tr_{\Lambda\chi} \ln(-i\partial_\mu\gamma_\mu + m + \Gamma_a\varphi_a) + \int d^4x \frac{1}{2G_S}\chi^2\varphi_a\varphi_a$$
$$+ \int d^4x \mathcal{L}_\chi + \int d^4x \mathcal{L}_{U_A(1)} \qquad (17)$$

where

$$\mathcal{L}_\chi = \frac{1}{2}(\partial_\mu\chi)^2 + \frac{1}{16}b^2\left(\chi^4\ln\frac{\chi^4}{\chi_G^4} - (\chi^4 - \chi_G^4)\right) \qquad (18)$$

$$\mathcal{L}_{U_A(1)} = \frac{1}{2G_S}\xi\chi^2\pi_0^2. \qquad (19)$$

with $\varphi_a = (\sigma_a, \pi_a)$ and $\Gamma_a = (\lambda_a, i\gamma_5\lambda_a)$. The usual SU(3) NJL contains five parameters, four of which being fixed to reproduce m_π, m_K, m'_η and f_π: the NJL model then contains one free parameter. In this version, one has two additional parameters : one is fixed to reproduce the mass of the glueball (associated with the χ field), the other one remaining free.

The particularities of this model have been used in two different directions. Firstly, we have studied the thermodynamics of a u-, d- and s-quark system [14]. Due to the additional free parameter, one can play with the order of the chiral transition in density or(and) temperature. Stability of a strange quark matter can also be investigated within this model [15]: it is found that as far as the constituent quark mass is sufficiently large in order to mimic confinement, a strange quark matter is stable in the chiral limit. The stability is obtained for $\rho_s = \rho_u = \rho_d$ in agreement with the chemical equilibrium. [ρ_i ($i = u, s, d$) is the density of the i-quark]. The stability is lost when finite quark masses are taken into account, confirming the results of ref. [16]. This is illustrated in Fig.5 where we have plotted the binding energy per baryon number versus the quantity $r_s = \rho_s/(\rho_u + \rho_d + \rho_s)$ for $M_u = 600$ MeV and $\chi_0 = 125$ MeV (dotted curves) or $\chi_0 = 80$ MeV (full curves). Another interesting feature of the model lies in the fact that it provides a mixing between the three scalar isoscalar fields σ_0, σ_8 and χ. The consequence is that the scalar mesons do not appear any longer as pure $q\bar{q}$ excitations but exhibit some glue content. In the same way, the glueball has now a $q\bar{q}$ content. The scalar glueball can then decay into two pseudoscalar mesons via a triangle quark loop diagram [17]. Its strong decay width receives two additional local contributions coming from the requirement of the scale invariance of the original NJL action : (i)the cutoff has to be multiplied by the χ field, (ii) the strength of the 4-quark coupling is now divided by χ^2 [G_s/χ^2]. However studying the scalar mesons puzzle is all but simple as reflected by a literature full of contradictory statements. The problem is still complicated by the presence of glueballs. Fig.6 shows results for the 2π decay[2] of

---

[1] one usually works in Euclidean space using an Euclidean metric.
[2] Results for the decay into the other pseudoscalars can be found in [17].

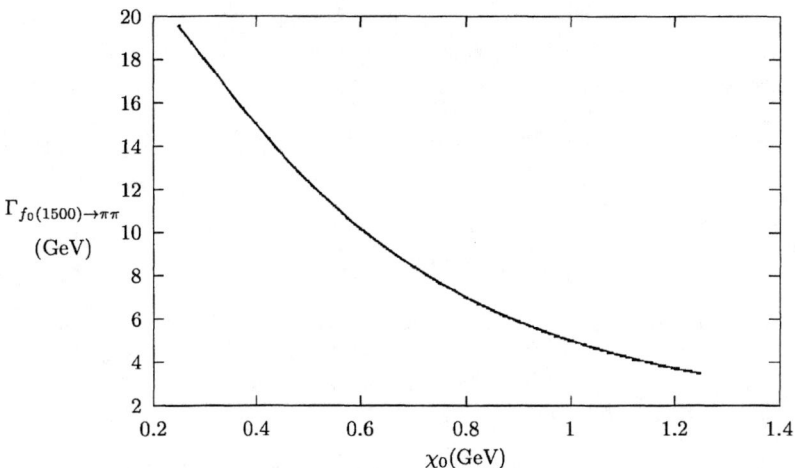

FIGURE 6. Decay width of the glueball $f_0(1500)$ into 2π.

the $f_0(1500)$ identified with the first scalar glueball. Indeed, the small value of the total width of $f_0(1500)$ ($\Gamma_{f_0(1500)} = 112 \pm 10$ MeV) seems incompatible with a nonet structure. Moreover, Amsler and Close [18] have argued that its strong branching ratios can not be explained with $f_0(1500)$ if it is assumed to be a pure $q\bar{q}$ state. Fig.6 corresponds to the following assumption for the scalar nonet:

$$I = 0 : f_0(1370), f_0(1710) \text{ mixed with the glueball } f_0(1500);$$
$$I = 1 : a_0(1450);$$
$$I = 1/2 : K_0^*(1430). \qquad (20)$$

One observes that the model yields catastrophic result for the decay into two pions. One can think that this comes from the large value of M_u ($M_u = 725$ MeV) we have to take to reproduce the large mass of the scalar nonet. One way to get a way out would be to introduce confinement [19]. Indeed, such a confinement has as effect to increase the scalar meson mass with regards to the nonconfining model [20]. The masses of the scalars can then be reproduced with a smaller value of M_u which could provide better agreement with the experimental value.

However one can work in another way, thinking that the identification of the scalar nonet + glueball is not the right one. For instance, one could assume:

$$I = 0 : f_0(980), f_0(1370) \text{ mixed with the glueball } f_0(1500);$$
$$I = 1 : a_0(980);$$
$$I = 1/2 : K_0^*(1430). \qquad (21)$$

Such masses can be reproduced in a confining model with a quark mass around 350 MeV. Moreover the two first excited states of the pion are well reproduced in this scheme. Work is in progress in this direction. Finally, following Lee and Weingarten [21], one could identify the glueball with $f_0(1710)$. In that case, our $f_0(1500)$, even if mixed with the others "hybrids", would have its larger component in the $s\bar{s}$ channel and its 2π decay width should be spectacularly reduced.

REFERENCES

1. Y. Nambu and G. Jona-Lasinio, *Phys. ReV.* **122**, 345 (1961)
2. H. Kleinert and B. Van den Bossche, *preprints* hep-ph/9907274; hep-ph/9908284
3. M. Jaminon and B. Van den Bossche, *Nucl. Phys.A* **582**, 517 (1995)
4. A. Polleri, R.A. Broglia, P.M. Pizzochero and N.N. Scoccola, *Z. Phys. A* **357**, 325 (1997)
5. R.M. Davidson and E. Ruiz Arriola, *Phys. Lett. B* **359**, 273 (1995)
6. M. Jaminon, M.C. Ruivo and C.A. de Sousa, *to be published*
7. M. Jaminon and E. Ruiz Arriola, *Phys. Lett. B* **443**, 33 (1998)
8. W. Broniowski, M. Polyakov, H.-C. Kim and K. Goeke, *Phys. Lett. B* **438**, 242 (1998)
9. S.R. Amendolia et al., *Nucl. Phys. B* **277**, 168 (1986), C.J. Bebek et al. *Phys. ReV. D* **13**; 25 (1976), A. Quenzer et al.*Phys. Lett. B* **76**, 512 (1978)
10. M. Jaminon and E. Ruiz Arriola, *to be published*
11. M. Buballa and M. Oertel, *Nucl. Phys. A* **642**, 39 (1998)
12. G. Ripka and M. Jaminon, *Ann. Phys.* **218**, 51 (1992)
13. M. Jaminon and G. Ripka, *Nucl. Phys A* **564**, 505 (1993)
14. J. Cugnon, M. Jaminon and B. Van den Bossche, *Nucl. Phys A* **598**, 515 (1996)
15. M. Jaminon and B. Van den Bossche, *to be published*
16. M. Buballa and M. Oertel, *Phys. Lett. B* **457**, 261 (1999)
17. M. Jaminon, M. Mathot and B. Van den Bossche, *Nucl. Phys. A*, in press
18. C. Amsler and F.E. Close, *Phys. Lett. B* **353**, 385 (1995); *Phys. ReV. D* **53**, 295 (1996)
19. M. Jaminon, M. Mathot and B. Van den Bossche, *to be published*; M. Jaminon and M. Mathot, *Proceedings of "International Workshop on Hadron Physics- Effective Theories of Low Energy QCD", Coimbra 1999*
20. L.S. Celenza, Xiang-Dong Li and C.M. Shakin, *Phys. ReV. C* **55**, 3083 (1997)
21. W. Lee and D. Weingarten, *Nucl. Phys. B (Proc. Suppl.)* **63**, 194 (1998)

Invariant Regularization of One-Loop Determinant in Non-Renormalizable Models

A.A. Osipov, B. Hiller and A.H. Blin

Centro de Física Teórica, Departamento de Física da Universidade de Coimbra, 3004-516 Coimbra, Portugal

Abstract. We use proper-time regularizations to define the one-loop fermion determinant in the form suggested by Gasser and Leutwyler some years ago. We show how to obtain the polynomial by which this definition of $\ln \det D$ needs to be modified in order to arrive at the fermion determinant whose modulus is invariant under chiral transformations. As an example it is shown how the fundamental symmetries associated with the NJL model are preserved in a consistent way.

I INTRODUCTION

We start with the definition

$$\ln \det D = -\frac{1}{2}\int_0^\infty \frac{dT}{T}\rho(T,\Lambda^2)\mathrm{Tr}\left(e^{-T\bar{D}^2}\right) - \int d^4x P(v,a,\sigma,\pi). \tag{1}$$

Here the operator $\bar{D} \equiv \gamma_5 D$ has been introduced [1], D is the Dirac operator in the presence of external vector (v), axial-vector (a), scalar (σ) and pseudoscalar (π) sources. The Schwinger proper-time method cannot be applied directly to fermions. This is why one has to square the operator $\bar{D}$. In this way the real and imaginary parts of $\ln \det D$ are treated on equal footing, as opposite to the $D^\dagger D$ definition. As an alternative one can use the integral representation of the complex power for the pseudo differential operator [2]. In the latter case an unambiguous definition of the determinant of the Dirac operator is obtained. The determinant is shown to be vector gauge invariant and to yield the correct axial and scale anomalies.

We consider a class of regularization schemes (proper-time regularizations) which can be incorporated in this expression through the kernel $\rho(T,\Lambda^2)$. These regularizations allow to shift in loop momenta. A typical example is the proper-time cutoff where the kernel ${}^t\rho(T,\Lambda^2)$ is equal to

$${}^t\rho(T,\Lambda^2) = \Theta\left(T - \frac{1}{\Lambda^2}\right). \tag{2}$$

Another choice for the kernel can be the covariant Pauli-Villars cutoff [3]

$$^c\rho(T,\Lambda^2) = 1 - (1 + T\Lambda^2)e^{-T\Lambda^2} \tag{3}$$

which leads to the well-known effective potential of the Nambu – Jona-Lasinio (NJL) model [4]. The result is[1]

$$V(m) = \frac{m^2}{2G}\left(1 - \frac{N_c G \Lambda^2}{4\pi^2}\right) + \frac{N_c}{8\pi^2}\left[m^4\ln\left(1 + \frac{\Lambda^2}{m^2}\right) - \Lambda^4\ln\left(1 + \frac{m^2}{\Lambda^2}\right)\right]. \tag{4}$$

Both of the kernels (2) and (3) have been used in many papers, for example, see papers [6,7] and [8] correspondingly. A wide set of possibilities for the kernel $\rho(T,\Lambda^2)$ have been considered in the papers [9,10].

The counterterms $P(v,a,\sigma,\pi)$ in formula (1) can be fixed from the transformation properties of $\ln\det D$. We consider here the case of chiral gauge theories with the $SU(2)_L \times SU(2)_R \times U(1)_V$ chiral symmetry. Explicitly, let D be equal to

$$D = \gamma^\mu(i\partial_\mu + v_\mu + a_\mu\gamma_5) - \sigma + i\gamma_5\pi \tag{5}$$

where $v_\mu = v_\mu^i \tau_i$, $a_\mu = a_\mu^i \tau_i$, $\pi = \pi^a \tau_a$, $\sigma = \sigma^a \tau_a$, $\tau_a = (1,\tau_i)$, $[\tau_i, \tau_j] = 2i\epsilon_{ijk}\tau_k$, $i = 1,2,3$. The corresponding chiral transformations of the external fields are given by

$$\delta v_\mu = \partial_\mu \alpha + i[\alpha, v_\mu] + i[\beta, a_\mu] \tag{6}$$

$$\delta a_\mu = \partial_\mu \beta + i[\alpha, a_\mu] + i[\beta, v_\mu] \tag{7}$$

$$\delta\sigma = i[\alpha, \sigma] - \{\beta, \pi\} \tag{8}$$

$$\delta\pi = i[\alpha, \pi] + \{\beta, \sigma\}. \tag{9}$$

Here $\alpha = \alpha_i \tau_i$ is the infinitesimal transformation generated by the vector currents and $\beta = \beta_i \tau_i$ is a chiral transformation. The transformation law of $\ln\det D$ in this case is known explicitly [11]:

$$\delta\ln\det D = \frac{iN_c}{(4\pi)^2}\int dx \mathrm{Tr}_f(\beta\Omega) \tag{10}$$

where

$$\Omega = \varepsilon^{\alpha\beta\mu\nu}\left[v_{\alpha\beta}v_{\mu\nu} + \frac{4}{3}\nabla_\alpha a_\beta \nabla_\mu a_\nu + \frac{2i}{3}\{v_{\alpha\beta}, a_\mu a_\nu\} \right.$$
$$\left. + \frac{8i}{3}a_\mu v_{\alpha\beta} a_\nu + \frac{4}{3}a_\alpha a_\beta a_\mu a_\nu\right]. \tag{11}$$

---

[1] See, for instance, [5] and references in it.

The field strength tensor $v_{\mu\nu}$ associated with v_μ is defined as

$$v_{\mu\nu} = \partial_\mu v_\nu - \partial_\nu v_\mu - i[v_\mu, v_\nu] \tag{12}$$

and $\nabla_\mu a_\nu$ stands for

$$\nabla_\mu a_\nu = \partial_\mu a_\nu - i[v_\mu, a_\nu]. \tag{13}$$

Our aim now is to calculate the polynomial $P(v,a,\sigma,\pi)$ in the framework of a nonrenormalizable aproach. Let us note that $P(v,a,\sigma,\pi)$ is unique up to a chirally invariant polynomial. One can always chose P in such a manner that the determinant is not modified if the external fields a_μ and π are switched off. In paper [1] it has been shown how to do this for renormalizable theories. There are two essential differences in our case. The first one is that we have to use a regularization with finite cutoff Λ. The ζ-function technique is not good for that because it does not lead to the correct description of the spontaneous chiral symmetry breaking phenomena. The second one is also related to the cutoff dependense of the result. As we shall show, the polynomial $P(v,a,\sigma,\pi)$ gets now systematically contributions from the terms which would vanish in the limit $\Lambda \to \infty$. This fact renders its evaluation rather technical.

II COUNTERTERMS AND SYMMETRY

To illustrate our consideration we shall discuss the NJL model with the $SU(2)_L \times SU(2)_R$ chiral symmetry. We use the model version with only the scalar-scalar and pseudoscalar-pseudoscalar type of four quark interactions. Integrating out the quark fields one obtains the action of the model in terms of scalar $\sigma \times 1$ and pseudoscalar $\pi = \pi_i \tau^i$ collective mesonic degrees of freedom

$$S_{coll} = -i \ln \det D - \int d^4x \frac{(\sigma + m)^2 + \vec{\pi}^2}{2G}. \tag{14}$$

The Dirac operator D is given by

$$D = i\gamma^\mu \partial_\mu - m - \sigma + i\gamma_5 \pi, \tag{15}$$

where m denotes the constituent quark mass generated in the process of spontaneous chiral symmetry breaking. In the phase with broken chiral symmetry the transformations (8) and (9) become

$$\bar\delta \sigma = -\{\beta, \pi\} \tag{16}$$

$$\bar\delta \pi = i[\alpha, \pi] + \{\beta, \sigma + m\} \tag{17}$$

for the considered isospin content of scalar and pseudoscalar. Under global chiral transformations the corresponding change in the Dirac operator $\bar{D} = \gamma_5 D$ is given by

$$i\bar{\delta}\bar{D} = [\bar{D}, \alpha] + \{\bar{D}, \beta\gamma_5\}. \tag{18}$$

Therefore, to get the related polynomial $P(\sigma, \pi)$ for this case one has to integrate the equality

$$\bar{\delta} \ln \det D = 0 \tag{19}$$

where $\ln \det D$ is defined according to Eq.(1). The variation of $P(\sigma, \pi)$ has to cancel the symmetry breaking part coming from the proper-time integral. In this way one gets

$$\bar{\delta}P(\sigma, \pi) = \frac{-i}{8\pi^2} \sum_{n=0}^{\infty} R_n \mathrm{tr}(\beta\gamma_5 a_{n+1}) \tag{20}$$

where tr represents trace in internal space. In the case under consideration it includes summations over flavour, colour and Lorentz indexes: $\mathrm{tr} = \mathrm{tr}_f \mathrm{tr}_c \mathrm{tr}_L$. One can see that $P(\sigma, \pi)$ is not invariant under chiral transformations, picking up the contribution which is linear in β. The functions R_n represent the integrals which appear in the result of the asymptotic expansion of the heat kernel

$$R_n = -\int_0^\infty \rho(T, \Lambda^2) d\left[T^{n-1} e^{-Tm^2}\right]$$
$$= \int_0^\infty dT T^{n-2}[m^2 T - (n-1)] e^{-Tm^2} \rho(T, \Lambda^2). \tag{21}$$

These integrals yield the following expression for R_n

$$^c R_n = \frac{n! \Lambda^4}{(\Lambda^2 + m^2)^{n+1}}. \tag{22}$$

This result corresponds to the kernel (3). For the case of $\rho(T, \Lambda^2)$ being equal to (2) one gets

$$^t R_n = (\Lambda^2)^{1-n} \exp\left(-\frac{m^2}{\Lambda^2}\right). \tag{23}$$

In renormalizable theory the terms R_n with $n \geq 2$ vanish in the limit $\Lambda \to \infty$. The same is also true if one applies the ζ - function regularization. This property of renormalizable models extremely simplifies the problem. In non-renormalizable models all of R_n terms contribute to the result.

The coefficients $a_n \equiv a_n(x, x)$ are the coincidence limit of the Seeley – DeWitt coefficients [9]. For our illustration we shall need the first four of them

$$a_0 = 1,$$
$$a_1 = -Q,$$
$$a_2 = \frac{1}{2}Q^2 + \frac{1}{6}Q_{\mu\mu} + \frac{1}{12}F^2,$$
$$\begin{aligned}a_3 = &-\frac{1}{6}Q^3 - \frac{1}{12}(\{Q, Q_{\mu\mu}\} + Q_\mu^2) - \frac{1}{60}Q_{\mu\mu\nu\nu} + \frac{1}{60}[F_{\mu\alpha;\alpha}, Q_\mu] \\ &- \frac{1}{60}(2\{F^2, Q\} + F_{\mu\nu}QF_{\mu\nu}) - \frac{1}{45}F_{\mu\alpha;\alpha}F_{\mu\beta;\beta} - \frac{1}{180}F_{\mu\alpha;\beta}F_{\mu\alpha;\beta} \\ &- \frac{1}{60}\{F_{\mu\nu}, F_{\mu\nu;\alpha\alpha}\} + \frac{1}{30}F^3.\end{aligned} \quad (24)$$

Some comments are in order here. First, we deal in this case with the linear realization of chiral symmetry. It means that we have for $\bar{D}^2$ the following representation

$$\bar{D}^2 = \nabla_\mu \nabla^\mu + m^2 + Q \quad (25)$$

where

$$\nabla_\mu = \partial_\mu + A_\mu, \quad A_\mu = \gamma_\mu \gamma_5 \pi, \quad (26)$$

$$Q = (\sigma^2 + 2m\sigma) + i\gamma^\mu \partial_\mu \sigma - 2(m+\sigma)i\gamma_5 \pi + 3\vec{\pi}^2. \quad (27)$$

Second, we wrote the coefficients (24) directly in Minkowski space. In this way one should understand summations over repeated Lorentz indexes to be implicit. We have used the following designations

$$F_{\mu\nu} = [\nabla_\mu, \nabla_\nu] = \gamma_\nu \gamma_5 \partial_\mu \pi - \gamma_\mu \gamma_5 \partial_\nu \pi + [\gamma_\nu, \gamma_\mu]\vec{\pi}^2, \quad (28)$$

$$F^2 = F_{\mu\nu}F^{\mu\nu}, \quad F^3 = F_{\mu\nu}F^{\nu\sigma}F_{\sigma\mu}, \quad (29)$$

$$Q_\mu = [\nabla_\mu, Q], \quad F_{\mu\nu;\nu} = [\nabla_\nu, F_{\mu\nu}], \quad (30)$$

One has to calculate traces $\text{tr}(\beta\gamma_5 a_n)$ and integrate Eq.(20). The first three non-zero contributions are given by

$$\text{tr}(\beta\gamma_5 a_1) = 4iN_c\bar{\delta}\vec{\pi}^2. \quad (31)$$

$$\text{tr}(\beta\gamma_5 a_2) = 4iN_c\bar{\delta}\left[\frac{1}{6}(\partial_\mu \vec{\pi})^2 - 2m\sigma\vec{\pi}^2 - \sigma^2\vec{\pi}^2 - \frac{2}{3}\vec{\pi}^4\right]. \quad (32)$$

$$\begin{aligned}\text{tr}(\beta\gamma_5 a_3) = 4iN_c\bar{\delta}&\left[\frac{1}{60}(\partial_\mu^2 \vec{\pi})^2 - \frac{1}{2}\vec{\pi}^2(\partial_\mu \sigma)^2 - \frac{1}{5}\vec{\pi}^2(\partial_\mu \vec{\pi})^2 - \frac{7}{60}(\partial_\mu \vec{\pi}^2)^2\right. \\ &- \frac{1}{3}(\sigma + m)\partial_\mu \sigma \partial_\mu \vec{\pi}^2 + \frac{1}{10}\vec{\pi}^6 \\ &+ \frac{1}{6}(\sigma^2 + 2m\sigma)[3\vec{\pi}^2(\sigma^2 + 2m\sigma) + 2\vec{\pi}^4 - (\partial_\mu \vec{\pi})^2] \\ &\left.- \frac{1}{3}m^2\vec{\pi}^4 - 2m\sigma\vec{\pi}^2 - \sigma^2\vec{\pi}^2 - \frac{2}{3}\vec{\pi}^4\right]. \quad (33)\end{aligned}$$

Let us note that the last four terms from a_3 (see Eq.(24)) do not contribute to $\text{tr}(\beta\gamma_5 a_3)$.

On the other side, the first term in formula (1) contributes to the Lagrangian of collective fields as

$$\mathcal{L}_{coll}^{(1)} = -\frac{1}{32\pi^2}\sum_{n=0}^{\infty} J_n \text{tr}(a_{n+1}), \tag{34}$$

where

$$J_n = \int_0^\infty \frac{dT}{T^{2-n}} e^{-Tm^2} \rho(T, \Lambda^2), \quad n = 0, 1, 2... \tag{35}$$

We have from (34)

$$\mathcal{L}_{coll}^{(1)} = \frac{N_c}{(2\pi)^2}\left\{\left(\sigma^2 + 2m\sigma + 3\vec{\pi}^2\right)J_0 \right.$$
$$\left. + \frac{1}{2}\left[(\partial_\mu\sigma)^2 + (\partial_\mu\vec{\pi})^2 + 4m^2\vec{\pi}^2 - (\sigma^2 + 2m\sigma + \vec{\pi}^2)^2\right]J_1 + ...\right\}. \tag{36}$$

Using the identity

$$m^2 J_n + (2-n)J_{n-1} = R_{n-1} \quad n = 1, 2, 3... \tag{37}$$

one can see, for instance, how symmetry breaking terms proportional to $\vec{\pi}^2$ are compensated in this expression by the contribution from (31). A fully chiral symmetric Lagrangian is therefore obtained at each order of the proper time expansion.

III CONCLUSIONS

We used the one-loop fermion determinant in the form suggested by Gasser and Leutwyler some years ago [1] to extend it to be applicable to non-renormalizable models. In this way the real and imaginary parts of $\ln \det D$ can be calculated with the same input. One obtains the correct description of the chiral anomaly when regularization is switched off. However it is necessary to correct the real part of $\ln \det D$ by the polynomial $P(v, a, \sigma, \pi)$ to get the chiral invariant result for this case. We have shown how to get the chiral symmetry restoring polynomial $P(v, a, \sigma, \pi)$. The simplest way to do this is to calculate in the phase with broken chiral symmetry, rewriting the symmetry transformations especially for this case. The result is an extension of the form presented in the paper [1], to incorporate explicitly the process of dynamical chiral symmetry breaking of the NJL model.

For simplicity we have considered the NJL model without vector and axial-vector degrees of freedom. However, the result can be easily extended to the more general case.

IV   ACKNOWLEDGEMENTS

This work is supported by grants provided by Fundação para a Ciência e a Tecnologia, PRAXIS XXI/BCC/7301/96, PRAXIS/C/FIS/12247/1998, PESO/P/PRO/15127/1999 and NATO "Outreach" Cooperation Program.

REFERENCES

1. Gasser, J., and Leutwyler, H., *Annals of Physics* **158**, 142-210 (1984).
2. Salcedo, L.L., and Ruiz Arriola, E., *Annals of Physics* **250**, 1-50 (1996).
3. Pauli, W., and Villars, F., *Reviews of Modern Physics* **21**, 434-444 (1949).
4. Nambu, Y., and Jona-Lasinio, G., *Physical Review* **122**, 345 (1961); **124**, 246 (1961).
5. Ying, S., *Annals of Physics* **250**, 69-111 (1996).
6. Bijnens, J., *Physics Reports* **265**, 369 (1996).
7. Nikolov, E.N., Broniowski, W., Christov, C.V., Ripka, G., and Goeke, K., *Nuclear Physics* **A 608**, 411-436 (1996).
8. Bernard, V., Osipov, A.A., and Meißner, U.-G., *Physics Letters* **B 285**, 119-125 (1992); Bernard, V., Blin, A.H., Hiller, B., Ivanov, Y.P., Osipov, A.A., and Meißner, U.-G., *Annals of Physics* **249**, 499-531 (1996).
9. Ball, R.D., *Physics Reports* **182**, 1-186 (1989).
10. Döring, F., Blotz, A., Schüren, C., Meissner, Th., Ruiz-Arriola, E., and Goeke, K., *Nuclear Physics* **A 536**, 548-572 (1992).
11. Bardeen, W.A., *Physical Review* **184**, 1848 (1969).

Light Scalar Mesons

Deirdre Black, Amir H. Fariborz and Joseph Schechter

Department of Physics, Syracuse University, Syracuse, New York 13244-1130, USA.

Abstract.
We review how a certain effective chiral Lagrangian approach to $\pi\pi$ scattering, πK scattering and $\eta' \to \eta\pi\pi$ decay provides evidence for the existence of light scalars $\sigma(550)$ and $\kappa(900)$ as well as describing the $f_0(980)$ and the $a_0(980)$. An attempt to fit these into a nonet suggests that their structure is closer to a dual quark–dual antiquark than to a quark–antiquark. A possible mechanism to explain the next higher mass scalar nonet is also proposed.

I INTRODUCTION

The possible existence of light scalar mesons (with masses less than about 1 GeV) has been a controversial subject for roughly forty years. There are two aspects: the extraction of the scalar properties from experiment and their underlying quark substructure. Because the $J = 0$ channels may contain strong competing contributions, such resonances may not necessarily dominate their amplitudes and could be hard to "observe". In such an instance their verification would be linked to the model used to describe them. The last few years have seen a revival of interest in this area. As examples, three models for the underlying quark structure have been discussed by many authors, including other contributors to this workshop [1]: i) the $K\bar{K}$ molecule model [2], ii) the $q\bar{q}$ model with strong meson-meson interactions (or "unitarized quark model") [3], iii) the intrinsic $qq\bar{q}\bar{q}$ model (Jaffe type [4]). These models have the common feature that four quarks are involved in some form; all are different from the "simple" $q\bar{q}$ model. Clearly, the elucidation of the structure of unusual low lying states can be expected to increase our understanding of non-perturbative QCD.

The present approach is based on comparing with experiment, the predictions for $\pi\pi$ scattering, πK scattering and $\eta' \to \eta\pi\pi$ decay from a phenomenological chiral Lagrangian containing particles of mass comparable to the energy regime of interest. These studies seem to require for consistency the existence of two isoscalars $\sigma(550)$ and $f_0(980)$, an isospinor $\kappa(900)$ and an isovector $a_0(980)$ with given properties. Note that in the effective Lagrangian approach, the quark substructure of the scalars is not specified. In particular a nonet field can *a priori* represent either

$q\bar{q}$ or $qq\bar{q}\bar{q}$ (or even more complicated) states. From this point of view our approach is "model independent".

Section II contains a brief summary of our scattering model in the $\pi\pi$ case. The generalization to the πK case and to $\eta' \to \eta\pi\pi$ decay is even more briefly summarized in section III. Section IV deals with the "family properties" of the nonet made up from the scalars we need. The model of "ideal mixing" for meson nonets is reviewed and generalized to include "dual ideal mixing". The realistic situation is noted to be closer to dual rather than ordinary ideal mixing. Finally, in section V, a possible mechanism is proposed to explain some puzzling features of a presumably more conventional next–to–lowest–lying scalar nonet.

II PI PI SCATTERING

The most difficult partial wave amplitude to explain is just the scalar channel with $I = J = 0$. Our notation for the partial wave is $T_J^I(s) = R_J^I + iI_J^I$. The complicated shape of the experimentally obtained $R_0^0(s)$ shown in Figs. 2 and 3 below suggests that resonances are present. Close to threshold, the chiral perturbation theory approach, which essentially supplies a Taylor expansion of the amplitude, is very accurate. However explaining the data shown to about 1.2 GeV would appear to require a prohibitively high order of expansion in this scheme. Thus we sacrifice some accuracy near threshold and use instead an expansion of the invariant amplitude in terms of resonance exchange diagrams (including contact terms needed for chiral symmetry). This holds the possibility of achieving a fit to experiment over a larger energy regime. Some theoretical support for such an approach comes from the leading order in $1/N_c$ approximation to QCD, which features just tree diagrams.

In detail, we calculate the tree diagrams of our model from an effective chiral Lagrangian which contains resonances but has interactions with a minimal (for simplicity) number of derivatives. Hence the initial computed amplitude will be (as in the $1/N_c$ expansion) purely real. This suggests that it is most sensible in the present approach to compare with the real part of the experimental amplitude. Of course we still must find a way to "regularize" the infinities which arise at the direct channel poles. We interpret the "regularization" procedure as equivalent to enforcing unitarity in the vicinity of the direct channel pole. In the case of a narrow isolated resonance we adopt the usual Breit–Wigner procedure in which the offending term in the amplitude is replaced as

$$\frac{MG}{M^2 - s} \to \frac{MG}{M^2 - s - iMG}. \qquad (1)$$

In the case of a very broad resonance we instead replace

$$\frac{MG}{M^2 - s} \to \frac{MG}{M^2 - s - iMG'}, \qquad (2)$$

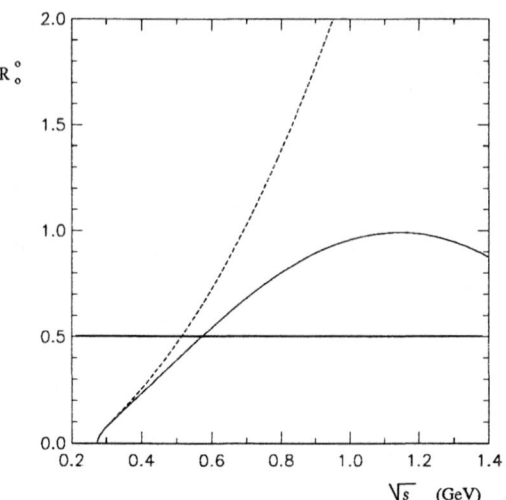

FIGURE 1. The solid line which shows the current algebra $+\rho$ result is much closer to the unitary bound of 0.5 than the dashed line which shows the current algebra result alone.

where G', which is not required to equal G, is taken as a fitting parameter. Finally, in the case of a "narrow" resonance in a non trivial background (characterized by a phase shift δ in the resonance partial wave) we replace,

$$\frac{MG}{M^2 - s} \to e^{2i\delta}\frac{MG}{M^2 - s - iMG}. \tag{3}$$

This method can be trivially modified to give a crossing symmetric invariant amplitude but unitarity may easily be violated in general. We thus choose the parameters for a putative σ meson represented by (2) to fit experiment. We then end up with an amplitude [5] which approximately satisfies unitarity and crossing symmetry. This is illustrated in a step by step manner in Figs. 1, 2 and 3.

The real part of the partial wave amplitude R_0^0 is obtained by projecting out the real part of the s-wave $I = 0$ component of the chiral invariant and crossing symmetric invariant amplitude. We see in Fig. 1 that the "current algebra" piece starts violating the unitarity bound, $|R_0^0| \leq 1/2$ at about 0.5 GeV and then runs away. However the inclusion of the ρ meson exchange diagrams turns the curve in the right direction and improves, but does not completely cure, the unitarity violation. This feature, which does not involve any unknown parameters, gives encouragment to our hope that the cooperative interplay of various pieces at tree level can explain the low energy scattering. In order to fix up Fig. 1 we note that the real part of a resonance contribution vanishes at the pole, is positive before the pole and *negative* above the pole. Thus a scalar resonance with a pole about 0.5 GeV (where R_0^0 in Fig. 1 needs a negative contribution to stay below 1/2) should do the job. The result of including such a σ meson, parametrized as in (2), is shown in Fig. 2. Now note that the predicted $R_0^0(s)$ in Fig. 2 vanishes around 1 GeV. Thus

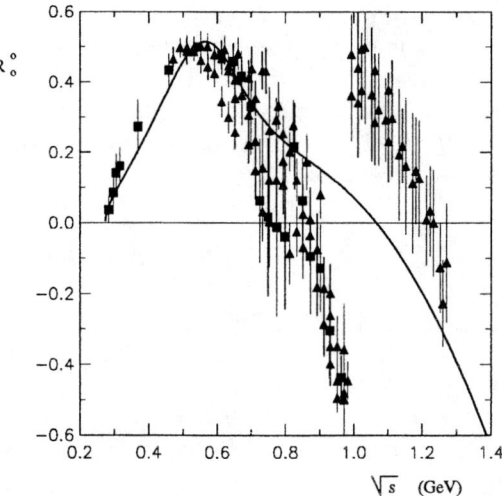

FIGURE 2. The sum of current algebra $+\rho + \sigma$ contributions compared to data.

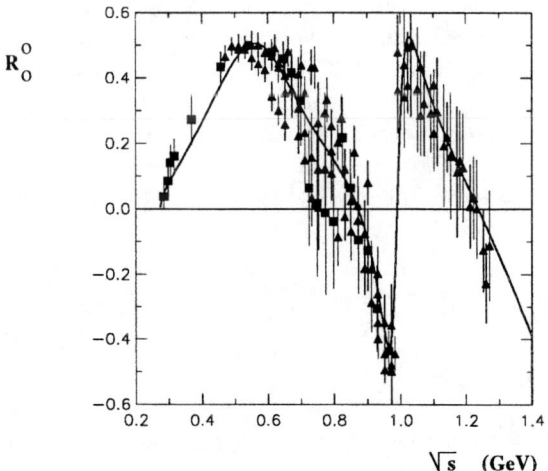

FIGURE 3. The sum of current algebra $+\rho + \sigma + f_0(980)$ contributions compared to data.

the phase δ at $1 Gev$ (assumed to keep rising) is about $90°$ there. Considering this as a background phase for the known $f_0(980)$, Eq(3) shows us that the real part of the $f_0(980)$ contribution will get reversed in sign (Ramsauer–Townsend effect). This is the missing piece in the jig–saw puzzle as Fig. 3 shows. Up to about $1.2 GeV$, the amplitude R_0^0 can be explained as the sum of current algebra, ρ exchange, $\sigma(550)$ exchange and $f_0(980)$ exchange pieces.

III PI K SCATTERING AND $\eta' \to \eta\pi\pi$

A similar treatment was carried out for the $J = 0$ partial wave amplitudes of πK scattering [6]. In this case the low energy amplitude is taken to correspond to the sum of a current algebra contact diagram, vector ρ and K^* exchange diagrams and scalar $\sigma(550)$, $f_0(980)$ and $\kappa(900)$ exchange diagrams. The situation in the interesting $I = 1/2$ channel turns out to be very analogous to the $I = 0$ channel of s-wave $\pi\pi$ scattering. Now a $\kappa(900)$ parametrized as in (2) is required to restore unitarity; it plays the role of the $\sigma(550)$ in the $\pi\pi$ case. Following our criterion we expect that to extend this treatment to the 1.5 GeV region, one should include the many possible exchanges of particles with masses up to about 1.5 GeV. Nevertheless we found that a satisfactory description of the 1-1.5 GeV s-wave region is obtained simply by including the well known $K_0^*(1430)$ scalar resonance, which plays the role of the $f_0(980)$ in the $\pi\pi$ calculation.

The $\eta' \to \eta\pi\pi$ process is a strong decay which yields information about the properties of the scalar $a_0(980)$ isovector resonance. The tree diagrams, which are similar to those of $\pi\eta$ scattering in our model [7], include $a_0(980)$, $\sigma(550)$ and $f_0(980)$ exchanges. Compared to the $\pi\pi$ and πK scatterings there is a simplification in that G-parity invariance prevents vector meson exchange diagrams from contributing. The associated "current algebra" contact diagrams also vanish. It was found that fitting the model to the experimental Dalitz plot and the rate gave $a_0(980)$ properties consistent with the recent experimental ones.

IV SCALAR NONET "FAMILY" PROPERTIES

The nine states associated with the $\sigma(550)$, $\kappa(900)$, $f_0(980)$ and $a_0(980)$ are required in order to fit experiment in our model. What do their masses and coupling constants suggest about their quark substructure? (See [8] for more details.) Suppose we first try to assign them to a conventional $q\bar{q}$ nonet:

$$\sigma(550) \sim \frac{1}{\sqrt{2}}(u\bar{u} + d\bar{d}),$$
$$\kappa^+(900) \sim u\bar{s},$$
$$a_0^+(980) \sim u\bar{d},$$
$$f_0(980) \sim s\bar{s}. \qquad (4)$$

Then there are two puzzles. i) Why aren't the $a_0(980)$ and the $\sigma(550)$, which have the same number of non–strange quarks, degenerate? ii) Why aren't these particles, being p–wave states, in the same 1+ GeV energy region as the other p–wave states?

To study this, first note that most meson multiplets can be nicely understood using the concept of "ideal mixing". In Okubo's formulation [9], originally applied to the vector meson multiplet, the meson fields are grouped into a nonet matrix,

$$N_a^b = \begin{bmatrix} N_1^1 & a_0^+ & \kappa^+ \\ a_0^- & N_2^2 & \kappa^0 \\ \bar{\kappa}^+ & \bar{\kappa}^0 & N_3^3 \end{bmatrix}, \quad (5)$$

where the particle names have been chosen to fit the scalar mesons. The two $I=0$ states are the SU(3) singlet, $(N_1^1 + N_2^2 + N_3^3)/\sqrt{3}$ and the SU(3) octet member, $(N_1^1 + N_2^2 - 2N_3^3)/\sqrt{6}$. Okubo's ansatz for the mass terms was,

$$\mathcal{L}_{mass} = -a\text{Tr}(NN) - b\text{Tr}(NN\mathcal{M}), \quad (6)$$

where $a > 0$ and b are real constants and $\mathcal{M} = diag(1, 1, x)$ (with $x = m_s/m_u$) is the "spurion" matrix which breaks flavor SU(3) invariance. With (5) and (6) the $SU(3)$ singlet and SU(3) octet isoscalar states mix in such a way (ideal mixing) that the physical mass eigenstates emerge as $(N_1^1 + N_2^2)/\sqrt{2}$ and N_3^3. Furthermore there are two mass relations

$$m^2(a_0) = m^2(\frac{N_1^1 + N_2^2}{\sqrt{2}}),$$
$$m^2(a_0) - m^2(\kappa) = m^2(\kappa) - m^2(N_3^3). \quad (7)$$

Note that there are two different solutions depending on the sign of b. If $b > 0$ we get Okubo's original case where [with the identifications $a_0 \to \rho$, $\kappa \to K^*$, $(N_1^1 + N_2^2)/\sqrt{2} \to \omega$ and $N_3^3 \to \phi$] there is the conventional ordering

$$m^2(\phi) > m^2(K^*) > m^2(\rho) = m^2(\omega). \quad (8)$$

This agrees with counting the number of (heavier) strange quarks when we identify $N_a^b \sim q_a \bar{q}^b$.

On the other hand if $b < 0$ and we identify $N_3^3 \to \sigma$ and $(N_1^1 + N_2^2)/\sqrt{2} \to f_0$, the resulting ordering would be

$$m^2(f_0) = m^2(a_0) > m^2(\kappa) > m^2(\sigma), \quad (9)$$

which is in nice agreement with the present "observed" scalar spectrum. But this clearly does not agree with counting the number of strange quarks while assuming that the scalar mesons are simple quark anti-quark composites. This unusual ordering will agree with counting the number of strange quarks if we assume instead that the scalar mesons are schematically constructed as $N_a^b \sim T_a \bar{T}^b$ where $T_a \sim \epsilon_{acd} \bar{q}^c \bar{q}^d$ is a "dual" quark. Specifically

$$N_a^b \sim T_a \bar{T}^b \sim \begin{bmatrix} \bar{s}\bar{d}ds & \bar{s}\bar{d}us & \bar{s}\bar{d}ud \\ \bar{s}\bar{u}ds & \bar{s}\bar{u}us & \bar{s}\bar{u}ud \\ \bar{u}\bar{d}ds & \bar{u}\bar{d}us & \bar{u}\bar{d}ud \end{bmatrix} \qquad (10)$$

Note in particular that the light $\sigma \sim N_3^3$ contains no strange quarks. While this picture seems unusual, precisely the configuration (10) was found by Jaffe [4] in the framework of the MIT bag model. The key dynamical point is that the states in (10) receive (due to the spin and color spin recoupling coefficients) exceptionally large binding energy from the "hyperfine" piece of the gluon exchange interchange:

$$H_{hf} = -\Delta \sum_{i,j} (\mathbf{S}_i \cdot \mathbf{S}_j)(\mathbf{F}_i \cdot \mathbf{F}_j), \qquad (11)$$

wherein the sum goes over all pairs i, j while S_i and F_i are respectively the spin and color generators acting on the ith quark or antiquark.

While the picture above seems close to our expectations it is not quite right in detail. For example the masses do not exactly obey (7). Furthermore the simplest model for decay would give that $f_0 \to \pi\pi$ vanishes, in contradiction to experiment. Hence we add the extra mass terms

$$\mathcal{L}_{mass} = \text{Eq.}(6) - c\text{Tr}(N)Tr(N) - d/rmTr(N)\text{Tr}(N\mathcal{M}). \qquad (12)$$

The c and d terms give $f_0 - \sigma$ mixing. Now we solve for (a, b, c, d) in terms of the four masses m_σ =550 MeV, m_κ =900MeV, m_{a_0} =983.5 MeV and m_{f_0} =980 MeV. The solution boils down to a quadratic equation for (say) d. This gives two possible values for the mixing angle θ_s defined by,

$$\begin{pmatrix} \sigma \\ f_0 \end{pmatrix} = \begin{pmatrix} \cos\theta_s & -\sin\theta_s \\ \sin\theta_s & \cos\theta_s \end{pmatrix} \begin{pmatrix} N_3^3 \\ \frac{N_1^1 + N_2^2}{\sqrt{2}} \end{pmatrix}. \qquad (13)$$

The solution $\theta_s \approx -90^o$, giving $\sigma \approx (N_1^1 + N_2^2)/\sqrt{2}$ seems to correspond to restoring the $q\bar{q}$ model (4) for the scalars once more. The other solution $\theta_s \approx -20^o$ corresponds to σ being mainly N_3^3 which was just noted to be a characteristic signature of the $qq\bar{q}\bar{q}$ model (10). The very existence of these two different solutions highlights the fact that by just assuming a flavor transformation property for the scalars we are not forcing a particular identification of their underlying quark structure. Different substructures are naturally associated with different values of the parameters in the same effective Lagrangian. In any event, the extra terms in (12) have restored the ambiguity about the scalars' structure. We need more information to decide the issue. For this purpose we look at the trilinear couplings.

Using SU(3) invariance we write

$$\mathcal{L}_{N\phi\phi} = A\,\epsilon^{abc}\epsilon_{def}N_a^d\partial_\mu\phi_b^e\partial_\mu\phi_c^f + B\text{Tr}(N)\,\text{Tr}(\partial_\mu\phi\partial_\mu\phi)$$
$$+ C\text{Tr}(N\partial_\mu\phi)\,\text{Tr}(\partial_\mu\phi) + D\text{Tr}(N)\,\text{Tr}(\partial_\mu\phi)\,\text{Tr}(\partial_\mu\phi), \qquad (14)$$

where A, B, C, D are four real constants and ϕ represents the usual pseudoscalar nonet matrix. The derivatives stem from the requirement that (14) be the leading part of a chiral invariant object. If desired, we can rewrite the A term as a linear combination of the usual $\text{Tr}(N\partial_\mu \phi \partial_\mu \phi)$ and the three other terms. The motivation for the form given is that by itself the A term yields zero for $f_0 \to \pi\pi$ and $\sigma \to K\bar{K}$, both of which should vanish in the "fall apart" picture of a $T\bar{T}$ type scalar meson. Note that all the coupling constants which enter into our treatment of $\pi\pi$ and πK scattering depend on just A and B; C and D contribute only to the decays containing η or η' in the final state. For examples of couplings:

$$\gamma_{\kappa K\pi} = \gamma_{a_0 KK} = -2A,$$
$$\gamma_{\sigma\pi\pi} = 2B\sin\theta_s - \sqrt{2}(B-A)\cos\theta_s, etc. \quad (15)$$

The mixing angle solution which best fits the couplings needed to explain the $\pi\pi$ and πK scattering turns out to be $\theta_s \approx -20°$. Together with a suitable choice of C and D, the interactions involving η and η' are also consistently described (as mentioned in section III). Thus it seems that our results point to a picture in which the light scalars are mainly dual quark- dual antiquark rather than quark-antiquark type. Very recently Achasov [10] has argued that new experimental data from Novosibirsk on the radiative decay $\phi(1020) \to \pi^0 \eta \gamma$ are better fit with a $qq\bar{q}\bar{q}$ type model of the $a_0(980)$.

To sum up: assuming $\sigma(550)$, $\kappa(900)$, $f_0(980)$ and $a_0(980)$ to belong to a nonet N_a^b which is fitted into a chiral Lagrangian, we have found the parameters A, B, C, D which specify sixteen scalar-pseudoscalar-pseudoscalar coupling constants. These couplings and masses are used to explain $\pi\pi$ scattering (σ, f_0), πK scattering (κ) and $\eta' \to \eta\pi\pi$ (a_0) with regularized tree amplitudes. Furthermore, a small $\sigma - f_0$ mixing angle in (13) suggests that N_a^b is describing a structure similar to a dual quark dual antiquark. If this picture is correct there are many interesting applications and questions.

V POSSIBLE MECHANISM FOR NEXT LOWEST-LYING SCALARS

Of course, the success of the phenomenological quark model suggests that there exists a nonet of "conventional" $q\bar{q}$ scalars in the 1+ GeV range. What are the experimental candidates for these? [11] The situation for the isoscalar candidates is presently confusing. The $f_0(1370)$ may actually correspond to two different states. The $f_0(1500)$ may be a glueball while the $f_J(1710)$ does not necessarily have spin zero. Thus we will not focus on the isoscalars now. On the other hand the Review of Particle Physics "endorses" the isovector and isospinor candidates

$$a_0(1450): M = 1474 \pm 19\text{MeV}, \quad \Gamma = 265 \pm 13\text{MeV},$$
$$K_0^*(1450): M = 1429 \pm 6\text{MeV}, \quad \Gamma = 287 \pm 23\text{MeV}.$$

On the way to taking these states seriously as members of an ordinary p-wave nonet we encounter three puzzles. i) The mass of the $a_0^+(1450)$ (presumably a

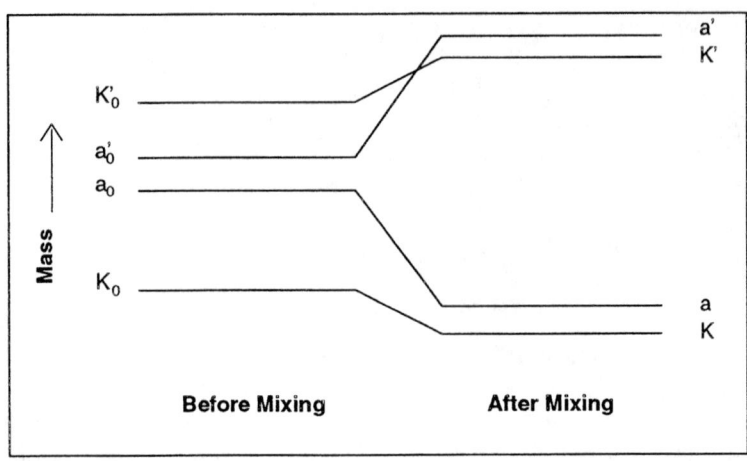

FIGURE 4. Mixing of two nonets-a',K',a and K stand respectively for the "physical" states $a_0(1450), K_0^*(1430), a_0(980)$ and $\kappa(900)$. K_0 and a_0 are the unmixed isospinor and isovector $qq\bar{q}\bar{q}$ states, while K_0' and a_0' are the corresponding unmixed $q\bar{q}$ states.

$u\bar{d}$ state is greater than that of the $K_0^{*+}(1430)$ (presumably a $u\bar{s}$ state). ii) The $a_0(1450)$ and $K_0^*(1430)$ are not less massive than the corresponding p-wave tensor mesons $a_2(1320)$ and $K_2^*(1430)$, as expected from an $L \cdot S$ interaction (e.g. $m[\chi_{c2}(1p)] > m[\chi_{c0}(1p)]$). iii) Assuming the known decay modes $K_0^*(1430) \to K\pi$ and $a_0(1450) \to \pi\eta, K\bar{K}, \pi\eta'$ saturate the total widths, we have from SU(3) flavor invariance that $\Gamma[a_0(1450)] = 1.51\Gamma[K_0^*(1430)]$. However, experimentally it is $(0.92 \pm 0.12)\Gamma[K_0(1430)]$ instead.

These puzzles can be simply resolved [12] if we assume that an ideally mixed heavier $q\bar{q}$ nonet N' in turn mixes with an ideally mixed $T\bar{T}$ nonet N (as in (10)) via

$$\mathcal{L}' = -\gamma \text{Tr}(NN'). \qquad (16)$$

Actually the assumption of exact ideal mixings is a simplification which can be relaxed. The mechanism is driven by the fact that $m(a_0') < m(K_0')$ while $m(a_0) > m(K_0)$. Here the subscript zero refers to the unmixed N and N' members. The splittings are summarized in Fig. 4.

The explanations are: i)Think of a perturbation theory approach. There is a smaller "energy denominator" for $a_0 - a_0'$ mixing than for $K_0 - K_0'$ mixing. Thus there is more $a_0 - a_0'$ repulsion as shown in Fig. 4. ii) Since the mixing of two levels "repels" them, both $a_0(1450)$ and $K_0^*(1430)$ are heavier than would be expected otherwise. Similarly the light scalars $a_0(980)$ and $\kappa(900)$ are lighter than they would be without the mixing (16). iii) The difference between the $a_0(1450)$ and

$K_0^*(1430)$ decay coupling constants can be understood from the necessarily greater mixture of the $qq\bar{q}\bar{q}$ component in the $a_0(1450)$ than in the $K_0^*(1430)$.

Clearly, looking at the isoscalars will be especially interesting when the experimental situation becomes clearer.

We would like to thank Francesco Sannino and Masayasu Harada for fruitful collaboration. One of us (J.S.) would like to thank the organizers for arranging a stimulating and enjoyable conference. The work has been supported in part by the US DOE under contract DE-FG-02-85ER40231.

REFERENCES

1. See the write-ups of V. A. Andrianov, W. Florkowski, S. Gerasimov, M. Jaminon, T. Kunihiro, M. Nemes, S. Nishiyama, G. Rupp, F. Stancu and M. Volkov.
2. N. Isgur and J. Weinstein, Phys. Rev. Lett.**48**, 659(1982).
3. N. Tornqvist, Z. Phys. **C68**, 647(1995); E. van Beveren et al, Z. Phys. **C30**, 615(1986).
4. R. Jaffe, Phys. Rev.**D15**, 267 (1977).
5. M. Harada, F. Sannino and J. Schechter, Phys. Rev. **D54**,1991(1996).
6. D. Black, A. H. Fariborz, F. Sannino and J. Schechter, Phys. Rev. **D58**, 054012 (1998).
7. A.H. Fariborz and J. Schechter, Phys. Rev. **D60**, 034002 (1999).
8. D. Black, A. H. Fariborz, F. Sannino and J. Schechter, Phys. Rev. **D59**, 074026 (1999).
9. S. Okubo, Phys. Lett. **5**, 165 (1963).
10. N. Achasov hep-ph/9904223.
11. Review of Particle Physics, Euro. Phys. J. **C3** (1999).
12. D. Black, A. H. Fariborz and J. Schechter, hep-ph/9907516.

First Radial Excitations of Scalar Meson Nonet

Mikhail K. Volkov[1] and Valeri L. Yudichev [2]

Bogoliubov Laboratory of Theoretical Physics,
Joint Institute for Nuclear Research, Dubna 141980, Russia

Abstract. First radial excitations of scalar and pseudoscalar meson nonets as well as their ground states are described as quark-antiquark bound systems in the framework of a nonlocal chiral quark model of the Nambu–Jona-Lasinio (NJL) type. Glueball states and their possible mixing with members of the $\bar{q}q$ multiplets are not considered.

Radial excitations are described by means of simple polynomial form factors in the momentum space which are introduced into the four-quark vertices of the quark Lagrangian. The form factors are chosen in the regime of spontaneous chiral symmetry breaking so that the gap equations do not change their standard form, which provides fulfillment of the low energy theorems. All free parameters of the form factors are fixed by experimentally observed masses of pseudoscalar mesons. The masses of scalar mesons are predicted.

The local six-quark 't Hooft interaction is introduced to solve the so-called $U_A(1)$-problem. The singlet-octet mixing for the ground and radially excited states of scalar and pseudoscalar mesons is taken into account.

The mass spectrum of scalar states is obtained. Classification of the considered meson nonets is given. It is shown that the meson states $a_0(1450)$, $f_0(1370)$, $f_J(1710)$ and $K_0^*(1430)$ are the first radial excitations of $a_0(980)$, $f_0(400-1200)$, $f_0(980)$, $K_0^*(960)$.

1. INTRODUCTION

A correct description of both the ground and excited states of scalar mesons encounters a variety of complex problems. Let us point out some of them. i) For a long time, the experimental status of the lightest scalar isoscalar singlet meson was unclear. In some papers, the resonance $f_0(1370)$ was considered the lowest resonance [1–3], and only in 1998 the state $f_0(400-1200)$ was included into the summary tables of PDG review[3] [4]. ii) Having quantum numbers of vacuum, the scalar isoscalar states are most probably get mixed with glueballs [5–8]. iii) There is also a lot of problems related to the description of $f_0(980)$ and $a_0(980)$.

---

[1] volkov@thsun1.jinr.ru
[2] yudichev@thsun1.jinr.ru
[3] However, in earlier editions of PDG the light σ state still could be found; it was excluded later.

Their unusual experimental branching ratios for several decays have brought forth different ideas concerning the structure of the mesons. Among them, there are the quark-antiquark model [1–3,7,9,10], the four-quark model [11,12] and the kaon molecule model [13,14]. iv) The strange meson $K_0^*(1430)$ seems too heavy to be the ground state: 1 GeV is more characteristic of the ground meson states (see [15–17]).

The description of ground and excited states of the pion, kaon and the vector meson nonet in the framework of a nonlocal version of the NJL model has been done in our earlier papers [18–20]. Here we intend to study the ground and first radially excited states of the scalar meson nonet and η, η' mesons.

To produce correct masses for the ground states of η and η', we, as usual, introduce the 't Hooft interaction [21–23]. Although our model is nonlocal, which is reflected in the presence of form factors at four-quark vertices, we, nevertheless, assume the 't Hooft term local. The form factors in scalar channels of quark current-current interaction are chosen identical with those in the pseudoscalar channel. This is the requirement of global chiral symmetry of quark interaction. With that assumption, there is no need for additional parameters in the form factors of scalar quark vertices. Therefore, the masses of scalar mesons can be immediately predicted after fixing the form factor parameters by the pseudoscalar meson masses from experiment. As a result, we have found that the model masses of radial excitations of scalar mesons are close to the experimentally observed $f_0(1370)$, $f_J(1710)$[4], $a_0(1450)$ and $K_0^*(1430)$ mesons. This allows us to interpret them as the first radial excitations of mesons $f_0(400-1200)$, $f_0(980)$, $a_0(980)$ and $K_0^*(960)$.

In view of the analysis given in papers [5,6], we consider $f_0(1500)$ as a $\bar{q}q$ state essentially mixed with a glueball. Insofar as our effective Lagrangian does not contain the glueball state, we do not pretend to describe $f_0(1500)$. However, we are going to include the glueball in our further work and take into account possible mixing with isoscalar $\bar{q}q$ states [7,8].

Concerning the strange scalar K_0^*, we think that the state with mass 1430 MeV is much likely a radial excitation of a light and wide resonance with mass about 960 MeV (see [15,16]). Further discussion on this problem is given in the conclusion.

As to the ground states $a_0(980)$ and $f_0(980)$, a detailed discussion on their internal structure and properties is beyond the scope of our paper.

Our report is organized as follows. In Sec. 2, we introduce the chiral quark Lagrangian with nonlocal four-quark vertices and local 't Hooft interaction. In Sec. 3, we calculate the effective Lagrangian for isovector and strange mesons in the one-loop approximation. There we renormalize meson fields and transform the free part of the Lagrangian to the diagonal form and obtain meson mass formulae. Section 4 is devoted to isoscalar mesons where we find masses and mixing coefficients. The model parameters are discussed in Sec. 5. The results of our work and possible ways to improve the model are discussed in Sec. 6.

---

[4] We assume hereafter $f_J(1710)$ is an isoscalar ($J = 0$).

2. $U(3) \times U(3)$ CHIRAL LAGRANGIAN WITH EXCITED MESON STATES AND 't HOOFT INTERACTION

We use a nonlocal separable four-quark interaction of a current-current form which admits nonlocal vertices (form factors) in the quark currents and the pure local six-quark 't Hooft interaction [21–23]:

$$\mathcal{L}(\bar{q},q) = \int d^4x \, \bar{q}(x)(i\not{\partial} - m^0)q(x) + \mathcal{L}_{\text{int}}^{(4)} + \mathcal{L}_{\text{int}}^{(6)}, \tag{1}$$

$$\mathcal{L}_{\text{int}}^{(4)} = \int d^4x \sum_{a=0}^{8} \sum_{i=1}^{N} \frac{G}{2} [j_{S,i}^a(x) j_{S,i}^a(x) + j_{P,i}^a(x) j_{P,i}^a(x)], \tag{2}$$

$$\mathcal{L}_{\text{int}}^{(6)} = -K \left[\det \left[\bar{q}(1+\gamma_5)q \right] + \det \left[\bar{q}(1-\gamma_5)q \right] \right]. \tag{3}$$

Here, m^0 is the current quark mass matrix ($m_u^0 \approx m_d^0$) and $j_{S(P),i}^a$ denotes the scalar (pseudoscalar) quark currents

$$j_{S(P),i}^a(x) = \int d^4x_1 d^4x_2 \, \bar{q}(x_1) F_{S(P),i}^a(x; x_1, x_2) q(x_2) \tag{4}$$

where $F_{S(P),i}^a(x; x_1, x_2)$ are the scalar (pseudoscalar) nonlocal quark vertices. To describe the first radial excitations of mesons, we take the form factors in the momentum space as follows (see [18–20]),

$$F_{S,j}^a(\mathbf{k}) = \lambda^a f_j^a, \qquad F_{P,j}^a = i\gamma_5 \lambda^a f_j^a, \tag{5}$$

$$f_1^a \equiv 1, \qquad f_2^a \equiv f_a(\mathbf{k}) = c_a(1 + d_a \mathbf{k}^2), \tag{6}$$

where λ^a are Gell-Mann matrices, $\lambda^0 = \sqrt{\frac{2}{3}}\mathbf{1}$, with $\mathbf{1}$ being the unit matrix. Here, we consider the form factors in the rest frame of mesons [5].

The part of Lagrangian (1), describing the ground states and first radial excitations, can be rewritten in the following form (see [21–23]):

$$\mathcal{L} = \int d^4x \left\{ \bar{q}(x)(i\not{\partial} - m^0)q(x) + \frac{G}{2} \sum_{a=0}^{8} \left[\left(j_{S,2}^a\right)^2 + \left(j_{P,2}^a\right)^2 \right] + \right.$$

$$\frac{1}{2} \sum_{a=1}^{9} \left[G_a^{(-)} (\bar{q}(x)\tau_a q(x))^2 + G_a^{(+)} (\bar{q}(x)i\gamma_5 \tau_a q(x))^2 \right] + \tag{7}$$

$$\left. G_{us}^{(-)} (\bar{q}(x)\lambda_u q(x))(\bar{q}(x)\lambda_s q(x)) + G_{us}^{(+)} (\bar{q}(x)i\gamma_5 \lambda_u q(x))(\bar{q}(x)i\gamma_5 \lambda_s q(x)) \right\},$$

---

[5] The form factors depend on the transversal parts of the relative momentum of quark-antiquark pairs $k_\perp = k - \frac{k \cdot P}{P^2} P$, where k and P are the relative and total momenta of a quark-antiquark pair, respectively. Then, in the rest frame of mesons, $\mathbf{P}_{\text{meson}} = 0$, the transversal momentum is $k_\perp = (0, \vec{k})$, and we can define the form factors as depending on the 3-dimensional momentum $\vec{k}$ alone.

where

$$\tau_i = \lambda_i \ (i=1,...,7), \quad \tau_8 = \lambda_u = (\sqrt{2}\lambda_0 + \lambda_8)/\sqrt{3},$$
$$\tau_9 = \lambda_s = (-\lambda_0 + \sqrt{2}\lambda_8)/\sqrt{3}, \tag{8}$$
$$G_1^{(\pm)} = G_2^{(\pm)} = G_3^{(\pm)} = G \pm 4Km_s I_1(m_s),$$
$$G_4^{(\pm)} = G_5^{(\pm)} = G_6^{(\pm)} = G_7^{(\pm)} = G \pm 4Km_u I_1(m_u),$$
$$G_u^{(\pm)} = G \mp 4Km_s I_1(m_s), \quad G_s^{(\pm)} = G, \quad G_{us}^{(\pm)} = \pm 4\sqrt{2}Km_u I_1(m_u).$$

Here m_u and m_s are the constituent quark masses and $I_1(m_q)$ is the integral defined for an arbitrary n as follows

$$I_n(m_q) = \frac{-iN_c}{(2\pi)^4} \int_{\Lambda_3} d^4k \frac{1}{(m_q^2 - k^2)^n}. \tag{9}$$

The 3-dimensional cut-off Λ_3 in (9) is implemented to regularize the divergent integrals[6].

3. THE MASSES OF ISOVECTOR AND STRANGE MESONS (GROUND AND EXCITED STATES)

After bosonization, the part of Lagrangian (7), describing the isovector and strange mesons, takes the form

$$\mathcal{L}(a_{0,1}, K_{0,1}^*, \pi_1, K_1, a_{0,2}, K_{0,2}^*, \pi_2, K_2) = \frac{a_{0,1}^2}{2G_{a_0}} - \frac{K_{0,1}^{*}{}^2}{G_{K_0^*}} - \frac{\pi_1^2}{2G_\pi} - \frac{K_1^2}{G_K} -$$
$$\frac{1}{2G}(a_{0,2}^2 + 2(K_{0,2}^*)^2 + \pi_2^2 + 2K_2^2) -$$
$$iN_c \text{Tr} \ln \left[1 + \frac{1}{i\partial\!\!\!/ - m} \sum_{a=1}^{7} \sum_{j=1}^{2} \lambda_a \left(\sigma_j^a + i\gamma_5 \varphi_j^a\right) f_j^a \right] \tag{10}$$

where $m = \text{diag}(m_u, m_d, m_s)$ is the matrix of constituent quark masses ($m_u \approx m_d$), σ_j^a and φ_j^a are the scalar and pseudoscalar fields: $\sum_{a=1}^{3}(\sigma_j^a)^2 \equiv a_{0,j}^2 = (a_{0,j}^0)^2 + 2a_{0,j}^+ a_{0,j}^-$, $\sum_{a=4}^{7}(\sigma_j^a)^2 \equiv 2K_{0,j}^{*}{}^2 = 2(\bar{K}_{0,j}^*)^0 (K_{0,j}^*)^0 + 2(K_{0,j}^*)^+ (K_{0,j}^*)^-$, $\sum_{a=1}^{3}(\varphi_j^a)^2 \equiv \pi_j^2 = (\pi_j^0)^2 + 2\pi_j^+ \pi_j^-$, $\sum_{a=4}^{7}(\varphi_j^a)^2 \equiv 2K_j^2 = 2\bar{K}_j^0 K_j^0 + 2\bar{K}_j^+ K_j^-$. As to the coupling constants G_a, they will be defined later (see Sect. 5 and (8)).

The free part of Lagrangian (10) has the following form

$$\mathcal{L}^{(2)}(\sigma, \varphi) = \frac{1}{2} \sum_{i,j=1}^{2} \sum_{a=1}^{7} \left(\sigma_i^a K_{\sigma,ij}^a(P) \sigma_j^a + \varphi_i^a K_{\varphi,ij}^a(P) \varphi_j^a\right) \tag{11}$$

---

[6] For instance, $I_1(m) = \frac{N_c m^2}{8\pi^2} \left[x\sqrt{x^2+1} - \ln(x + \sqrt{x^2+1})\right]\big|_{x=\Lambda_3/m}$.

where the coefficients $K^a_{\sigma(\varphi),ij}(P)$ are given below,

$$K^a_{\sigma(\varphi),ij}(P) = -\delta_{ij}\left[\frac{\delta_{i1}}{G_a^{(\mp)}} + \frac{\delta_{i2}}{G}\right] -$$
$$iN_c\text{Tr}\int_{\Lambda_3}\frac{d^4k}{(2\pi)^4}\frac{1}{\not{k}+\not{P}/2-m_q^a}r^{\sigma(\varphi)}f_i^a\frac{1}{\not{k}-\not{P}/2-m_{q'}^a}r^{\sigma(\varphi)}f_j^a, \quad (12)$$

$$r^\sigma = 1, \quad r^\varphi = i\gamma_5, \quad (13)$$

$$m_q^a = m_u \;\; (a=1,...,7), \quad m_{q'}^a = m_u \;\; (a=1,...,3), \quad m_{q'}^a = m_s \;\; (a=4,...,7), \quad (14)$$

with m_u and m_s being the constituent quark masses and f_j^a defined in (6). Integral (12) is evaluated by expanding in the meson field momentum P. To order P^2, one obtains

$$K^a_{\sigma(\varphi),11}(P) = Z^a_{\sigma(\varphi),1}[P^2 - (m_q^a \pm m_{q'}^a)^2 - M^2_{\sigma^a(\varphi^a),1}],$$
$$K^a_{\sigma(\varphi),22}(P) = Z^a_{\sigma(\varphi),2}[P^2 - (m_q^a \pm m_{q'}^a)^2 - M^2_{\sigma^a(\varphi^a),2}],$$
$$K^a_{\sigma(\varphi),12}(P) = K^a_{\sigma(\varphi),21}(P) = \gamma^a_{\sigma(\varphi)}[P^2 - (m_q^a \pm m_{q'}^a)^2], \quad (15)$$

where

$$Z^a_{\sigma,1} = 4I_2^a, \quad Z^a_{\sigma,2} = 4I_2^{ffa}, \quad \gamma^a_\sigma = 4I_2^{fa}, \quad (16)$$

$$Z^a_{\varphi,1} = ZZ^a_{\sigma,1}, \quad Z^a_{\varphi,2} = Z^a_{\sigma,2}, \quad \gamma^a_\varphi = Z^{1/2}\gamma^a_\sigma \quad (17)$$

and

$$M^2_{\sigma^a(\varphi^a),1} = (Z^a_{\sigma(\varphi),1})^{-1}\left[1/G_a^{(\mp)} - 4(I_1(m_q^a) + I_1(m_{q'}^a))\right], \quad (18)$$
$$M^2_{\sigma^a(\varphi^a),2} = (Z^a_{\sigma(\varphi),2})^{-1}\left[1/G - 4(I_1^{ffa}(m_q^a) + I_1^{ffa}(m_{q'}^a))\right]. \quad (19)$$

The factor Z appears here due to $\pi - a_1$-transitions [9,10,19],

$$Z = 1 - 6m_u^2/M^2_{a_1}, \quad (20)$$

and the integrals $I_2^{f..f}$ contain form factors:

$$I_2^{f..f_a}(m_q^a, m_{q'}^a) = \frac{-iN_c}{(2\pi)^4}\int_{\Lambda_3}d^4k\frac{f_a(\mathbf{k})..f_a(\mathbf{k})}{[(m_q^a)^2-k^2][(m_{q'}^a)^2-k^2]}. \quad (21)$$

Further, we consider only the scalar isovector and strange mesons because the masses of the pseudoscalar mesons have been already described in [19].

After the renormalization of the scalar fields

$$\sigma_i^{ar} = \sqrt{Z_{\sigma,i}^a}\,\sigma_i^a \tag{22}$$

the part of Lagrangian (11), describing for example the scalar isovector mesons, takes the form

$$\mathcal{L}_{a_0}^{(2)} = \tfrac{1}{2}\left(P^2 - 4m_u^2 - M_{a_0,1}^2\right)a_{0,1}^2 + \Gamma_{a_0}\left(P^2 - 4m_u^2\right)a_{0,1}a_{0,2} + \tfrac{1}{2}\left(P^2 - 4m_u^2 - M_{a_0,2}^2\right)a_{0,2}^2, \tag{23}$$

where

$$\Gamma_{\sigma^a} = I_2^{fa}/\sqrt{I_2 I_2^{ffa}}. \tag{24}$$

After the transformations of meson fields

$$\sigma^a = \cos(\theta_{\sigma,a} - \theta_{\sigma,a}^0)\sigma_1^{ar} - \cos(\theta_{\sigma,a} + \theta_{\sigma,a}^0)\sigma_2^{ar},$$
$$\hat{\sigma}^a = \sin(\theta_{\sigma,a} - \theta_{\sigma,a}^0)\sigma_1^{ar} - \sin(\theta_{\sigma,a} + \theta_{\sigma,a}^0)\sigma_2^{ar}, \tag{25}$$

Lagrangian (23) takes the diagonal form:

$$L_{a_0}^{(2)} = \tfrac{1}{2}(P^2 - M_{a_0}^2)a_0^2 + \tfrac{1}{2}(P^2 - M_{\hat{a}_0}^2)\hat{a}_0^2. \tag{26}$$

Here we have

$$M_{(a_0,\hat{a}_0)}^2 = \frac{1}{2(1 - \Gamma_{a_0}^2)}\bigg[M_{a_0,1}^2 + M_{a_0,2}^2 \pm \sqrt{(M_{a_0,1}^2 - M_{a_0,2}^2)^2 + (2M_{a_0,1}M_{a_0,2}\Gamma_{a_0})^2}\bigg] + 4m_u^2, \tag{27}$$

and

$$\tan 2\bar{\theta}_{\sigma,a} = \sqrt{\frac{1}{\Gamma_{\sigma^a}^2} - 1}\left[\frac{M_{\sigma^a,1}^2 - M_{\sigma^a,2}^2}{M_{\sigma^a,1}^2 + M_{\sigma^a,2}^2}\right], \quad 2\theta_{\sigma,a} = 2\bar{\theta}_{\sigma,a} + \pi, \tag{28}$$

$$\sin\theta_{\sigma,a}^0 = \sqrt{(1 + \Gamma_{\sigma^a})/2}. \tag{29}$$

Transformations (25) express the "physical" fields σ and $\hat{\sigma}$ through the "bare" ones σ_i^{ar}, and for calculations, these equations must be inverted. One will find the values of mixing coefficients for the scalar and pseudoscalar fields in [24–26].

An analogous procedure was implemented in the case of strange scalar mesons but is omitted here to make the paper short (see [24–26]).

4. THE MASSES OF ISOSCALAR MESONS (GROUND AND EXCITED STATES)

The 't Hooft interaction effectively gives rise to additional four-quark vertices in the isoscalar part of Lagrangian (7):

$$\mathcal{L}_{\text{isosc}} = \sum_{a,b=8}^{9} \left[(\bar{q}\tau_a q) T^S_{ab} (\bar{q}\tau_b q) + (\bar{q} i\gamma_5 \tau_a q) T^P_{ab} (\bar{q} i\gamma_5 \tau_b q) \right] \tag{30}$$

where $T^{S(P)}$ is a matrix with elements defined as follows (for definition of $G_u^{(\mp)}$, $G_s^{(\mp)}$ and $G_{us}^{(\mp)}$ see (8))

$$\begin{aligned} T^{S(P)}_{88} &= G_u^{(\mp)}/2, & T^{S(P)}_{89} &= G_{us}^{(\mp)}/2, \\ T^{S(P)}_{98} &= G_{us}^{(\mp)}/2, & T^{S(P)}_{99} &= G_s^{(\mp)}/2. \end{aligned} \tag{31}$$

This leads to nondiagonal terms in the free part of the effective Lagrangian for isoscalar scalar and pseudoscalar mesons after bosonization

$$\mathcal{L}_{\text{isosc}}(\sigma,\varphi) = -\frac{1}{4} \sum_{a,b=8}^{9} \left[\sigma_1^a (T^S)^{-1}_{ab} \sigma_1^b + \varphi_1^a (T^P)^{-1}_{ab} \varphi_1^b \right] - \frac{1}{2G} \sum_{a=8}^{9} \left[(\sigma_2^a)^2 + (\varphi_2^a)^2 \right] - $$
$$i \, \text{Tr} \ln \left\{ 1 + (i\slashed{\partial} - m)^{-1} \sum_{a=8}^{9} \sum_{j=1}^{2} \tau^a [\sigma_j^a + i\gamma_5 \varphi_j^a] f_j^a \right\}, \tag{32}$$

where $(T^{S(P)})^{-1}$ is the inverse of $T^{S(P)}$. From (32), in the one-loop approximation, one obtains the free part of the effective Lagrangian

$$\mathcal{L}^{(2)}(\sigma,\varphi) = \frac{1}{2} \sum_{i,j=1}^{2} \sum_{a,b=8}^{9} \left(\sigma_i^a K^{[a,b]}_{\sigma,ij}(P) \sigma_j^b + \varphi_i^a K^{[a,b]}_{\phi,ij}(P) \varphi_j^b \right). \tag{33}$$

(For the definition of $K^{[a,b]}_{\sigma(\varphi),i}$ and details see Appendix in [24–26].)

After the renormalization of both the scalar and pseudoscalar fields, analogous to (22), we come to the Lagrangian which can be represented in a form slightly different from that of (33). It is convenient to introduce 4-vectors of "bare" fields

$$\Phi = (\varphi_1^{8r}, \varphi_2^{8r}, \varphi_1^{9r}, \varphi_2^{9r}), \qquad \Sigma = (\sigma_1^{8r}, \sigma_2^{8r}, \sigma_1^{9r}, \sigma_2^{9r}). \tag{34}$$

Thus, we have

$$\mathcal{L}^{(2)}(\Sigma,\Phi) = \frac{1}{2} \sum_{i,j=1}^{4} \left(\Sigma_i \mathcal{K}_{\Sigma,ij}(P) \Sigma_j + \Phi_i \mathcal{K}_{\Phi,ij}(P) \Phi_j \right) \tag{35}$$

where the new functions $\mathcal{K}_{\Sigma(\Phi),ij}(P)$ have been introduced (see Appendix in [24–26]).

Up to this moment one has four pseudoscalar and four scalar meson states which are the octet and nonet singlets. The mesons of the same parity have the same quantum numbers and, therefore, they are expected to be mixed. In our model, the mixing is represented by 4×4 matrices $R^{\sigma(\varphi)}$ which transform the "bare" fields φ_i^{8r}, φ_i^{9r}, σ_i^{8r} and σ_i^{9r} entering into 4-vectors Φ and Σ to the "physical" ones η, η', $\hat{\eta}$, $\hat{\eta}'$, σ, $\hat{\sigma}$, f_0 and $\hat{f}_0$ represented as components of vectors $\Phi_{\rm ph}$ and $\Sigma_{\rm ph}$:

$$\Phi_{\rm ph} = (\eta, \hat{\eta}, \eta', \hat{\eta}'), \qquad \Sigma_{\rm ph} = (\sigma, \hat{\sigma}, f_0, \hat{f}_0) \qquad (36)$$

where a caret over a meson field stands for the first radial excitation of the meson. The transformation $R^{\sigma(\varphi)}$ is linear and nonorthogonal:

$$\Phi_{\rm ph} = R^\varphi \Phi, \qquad \Sigma_{\rm ph} = R^\sigma \Sigma. \qquad (37)$$

In terms of "physical" fields, the free part of the effective Lagrangian is of the conventional form, and the coefficients of matrices $R^{\sigma(\varphi)}$ give mixing of the $\bar{u}u$ and $\bar{s}s$ components (see [24–26]), with and without form factors.

5. MODEL PARAMETERS AND MESON MASSES

In our model, we have five basic parameters: the masses of the constituent $u(d)$ and s quarks, $m_u = m_d$ and m_s, the cut-off parameter Λ_3, the four-quark coupling constant G and the 't Hooft coupling constant K. We have fixed these parameters with the help of input parameters: the pion decay constant $F_\pi = 93$ MeV, the ρ-meson decay constant $g_\rho = 6.14$ (decay $\rho \to 2\pi$)[7], the masses of pion and kaon and the mass difference of η and η' mesons (for details of these calculations, see [19,20,23]). Here we give only numerical estimates of these parameters:

$$m_u = 280 \text{ MeV}, \quad m_s = 405 \text{ MeV}, \quad \Lambda_3 = 1.03 \text{ GeV},$$
$$G = 3.14 \text{ GeV}^{-2}, \quad K = 6.1 \text{ GeV}^{-5}. \qquad (38)$$

We also have a set of additional parameters $c_{qq}^{\sigma^a(\varphi^a)}$ in form factors f_2^a. These parameters are defined by masses of excited pseudoscalar mesons, $c_{uu}^{\pi,a_0} = 1.44$, $c_{uu}^{\eta,\eta',\sigma,f_0} = 1.5$, $c_{us}^{K,K_0^*} = 1.59$, $c_{ss}^{\eta,\eta',\sigma,f_0} = 1.66$. The slope parameters d_{qq} are fixed by special conditions satisfying the standard gap equation, $d_{uu} = -1.78$ GeV^{-2}, $d_{us} = -1.76$ GeV^{-2}, $d_{ss} = -1.73$ GeV^{-2} (see [19]). Using these parameters, we obtain masses of pseudoscalar and scalar mesons that are listed in Table 1 together with experimental values.

From our calculations we suggest the following interpretation of $f_0(1370)$, $f_J(1710)$ and $a_0(1470)$ mesons: we consider them as the lowest radial excitations of the ground states $f_0(400 - 1200)$, $f_0(980)$ and $a_0(980)$.

---

[7] Here we do not consider vector and axial-vector mesons, however, we have used the relation $g_\rho = \sqrt{6} g_\sigma$ together with the Goldberger–Treiman relation $g_\pi = m/F_\pi = Z^{-1/2} g_\sigma$ to fix the parameters m_u and Λ_3 (see [19]).

TABLE 1. The model masses of mesons, MeV

	GR	EXC	GR(Exp.) [4]	EXC(Exp.) [4]
M_σ	530	1330	400-1200	1200-1500
M_{f_0}	1070	1600	980 ± 10	1712 ± 5
M_{a_0}	830	1500	983.4 ± 0.9	1474 ± 19
$M_{K_0^*}$	960	1500	905 ± 50 [16]	1429 ± 12
M_π	140	1300	139.56995 ± 0.00035	1300 ± 100
M_K	490	1300	497.672 ± 0.031	1460(?)
M_η	520	1280	547.30 ± 0.12	1297.8 ± 2.8
$M_{\eta'}$	910	1470	957.78 ± 0.14	1440-1470

6. DISCUSSION AND CONCLUSION

Our calculations have shown that we can interpret the scalar states $f_0(1370)$, $a_0(1450)$ and $f_0(1710)$ as the first radial excitations of $f_0(400-1200)$, $a_0(980)$ and $f_0(980)$. We estimated their masses in the framework of a nonlocal chiral quark model.[8] We would like to emphasize that we did not use additional parameters except those necessary to fix the mass spectrum of pseudoscalar mesons. We used the same form factors both for the scalar and pseudoscalar mesons, which is the requirement of global chiral symmetry.

We assumed[9] that the state $f_0(1500)$ is essentially mixed with the glueball state [5–8], and its probable mixing with $f_0(980)$, $f_0(1370)$ and $f_J(1710)$ may provide us with a more correct description of the masses of these states[10] (see Table 1). We are going to consider this problem in a subsequent work.

A more complicated situation takes place for the ground state $a_0(980)$. In the framework of our quark-antiquark model, we have a mass deficit for this meson, 830 MeV instead of 980 MeV. We suspect that this drawback is caused by four-quark component in this state which we did not take into account [11,12].

There is also some mystery about the strange scalar meson K_0^*. Its experimental mass is large enough, $M_{K_0^*} = 1430$ MeV. In our model, there are two strange scalars, $K_0^*(960)$ and $K_0^*(1500)$. Thus, together with [15,16] we suppose that it is possible for a strange resonance, $K_0^*(960)$ to exist in nature still missed in detectors as the ground state whereas the resonance $K_0^*(1430)$ is its radial excitation. Our model gives for the excited meson K_0^*: $M_{\hat{K}_0^*} \approx 1500$ MeV.

One will find the estimates of scalars' decay widths and detailed calculations of the mass spectrum in [24–26].

---

[8] The estimates of widths of their major decay modes are given in [24–26].
[9] Arguments for this choice come from the analysis of γdecays. The reader is referred to [24–26] for details.
[10] Our estimates for the masses of f_0 and $\hat{f}_0$: $M_{f_0} = 1070$ MeV and $M_{\hat{f}_0} = 1600$ MeV are expected to shift to $M_{f_0} = 980$ MeV and $M_{\hat{f}_0} = 1710$ MeV after mixing with the glueball $f_0(1500)$.

This work has been supported by RFBR Grant N 98-02-16135 and by Heisenberg–Landau Program 1999.

REFERENCES

1. Törnqvist, N., *Phys. Rev. Lett.* **49**, 624–627 (1982).
2. Lanik, L., *Phys. Lett. B* **306**, 139–144 (1993).
3. Dmitrašinović, V., *Phys. Rev. C* **53**, 1383–1396 (1996).
4. Review of Particle Physics, *Europ. Phys. J. C* **3**, 353–611 (1998).
5. Anisovich, A.A., Bugg, D.V., and Sarantsev A.V., *Phys. Rev. D* **58**, 111503-1–111503-4 (1998).
6. Narison, S., *Nucl. Phys. B* **509**, 312–356 (1998).
7. Kusaka, K., Volkov, M.K., and Weise, W., *Phys. Lett. B* **302**, 145–150 (1993).
8. Jaminon, M., and Bossche, B. Van den, *Nucl. Phys. A* **619**, 285–294 (1997).
9. Volkov, M.K., *Ann. Phys.* **157**, 282–303 (1984).
10. Volkov, M.K., *Sov. J. Part. Nucl.* **17**, 186–203 (1986).
11. Jaffe, .L., *Phys. Rev. D* **15**, 267–280 (1977).
12. Achasov, N.N., Devyanin, S.A., and Shestakov, G.N., *Usp. Fiz. Nauk.* **142**, 361–393 (1984).
13. Weinstein, J., and Isgur, N., *Phys. Rev. Lett.* **48**, 659–661 (1982).
14. Weinstein, J., and Isgur, N., *Phys. Rev. D* **27**, 588–599 (1983).
15. Ishida, M.Y., and Ishida, S., "On existence of $\sigma(600)$ – Its physical implications and related problems", presented at HADRON'97, 4th Int. Conf. on Hadr. spectr., BNL, 1997, hep-ph/9712231.
16. Ishida, S., Ishida, M., Ishida, T., Takamatsu, K., and Tsuru, T., *Prog. Theor. Phys.* **98**, 621–629 (1997).
17. Scadron, M.D., *Phys. Rev. D* **26**, 239–247 (1982).
18. Volkov, M.K., and Weiss, ., *Phys. Rev. D* **56**, 221–229 (1997).
19. Volkov, M.K., *Phys. At. Nucl.* **60**, 1920–1929 (1997).
20. Volkov, M.K., Ebert, D., and Nagy M., *Int. J. Mod. Phys. A* **13**, 5443–5457 (1998).
21. Vogl, H., and Weise, W., *Progr. Part. Nucl. Phys.* **27**, 195–272 (1991).
22. Klevansky, S.P., *Rev. Mod. Phys.* **64**, 649–708 (1992).
23. Volkov, M.K., Nagy, M., and Yudichev, V.L., *Nuovo Cim. A* **112**, 225–232 (1999).
24. Volkov, M.K., and Yudichev, V.L., "Excited scalar mesons in a chiral quark model", to be published in *Int.J.Mod.Phys. A* **14**, (1999); hep-ph/9904226.
25. Volkov, M.K., and Yudichev, V.L., "Radial excitations of scalar and η, η' mesons in a chiral quark model", to be published in *Phys.At.Nucl.* **63** No. 8 (2000).
26. Volkov, M.K., and Yudichev, V.L., "Radially excited scalar, pseudoscalar, and vector meson nonets in a chiral quark model", to be published in *Phys.Part.Nucl.* **31**, No. 3 (2000); hep-ph/9906571.

Scalar Mesons as "Simple" $Q\bar{Q}$ States

Eef van Beveren[*] and George Rupp[†]

[*]*Departamento de Física, Universidade de Coimbra, 3000 Coimbra, Portugal (eef@teor.fis.uc.pt)*
[†]*Centro de Física das Interacções Fundamentais, Instituto Superior Técnico, Edifício Ciência, 1049-001 Lisboa, Portugal (george@ajax.ist.utl.pt)*

Abstract. The Nijmegen Unitarized Meson Model and its application to scalar mesons is briefly revisited. It is shown that all scalar states up to 1.5 GeV can be described as $^3P_0$ $q\bar{q}$ states coupled to the OZI-allowed open and closed two-meson channels consisting of pseudoscalar and vector mesons. Crucial are the manifestation of a resonance-doubling phenomenon, typical for strong S-wave decay, and the employment of truly flavor-symmetric coupling constants. Also S-wave meson-meson scattering is thus reasonably well described, without any parameter fit.

INTRODUCTION

The scalar mesons have been posing a serious problem to hadron spectroscopists over the past three decades, since it seems to be impossible to group these particles into standard nonets typical for mesonic $q\bar{q}$ systems. Among the various apparent inconsistencies with standard mesonic states, we should mention the enigmatic light and broad σ meson alias f_0(400-1200), the light and narrow $f_0(980)$ (old S^*) and $a_0(980)$ (old δ), and the excess of experimental candidates to constitute one ground-state scalar nonet. Therefore, a variety of alternative descriptions and mechanisms have been proposed, such as multiquark ($q^2\bar{q}^2$) configurations, glueballs, $K\bar{K}$ molecules, and instanton contributions.

In this talk, we shall demonstrate that no such exotic approaches are needed to obtain a satisfactory description of the scalar-meson sector, provided one works in a unitarized framework such as the Nijmegen unitarized meson model (NUMM) [1]. In particular, the doubling of resonance poles for the ground states as predicted by the NUMM [2,3], which is typical for S-wave scattering channels strongly coupled to confined channels [4], allows for *two complete* scalar nonets, thus accommodating all experimentally observed states up to 1.5 GeV. Such a resonance doubling was recently also observed in Refs. [5,6], employing a revised version of the Helsinki unitarized quark model (HUQM). However, in the latter work this doubling only occurs for some states, which precludes describing, for instance, an as yet to be confirmed light K_0^* (old κ) and the established $f_0(1500)$ resonance within the very same framework, contrary to the NUMM. We ascribe this failure to the use of coupling constants for the three-meson vertices that are not flavor independent [3,7], thus leading to a breaking of the usual nonet pattern for mesons.

In the following, we shall very briefly review the essence of the NUMM, present the results for the scalar mesons, and make some concluding remarks in a perspective of future work.

NIJMEGEN UNITARIZED MESON MODEL

The basic unitarization philosophy underlying the NUMM stems from the observation that most mesons are resonances, some of which so broad that their very existence seems doubtful, as for example the $f_0(400\text{-}1200)$. So it makes no sense to treat such states as stable $q\bar{q}$ systems, even if *a posteriori* hadronic decays are dealt with in perturbation theory. The problem is that such an approach ignores the possible real mass shifts due to strong decay, which, at least in principle, could be of the same order of magnitude as the resonance widths. To make things worse, there is no reason to presume beforehand that the effects of closed thresholds, corresponding to *virtual* two-meson decays or, in diagrammatic language, to *mesonic loops*, are negligible.

In order to meet these objections, in the NUMM the valence $q\bar{q}$ system describing a stable or "bare" meson and the various OZI-allowed two-meson decay channels are treated on an equal footing. To achieve this, use is made of a coupled-channel Schrödinger-type formalism, in which a physical meson is represented by a long state vector, for example in the case of the f_0 meson given by

$$|f_0> \;=\; \begin{pmatrix} n\bar{n} & (l=1) \\ s\bar{s} & (l=1) \\ \pi\pi & (l=0) \\ \eta_n\eta_n & (l=0) \\ \eta_s\eta_s & (l=0) \\ KK & (l=0) \\ \rho\rho & (l=0) \\ \rho\rho & (l=2) \\ \omega\omega & (l=0) \\ \omega\omega & (l=2) \\ \phi\phi & (l=0) \\ \phi\phi & (l=2) \\ K^*K^* & (l=0) \\ K^*K^* & (l=2) \end{pmatrix} \;,\; \begin{cases} V_{q\bar{q}} = \tfrac{1}{2}\mu_q\omega^2 r^2 \\ V_{M_1 M_2} = 0 \\ V_{qM} = \tilde{g}\dfrac{r}{r_0}e^{-\frac{1}{2}(\frac{r}{r_0})^2} \end{cases} . \qquad (1)$$

Since the f_0 is a scalar isosinglet, we take two P-wave $q\bar{q}$ channels that can mix, coupled to a series of S- and D-wave two-meson channels consisting of pseudoscalar and vector mesons. The mixing takes place via the meson-meson channels to which both $n\bar{n}$ and $s\bar{s}$ couple, i.e., the channels with kaons. Note that n is shorthand for

u or d quark, so η_n and η_s stand for the ideally mixed, non-strange and strange isosinglet pseudoscalar, respectively. In Eq. 1, we have also given the used potentials, which amount to a harmonic oscillator with constant frequency in the $q\bar{q}$ channels, and a peaked function vanishing at the origin for the transitions between $q\bar{q}$ and meson-meson channels, which should mimic the $^3P_0$ mechanism (see Ref. [1] for reasons and details). No direct interaction between the two $q\bar{q}$ channels is assumed, so the mixing takes place via the mesonic channels that couple to both. Also, no meson-meson or "final-state" interactions are included for simplicity. These assumptions are not strictly necessary, but facilitate the numerical tractability of the equations [1] and, moreover, allow a cleaner view on what are the pure unitarization effects. A possible relaxation of these restrictions will be discussed in the concluding remarks. As to the generic coupling constant $\tilde{g}$ in Eq. (1), it should be noted that it includes phenomenological factors for one-gluon exchange and closed-threshold suppression [1], besides flavor-symmetric coupling constants for the various $^3P_0$ three-meson vertices [7].

The NUMM has been applied to heavy quarkonia [8], pseudoscalar and vector mesons [1], and scalar mesons [2], with generally good results for the mesonic spectra and meson-meson phase shifts. Especially the predictions in the scalar sector are of a remarkable quality, considering that all model parameters have been previously fixed in a fit to the light and heavy pseudoscalar and vector mesons, without any additional adjustment to the scalars. However, before examining in detail the scalar sector, let us first try to get a feeling for the possible effects of unitarization in a very simple toy model (see also Ref. [9]).

To study qualitatively the influence of mesonic decay on the bare spectrum of a confining $q\bar{q}$ potential, let us consider the two-channel Schrödinger equation

$$\begin{pmatrix} H_{q\bar{q}} & \lambda V_{qM} \\ \lambda V_{qM} & H_{MM} \end{pmatrix} \begin{pmatrix} \Psi_{q\bar{q}} \\ \Psi_{MM} \end{pmatrix} = E \begin{pmatrix} \Psi_{q\bar{q}} \\ \Psi_{MM} \end{pmatrix} . \quad (2)$$

Here, $H_{q\bar{q}}$ contains a confining potential, that is, a harmonic oscillator, H_{MM} is taken to be a free S-wave Hamiltonian, and λV_{qM} simulates the transitions between the $q\bar{q}$ and meson-meson channel through $^3P_0$ quark-pair creation. In the case that the transition strength $\lambda = 0$, Eq. (2) just yields two disconnected spectra, a discrete one for the bare $q\bar{q}$ state, and a continuous one for the free two-meson system. Once $\lambda \neq 0$, one unique spectrum emanates, which amounts to a number of resonances resulting from the possibility for the confined $q\bar{q}$ system to decay into two mesons. In Fig. 1 [9], we plot the total meson-meson cross section for the case of small λ. We clearly see that there is an evident correspondence between the found peaks and the discrete $q\bar{q}$ energy levels indicated with crosses, the central resonance positions almost exactly coinciding with the bound-state energies. This is a situation one typically would find in atomic physics. However, in hadronic physics the state of affairs is usually very different, where the now large coupling λ reflects the possibility of strong decay. Such a situation is depicted in Fig. 2 [9], where the resonance peaks and bumps are not only at energies quite off the values in the bare spectrum, but even different in number. It is obvious that, if anything similar were to happen in real hadron spectroscopy, the consequences would be dramatic, since then hardly any inference could be drawn from the physical spectrum concerning the underlying confining $q\bar{q}$ potential. In the following application of the NUMM to scalar mesons, we shall verify that this indeed occurs.

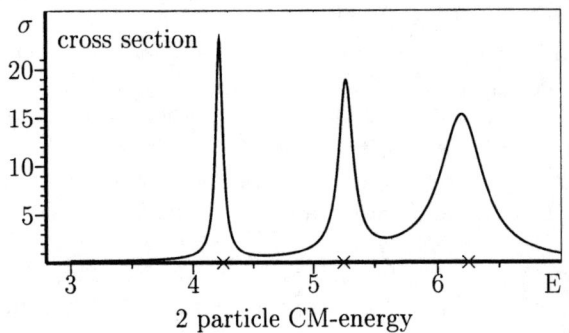

FIGURE 1. Elastic scattering cross section for small transition strength (arbitrary units).
Figure reprinted from: E. van Beveren, *Scalar mesons as $q\bar{q}$ systems with meson-meson admixtures,* Nucl. Phys. B (Proc. Suppl.) **21**, Page no. 44, Copyright (1991), with permission from Elsevier Science.

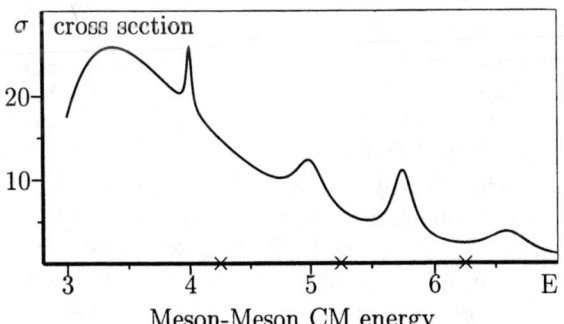

FIGURE 2. Elastic cross section for large transition strength (arbitrary units).
Figure reprinted from: E. van Beveren, *Scalar mesons as $q\bar{q}$ systems with meson-meson admixtures,* Nucl. Phys. B (Proc. Suppl.) **21**, Page no. 44, Copyright (1991), with permission from Elsevier Science.

SCALAR MESONS

As mentioned before, the application of the full NUMM to scalar mesons is straightforward, not involving any additional fit or alteration of the interactions. The only approximation is the omission of color splitting, which we verified to be negligible anyhow, due to the $^3P_0$ nature of the scalars themselves. The thus obtained S-matrices for all mesonic $J^{PC} = 0^{++}$ states have been searched for poles in the second Riemann sheet. The results for the real parts of the found poles are given in Table 1 [3], together with the predictions of model [5,6] (HUQM), the experimentally confirmed candidates, and the respective interpretations in terms of $q\bar{q}$ states. The most striking feature of the NUMM results is a doubling of states with respect to the bare $q\bar{q}$ spectrum, which was already forboded somehow by the toy model presented above. Here, we restrict ourselves to note that this dynamical phenomenon is typical for S-wave scattering channels strongly coupled to confined channels, and refer to Refs. [3,4] for more details and discussion. The resonance doubling allows for an identification of all observed scalar states up to 1.5 GeV, even obtaining one extra resonance not yet confirmed by experiment, namely a light K_0^* (old κ). But also this state has recently received renewed phenomenological and theoretical support [11–15]. As to the model of Ref. [5,6], we observe from Table 1 that such a resonance is not predicted, nor the established $f_0(1500)$. Moreover, the $f_0(1300)$ is interpreted as mainly $s\bar{s}$, while we claim it is predominantly $n\bar{n}$ and the $f_0(1500)$ mainly $s\bar{s}$. Apart from considering our interpretation more natural and favored by the known decay rates [3], we should mention a very recent lattice calculation largely supporting our model prediction [16]. The fact that model [5,6] fails to find two complete scalar nonets we ascribe to the use of coupling constants for the three-meson vertices that are not flavor independent [3]. Here, the crucial point is a point-particle approach in the derivation of the couplings, which leads to a wrong normalization in the case of the scalar mesons, being $^3P_0$ states themselves just as the created $q\bar{q}$ pairs [7,17].

Another feature of the NUMM, and also of the HUQM for that matter, is the automatic obtainment of a unitary, analytic S-matrix, which allows for a straightforward calculation of partial cross sections and phase shifts. Thus, we present our results for the elastic S-wave $\pi\pi$ and $K\pi$ phase shifts, in Figs. 3 and 4, respectively. We must reemphasize that these are model *predictions* and not the result of a fit. In that perspective, the results are surprisingly good, reproducing the bulk features of the experimental phases, including the resonant structures. In the $\pi\pi$ case, both the broad structure from roughly 400 to 950 MeV, owing to a $f_0(400\text{-}1200)$ (or σ) pole at $470 - 208i$ MeV, and the sharp resonance close to 1 GeV, due to a $f_0(980)$ pole at $994 - 17i$ MeV, are reasonably well described. Of course, some background structure is clearly lacking, which is no surprise in view of the neglect of final-state interactions in the meson-meson channels. Also notice that in the $K\pi$ case, where final-state interactions from t-channel meson exchanges are expected to be less important, the phase shifts are extremely well reproduced in the energy region 0.7–1.2 GeV, exactly where we find the lowest K_0^* pole, i.e., at $727 - 263i$ MeV. So

TABLE 1. Scalar-meson predictions and $q\bar{q}$ interpretations for the HUQM and NUMM, together with experimentally established states.
Table reprinted from: E. van Beveren and G. Rupp, *Comment on "Understanding the scalar meson $q\bar{q}$ nonet"*, Eur. Phys. J. C **10**, Page no. 471, Copyright (1999), with permission from Springer-Verlag.

	HUQM [5,6]		NUMM [2,3]		Exp. [10]
Resonance	ReE_{pole}	$q\bar{q}$ configuration	ReE_{pole}	$q\bar{q}$ configuration	Mass
$\sigma/f_0(400\text{--}1200)$	470	$1^{st} \approx \frac{1}{\sqrt{2}}(u\bar{u}+d\bar{d})$	470	$1^{st} \approx \frac{1}{\sqrt{2}}(u\bar{u}+d\bar{d})$	400–1200
$S^*/f_0(980)$	1006	$1^{st} \approx s\bar{s}$	994	$1^{st} \approx s\bar{s}$	980 ± 10
$\delta/a_0(980)$	1094	$1^{st} \frac{1}{\sqrt{2}}(u\bar{u}-d\bar{d})$	968	$1^{st} \frac{1}{\sqrt{2}}(u\bar{u}-d\bar{d})$	983 ± 1
κ/K_0^*	-	-	727	$1^{st}\ s\bar{d}$	?
$f_0(1370)$	1214	$2^{nd} \approx s\bar{s}$	1300	$2^{nd} \approx \frac{1}{\sqrt{2}}(u\bar{u}+d\bar{d})$	1200–1500
$f_0(1500)$	-	-	1500	$2^{nd} \approx s\bar{s}$	1500 ± 10
$a_0(1450)$	1592	$2^{nd} \frac{1}{\sqrt{2}}(u\bar{u}-d\bar{d})$	1300	$2^{nd} \frac{1}{\sqrt{2}}(u\bar{u}-d\bar{d})$	1474 ± 19
$K_0^*(1430)$	1450	$1^{st}\ s\bar{d}$	1400	$2^{nd}\ s\bar{d}$	1429 ± 6

it is unmistakably demonstrated that this very controversial resonance is perfectly compatible with a non-resonant behavior of the phases in the same region.

CONCLUDING REMARKS

In the foregoing, we hope to have made it clear that, for a reliable and detailed description of meson spectra in general, the effects of mesonic loops and decay should be taken into account. In particular, this is *a fortiori* true in the case of the scalar mesons, where, from a theoretical point of view, one is faced with very large couplings to S-wave two-meson channels, and, on the experimental side, a confusing picture of many broad as well as narrow resonances of very disparate masses emerges. Whatever the used approach, however, extreme care is required in the choice of the classes of the to be included decay channels, and in the computation of the respective coupling constants, lest one introduce explicitly flavor-breaking mechanisms that may distort the spectra in an unrealistic fashion.

The NUMM employed here may, of course, be subject to improvements. As indicated before, final-state interactions from t-channel meson exchanges should be included, aiming at restoring crossing symmetry to some degree, and in order to allow for a more accurate reproduction of the experimental meson-meson phase shifts. Furthermore, relativity should be addressed in a thoroughter way than just by some relativistic kinematics, preferably in a covariant quasipotential framework. This would require a profound overhaul of the mathematical formulation of the model, but could then make further refinements of the used interactions feasible.

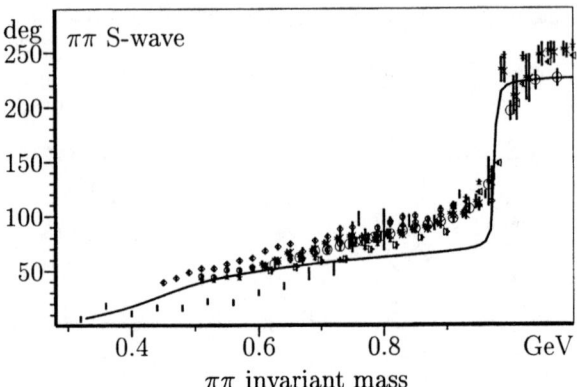

FIGURE 3. Model [2] results for $\pi\pi$ elastic S-wave phase shifts. The various sets of data are taken from (⊙, [18]), (∗, [19]), (⋆, ×, ◇, ◁, ▷ respectively for analyses A, B, C, D, and E of [20]), (∘, [21]), and (·, [22]).
Figure reprinted from: E. van Beveren, *Scalar mesons as $q\bar{q}$ systems with meson-meson admixtures*, Nucl. Phys. B (Proc. Suppl.) **21**, Page no. 47, Copyright (1991), with permission from Elsevier Science.

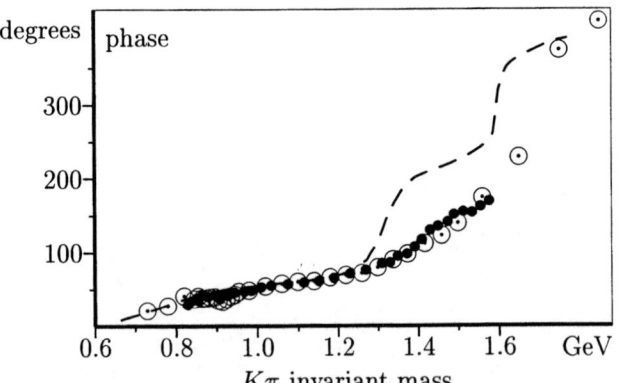

FIGURE 4. Kaon-pion $I = \frac{1}{2}$ S-wave phase shifts. The data indicated by ⊙ are taken from Ref. [23] and by • from Ref. [24]. The model results (dashed line) are taken from Ref. [2].
Figure reprinted from: E. van Beveren and G. Rupp, *Comment on "Understanding the scalar meson $q\bar{q}$ nonet"*, Eur. Phys. J. C **10**, Page no. 470, Copyright (1999), with permission from Springer-Verlag.

REFERENCES

1. E. van Beveren, G. Rupp, T. A. Rijken, and C. Dullemond, Phys. Rev. D **27**, 1527 (1983).
2. E. van Beveren, T. A. Rijken, K. Metzger, C. Dullemond, G. Rupp, and J. E. Ribeiro, Z. Phys. C **30**, 615 (1986).
3. Eef van Beveren and George Rupp, Eur. Phys. J. C **10**, 469 (1999).
4. E. van Beveren, C. Dullemond, T. A. Rijken, and G. Rupp, Lect. Notes Phys. **211**, 331 (1984).
5. Nils A. Törnqvist, Z. Phys. C **68**, 647 (1995).
6. Nils A. Törnqvist and Matts Roos, Phys. Rev. Lett. **76**, 1575 (1996).
7. Eef van Beveren and George Rupp, *Flavour symmetry of mesonic decay couplings*, hep-ph/9806248, to appear in Eur. Phys. J. C (1999).
8. E. van Beveren, C. Dullemond, and G. Rupp, Phys. Rev. D **21**, 772 (1980); D **22**, 787 (E) (1980).
9. E. van Beveren, Nucl. Phys. B **21** (Proc. Suppl.), 43 (1991).
10. C. Caso et al, Eur. Phys. J. C **3**, 1 (1998).
11. S. Ishida, M. Ishida, T. Ishida, K. Takamatsu, and T. Tsuru, Prog. Theor. Phys. **98**, 621 (1997).
12. Deirdre Black, Amir H. Fariborz, Francesco Sannino, and Joseph Schechter, Phys. Rev. D **58**, 054012 (1998).
13. Th. A. Rijken, V. G. J. Stoks, and Y. Yamamoto, Phys. Rev. C **59**, 21 (1999).
14. J. A. Oller, E. Oset, and J. R. Peláez, Phys. Rev. D **59**, 074001 (1999).
15. M. Napsuciale, *Scalar meson masses and mixing angle in a $U(3) \times U(3)$ linear sigma model*, hep-ph/9803396.
16. W. Lee and D. Weingarten, *Scalar quarkonium masses and mixing with the lightest scalar glueball*, hep-lat/9910008, to appear in Phys. Rev. D.
17. Eef van Beveren and George Rupp, Phys. Lett. B **454**, 165 (1999).
18. S. D. Protopopescu et al., Phys. Rev. D **7**, 1279 (1973).
19. B. Hyams et al., Nucl. Phys. B **64**, 134 (1973); Nucl. Phys. B **100**, 205 (1975).
20. G. Grayer et al., Nucl. Phys. B **75**, 189 (1974).
21. P. Estabrooks and A. D. Martin, Nucl. Phys. B **79**, 301 (1974).
22. N. N. Biswas et al., Phys. Rev. Lett. **47**, 1378 (1981).
23. P. Estabrooks et al., Nucl. Phys. B **133**, 490 (1978); P. Estabrooks, Phys. Rev. D **19**, 2678 (1979).
24. D. Aston, Nucl. Phys. B **296**, 493 (1988).

The Resonating Mean-Field Theoretical Approach to the Nambu-Jona-Lasinio Model
- $\sigma - \pi$ Sector -

Seiya Nishiyama*[1], João da Providência†[2] and Osamu Ohno‡[3]

*† *Centro de Física Teórica, Universidade de Coimbra, 3000-Coimbra,*
*‡ *Department of Physics, Kochi University, Kochi 780-8520, Japan*

Abstract. Using the Nambu-Jona-Lasinio (NJL) model, dynamical chiral-symmetry breaking, light-mesons spectra and properties of the mesons have been investigated by a conventional Hartree-Fock approach. To show the advantage of the resonating (Res) mean-field theory for a fermion system with large quantum fluctuations over the usual mean-field theory, we have applied it successfully to the NJL model to describe phenomena associated with a pionic excitation. Further we use it to describe precisely the phenomena including a sigma mesonic excitation. A state with large quantum fluctuations is approximated by superposition of three Dirac seas, non-orthogonal Slater determinants (S-dets) with different correlation structures. Suppose the Dirac seas are composed of "constituent quark masses". We make a direct optimization of Res-mean-field energy functional, i.e., a variation of Res-mean-field ground-state energy with respect to Res-mean-field parameters, the "constituent quark masses". The Res-mean-field ground and excited states generated with the S-dets explain most of mass spectra and associated properties of the pion and sigma meson.

INTRODUCTION

The resonating mean-field (Res-mean-field) thoery has been successfully applied by two of the present authors (S. N.) and (J. da P.) [1] (referred to as I) for description of mass spectrum and associated properties of the pion. In I we could reproduce a good numerical result of the pion mass as a "resonon" excitation energy of the Res-mean-field [2,3]. For the past decade, Fukutome and one of the present authors (S. N.) have proposed theories called the resonating Hartree-Fock

---

[1] Permanet address: Department of Physics, Kochi University, Kochi 780-8520, Japan
 E-mail address: nisiyama@fteor6.fis.uc.pt; nisiyama@cc.kochi-u.ac.jp
[2] E-mail address: providencia@gemini.ci.uc.pt
[3] E-mail address: kusamu@cc.kochi-u.ac.jp

(Res-HF) theory [2] and the resonating Hartree-Bogoliubov (Res-HB) theory [3] to treat problems of large quantum fluctuations in normal fermion systems and superconducting fermion systems. In the Res-HF and Res-HB theories, the ground state is assumed to be superposition of multiple mean-field wave functions with different correlation structures. The Res-HF theory had been applied to a well-known solvable model in order to clarify essential features of the theory and to show its advantage over the usual HF theory. The Res-HF ground state with a few Slater determinants (S-dets) reproduces the exact ground-state energy and explains most of the ground-state correlation energy in all the correlation regimes [4,5]. The Res-HF approach has turned out to be a promising tool and provides a method to work better than the usual HF. The Res-HB theory had been applied to describe resonance of shapes in nuclei by a schematic model with pairing correlations [6].

It is now widely recognized that quantum chromo dynamics (QCD) endows the fundamental theory of strong interactions. In QCD, spontaneous breaking of chiral symmetry, leading to a condensate of quark-antiquark pairs in the QCD vacuum, plays a crucial role in description of low-energy characteristics of hadrons. The quark-antiquark condensation caused by dynamical chiral-symmetry breaking was first proposed by Nambu and Jona-Lasinio (NJL) [7]. The NJL model containing a chiral effective interaction between quarks provides a basic framework for spontaneous realization of chiral-symmetry breaking. It allows for collective mesonic excitations whose properties have characteristics of physical particles. Then the NJL model has been regarded as a reasonable and manageable approximation to QCD for intermediate length scales and to dynamical chiral-symmetry breaking [8].

Using the NJL model Hamiltonian, one of the present authors (J. da P.), Ruivo and Sousa have investigated dynamical chiral-symmetry breaking, light-mesons spectra and properties of the mesons on the basis of the conventional HF and time-dependent HF (TDHF) approaches [9]. This theoretical framework and technique in many-body problems, very familiar to nuclear physicists, turned out to be appropriate to describe the properties of hadrons in an exceedingly intuitive way in the spirit of a mean-field approximation. The HF procedure describes the realization of a stable equilibrium configuration in the context of the mean-field approximation. The TDHF method gives the random phase approximation (RPA) collective excitations with small amplitudes which allow to interpret both the low-energy light-mesons mass spectra and properties of the mesons [10].

The Res-mean-field theories are able to treat in a rigorous manner large quantum fluctuations and strong correlation effects due to quantum- and dynamical- tunneling effects falling outside the scope of the RPA and of mode-mode coupling theories. Thus, they are possible to reveal surprisingly dramatic aspects of the physics of fermion systems with large quantum fluctuations. The Res-mean-field methods have been constructed on the basis of a group theoretical deduction starting from the fact that they are based on the Lie algebras of the fermion pair operators arising from the canonical anti-commutation relation of the fermion. Therefore, they should have a universal applicability to problems of current topics in wider fields of physics, from mesoscopic physics to hadron physics.

Extending the method developed in I, we intend to describe more precisely phenomena of the pion and sigma meson. A state $|\Psi\rangle$ with large quantum fluctuations is approximated by superposition of three Dirac seas $|\Psi\rangle = \sum_{i=1}^{3} c_i |u_i\rangle$. The $|\Psi\rangle$ is a coherent state representation (CS rep) of fermion state vectors represented by a function $\Psi(g) = \langle g|\Psi\rangle$ on U(4). The ket $|g\rangle$ is given as $|g\rangle = U(g)|\phi\rangle$ in which the unitary operator $U(g)$ induces a Thouless transformation of a reference S-det $|\phi\rangle$ [11]. The U(4) CS rep of a spinor state is a representation on the coset U(4)/U(2) denoted as u which is an 4×2 sub-matrix of a U(4) matrix.

MODEL AND RES-MEAN-FIELD APPROXIMATION

We consider a naive NJL model which describes a system of many quarks interacting via a two-body force according to the chiral-invariant Hamiltonian

$$H = \sum_{i=1}^{N} p_i \cdot \alpha_i - g \sum_{i \neq j}^{N} \delta(r_i - r_j) \left[\beta_i \beta_j - \beta_i \gamma_i^5 \beta_j \gamma_j^5 \right], \tag{1}$$

where α_i, β_i and γ_i^5 stand for the standard Dirac matrices acting on the degrees of freedom of the quark i and g is the coupling constant. We use the NJL model with only one single flavor but without isospin [7]. If $g = 0$, the above Hamiltonian describes a massless Dirac free quark and a Dirac sea (the S-det) of massless quark having zero chirality is given as $|\Phi_0\rangle = \prod_{i=1}^{N} d_i^{\dagger(0)} |0\rangle$, where $d_i^{\dagger(0)}$ is the creation operator of a massless negative-energy state and $|0\rangle$ is the absolute vacuum. The index i stands for the momentum p and helicity s and the absolute value of p_i satisfies $|p_i| \leq \Lambda$ (the highest momentum of the occupied states). When g is switched on, the system undergoes a phase transition into a state of chiral broken symmetry above a critical value of the coupling strength $g_{\rm cr}$. As the consequence of the chiral-symmetry breaking, the quarks acquire a dynamical mass. A Dirac sea (the S-det) of massive quarks is written as $|\Phi\rangle = \prod_{i=1}^{N} d_i^{\dagger} |0\rangle$, where $d_i^{\dagger}$ is the creation operator of a massive negative-energy state. According to Refs. [7,9], the operators b_i ($b_i^{\dagger}$) and d_i ($d_i^{\dagger}$) are related to $b_i^{(0)}$ ($b_i^{\dagger(0)}$) and $d_i^{(0)}$ ($d_i^{\dagger(0)}$) as

$$\begin{aligned}
\left[b_{-p,s}, b_{p,s}, d_{-p,s}^{\dagger}, d_{p,s}^{\dagger} \right] &= U(g) \left[b_{-p,s}^{(0)}, b_{p,s}^{(0)}, d_{-p,s}^{\dagger(0)}, d_{p,s}^{\dagger(0)} \right] U^{\dagger}(g) \\
&= \left[b_{-p,s}^{(0)}, b_{p,s}^{(0)}, d_{-p,s}^{\dagger(0)}, d_{p,s}^{\dagger(0)} \right] g.
\end{aligned} \tag{2}$$

Using $\det g$ (determinant of the matrix g) and I_4 (4 dimensional unit matrix), the 4×4 matrix g is given in terms of $\beta_p = |p|/E_p$ and $E_p = \sqrt{p^2 + M^2}$ as

$$g = \sqrt{(1+\beta_p)/2} \cdot I + \sqrt{(1-\beta_p)/2} \cdot \gamma^5 \beta \Sigma_1, \quad gg^{\dagger} = g^{\dagger}g = I_4, \quad \det g = 1, \tag{3}$$

where by Σ we denote the 4×4 matrix-valued vector $\Sigma = (\Sigma_1, \Sigma_2, \Sigma_3)$ defined as $\Sigma = \begin{bmatrix} \sigma & 0 \\ 0 & \sigma \end{bmatrix}$ with $\sigma = (\sigma_1, \sigma_2, \sigma_3)$. Here σ denotes the vector having the Pauli matrices for components. In the Res-mean-field approximation, the "constituent quark mass" M is treated as a variational parameter. The total number of quarks in negative-energy states is N and a momentum cutoff Λ is used to regularize the Res-mean-field theory [9]. The S-det of massive quarks with negative energy is a coherent state of massless quark-antiquark pairs with negative energy.

RES-MEAN-FIELD EQUATION FOR CONSTITUENT QUARKS

Following I, we introduce three 4×2 isometric matrices $u_{i,\,ip,\,r}$ ($i = 1 \sim 3$) with "constituent quark masses" M_i ($i = 1 \sim 3$) and 2×2 $z_{ij,\,p}$ ($i \neq j = 1 \sim 3$) by

$$\left. \begin{array}{l} u_{1,\,1p,\,r} = \begin{bmatrix} -\sqrt{(1-\beta_{1p})/2}\,\sigma \cdot n \cdot \chi_r \\ \sqrt{(1+\beta_{1p})/2} \cdot \chi_r \end{bmatrix}, \; \chi_1 = \begin{bmatrix} 1 \\ 0 \end{bmatrix}, \; \chi_2 = \begin{bmatrix} 0 \\ 1 \end{bmatrix}, \\[2ex] u_{2,\,2p,\,r} = \gamma^5 u_{1,\,2p,\,r}\,, \quad u_{3,\,3p,\,r} = u_{1,\,3p,\,r}\,, \quad (r = 1, 2) \end{array} \right\} \quad (4)$$

$$u^{\dagger}_{i,\,ip,\,r} u_{i,\,ip,\,r'} = \delta_{rr'}\,, \quad z_{ij,\,p,\,rr'} = u^{\dagger}_{i,\,ip,\,r} u_{j,\,jp,\,r'}\,. \quad (r, r' = 1, 2) \quad (5)$$

Substituting Eq. (4) into Eq. (5), for $r, r' = 1, 2$ we have

$$\left. \begin{array}{l} z_{12,\,p,\,rr'} = -\frac{1}{2}(\sqrt{1-\beta_{1p}}\sqrt{1+\beta_{2p}} + \sqrt{1+\beta_{1p}}\sqrt{1-\beta_{2p}}\,)\chi^{\dagger}_r \sigma \cdot p \chi_{r'}, \\[2ex] z_{13,\,p,\,rr'} = \frac{1}{2}(\sqrt{1+\beta_{1p}}\sqrt{1+\beta_{3p}} + \sqrt{1-\beta_{1p}}\sqrt{1-\beta_{3p}}\,)\delta_{rr'}, \end{array} \right\} \quad (6)$$

where $\beta_{ip} = M_i/E_{ip}$ and $E_{ip} = \sqrt{p^2 + M_i^2}$ ($i = 1 \sim 3$). From (6) their matrix- and inverse matrix- forms together with the determinants are obtained, respectively as

$$z_{12,\,p} = -\frac{1}{\widetilde{E}_p}\sigma \cdot p\,, \; z^{-1}_{12,\,p} = -\frac{\widetilde{E}_p}{p^2}\sigma \cdot p\,, \; \det z_{12,\,p} = -\frac{1}{2}\left(1 + \frac{p^2 - M_1 M_2}{E_{1p} E_{2p}}\right), \quad (7)$$

$$\widetilde{E}_p = \frac{2\sqrt{E_{1p}}\sqrt{E_{2p}}}{\sqrt{E_{1p}/p - M_1/p}\sqrt{E_{2p}/p + M_2/p} + \sqrt{E_{1p}/p + M_1/p}\sqrt{E_{2p}/p - M_2/p}}, \quad (8)$$

$$z_{13,\,p} = \frac{1}{\widehat{E}_p} p \cdot I_2\,, \; z^{-1}_{13,\,p} = \frac{\widehat{E}_p}{p} \cdot I_2\,, \; \det z_{13,\,p} = \frac{1}{2}\left(1 + \frac{p^2 + M_1 M_3}{E_{1p} E_{3p}}\right), \quad (9)$$

$$\widehat{E}_p = \frac{2\sqrt{E_{1p}}\sqrt{E_{3p}}}{\sqrt{E_{1p}/p + M_1/p}\sqrt{E_{3p}/p + M_3/p} + \sqrt{E_{1p}/p - M_1/p}\sqrt{E_{3p}/p - M_3/p}}. \quad (10)$$

The corresponding formulas for $z_{32,\,p,\,rr'}$ are obtained if we replace the index 1 by 3 in Eq. (7). Using new variables $x_i = \frac{M_i}{\Lambda}$ ($i = 1 \sim 3$), the overlap integrals $\det z_{12}$ and $\det z_{13}$ are expressed as

$$\det z_{12} = \prod_p \det z_{12,\,p} \cdot \theta(\Lambda^2 - p^2) = \exp\left(\sum_p \ln \det z_{12,\,p} \cdot \theta(\Lambda^2 - p^2)\right), \quad (11)$$

$$\det z_{13} = \prod_p \det z_{13,\,p} \cdot \theta(\Lambda^2 - p^2) = \exp\left(\sum_p \ln \det z_{13,\,p} \cdot \theta(\Lambda^2 - p^2)\right), \quad (12)$$

from which we get $\det z_{12} = e^{a\widetilde{z}(x_1,x_2)}$ and $\det z_{13} = e^{a\widehat{z}(x_1,x_3)}$. Here we have used $\sum_p = V \int d^3p/(2\pi)^3$ and $V = 6\pi^2 a/\Lambda^3$. We had to introduce the dimensionless volume parameter a to get a finite value of the overlap integrals. The a gives a "confinement volume of quarks". As already stressed in I, the deficiency of unstableness of the NJL vacuum may be a manifestation of the lack of confinement in the NJL model. This is the main drawback of the NJL model. The introduction of a "confinement volume of quarks" in I and the present calculation is one of possible choice of treatment to give a stabilizing mechanism of the NJL vacuum against a chiral rotation. The step function θ is defined as $\theta(x) = \begin{cases} 1, & x \geq 0, \\ 0, & x < 0. \end{cases}$ The overlap functions $\widetilde{z}(x_1, x_2)$ ($x_1 \geq x_2$) and $\widehat{z}(x_1, x_3)$ ($x_1 \geq x_3$) are given as

$$\begin{aligned}\widetilde{z}(x_1, x_2) &= \tfrac{1}{3} - \ln 2 - (x_1^2 + x_2^2) + (x_1^3 \tan^{-1} \tfrac{1}{x_1} + x_2^3 \tan^{-1} \tfrac{1}{x_2}) \\ &+ \ln(1 + \tfrac{1-x_1 x_2}{\sqrt{1+x_1^2}\sqrt{1+x_2^2}}) - \tfrac{1}{3}(1 - 2x_1^2 + 3x_1 x_2 - x_2^2)\sqrt{\tfrac{1+x_1^2}{1+x_2^2}} \\ &- \tfrac{1}{3}x_1(2x_1^2 - 3x_1 x_2 + 2x_2^2)E(\tan^{-1} \tfrac{1}{x_2}, \sqrt{1 - \tfrac{x_2^2}{x_1^2}}) \\ &+ \tfrac{1}{3}x_1 x_2^2 F(\tan^{-1} \tfrac{1}{x_2}, \sqrt{1 - \tfrac{x_2^2}{x_1^2}}),\end{aligned} \quad (13)$$

$$\begin{aligned}\widehat{z}(x_1, x_3) &= \tfrac{1}{3} - \ln 2 - (x_1^2 + x_3^2) + (x_1^3 \tan^{-1} \tfrac{1}{x_1} + x_3^3 \tan^{-1} \tfrac{1}{x_3}) \\ &+ \ln(1 + \tfrac{1+x_1 x_3}{\sqrt{1+x_1^2}\sqrt{1+x_3^2}}) - \tfrac{1}{3}(1 - 2x_1^2 - 3x_1 x_3 - x_3^2)\sqrt{\tfrac{1+x_1^2}{1+x_3^2}} \\ &- \tfrac{1}{3}x_1(2x_1^2 + 3x_1 x_3 + 2x_3^2)E(\tan^{-1} \tfrac{1}{x_3}, \sqrt{1 - \tfrac{x_3^2}{x_1^2}}) \\ &+ \tfrac{1}{3}x_1 x_3^2 F(\tan^{-1} \tfrac{1}{x_3}, \sqrt{1 - \tfrac{x_3^2}{x_1^2}}),\end{aligned} \quad (14)$$

where $E(\phi, k)$ and $F(\phi, k)$ are incomplete elliptic integrals. We also get the function $\widetilde{z}(x_3, x_2)$ if the index 1 is replaced by 3 in (13). Using (4), (7) and (9), the density matrices $W_{ii,\,p}$ ($i = 1 \sim 3$) for negative-energy states are given as

$$W_{11(33),\,p} = \frac{1}{2}\left(1 - \beta\frac{M_{1(3)}}{E_{1(3)p}} - \gamma^5\frac{\Sigma p}{E_{1(3)p}}\right)\theta(\Lambda^2 - p^2), \quad (15)$$

$$W_{22,\,p} = \frac{1}{2}\left(1 + \beta\frac{M_2}{E_{2p}} - \gamma^5\frac{\Sigma p}{E_{2p}}\right)\theta(\Lambda^2 - p^2) = \gamma^5 W_{11,\,p}(M_1 \to M_2)\gamma^5. \quad (16)$$

The interstate density matrices $W_{12,\,p}$ and $W_{13,\,p}$ are calculated to be

$$W_{12,\,p} = \frac{1}{2}\left(1 + \beta\frac{\widetilde{A}_p - \widetilde{B}_p}{\widetilde{A}_p + \widetilde{B}_p} - \gamma^5\frac{\widetilde{C}_p + \widetilde{D}_p}{\widetilde{A}_p + \widetilde{B}_p}\Sigma p - \beta\gamma^5\frac{\widetilde{C}_p - \widetilde{D}_p}{\widetilde{A}_p + \widetilde{B}_p}\Sigma p\right)\theta(\Lambda^2 - p^2), \quad (17)$$

$$W_{13,\,p} = \frac{1}{2}\left(1 + \beta\frac{\widehat{A}_p - \widehat{B}_p}{\widehat{A}_p + \widehat{B}_p} - \gamma^5\frac{\widehat{C}_p + \widehat{D}_p}{\widehat{A}_p + \widehat{B}_p}\Sigma p - \beta\gamma^5\frac{\widehat{C}_p - \widehat{D}_p}{\widehat{A}_p + \widehat{B}_p}\Sigma p\right)\theta(\Lambda^2 - p^2), \quad (18)$$

$$\left.\begin{array}{l}\widetilde{A}_p \equiv \sqrt{\frac{E_{1p}}{p} - \frac{M_1}{p}}\sqrt{\frac{E_{2p}}{p} + \frac{M_2}{p}}, \quad \widehat{A}_p \equiv \sqrt{\frac{E_{1p}}{p} - \frac{M_1}{p}}\sqrt{\frac{E_{3p}}{p} - \frac{M_3}{p}}, \\[4pt] \widetilde{B}_p \equiv \sqrt{\frac{E_{1p}}{p} + \frac{M_1}{p}}\sqrt{\frac{E_{2p}}{p} - \frac{M_2}{p}}, \quad \widehat{B}_p \equiv \sqrt{\frac{E_{1p}}{p} + \frac{M_1}{p}}\sqrt{\frac{E_{3p}}{p} + \frac{M_3}{p}}, \\[4pt] \widetilde{C}_p \equiv \sqrt{\frac{E_{1p}}{p} + \frac{M_1}{p}}\sqrt{\frac{E_{2p}}{p} + \frac{M_2}{p}}\frac{1}{p}, \quad \widehat{C}_p \equiv \sqrt{\frac{E_{1p}}{p} + \frac{M_1}{p}}\sqrt{\frac{E_{3p}}{p} - \frac{M_3}{p}}\frac{1}{p}, \\[4pt] \widetilde{D}_p \equiv \sqrt{\frac{E_{1p}}{p} - \frac{M_1}{p}}\sqrt{\frac{E_{2p}}{p} - \frac{M_2}{p}}\frac{1}{p}, \quad \widehat{D}_p \equiv \sqrt{\frac{E_{1p}}{p} - \frac{M_1}{p}}\sqrt{\frac{E_{3p}}{p} + \frac{M_3}{p}}\frac{1}{p}.\end{array}\right\} \quad (19)$$

To get the correct density matrix W_p we replace $\beta_p = |p|/E_p$ with $\beta_p = M/E_p$.

The Res-mean-field configuration-interaction (CI) equation to determine the mixing coefficients c_i ($i = 1 \sim 3$) is written simply in terms of the real quantities as

$$\begin{bmatrix} H[W_{11}] - E & (H[W_{12}] - E)\det z_{12} & (H[W_{13}] - E)\det z_{13} \\ (H[W_{12}] - E)\det z_{12} & H[W_{22}] - E & (H[W_{32}] - E)\det z_{32} \\ (H[W_{13}] - E)\det z_{13} & (H[W_{32}] - E)\det z_{32} & H[W_{33}] - E \end{bmatrix}\begin{bmatrix} c_1 \\ c_2 \\ c_3 \end{bmatrix} = 0$$

and the normalization $c_1^2 + c_2^2 + c_3^2 + 2c_1c_2\det z_{12} + 2c_1c_3\det z_{13} + 2c_3c_2\det z_{32} = 1$.
The usual mean-field energy functional $\langle u_i|H|u_i\rangle (= H[W_{ii}],\ i = 1 \sim 3)$ is given by

$$H[W_{ii}] = -\frac{\Lambda^4}{4\pi^2}\left[\sqrt{1 + x_i^2} - \frac{3}{2}x_i^2 v(x_i) + \frac{g\Lambda^2}{\pi^2}x_i^2 v^2(x_i) + \frac{2}{9}\frac{g\Lambda^2}{\pi^2}\right]V, \quad (20)$$

where $v(x) \equiv \sqrt{1 + x^2} - x^2\ln f(x)$ and $f(x) \equiv \frac{1}{x} + \sqrt{1 + \frac{1}{x^2}}$. Using (17) and (18), the matrix elements of the Hamiltonian $\langle u_1|H|u_2\rangle$ and $\langle u_1|H|u_3\rangle$ are calculated as

$\langle u_1|H|u_2\rangle = H[W_{12}]\cdot\det z_{12}$,
$H[W_{12}] = -\frac{\Lambda^4}{4\pi^2}\left[\frac{1}{x_1+x_2}\left[x_1\{\sqrt{1+x_2^2} + \frac{1}{2}x_2^2 v(x_2)\} + x_2\{\sqrt{1+x_1^2} + \frac{1}{2}x_1^2 v(x_1)\}\right]\right.$ (21)
$+ \frac{1}{4}\frac{g\Lambda^2}{\pi^2}\frac{1}{(x_1+x_2)^2}\left[\sqrt{1+x_1^2} + \frac{1}{2}x_1^2 v(x_1) - \sqrt{1+x_2^2} - \frac{1}{2}x_2^2 v(x_2)\right]^2 + \left.\frac{2}{9}\frac{g\Lambda^2}{\pi^2}\right]V,$

$\langle u_1|H|u_3\rangle = H[W_{13}]\cdot\det z_{13}$,
$H[W_{13}] = -\frac{\Lambda^4}{4\pi^2}\left[\frac{1}{x_1-x_3}\left[x_1\{\sqrt{1+x_3^2} + \frac{1}{2}x_3^2 v(x_3)\} - x_3\{\sqrt{1+x_1^2} + \frac{1}{2}x_1^2 v(x_1)\}\right]\right.$ (22)
$+ \frac{1}{4}\frac{g\Lambda^2}{\pi^2}\frac{1}{(x_1-x_3)^2}\left[\sqrt{1+x_1^2} + \frac{1}{2}x_1^2 v(x_1) - \sqrt{1+x_3^2} - \frac{1}{2}x_3^2 v(x_3)\right]^2 + \left.\frac{2}{9}\frac{g\Lambda^2}{\pi^2}\right]V,$

and $\langle u_3|H|u_2\rangle$ can be obtained by replacing the index 1 by 3 in $\langle u_1|H|u_2\rangle$. By solving the Res-mean-field CI equation, we can get the Res-mean-field energy E and the corresponding mixing coefficients c_i ($i = 1 \sim 3$).

Direct optimization of the Res-mean-field energy is achieved by a variation of the Res-mean-field ground-state energy $\frac{\partial}{\partial x_i} E_{\text{gr}}^{\text{Res}}(x_1, x_2, x_3) = 0$ ($i = 1 \sim 3$). This is the self-consistency condition in the Res-mean-field theory. Similarity between this condition and the gap equation in the BCS theory [12] and the Res-HB theory [3,6] is a manifestation of the analogy of the Res-mean-field theory with the BCS theory which motivated the NJL model [7].

RES-GROUND-STATE ENERGY AND $\pi - \sigma$ MASS SPECTRA

There are three parameters Λ (the cutoff parameter), g (the coupling constant) and a (the volume parameter) in the present NJL model. According to [10], we use the same value of the cutoff parameter $\Lambda = 631.0$ [MeV] as the one in I.

Numerical calculations are carried out at $g\Lambda^2 = 12.0 \sim 16.0$ and $a = 5.0 \sim 9.0$. We can find a very interesting and good numerical result for the parameters $g\Lambda^2 = 14.5$ and $a = 7.5$ which gives the "constituent quark masses" $M_1 = 378.6$ [MeV], $M_2 = 246.1$ [MeV] and $M_3 = 6.3$ [MeV], the mixing coefficients $c_{1,\text{gr}} = 0.837$ ($c_{1,\text{gr}}^2 = 0.701$), $c_{2,\text{gr}} = 0.269$ ($c_{2,\text{gr}}^2 = 0.072$) and $c_{3,\text{gr}} = 0.182$ ($c_{3,\text{gr}}^2 = 0.033$), and the overlap integrals $\det z_{12} = 0.040$, $\det z_{13} = 0.392$ and $\det z_{32} = 0.567$. The value $a = 7.5$ corresponds to the "confinement volume" of a cube with sides $L = 2.39$ [fm] long.

Using these values, from the excitation energies $E_{\text{excitation},1}^{\text{Res}} = E_{\text{ex},1}^{\text{Res}} - E_{\text{gr}}^{\text{Res}}$ and $E_{\text{excitation},2}^{\text{Res}} = E_{\text{ex},2}^{\text{Res}} - E_{\text{gr}}^{\text{Res}}$ (the "resonon" excitation energies [2,3]) we can obtain also the very good pion and sigma meson mass spectra $m_\pi = 139.8$ [MeV] and $m_\sigma = 674.8$, respectively, in excellent agreement with the experimental values of the pion and sigma meson mass spectra $m_\pi = 139.6$ [MeV] and $m_\sigma = 670.0$ [MeV]. As emphasized in I, the idea of introduction of the "confinement volume of quarks" leads automatically to a non vanishing pion mass also in the present calculation.

ORDER PARAMETER AND PION DECAY CONSTANT

We consider the solution for "constituent quark masses". Using the Res-mean field ground-state wave function $|\Psi_{\text{gr}}^{\text{Res}}\rangle = \sum_{i=1}^{3} c_{i,\text{gr}} |u_i\rangle$, the order parameter for the quarks is expressed by the formula

$$\begin{aligned}
\langle \bar{\psi}\psi \rangle &= \langle \Psi_{\text{gr}}^{\text{Res}} | \bar{\psi}\psi | \Psi_{\text{gr}}^{\text{Res}} \rangle \\
&= c_{1,\text{gr}}^2 \sum_p \text{Tr}[\beta W_{11,\, p}] + c_{2,\text{gr}}^2 \sum_p \text{Tr}[\beta W_{22,\, p}] + c_{3,\text{gr}}^2 \sum_p \text{Tr}[\beta W_{33,\, p}] \\
&+ c_{1,\text{gr}} c_{2,\text{gr}} \sum_p (\text{Tr}[\beta W_{12,\, p}] + \text{Tr}[\beta W_{12,\, p}^\dagger]) \cdot \Pi_p \det z_{12,\, p} \cdot \theta(\Lambda^2 - p^2) \\
&+ c_{1,\text{gr}} c_{3,\text{gr}} \sum_p (\text{Tr}[\beta W_{13,\, p}] + \text{Tr}[\beta W_{13,\, p}^\dagger]) \cdot \Pi_p \det z_{13,\, p} \cdot \theta(\Lambda^2 - p^2) \\
&+ c_{3,\text{gr}} c_{2,\text{gr}} \sum_p (\text{Tr}[\beta W_{32,\, p}] + \text{Tr}[\beta W_{32,\, p}^\dagger]) \cdot \Pi_p \det z_{32,\, p} \cdot \theta(\Lambda^2 - p^2) ,
\end{aligned} \quad (23)$$

from which the quantity $-\{\langle \bar{\psi}\psi \rangle / V\}^{\frac{1}{3}}$ is evaluated as

$$
\begin{aligned}
-\left\{\frac{\langle\bar\psi\psi\rangle}{V}\right\}^{\frac{1}{3}} &= \frac{\Lambda}{(2\pi^2)^{1/3}}\Big[c_{1,\text{gr}}^2 x_1 v(x_1) - c_{2,\text{gr}}^2 x_2 v(x_2) + c_{3,\text{gr}}^2 x_3 v(x_3) \\
&+ c_{1,\text{gr}}c_{2,\text{gr}} \cdot \tfrac{\det z_{12}}{x_1+x_2}\left\{\sqrt{1+x_1^2} + \tfrac{1}{2}x_1^2 v(x_1) - \sqrt{1+x_2^2} - \tfrac{1}{2}x_2^2 v(x_2)\right\} \\
&+ c_{1,\text{gr}}c_{3,\text{gr}} \cdot \tfrac{\det z_{13}}{x_1-x_3}\left\{\sqrt{1+x_1^2} + \tfrac{1}{2}x_1^2 v(x_1) - \sqrt{1+x_3^2} - \tfrac{1}{2}x_3^2 v(x_3)\right\} \\
&+ c_{3,\text{gr}}c_{2,\text{gr}} \cdot \tfrac{\det z_{32}}{x_3+x_2}\left\{\sqrt{1+x_3^2} + \tfrac{1}{2}x_3^2 v(x_3) - \sqrt{1+x_2^2} - \tfrac{1}{2}x_2^2 v(x_2)\right\}\Big]^{\frac{1}{3}}.
\end{aligned}
\tag{24}
$$

Substituting the solution for the parameters $g\Lambda^2 = 14.5$ ($\Lambda = 631.0$ [MeV]) and $a = 7.5$, $x_1 = 0.60$, $x_2 = 0.39$, $x_3 = 0.01$, $c_{1,\text{gr}} = 0.837$, $c_{2,\text{gr}} = 0.269$, $c_{3,\text{gr}} = 0.182$, $\det z_{12} = 0.040$, $\det z_{13} = 0.392$ and $\det z_{32} = 0.567$ into Eq. (24), we are led to a good numerical result $-\{\langle\bar\psi\psi\rangle/V\}^{\frac{1}{3}} = 137.5$ [MeV]. This value, however, is still a little bit small when compared with the theoretical one and the experimental one given in Ref. [10].

Following I the pion decay constant f_π can be described as the time component of the axial-vector matrix element $\langle 0^{\text{Res}}|j_5|\pi_p^{\text{Res}}\rangle = f_\pi[E_{\text{excitation},1}^{\text{Res}}]^{\frac{1}{2}}$, $j_5 = \Sigma_{j=1}^N \gamma_j^5 e^{-ip\cdot r_j}$, for the Res-mean-field RPA [13,14] pionic state $|\pi_p^{\text{Res}}\rangle$ of momentum p and vacuum $|0^{\text{Res}}\rangle$. With the aid of the interstate density matrix $W_{12,p}$, the approximate Res-mean-field RPA pionic state is given by $|\pi_p^{\text{Res}}\rangle \simeq \mathcal{N}W_{12,p}^\dagger|0^{\text{Res}}\rangle$ where $\mathcal{N}$ is the normalization factor. Then by using the Res-mean-field excited-state wave function $|\Psi_{\text{ex},1}^{\text{Res}}\rangle = \Sigma_{i=1}^3 c_{i,\text{ex},1}|u_i\rangle$ and the helicity operator S_p $(= \frac{\Sigma\cdot p}{|p|})$ we get

$$
\begin{aligned}
f_\pi &= \langle 0^{\text{Res}}|j_5|\pi_p^{\text{Res}}\rangle[E_{\text{excitation},1}^{\text{Res}}]^{-\frac{1}{2}} \simeq \mathcal{N}\langle 0^{\text{Res}}|j_5 W_{12,p}^\dagger|0^{\text{Res}}\rangle[E_{\text{excitation},1}^{\text{Res}}]^{-\frac{1}{2}} \\
&\simeq \mathcal{N}\langle\Psi_{\text{gr}}^{\text{Res}}|\Sigma_{i=1}^N(\gamma_i^5 S_{p_i})|\Psi_{\text{ex},1}^{\text{Res}}\rangle[E_{\text{excitation},1}^{\text{Res}}]^{-\frac{1}{2}}.
\end{aligned}
\tag{25}
$$

The mixing coefficients $c_{1,\text{ex},1(2)}$, $c_{2,\text{ex},1(2)}$ and $c_{3,\text{ex},1(2)}$ are determined through the orthogonalities $\langle\Psi_{\text{gr}}^{\text{Res}}|\Psi_{\text{ex},1}^{\text{Res}}\rangle = 0$, $\langle\Psi_{\text{gr}}^{\text{Res}}|\Psi_{\text{ex},2}^{\text{Res}}\rangle = 0$ and $\langle\Psi_{\text{ex},1}^{\text{Res}}|\Psi_{\text{ex},2}^{\text{Res}}\rangle = 0$. Finally we obtain the ratio of the pion decay constant f_π to the $\sqrt{V}$

$$
\begin{aligned}
\frac{f_\pi}{\sqrt{V}} &= -\frac{2\Lambda}{\sqrt{6\sqrt{2}\pi}} \cdot \left[\frac{E_{\text{excitation},1}^{\text{Res}}}{\Lambda}\right]^{-\frac{1}{2}} \times \Big[1 - (1 - 2x_1^2 - 3x_1 x_2 - x_2^2)\sqrt{\tfrac{1+x_1^2}{1+x_2^2}} \\
&\quad -x_1(2x_1^2 + 3x_1 x_2 + 2x_2^2)E(\tan^{-1}\tfrac{1}{x_2},\sqrt{1-\tfrac{x_2^2}{x_1^2}}) \\
&\quad +x_1 x_2^2 F(\tan^{-1}\tfrac{1}{x_2},\sqrt{1-\tfrac{x_2^2}{x_1^2}})\Big]^{-\frac{1}{4}} \times \\
&\Big\{c_{1,\text{gr}}c_{1,\text{ex},1}[\sqrt{1+x_1^2}(1-2x_1^2) + 2x_1^3] + c_{2,\text{gr}}c_{2,\text{ex},1}[\sqrt{1+x_2^2}(1-2x_2^2) + 2x_2^3] \\
&+ c_{3,\text{gr}}c_{3,\text{ex},1}[\sqrt{1+x_3^2}(1-2x_3^2) + 2x_3^3] \\
&+ (c_{1,\text{gr}}c_{2,\text{ex},1} + c_{2,\text{gr}}c_{1,\text{ex},1})\cdot \det z_{12} \\
&\quad \cdot \tfrac{1}{x_1+x_2}[x_1(1+x_2^2)^{\frac{3}{2}} + x_2(1+x_1^2)^{\frac{3}{2}} - x_1 x_2(x_1^2+x_2^2)] \\
&- (c_{1,\text{gr}}c_{3,\text{ex},1} + c_{3,\text{gr}}c_{1,\text{ex},1})\cdot \det z_{13} \\
&\quad \cdot \tfrac{1}{x_1-x_3}[x_1(1+x_3^2)^{\frac{3}{2}} - x_3(1+x_1^2)^{\frac{3}{2}} + x_1 x_3(x_1^2-x_3^2)] \\
&+ (c_{3,\text{gr}}c_{2,\text{ex},1} + c_{2,\text{gr}}c_{3,\text{ex},1})\cdot \det z_{32} \\
&\quad \cdot \tfrac{1}{x_3+x_2}[x_3(1+x_2^2)^{\frac{3}{2}} + x_2(1+x_3^2)^{\frac{3}{2}} - x_3 x_2(x_3^2+x_2^2)]\Big\}. \quad (x_1 > x_2 > x_3)
\end{aligned}
\tag{26}
$$

Substituting the previous solutions and $c_{1,\text{ex},1} = 0.392$, $c_{2,\text{ex},1} = -0.952$ and $c_{3,\text{ex},1} = 0.040$ into Eq. (26), we are able to get a good numerical result $f_\pi/\sqrt{V} = 43.8$ [MeV]. However, this value is also still a little bit small when compared with the theoretical one and the experimental one given in Ref. [10] as well as the value of chiral deformation. Therefore, if we take the degrees of freedom of the isospin, flavor and color into account fully, we can expect to get much improved values for the $f_\pi/\sqrt{V}$ and $-\{\langle\bar{\psi}\psi\rangle/V\}^{\frac{1}{3}}$.

SUMMARY AND CONCLUDING REMARKS

To show the advantage of the Res-mean-field theory for a fermion system with large quantum fluctuations over the usual mean-field theory, we have applied it to the naive NJL model with single flavor but without isospin. A state with large quantum fluctuations is approximated by superposition of three Dirac seas, i.e., the non-orthogonal S-dets with different correlation structures. We have optimized directly the Res-mean-field energy functional by variations of the Res-mean-field ground-state energy with respect to the Res-mean-field parameters "constituent quark masses". The Res-mean-field ground and excited states generated with the three S-dets explain most of the pion and sigma meson mass spectra.

The present treatment has following characteristic points: (i) There are three model parameters Λ (cutoff parameter), g (coupling constant) and a (dimensionless volume parameter). (ii) The g does not enter the vacuum properties. (iii) The regime of the results depends only on the dimensionless ratio M_i/Λ and only the M_i determine the energy scale through $\frac{\partial}{\partial x_i} E_{\text{gr}}^{\text{Res}}(x_1, x_2, x_3) = 0$. We interpret M_i as the "constituent quark masses" and use as the variational parameters. But only the heavier one has a significant physical meaning because the magnitude of square of the mixing coefficient $c_{1,\text{gr}}^2$ is much larger than those of the others, $c_{2,\text{gr}}^2$ and $c_{3,\text{gr}}^2$.

TABLE 1. Pionic and sigma mesonic properties for "constituent quark masses"

		NJL MODEL		
		Res-MFT	TDHF[*a]	Exp.-Value
Inputs	a	7.5	-	-
	Λ[MeV]	631.0	631.0[*]	-
Outputs	$g\Lambda^2$	14.5	12.0	-
	M_1, M_2, M_3 [MeV]	378.6, 246.1, 6.31	335.0[*]	350
	$-\langle\bar{\psi}\psi\rangle^{\frac{1}{3}}$ [MeV]	137.5	246.6[*]	225±25
	m_π [MeV]	139.8	138.0[*]	139.6
	m_σ [Mev]	674.8	626.6[*]	670.0
	f_π[MeV]	43.8	93.0[*]	93

[a] The asterisk denotes results taken from [10]. The Res-MFT means the resonating mean-field theory. The output value 14.5 of $g\Lambda^2$ in the column TDHF corresponds to the output value 2.41 of the same quantity in [10] if we take the degrees of freedom of the flavor $N_f = 2$ and of the color $N_c = 3$.

Finally, we present in Table [1] numerical results for various physical quantities. Particularly, the numerical values of $g\Lambda^2$, "constituent quark masses" M_i ($i = 1 \sim 3$), pion mass m_π, order parameter $-\{\langle\bar{\psi}\psi\rangle/V\}^{\frac{1}{3}}$ and pion decay constant $f_\pi/\sqrt{V}$ tabulated in Table [1] compare comparatively well with the experimental datas.

The radical spirit of the Res-mean-field theory may be expected to open a new field for the exploration of the low-energy hadron physics using the strong analogy between a chiral effective Hamiltonian with four-fermion interaction and a familiar non-relativistic fermion two-body Hamiltonian. The present calculations are not so realistic as far as we consider only the three S-dets. Then we have the following interesting problems to be solved: (1) The consideration of the degrees of freedom of the isospin, flavor and color. (2) The inclusion of collective f_0, ρ, ω and a_1 mesonic states. (3) The evaluation of leptonic coupling constant g_π.

ACKNOWLEDGMENTS

One of the authors (S. N.) would like to express his sincere thanks to thank Professor J. da Providência for kind and warm hospitality extended to him at the Centro de Física Teórica, Universidade de Coimbra, Portugal. This work was supported by the Portguese Project PRAXIS XXI. S. N. was supported by the Portguese program PRAXIS XXI/BCC/4270/94.

REFERENCES

1. S. Nishiyama and J. da Providência, Phys. Rev. **C60** (1999);
   S. Nishiyama, Invited talk at the 23rd International Workshop on *Condensed Matter Theories* held at Ithaca, GREECE, 17-23 June, 1999.
2. H. Fukutome, Prog. Theor. Phys. **80** (1988) 417.
3. S. Nishiyama and H. Fukutome, Prog. Theor. Phys. **85** (1991) 1211.
4. S. Nishiyama, Nucl. Phys. **A576** (1994) 317.
5. S. Nishiyama, M. Ido and K. Ishida, *New Perspectives in Nuclear Structure, Proceedings of the 5th International Spring Seminar on Nuclear Physics*, (World Scientific Publishing Co. Pte. Ltd. 1996) p.121-p.130.
   S. Nishiyama, M. Ido and K. Ishida, to appear in Int. J. Mod. Phys. **E** (1999).
6. S. Nishiyama and H. Fukutome, J. Phys. **G18** (1992) 317.
7. Y. Nambu and G. Jona-Lasinio, Phys. Rev. **122** (1961), 345; **124** (1961), 246.
8. T. Hatsuda and T. Kunihiro, Phys. Rep. **247** (1994), 221; See also references therein.
9. J. da Providência, M. C. Ruivo and C. A. de Sousa, Phys. Rev. **D36** (1987) 1882.
10. C. A. de Sousa, Z. Phys. **C43** (1989) 503; **C49** (1991) 619.
11. D. J. Thouless, Nucl. Phys. **21** (1960) 225.
12. J. Bardeen, L. N. Cooper and J. R. Schrieffer, Phys. Rev. **108** (1957) 1175.
13. H. Fukutome, Prog. Theor. Phys. **81** (1989) 342.
    S. Nishiyama and H. Fukutome, Prog. Theor. Phys. **86** (1991) 371.
14. S. Nishiyama, K. Ishida and M. Ido, Nucl. Phys. **A599** (1996) 457.

Extended non-chiral quark models confronting QCD

A.A.Andrianov*,†[1] and V.A.Andrianov†[2]

*Departament d'ECM, Universitat de Barcelona
08028 Barcelona, Spain
†Department of Theoretical Physics, St.-Petersburg State University,
198904 St.-Petersburg, Russia

Abstract. We discuss the low energy effective action of QCD in the quark sector. When it is built at the CSB (chiral symmetry breaking) scale by means of perturbation theory it has the structure of a generalized Nambu-Jona-Lasinio (NJL) model with CSB due to attractive forces in the scalar channel. We show that if the lowest scalar meson state is sufficiently lighter than the heavy pseudoscalar π' then QCD favors a low-energy effective theory in which higher dimensional operators (of the Nambu-Jona-Lasinio type) are dominated and relatively strong. A light scalar quarkonium ($m_\sigma = 500 \div 600$ MeV) would provide an evidence in favor to this NJL mechanism.

Thus the non-chiral Quasilocal Quark Models (QQM) in the dynamical symmetry-breaking regime are considered as approximants for low-energy action of QCD. In the mean-field (large-N_c) approach the equation on critical coupling surface is derived. The mass spectrum of scalar and pseudoscalar excited states is calculated in leading-log approach which is compatible with the truncation of the QCD effective action with few higher-dimensional operators. The matching to QCD based on the Chiral Symmetry Restoration sum rules is performed and it helps to select out the relevant pattern of CSB as well as to enhance considerably the predictability of this approach.

INTRODUCTION: DEFINITION OF QQM

The low energy effective action of QCD in the quark sector has a qualitatively different structure depending on whether it is built at the CSB scale by means of perturbation theory or below the CSB scale when the major chiral symmetry breaking effect - the formation of light pseudoscalar mesons - is implemented manifestly. In the first case the models [1] extend the Nambu-Jona-Lasinio one [2] (see the reviews [3–11]) with chiral symmetry broken due to strong attractive 4-fermion forces in the color-singlet scalar channel. In the second case the resulting model [12] is a generalization of the chiral quark model with a built-in constituent quark

---
[1]) Supported by Grant GRACENAS 6-19-97 and by Generalitat de Catalunya, Grant PIV 1999.
[2]) Supported by Grant RFBR 98-02-18137, Travel Grant of Russian Academy of Science and by Funds of the Workshop Hadron 99, Coimbra.

mass and the non-linear realization of chiral symmetry. This type of QCD effective action is not discussed in our talk although it may be more relevant [12–14] if the quarkonium scalar meson is sufficiently heavy [15,16].

The quasilocal approach of [1] (see also [17–19]) represents a systematic extension of the NJL model towards the complete effective action of QCD where many-fermion vertices with derivatives are incorporated with the manifest chiral symmetry of interaction, motivated by the soft momentum expansion of the perturbative QCD effective action. For sufficiently strong couplings, the new operators promote the formation of additional scalar and pseudoscalar states. These models allow an extension of the linear σ model provided by the NJL model, with the pion being a broken symmetry partner of the lightest scalar meson just as before, and with excited pions and scalar particles coming in pairs. In particular, when only scalar and pseudoscalar color-singlet channels are examined and dynamical quark masses are supposed to be sufficiently smaller than the CSB cutoff one may derive the minimal two-channel lagrangian of the QQM in the separable form [1,19]:

$$\mathcal{L}^{QQM} = \bar{q}i\partial\!\!\!/q + \mathcal{L}^I;$$
$$\mathcal{L}^I = \frac{1}{4N_f N_c \Lambda^2} \sum_{k,l=1}^{2} a_{kl} \left[\bar{q}f_k(\hat{s})q\,\bar{q}f_l(\hat{s})q - \bar{q}f_k(\hat{s})\tau^a\gamma_5 q\,\bar{q}f_l(\hat{s})\tau^a\gamma_5 q \right], \quad (1)$$

where $\partial\!\!\!/ \equiv \gamma^\mu \partial_\mu$, a_{kl} represents a symmetric matrix of real coupling constants and polynomial formfactors are chosen as follows:

$$f_1(\hat{s}) = 2 - 3\hat{s}; \qquad f_2(\hat{s}) = -\sqrt{3}\hat{s}; \qquad \hat{s} \equiv -\frac{\partial^2}{\Lambda^2}. \quad (2)$$

As this model interpolates the low-energy QCD action it is supplied with the cutoff $\Lambda \sim 1$ GeV which bounds virtual quark momenta in quark loops. We restrict ourselves with consideration of two-flavor case, thus τ_a denote Pauli matrices.

A somewhat different, nonlocal approach to describe excited meson states was developed in [20,21]

EFFECTIVE POTENTIAL AND MESON SPECTRUM

For strong four-fermion coupling constants $a_{kl} \sim 8\pi^2 \delta_{kl}$ the lagrangian (1) reveals the phenomenon of dynamical chiral symmetry breaking. This phenomenon can be described with the help of the effective potential for the attractive scalar channel where scalar mesons arise as composite states. Indeed its non-trivial minimum gives rise to a dynamical quark mass and the perturbative fluctuations around this minimum characterize the mass spectrum of meson states. To derive the required effective potential one should bosonize the quark action, i.e. incorporate auxiliary bosonic variables: $\sigma_k \sim i\bar{q}f_k(\hat{s})q$, $\pi_k^a \sim \bar{q}f_k(\hat{s})\tau^a\gamma_5 q$ and integrate out fermionic degrees of freedom. At the first step we introduce the bosonic variables in two channels:

$$\mathcal{L}_I = \sum_{k=1}^{2} i\bar{q}\left(\sigma_k + i\gamma_5 \pi_k^a \tau^a\right) f_k(\hat{s})q + N_f N_c \Lambda^2 \sum_{k,l=1}^{2} \left(\sigma_k a_{kl}^{-1} \sigma_l + \pi_k^a a_{kl}^{-1} \pi_l^a\right). \quad (3)$$

Let us parametrize the matrix of coupling constants in a close vicinity of tricritical point:

$$8\pi^2 a_{kl}^{-1} = \delta_{kl} - \frac{\Delta_{kl}}{\Lambda^2}, \qquad |\Delta_{kl}| \ll \Lambda^2. \quad (4)$$

The last inequality turns out to be equivalent to require the dynamical mass to be essentially less than the cutoff.

After integrating out the quark fields one comes to the bosonic effective action $\mathcal{W}(\sigma_k, \pi_k^a)$ and therefrom, for constant meson variables, to the effective potential:

$$V_{eff} = \frac{N_c N_f}{8\pi^2}\Bigg(-\sum_{k,l=1}^{2}\sigma_k\sigma_l\Delta_{kl} - (\pi_2^a)^2\Delta_{22} + 8(\sigma_1)^4\left(\ln\frac{\Lambda^2}{4(\sigma_1)^2} + \frac{1}{2}\right)$$
$$-\frac{159}{8}(\sigma_1)^4 - \frac{5\sqrt{3}}{2}\sigma_1^3\sigma_2 + \frac{9}{4}\sigma_1^2\sigma_2^2 + \frac{\sqrt{3}}{2}\sigma_1\sigma_2^3 + \frac{9}{8}(\sigma_2)^4$$
$$+\left(\frac{3}{4}\sigma_1^2 + \frac{\sqrt{3}}{2}\sigma_1\sigma_2 + \frac{9}{4}\sigma_2^2\right)(\pi_2^a)^2 + \frac{9}{8}(\pi_2^a)^4\Bigg) + O\left(\frac{\ln\Lambda}{\Lambda^2}\right), \quad (5)$$

for the fixed direction of chiral symmetry breaking $\pi_1^a = 0$.

The QCD inspired action should not, of course, induce the isospin symmetry breaking and therefore a non trivial solution for v.e.v is expected to be in the scalar channel, $<\pi_2^a>= 0$. It implies the following inequality to hold:

$$\frac{3}{4}\sigma_1^2 + \frac{\sqrt{3}}{2}\sigma_1\sigma_2 + \frac{9}{4}\sigma_2^2 > \Delta_{22}.$$

The conditions on extremum of the effective potential (5), the mass-gap equations,

$$\Delta_{11}\sigma_1 + \Delta_{12}\sigma_2 = 16\sigma_1^3\ln\frac{\Lambda^2}{4\sigma_1^2} - \frac{159}{4}\sigma_1^3 - \frac{15\sqrt{3}}{4}\sigma_1^2\sigma_2 + \frac{9}{4}\sigma_1\sigma_2^2 + \frac{\sqrt{3}}{4}\sigma_2^3$$
$$\Delta_{12}\sigma_1 + \Delta_{22}\sigma_2 = -\frac{5\sqrt{3}}{4}\sigma_1^3 + \frac{9}{4}\sigma_1^2\sigma_2 + \frac{3\sqrt{3}}{2}\sigma_1\sigma_2^2 + \frac{9}{4}(\sigma_2)^3, \quad (6)$$

allow to find the relations between the components of dynamical mass function and (reduced) coupling constants Δ_{kl}. In practice, one uses the v.e.v.'s of scalar fields as input parameters, in particular, $2\sigma_1 = m_{dyn} = 200 \div 300$ MeV, and determines the required Δ_{kl}.

The second variation of effective action in the vicinity of above v.e.v.,

$$\frac{\delta^2\mathcal{W}}{\delta\sigma_k(p)\delta\sigma_l(p')} = \frac{N_c N_f}{8\pi^2}(A_{kl}^\sigma p^2 + B_{kl}^\sigma)\delta^{(4)}(p+p');$$
$$\frac{\delta^2\mathcal{W}}{\delta\pi_k(p)\delta\pi_l(p')} = \frac{N_c N_f}{8\pi^2}(A_{kl}^\pi p^2 + B_{kl}^\pi)\delta^{(4)}(p+p'), \quad (7)$$

brings both the kinetic terms $\sim A_{kl}^{\sigma,\pi}$ and the mass matrix $B_{kl}^{\sigma,\pi}$ which represents the second derivative of the effective potential (5) (see their general structure in [22]).

The kinetic matrices $A_{k,l}^{\sigma,\pi}$ take the form:

$$A_{kl}^{\sigma} \simeq A_{kl}^{\pi} \simeq \begin{pmatrix} \left(4\ln\frac{\Lambda^2}{4\sigma_1^2} - \frac{23}{2}\right) & -\frac{\sqrt{3}}{2} \\ -\frac{\sqrt{3}}{2} & \frac{3}{2} \end{pmatrix}. \tag{8}$$

Let us now display the matrix of second variations $B_{k,l}^{\sigma,\pi}$:

$$B_{11}^{\sigma} = -2\Delta_{11} + 96\sigma_1^2 \ln\left(\frac{\Lambda^2}{4\sigma_1^2}\right) - \frac{605}{2}\sigma_1^2 - 15\sqrt{3}\sigma_1\sigma_2 + \frac{9}{2}\sigma_2^2,$$

$$B_{12}^{\sigma} = -2\Delta_{12} - \frac{15\sqrt{3}}{2}\sigma_1^2 + 9\sigma_1\sigma_2 + \frac{3\sqrt{3}}{2}\sigma_2^2,$$

$$B_{22}^{\sigma} = -2\Delta_{22} + \frac{9}{2}\sigma_1^2 + 3\sqrt{3}\sigma_1\sigma_2 + \frac{27}{2}\sigma_2^2,$$

$$B_{11}^{\pi} = -2\Delta_{11} + 32\sigma_1^2 \ln\left(\frac{\Lambda^2}{4\sigma_1^2}\right) - \frac{159}{2}\sigma_1^2 - 5\sqrt{3}\sigma_1\sigma_2 + \frac{3}{2}\sigma_2^2,$$

$$B_{12}^{\pi} = -2\Delta_{12} - \frac{5\sqrt{3}}{2}\sigma_1^2 + 3\sigma_1\sigma_2 + \frac{\sqrt{3}}{2}\sigma_2^2,$$

$$B_{22}^{\pi} = -2\Delta_{22} + \frac{3}{2}\sigma_1^2 + \sqrt{3}\sigma_1\sigma_2 + \frac{9}{2}\sigma_2^2. \tag{9}$$

Their diagonalization allows to find the masses of meson states. One recovers two states in the scalar channel and two triplet states in the pseudoscalar one. The lightest multiplet consists of the massless pion, $m_\pi = 0$ and the NJL scalar meson, $m_\sigma = 4\sigma_1 = 2m_{dyn}$ in the leading-log approach. The masses of heavier mesons depend essentially on the pattern of CSB in the vicinity of tricritical point. But in the leading-log approach they are approximately equal, $m_{\sigma'}^2 \sim m_{\pi'}^2 \sim -\frac{4}{3}\Delta_{22} = \mathcal{O}(\log\Lambda)$. The last estimation follows from the mass-gap eqs.(6). A more precise relation takes places:

$$m_{\pi'}^2 \simeq -\frac{4}{3}\Delta_{22} + \sigma_1^2 + \frac{2\sqrt{3}}{3}\sigma_1\sigma_2 + 3\sigma_2^2;$$

$$m_{\sigma'}^2 - m_{\pi'}^2 \simeq 2\sigma_1^2 + \frac{4\sqrt{3}}{3}\sigma_1\sigma_2 + 6\sigma_2^2 > 0. \tag{10}$$

CSR RULES

Let us employ the constraints based on Chiral Symmetry Restoration (CSR) in QCD at high energies. We consider two-point correlators of color-singlet quark currents in Euclidean space-time,

$$\Pi_C(p^2) = \int d^4x \, \exp(ipx) \, \langle T\left(\bar{q}\Gamma q(x) \, \bar{q}\Gamma q(0)\right)\rangle, \tag{11}$$

restricting ourselves in this talk with

$$C \equiv S, P; \qquad \Gamma = i, \, \gamma_5 \tau^a. \tag{12}$$

In the chiral limit the scalar correlator Π_S and the pseudoscalar one Π_P^{aa} approach to each other rapidly as their O.P.E.'s [23,24] coincide at all orders in perturbation theory and, as well, in the non-perturbative, purely gluonic part [12,25],

$$\left(\Pi_P^{aa}(p^2) - \Pi_S(p^2)\right)_{p^2 \to \infty} \equiv \frac{\Delta_{SP}}{p^4} + \mathcal{O}\left(\frac{1}{p^6}\right), \qquad \Delta_{SP} \simeq 24\pi\alpha_s <\bar{q}q>^2, \tag{13}$$

where $<\bar{q}q>$ stands for the quark condensate and the vacuum dominance hypothesis [23] is exploited for the estimation of four-quark condensates as we follow the large-N_c limit. Meantime, in the latter limit the correlators are well saturated by narrow resonances,

$$\Pi_P^{aa}(p^2) - \Pi_S(p^2) = \sum_n \left[\frac{Z_n^P}{p^2 + m_{P,n}^2} - \frac{Z_n^S}{p^2 + m_{S,n}^2}\right]. \tag{14}$$

As the difference decreases rapidly, one can assume that the lower lying resonances will dominate in the above sum.

The outcoming CSR rules in the two-channel model (1) read:

$$Z_\sigma + Z_{\sigma'} = Z_\pi + Z_{\pi'}; \qquad Z_\sigma m_\sigma^2 + Z_{\sigma'} m_{\sigma'}^2 = Z_{\pi'} m_{\pi'}^2 + \Delta_{SP}. \tag{15}$$

The first relation can be fulfilled in the (one-channel) NJL model which corresponds to the one-resonance ansatz, $Z_{\sigma',\pi'} = 0$, whereas the last one can be valid only in a two-resonance model, for the Δ_{SP} defined in (13) (see [13]).

CONSTRAINTS ON MESON PARAMETERS

The relevant correlators and the values of residues Z_i can be found by variation of the external sources S_k, P_k^a which couple to the scalar and pseudoscalar quark densities. The structure of the corresponding operators in the quark lagrangian is completely analogous to the Yukawa vertex in (3). Then the effect of external sources can be separated by shifting the scalar fields in the quark vertex of (3) and further on by integrating out the quark fields. As a result the effective action for generating of two-point correlators is parametrized in terms of the second variation matrix (7):

$$\mathcal{W}^{(2)} \simeq \frac{N_c N_f \Lambda^2}{8\pi^2} \sum_{k,l=1}^2 \left(S_k \Gamma_{kl}^{(\sigma)} S_l + P_k^a \Gamma_{kl}^{(\pi)} P_l^a\right),$$

$$\Gamma_{kl}^{(\sigma)} = \delta_{kl} - 2\Lambda^2 \left(A^\sigma p^2 + B^\sigma\right)_{kl}^{-1}; \qquad \Gamma_{kl}^{(\pi)} = \delta_{kl} - 2\Lambda^2 \left(A^\pi p^2 + B^\pi\right)_{kl}^{-1}. \tag{16}$$

In particular, the strictly local quark densities can be presented as a superposition of two currents:

$$\bar{q}\Gamma q = \frac{1}{2}\left(\bar{q}f_1\Gamma q - \sqrt{3}\bar{q}f_2\Gamma q\right); \quad \Gamma \equiv i, \gamma_5\tau^a. \tag{17}$$

Respectively, their two-point correlator in the scalar channel, $\Pi_S(p^2)$ reads,

$$\Pi_\sigma(p^2) = -\frac{N_c\Lambda^2}{2\pi^2} + \frac{Z_\sigma}{p^2 + m_\sigma^2} + \frac{Z_{\sigma'}}{p^2 + m_{\sigma'}^2};$$

$$Z_\sigma \simeq -\frac{N_c\Lambda^4}{12\pi^2 m_{\sigma'}^2 \ln\frac{\Lambda^2}{4\sigma_1^2}}\left[-48\sigma_1^2 \ln\frac{\Lambda^2}{4\sigma_1^2} + 3\Delta_{11} + 2\sqrt{3}\Delta_{12} + \Delta_{22}\right]$$

$$\simeq -\frac{N_c\Lambda^4 \Delta_{22}(\sigma_1 - \sqrt{3}\sigma_2)^2}{12\pi^2 m_{\sigma'}^2 \sigma_1^2 \ln\frac{\Lambda^2}{4\sigma_1^2}};$$

$$Z_{\sigma'} + Z_\sigma = \frac{N_c\Lambda^4}{2\pi^2} \equiv Z_0. \tag{18}$$

These relations are derived with the help of mass-gap eqs.(6).

To the first order in the leading-log approach, the weak decay coupling constant for pion can be found from the gauged second variation:

$$F_\pi^2 \simeq \frac{N_c\sigma_1^2}{\pi^2} \ln\frac{\Lambda^2}{4\sigma_1^2}, \tag{19}$$

and it coincides with that one of the NJL model. Respectively the value of quark condensate can be expressed in terms of v.e.v. of σ_i,

$$<\bar{q}q> \simeq -\frac{N_c\Lambda^2}{8\pi^2}(\sigma_1 - \sqrt{3}\sigma_2). \tag{20}$$

Thus taking these equations into account and remembering the leading order of the π'-meson mass (10) one arrives to the remarkable relation:

$$Z_\sigma \simeq 4\frac{<\bar{q}q>^2}{F_\pi^2}. \tag{21}$$

Let us examine the two-point correlator of local densities (17) in the pseudoscalar channel:

$$\Pi_\pi(p^2) = -\frac{N_c\Lambda^2}{2\pi^2} + \frac{Z_\pi}{p^2} + \frac{Z_{\pi'}}{p^2 + m_{\pi'}^2};$$

$$Z_\pi \simeq Z_\sigma \quad \text{for} \quad m_{\pi'}^2 \simeq m_{\sigma'}^2;$$

$$Z_{\pi'} + Z_\pi = Z_0; \quad Z_{\pi'} \simeq Z_{\sigma'}. \tag{22}$$

The equality of Z_π and Z_σ in eq.(21) realizes both the approximate restoration of chiral symmetry in each multiplet and the fulfillment of PCAC requirement (21) for the residue in the pion pole.

We stress that the residues in poles are of different order of magnitude:

$$Z_\sigma \sim Z_\pi = \mathcal{O}\left(\frac{Z_0}{\ln\frac{\Lambda^2}{4\sigma_1^2}}\right) \ll Z_{\sigma'} \sim Z_{\pi'} = \mathcal{O}(Z_0). \tag{23}$$

Now we are able to impose and check the CSR constraints (15). The leading asymptotics represents the generalized σ-model relation and is automatically fulfilled:

$$Z_\sigma + Z_{\sigma'} = Z_\pi + Z_{\pi'} = Z_0, \tag{24}$$

that in fact reflects the manifest chiral symmetry of the QQM lagrangian.

As to the second constraint the possibility to satisfy it depends on the value of the QCD coupling constants α_s. Indeed, it can be written by means of (10) as follows:

$$m_{\sigma'}^2 - m_{\pi'}^2 \simeq \frac{\Delta_{SP}}{Z_0};$$
$$\sigma_1^2 + \frac{2\sqrt{3}}{3}\sigma_1\sigma_2 + 3\sigma_2^2 \simeq \frac{3N_c\alpha_s}{8\pi}(\sigma_1 - \sqrt{3}\sigma_2)^2. \tag{25}$$

As the left part is always positive there exists a lower bound for $\alpha_s \geq \frac{8\pi}{9N_c}$ providing solutions of the constraint. The lowest value of $\alpha_s \simeq 0.9$ is given by $\sigma_1 = -\sqrt{3}\sigma_2$ and for these v.e.v.'s one obtains the following splitting between the σ'- and π'-meson masses:

$$m_{\sigma'}^2 - m_{\pi'}^2 \simeq \frac{8}{3}\sigma_1^2 = \frac{1}{6}m_\sigma^2; \tag{26}$$

i.e. for $m_\sigma = 500 \div 600 MeV$ these masses practically coincide, $m_{\sigma'} \simeq m_{\pi'} = 1300 MeV$ and such a σ'-meson may be identified [26] with $f_0(1300)$. The above value of α_s lies in the region of rather strong coupling where next-to-leading corrections to the anomalous dimension of four-quark operator in (13) are not negligible, $\sim \frac{\alpha_s}{\pi} \sim 0.3$ and should be systematically taken into account to obtain a reasonable precision. However the very fact that one has to match to QCD asymptotics at a scale $\mu \sim 600 MeV \sim m_\sigma$ lower than the masses of heavy resonances is not troublesome as it relates just coefficients of $1/p^2$ expansion irrespectively of how high is the momentum. On the other hand the matching should be performed in the region where one can neglect even more heavier resonances, i.e. at a scale ≤ 1 GeV.

CONCLUSIONS

1. We have shown that the quasilocal quark models truncating (perturbative) low-energy QCD effective action can serve to describe the physics of heavy meson resonances. The matching to nonperturbative QCD based on the chiral symmetry restoration at high energies improves the predictability of such models. QQM extend the NJL model and inevitably contain a rather light scalar meson which is however not excluded by the particle phenomenology [26].

2. The fast convergence in QQM of mass spectra and other characteristics of heavy parity doublers entail their decoupling from the low-energy pion physics. For instance, let us calculate the dim-4 chiral coupling constant [27,28],

$$L_8 = \frac{F_\pi^4}{64 <\bar q q>^2} \left(\frac{Z_\sigma}{m_\sigma^2} + \frac{Z_{\sigma'}}{m_{\sigma'}^2} - \frac{Z_{\pi'}}{m_{\pi'}^2} \right)$$
$$\simeq \frac{F_\pi^2}{16 m_\sigma^2} \left(1 - \frac{6\alpha_s \pi F_\pi^2 m_\sigma^2}{m_{\pi'}^4} \right). \tag{27}$$

The second term represents the net effect of heavy resonances after the CSR constraints (15) have been imposed. It is easy to find that its relative contribution is less than 2%. Therefore this constant is essentially determined in QQM by the lightest scalar meson. Its value, $L_8 = (0.9 \pm 0.4) \times 10^{-3}$ from [27,29,30] nearly accepts $m_\sigma \simeq 600 MeV$.

3. Some disadvantage of QQM as well as of the original NJL model is that they presuppose the large, critical values of four-quark coupling constants which is difficult to justify with perturbative calculations in QCD. As well the CSR matching has to be performed at a scale where the QCD coupling constant is rather large and the perturbation theory is unreliable. The Extended Chiral Quark Models [12,13] seem to be free of these shortcomings and are able to adjust the light scalar meson mass to be of order 1 GeV. However the final choice between them may be done by the fit of a larger variety of meson characteristics which is in progress.

4. Let us comment the approximations used to derive the meson characteristics: namely, the large N_c and leading-log approximations. The first one is equivalent [31,32] to the neglect of meson loops. The second one fits well the quarks confinement as quark-antiquark threshold contributions are suppressed in two-point functions in the leading log approximation. The accuracy of this approximation is controlled also by the magnitudes of higher dimensional operators neglected in QQM, *i.e.* by contributions of heavy mass resonances not included into QQM. All these approximations are mutually consistent. In particular, in the effective action without gluons the quark confinement should be realized with the help of an infinite number of quasilocal vertices with higher-order derivatives. Then the imaginary part of quark loops can be compensated and their momentum dependence can

eventually reproduce the infinite sum of meson resonances in the large-N_c limit. If the effective action is truncated with a finite number of vertices and thereby deals with only a few resonances one has to retain only a finite number of terms in the low-momentum expansion of quark loops in the CSB phase, with a non-zero dynamical mass.

5. There are more possibilities to bind the phenomenological constants of QQM based on CSR constraints for three- and four-point correlators and also in the vector and axial-vector channels. Some of these CSR rules have been explored in [12,33–37].

ACKNOWLEDGEMENTS

We express our gratitude to the organizers of the International Workshop on Hadron Physics 1999 in Coimbra and especially to Prof. J. da Providencia for hospitality and financial support of one of us (V.A.). We are also grateful to D. Espriu and R. Tarrach for useful discussion and attention to our work.

REFERENCES

1. Andrianov, A.A., and Andrianov, V.A., *Int. J. Mod. Phys.* **A8**, 1981 (1993); *Theor. Math. Phys.* **94**, 3 (1993); *Proc. School-Sem. "Hadrons and nuclei from QCD", Tsuruga/Vladivostok/Sapporo 1993*, Singapore: WSPC, 1994, pp. 341-353, hep-ph/9309297.
2. Nambu, Y., and Jona-Lasinio, G., *Phys. Rev.* **122**, 345 (1961).
3. Volkov, M.K.,*Ann. Phys. (N.Y.)* **157**, 282 (1984).
4. Meissner, U.-G.,*Phys. Rept.* **161**, 213 (1988).
5. H. Vogl, H., and W. Weise, W., Progr. *Part. Nucl. Phys.* **27**, 195 (1991).
6. Klevansky, S.,*Rev. Mod. Phys.* **64**, 649 (1992).
7. Andrianov, A.A., and Andrianov, V.A., *Z. Phys.* **C55** 435 (1992); *Theor. Math. Phys.* **93**, 1126 (1992).
8. Bijnens, J., Bruno, C., and de Rafael, E., *Nucl. Phys.* **B390**, 501 (1993).
9. Hatsuda, T., and Kunihiro, T., *Phys. Rept.* **247**, 221 (1994).
10. Ebert, D., Reinhardt, H., and Volkov, M. K., *Progr. Part. Nucl. Phys.* **33**, 1 (1994).
11. Bijnens, J.,*Phys. Rept.* **265**, 369 (1996).
12. Andrianov, A.A., Espriu, D., and Tarrach, R., *Nucl. Phys.***B533**, 429 (1998).
13. Andrianov, A.A., and Espriu, D., *JHEP* **10**, 022 (1999).
14. Andrianov, A.A., Espriu, D., and Tarrach, R., hep-ph/9909366.
15. J. A. Oller, E. Oset, *Phys.Rev.* **D60**, 074023 (1999).
16. Black, D., Amir H. Fariborz, A. H., and Schechter, J., hep-ph/9907516 .
17. Andrianov, A.A., and Andrianov, V.A., *Nucl. Phys. Proc. Suppl.* **39BC**, 257 (1995).
18. Pallante, E., and Petronzio, R., *Z. Phys.* **C65**,487 (1995).

19. Andrianov, A.A., Andrianov, V.A., and Yudichev, V.L., *Theor. Math. Phys.* **108**, 1069 (1996).
20. Volkov, M. K., and Weiss,C., *Phys. Rev.* **D56**, 221 (1997).
21. Volkov, M. K., Ebert, D., and Nagy, M. *Int. J. Mod. Phys.* **A13**, 5443 (1998).
22. Andrianov, A.A., Andrianov, V.A., and Rodenberg, R., *JHEP.* **06**, 003 (1999).
23. Shifman, M.A., Vainstein, A.I., and Zakharov, V.I., *Nucl. Phys.* **B147**, 385, 448 (1979).
24. L. J. Reinders, L.J., Rubinstein, H., and Yazaki, S., *Phys. Rept.* **127**, 1 (1985).
25. Andrianov, A.A., and Andrianov, V.A., *Zapiski Nauch. Sem. POMI (Proc. Steklov Math. Inst., St.Petersburg, Russia)*, **245**, No.14, 5 (1996), hep-ph/9705364.
26. Particle Data Group: Caso, C., *et al.*, *European Phys. J.* **C3**, 1 (1998).
27. J. Gasser, J., and Leutwyler, H., *Nucl. Phys.* **B250**, 465 (1985).
28. Bijnens, J., de Rafael, E., and Zheng, H., *Z. Phys.* **C62**, 437 (1994).
29. Ecker, G., *Progr. Part. Nucl. Phys.* **35**, 1 (1995).
30. Pich, A.,*Rept. Prog. Phys.* **58**, 563 (1995).
31. t'Hooft, G., *Nucl. Phys.* **B72**, 461 (1974).
32. Witten, E.,*Nucl. Phys.* **B160** 57 (1979).
33. Donoghue, J. F., and Golowich, E., *Phys. Rev.* **D49** 1513 (1994).
34. Moussallam, B., *Nucl. Phys.* **B504**, 381 (1997).
35. Knecht, M., and de Rafael, E., *Phys. Lett.* **B424**, 335 (1998).
36. Peris, S., Perrottet, M., and de Rafael, E.,*JHEP* **05** 011(1998).
37. Knecht, M., Peris, S., and de Rafael, E., *Phys. Lett.* **B443**, 255 (1998).

Strong decays of the scalar glueball in a confining effective Lagrangian of QCD

M. Jaminon, M. Mathot and B. Van den Bossche

*Université de Liège, Institut de Physique B5, Sart Tilman
B-4000 Liège 1, Belgium*

Abstract. We first calculate the masses and the coupling constants of the scalar and pseudoscalar mesons as well as the ones of the glueball. We use a confining model which implements the QCD trace anomaly and the axial anomaly. The idea would be at the end to calculate strong decay widths of the lightest scalar glueball into two pseudoscalar mesons using this confining model.

INTRODUCTION

In a previous paper [1] we calculated the strong decay widths of the scalar glueball into $\pi\pi$, $\eta\eta$, $\eta\eta'$, $K\bar{K}$ using a modified version of the NJL model which has been developed [2] in order to implement the QCD trace anomaly. This was performed by introducing one single point-like scalar dilaton field χ whose mean value χ_s is identified with the vacuum gluon condensate. Our bosonized action implements the QCD scale anomaly via [2]

$$L_\chi = \frac{1}{2}(\partial_\mu \chi)^2 + \frac{1}{16}b^2 \left[\chi^4 \ln \frac{\chi^4}{\chi_G^4} - (\chi^4 - \chi_G^4) \right], \tag{1}$$

with two additional parameters b^2 and χ_G with regards to the original NJL. The axial anomaly is implemented using a mass term for the field π_0

$$L_{U_A(1)} = \frac{1}{2G_s}\xi\chi^2\pi_0^2. \tag{2}$$

Due to the nonrenormalizability of the model, quark loops have to be regularized via a cut-off $\Lambda\chi$ where the dilaton field χ accounts for the scale invariance of the trace. The effective action can then be written:

$$I_{eff}(\varphi,\chi) = -\text{Tr}_{\Lambda\chi} \ln(-i\partial_\mu\gamma_\mu + m + \Gamma_a\varphi_a)$$
$$+ \int d^4x \frac{\chi^2}{2G_s}\varphi_a\varphi_a + \int d^4x\, L_\chi + \int d^4x\, L_{U_A(1)} \tag{3}$$

where $\varphi_a = (\sigma_a, \pi_a)$, $\Gamma_a = (\lambda_a, i\gamma_5\lambda_a)$ and $m = \text{diag}(m_u, m_d, m_s)$ stands for the current quark mass matrix.
This provides a mixing between the glueball and the scalar isoscalar mesons: the glueball then acquires a $q\bar{q}$ content and the scalar mesons a glue one. However, solving the scalar mesons puzzle is all but simple. Things are still complicated when the glueball is included into the game. In [1], we have assumed a scheme where the glueball, identified with $f_0(1500)$, is coupled to the $I = 0$ scalar mesons $f_0(1370)$ and $f_0(1710)$. Due to the mixing, the glueball can decay into two pseudoscalar mesons. The amplitudes receive two distinguishable contributions: one coming from the usual triangle diagram, the other one, called "local contribution", arising from the regularization and from the $\chi^2\varphi_a^2$ term. The results obtained for the decay into two pions were quite unrealistic ($\Gamma_{f_0(1500)\pi_0\pi_0} \approx 1 - 10$ GeV to be compared with the *total* width $\Gamma_{f_0(1500)} = 112\pm 10$ MeV). A way to explain this large value lies in the fact that the value of the u-quark mass has to be very large ($M_u \approx 725$ MeV) in order to reproduce the assumed scalar nonet. Hope is that confinement would improve the agreement with experiment. Indeed it is known that confinement will allow to reduce the value of M_u. In a first step we apply the confining model of refs. [3–6] to the scaled NJL and calculate the meson masses and coupling constants that are involved in the calculation of the strong decay widths.

THE CONFINEMENT MODEL

So far, our model lacks confinement, leading to unphysical $q\bar{q}$ decays when we stand above the quark-antiquark pair creation threshold $P^2 = 4M^2$, where P and M denote the quadri-momentum of the incoming meson and the constituent quark mass respectively. Since numerical results are no longer reliable for energies far above this threshold, we have to include confinement. This will be done using the model introduced in [3–6].
In this model, the Lagrangian contains a confining part L_{conf}. Different forms of L_{conf} can be considered [3,4]. We will restrict ourselves to the Lorentz-vector (L.V.) confinement

$$L_{conf} = \bar{q}(x)\gamma^\mu q(x) V^c(x-y) \bar{q}(y)\gamma_\mu q(y). \tag{4}$$

L.V. confinement is the most usual one and verifies chiral symmetry when $m = 0$. We adopt the following form for the confining potential:

$$V^c(r) = \kappa r \exp^{-\mu r}, \tag{5}$$

with κ the string tension and μ a small parameter introduced to soften the singularities of the Fourier transform of $V^c(r)$. If μ is small, the potential is approximatively linear over the considered range of interaction. One has

$$V^c(\vec{k} - \vec{k'}) = -4\pi\kappa \left\{ \frac{2}{[(\vec{k}-\vec{k'})^2 + \mu^2]^2} - \frac{8\mu^2}{[(\vec{k}-\vec{k'})^2 + \mu^2]^3} \right\} \tag{6}$$

when one neglects energy transfer via the confining field. In this work we take $\kappa = 0.0575$ GeV^2 and $\mu = 0.02$ GeV [6].

SELF-ENERGY

In this model, the k-dependent self energy of the quarks writes:

$$\Sigma(k) = B(k^2)\slashed{k} + M(k^2) \qquad (7)$$

$$M(k^2) = i \int \frac{d^4k'}{(2\pi)^4} \frac{4N_c n_f G_s/\chi_s^2 + 4V^c(k-k')}{(1-B(k'^2))^2 k'^2 - M^2(k'^2)} M(k'^2) \qquad (8)$$

$$B(k^2)k^2 = -i \int \frac{d^4k'}{(2\pi)^4} \frac{2V^c(k-k')(k.k')(1-B(k'^2))}{(1-B(k'^2))^2 k'^2 - M^2(k'^2)} \qquad (9)$$

Obviously, when $\kappa = 0$ (no confinement), one recovers the usual scaled NJL gap equation:

$$M = i \int \frac{d^4k'}{(2\pi)^4} \frac{4N_c n_f G_s/\chi_s^2}{k^2 - M^2} M \qquad (10)$$

It is shown in [3] that $B(k^2)$ keeps a very small value for all k^2 and that $M(k^2)$ smoothly depends on k^2. We will then assume in the following that $\Sigma(k) = C^{te} = M$, M being the constituent quark mass. This assumption is justified as long as κ keeps values lower than 0.1 GeV^2.

VERTEX FUNCTIONS AND VACUUM POLARIZATION INTEGRALS

The usual polarization integrals related to the scalar and pseudoscalar sectors write:

$$J_S(P^2) = iN_c n_f \text{Tr} \int \frac{d^4k}{(2\pi)^4} S(k+P/2) S(k-P/2) \qquad (11)$$

$$J_5(P^2) = -iN_c n_f \text{Tr} \int \frac{d^4k}{(2\pi)^4} \gamma_5 S(k+P/2) \gamma_5 S(k-P/2) \qquad (12)$$

with $S(k) = (\slashed{k} - M + i\epsilon)^{-1}$ the quark propagator.
They have unphysical cuts starting at $P^2 = 4M^2$ that corresponds to the quark and antiquark going on their mass shell. Above this $q\bar{q}$ creation threshold J_5 and

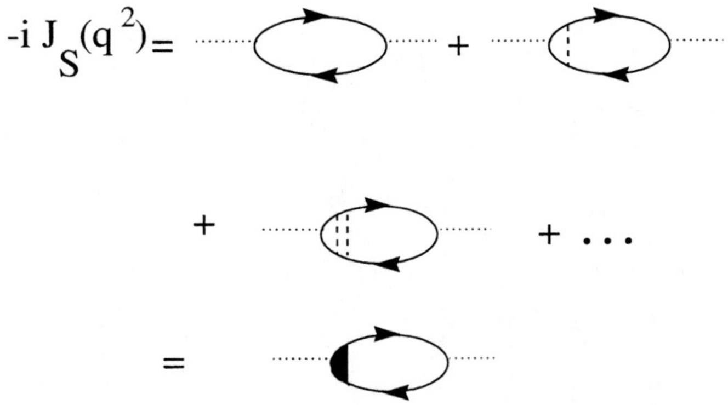

FIGURE 1. Quark-loop integral including a series of confining interactions (dashed line)

J_S become complex with a continuously increasing imaginary part leading to unreliable numerical results. This desease is eliminated when the confining interaction is included in the model. The polarization integrals take now the following form:

$$\hat{J}_S(P^2) = iN_c n_f \text{Tr} \int \frac{d^4k}{(2\pi)^4} S(k+P/2)\bar{\Gamma}_S(P,k)S(k-P/2) \quad (13)$$

$$\hat{J}_5(P^2) = -iN_c n_f \text{Tr} \int \frac{d^4k}{(2\pi)^4} \gamma_5 S(k+P/2)\bar{\Gamma}_5(P,k)S(k-P/2) \quad (14)$$

where

$$\bar{\Gamma}_S(P,k) = 1 - i \int \frac{d^4k'}{(2\pi)^4} \left[\gamma^\rho S(k'+P/2)\bar{\Gamma}_S(P,k')S(k'-P/2)\gamma_\rho V^c(\vec{k}-\vec{k}')\right] \quad (15)$$

$$\bar{\Gamma}_5(P,k) = \gamma_5 - i \int \frac{d^4k'}{(2\pi)^4} \left[\gamma^\rho S(k'+P/2)\bar{\Gamma}_5(P,k')S(k'-P/2)\gamma_\rho V^c(\vec{k}-\vec{k}')\right] \quad (16)$$

are vertex functions suming a series of confining interaction (see Fig. 1). Writing

$$S(\vec{k}) = \frac{M}{E(\vec{k})} \left[\frac{\Lambda^+(\vec{k})}{k^0 - E(\vec{k}) + i\epsilon} - \frac{\Lambda^-(-\vec{k})}{k^0 + E(\vec{k}) - i\epsilon}\right] \quad (17)$$

with the standard Dirac projection operators ($E(\vec{k}) = \sqrt{|\vec{k}|^2 + M^2}$):

$$\Lambda^+(\vec{k}) = \frac{\slashed{k}+M}{2M}, k^\mu = [E(\vec{k}), \vec{k}]$$

$$\Lambda^-(-\vec{k}) = \frac{\tilde{k} + M}{2M}, \tilde{k}^\mu = [-E(\vec{k}), \vec{k}],$$

the polarization functions can be written in terms of four vertex functions, noted $\Gamma_5^{+-}, \Gamma_5^{-+}, \Gamma_S^{+-}, \Gamma_S^{-+}$, all of them being derived from the original ones:

$$\Lambda^+(\vec{k})\Gamma_i(P,k)\Lambda^-(-\vec{k}) = \Gamma_i^{+-}(P,k)\Lambda^+(\vec{k})\gamma_i\Lambda^-(-\vec{k})$$

$$\Lambda^-(-\vec{k})\Gamma_i(P,k)\Lambda^+(\vec{k}) = \Gamma_i^{-+}(P,k)\Lambda^-(-\vec{k})\gamma_i\Lambda^+(\vec{k})$$

with $\gamma_i = 1, \gamma_5$ for $i = S, 5$. For an incoming meson at rest ($\vec{P} = 0$), they satisfy the following integral equations:

$$\Gamma_S^{+-}(P_0, |\vec{k}|) = 1 - 4\pi \int \frac{|\vec{k}'|^2 d|\vec{k}'|}{(2\pi)^3} \frac{-4|\vec{k}'|^2}{E(\vec{k}')} \frac{\Gamma_S^{+-}(P_0, |\vec{k}'|)}{(P^0)^2 - (2E(\vec{k}'))^2}$$
$$\times \left[\int_{-1}^1 V^c(\vec{k}-\vec{k}')d\cos\theta + \frac{M^2}{2|\vec{k}||\vec{k}'|} \int_{-1}^1 \cos\theta V^c(\vec{k}-\vec{k}')d\cos\theta \right] \quad (18)$$

$$\Gamma_S^{-+}(P_0, |\vec{k}|) = \Gamma_S^{+-}(P_0, |\vec{k}|) \quad (19)$$

$$\Gamma_5^{+-}(P_0, |\vec{k}|) = 1 - \int \frac{d^3k'}{(2\pi)^3} \left[\frac{M^2 - 2E(\vec{k})E(\vec{k}')}{E(\vec{k})E(\vec{k}')} \right] \frac{\Gamma_5^{+-}(P^0, |\vec{k}'|)V^c(\vec{k}-\vec{k}')}{P^0 - 2E(\vec{k}')} \quad (20)$$

$$\Gamma_5^{-+}(P_0, |\vec{k}|) = 1 + \int \frac{d^3k'}{(2\pi)^3} \left[\frac{M^2 - 2E(\vec{k})E(\vec{k}')}{E(\vec{k})E(\vec{k}')} \right] \frac{\Gamma_5^{-+}(P^0, |\vec{k}'|)V^c(\vec{k}-\vec{k}')}{P^0 + 2E(\vec{k}')} \quad (21)$$

In terms of these new vertex functions, the polarization functions have the following expressions, once the integration over k_0 has been performed:

$$\hat{J}_S(P_0) = -2N_c n_f \int \frac{d^3k}{(2\pi)^3} \left[\frac{\Gamma_S^{+-}(P_0, |\vec{k}|)}{P^0 - 2E(\vec{k})} - \frac{\Gamma_S^{-+}(P_0, |\vec{k}|)}{P^0 + 2E(\vec{k})} \right] \frac{|\vec{k}|^2}{E^2(\vec{k})} \quad (22)$$

$$\hat{J}_5(P_0) = -2N_c n_f \int \frac{d^3k}{(2\pi)^3} \left[\frac{\Gamma_5^{+-}(P_0, |\vec{k}|)}{P^0 - 2E(\vec{k})} - \frac{\Gamma_5^{-+}(P_0, |\vec{k}|)}{P^0 + 2E(\vec{k})} \right] \quad (23)$$

The main point is that $\Gamma_{S,5}^{+-}(P_0, |\vec{k}|) = 0$ when $P_0 - 2E(\vec{k}) = 0$. This feature suppresses the unphysical $q\bar{q}$ threshold and the polarization functions remain real.

MESON MASSES AND COUPLING CONSTANTS

Parameters

Our model contains nine parameters: the constituent quark masses (M_u, M_s), the strengths (G_s, b^2), the cut-off Λ, the axial anomaly strength ξ, the vacuum gluon condensate χ_s, and the confining parameters (κ, μ). As in [6], we use $\kappa = 0.0575$ GeV^2 and $\mu = 0.02$ GeV. The other parameters are adjusted to reproduce f_π, m_π, m_K, $m_{\eta'}$, m_{a_0}, and $m_{f_0(1500)}$.

$f_0(1500)$ is assumed to be the glueball while the $I = 1$ meson a_0 is considered as the $a_0(1450)$ or $a_0(980)$. The Goldberger-Treiman relation $M_u = g_{\pi q\bar{q}} f_\pi$ where

$$g_{\pi q\bar{q}}^{-2} = \frac{\partial}{\partial q^2}\left[\hat{J}_5(q^2)\right]_{q^2=m_\pi^2} \tag{24}$$

fixes $\Lambda\chi_s$. The masses m_π, m_{a_0} and m_K satisfy respectively:

$$1 - \frac{G_s}{\chi_s^2}\hat{J}_5(m_{\pi^2}^2) = 0 \tag{25}$$

$$1 - \frac{G_s}{\chi_s^2}\hat{J}_S(m_{a_0}^2) = 0 \tag{26}$$

and

$$1 - \frac{G_s}{\chi_s^2}\hat{J}_5^{us}(m_K^2) = 0 \tag{27}$$

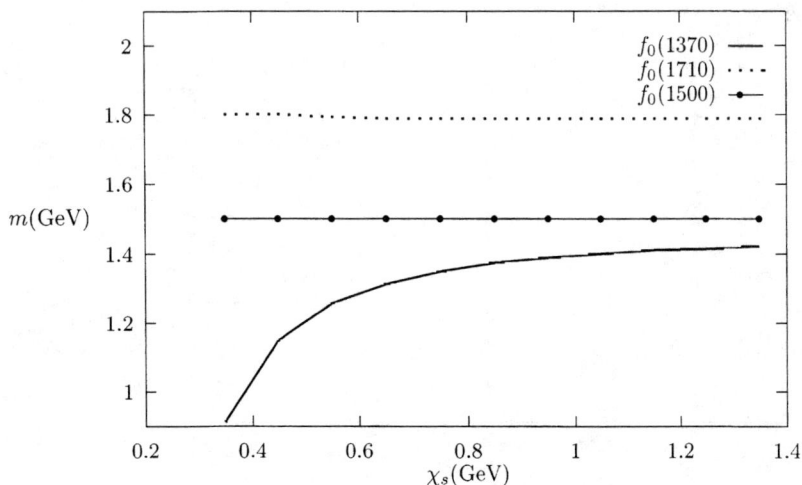

FIGURE 2. Scalar isoscalar meson masses in the confining model ($M_u = 0.605$ GeV)

$\hat{J}_5^{us}$ is obtained from $\hat{J}_5$ (Eq. 14) with one quark propagator with M_u, and the other one with M_s. The strengths b^2 and ξ of the scale and axial anomalies are adjusted on $m_{f_0(1500)}$ and $m_{\eta'}$ respectively. Since the glueball $f_0(1500)$ is coupled with the $I = 0$ scalar mesons, one has to diagonalize the inverse propagator matrix which comes from the expansion of the action up to the second order into the meson fields.

$$S^{-1} = \begin{pmatrix} -\frac{1}{3}(2\hat{J}_S^u + \hat{J}_S^s) + \frac{\chi^2}{G_s} & \frac{\sqrt{2}}{3}(\hat{J}_S^s - \hat{J}_S^u) & f_{0\chi}(\chi_s, \Lambda, G_s) \\ \frac{\sqrt{2}}{3}(\hat{J}_S^s - \hat{J}_S^u) & -\frac{1}{3}(\hat{J}_S^u + 2\hat{J}_S^s) + \frac{\chi^2}{G_s} & f_{8\chi}(\chi_s, \Lambda, G_s) \\ f_{0\chi}(\chi_s, \Lambda, G_s) & f_{8\chi}(\chi_s, \Lambda, G_s) & f_{\chi\chi}(\chi_s, \Lambda, G_s) \end{pmatrix} \quad (28)$$

where $f_{i\chi}$ ($i = 0, 8, \chi$) contains contributions coming from the Taylor expansion of the quark determinant into the χ field. These contributions are called "local contributions" [1]. $\hat{J}_S^u$ and $\hat{J}_S^s$ denote the polarization functions calculated with the constituent up and strange quark masses, respectively. The scalar masses correspond to the zeroes of the eigenvalues of S^{-1}.

We proceed in the same way for the pseudoscalar mesons. We perform the diagonalization of the matrix

$$\begin{pmatrix} -\frac{1}{3}(2\hat{J}_5^u + \hat{J}_5^s) + \frac{\chi^2}{G_s}(1+\xi) & \frac{\sqrt{2}}{3}(\hat{J}_5^s - \hat{J}_5^u) \\ \frac{\sqrt{2}}{3}(\hat{J}_5^s - \hat{J}_5^u) & -\frac{1}{3}(\hat{J}_5^u + 2\hat{J}_5^s) + \frac{\chi^2}{G_s} \end{pmatrix} \quad (29)$$

and assuming $m_{\eta'} = 0.958\ GeV$, we get the parameter ξ. We then remain with one free parameter χ_s.

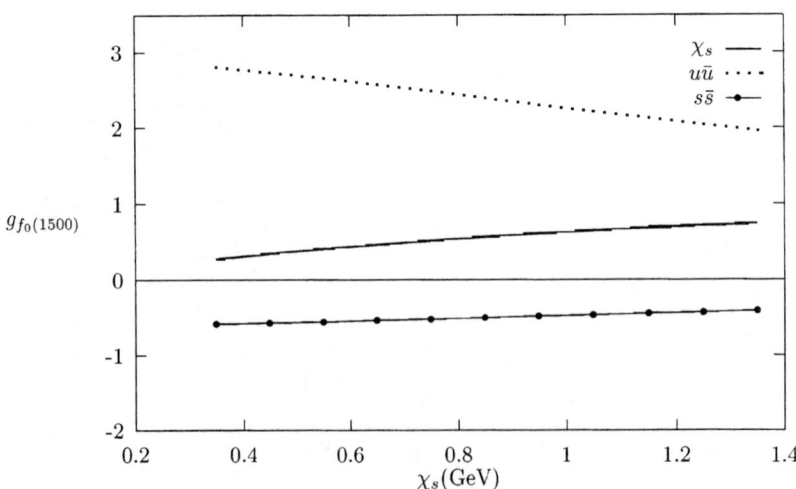

FIGURE 3. Glueball coupling constants in the confining model ($M_u = 0.605$ GeV)

TABLE 1. Vacuum parameters and pseudoscalar sector

	$a_0(1450)(d^4k)$	$a_0(1450)$	$a_0(980)$
$\kappa\ (GeV^2)$	0.0	0.0575	0.0575
$\mu\ (GeV)$		0.02	0.02
$M_u\ (GeV)$	0.725	0.605	0.368
$M_s\ (GeV)$	0.865	0.757	0.574
$\Lambda\chi_s\ (GeV)$	0.756	0.639	0.796
ξ	0.458	0.281	0.432
$M_{\eta'}(GeV)$	0.958	0.958	0.958
$M_\eta(GeV)$	0.482	0.475	0.469
$g_{\pi q\bar{q}}$	7.65	6.915	4.111
$g_{Kq\bar{q}}$	6.38	7.155	4.279
$g_{\eta' u\bar{u}}$	-1.82	-2.527	-1.181
$g_{\eta' s\bar{s}}$	-4.29	-5.434	-3.154
$g_{\eta u\bar{u}}$	-5.23	-5.404	-3.059
$g_{\eta s\bar{s}}$	4.08	4.304	2.743

Results

In Fig. 2 we plot the masses of the scalar isoscalar mesons and of the glueball versus the free parameter χ_s. $M_u = 605$ MeV is adjusted to reproduce $a_0(1450)$, with the confining parameters used in [6]. The glueball is identified with the $f_0(1500)$ and its mass is fixed at 1.5 GeV. Our results are compatible with scalars identified with $f_0(1710)$ and $f_0(1370)$, when $\chi_s \approx 0.6$ GeV. Without mixing with the glueball, $f_0(1710)$ would be a pure $s\bar{s}$ state while $f_0(1370)$ would represent the pure $u\bar{u}$ state (in the isospin symmetry). The mass of $f_0(1710)$ is slightly increased with regards to a nonconfining model [1].

Fig. 3 exhibits the glueball coupling constants (see [1] for their definition). With regards to [1], confinement slightly raises the coupling to the u-quark and reduces the one to the s-quark. The effect of confinement is not very important at that level.

We now turn to the case where M_u is chosen to reproduce $a_0(980)$. The nonconfining model would provide $M_u = 0.49$ GeV. However, calculating quantities such as the decay widths of the $f_0(1500)$ would not make sense since the glueball would stand far above the $q\bar{q}$ threshold. When confinement is included we get $M_u = 0.368$ GeV, leading to a value of the constituent mass in better agreement with the usual accepted one.

Masses and coupling constants are plotted in Figs. 4-5 respectively. Results are now compatible with scalar mesons identified with $f_0(1370)$ (pure $s\bar{s}$ state without mixing with the glueball) and $f_0(980)$ (pure $u\bar{u}$ state). The coupling constants are strongly lowered with regards to the previous scheme.

Table 1 collects the results for the pseudoscalar sector. These results are χ_s independent. The first column refers to the model without confinement [1]. The second one corresponds to the same scheme with confinement, the last one to the scheme where $a_0(980)$ is the scalar isovector meson.

In ref. [1], we concluded that our scaled non confining model would not be able to reproduce the decay of $f_0(1500)$ into two pions. Indeed, the required large value of M_u yields a large value of $g_{\pi q \bar{q}}$ (GT relation) and of $g_{f_0(1500)u\bar{u}}$ to which the decay

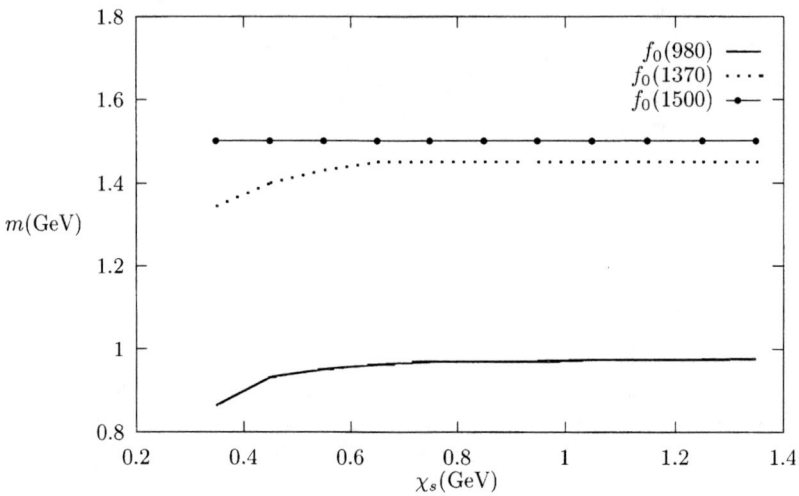

FIGURE 4. Scalar isoscalar meson masses with confinement ($M_u = 0.368\ GeV$)

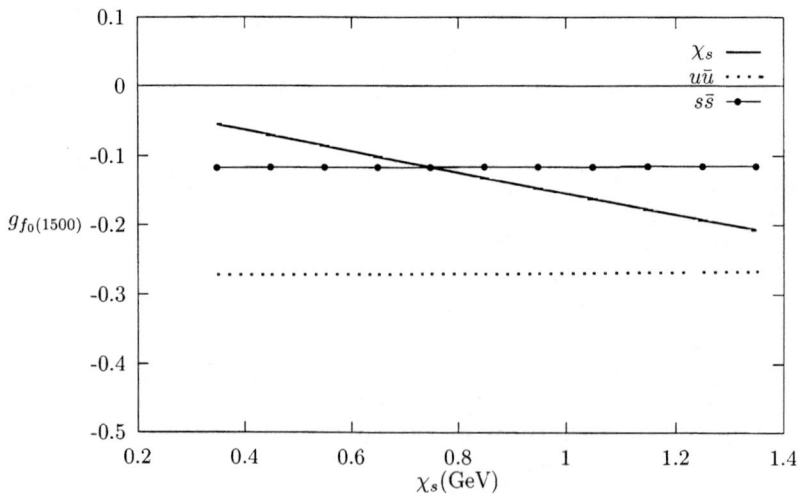

FIGURE 5. Glueball coupling constants ($M_u = 0.368\ GeV$)

width is proportional

$$\Gamma_{f_0(1500)\pi_0\pi_0} \propto \left[g_{f_0(1500)u\bar{u}}(g_{\pi u\bar{u}})^2\right]^2$$

One then found awfully large values for $\Gamma_{f_0(1500)\pi_0\pi_0}$ $(1-10\text{ GeV})$. The hope is that confinement will strongly improve the results. At this point of progress of our work, we can only stress the following remarks:

i) If one keeps the same scheme as in the non confining model, the value of M_u is not sufficiently reduced to decrease in a considerable way the values of $g_{\pi q\bar{q}}$ and $g_{f_0(1500)u\bar{u}}$. However the value of the triangle loop is also modified by confinement and the exact calculation has to be performed before drawing any conclusion.

ii) With the other scheme, all the coupling constants are strongly reduced. One can therefore expect a better agreement for the width of the glueball into two pions. Note that within this scheme the two first excited states of the pion are well reproduced. Work is in progress at that level [7].

REFERENCES

1. M. Jaminon, M.Mathot and B. Van den Bossche, *Nucl. Phys. A* (1999), in press
2. M. Jaminon and G. Ripka, *Ann. Phys.* **218**, 51 (1992).
3. L.S. Celenza, Xiang Dong Li and C.M. Shakin, *Phys. Rev C* **55**, 1492 (1997)
4. L.S. Celenza, Xiang Dong Li and C.M. Shakin, *Phys. Rev C* **56**, 3326 (1997)
5. L.S. Celenza, B. Huang and C.M. Shakin, *Phys. Rev C* **59**, 1041 (1999)
6. L.S. Celenza, B. Huang and C.M. Shakin, *Phys. Rev C* **59**, 1700 (1999)
7. M. Jaminon, M. Mathot and B. Van den Bossche, (1999), to be published

Models of Low Energy Effective Theory Applied to Kaon Non-Leptonic Decays and Other Matrix Elements[1]

Johan Bijnens*

*Department of Theoretical Physics 2, Lund University
Sölvegatan 14A, S 22362 Lund, Sweden*

Abstract. In this talk I describe work on computing non-leptonic matrix elements consistently with both long and short distance contributions included. On the simpler example of the π^+-π^0 mass difference I explain in detail the matching procedure and the difference between various low-energy models. I then explain the new difficulties in non-leptonic Kaon decays and how the matching here can in principle be done in the same way when scheme dependences are correctly accounted for. In the end I summarize the results J. Prades and I obtain for the $\Delta I = 1/2$ rule and B_6.

INTRODUCTION

The problem of describing non-leptonic decays is a very old one and is still not fully solved today. In this talk I will describe the large N_c method first suggested in a series of papers by Bardeen, Buras and Gérard [1] and later extended by several other authors. I will illustrate most of the problems and solutions on the example of the charged and neutral pion mass difference and afterwards show how this method can be extended systematically to the case of non-leptonic weak decays as well. The main results described there are those of [2] and also descibed in [3].

The subject of this meeting was hadronic physics, so why are we interested in these extra quantities. They provide a very strong test of our understanding of the strong interaction at all length scales. Our present knowledge of the strong interaction can be summarized as:

Short Distance: This is the perturbative QCD domain and here QCD has had many successes, we count this region as understood.

(Very) Long Distance: This is the Chiral Perturbation Theory (CHPT) regime [4]. Many successes again and basically understood.

---

[1] Work supported in part by TMR, EC-Contract No. ERBFMRX-CT980169 (EURODAΦNE).

Intermediate Distance: This is the domain of models supplemented with various arguments, sum rules, lattice QCD results, etc. and is the most difficult.

In the type of observables covered in this talk all three regimes are important. We consider processes with incoming and outgoing hadrons but with an internally exchanged photon or weak boson. The difficulty now resides in the fact that even if the external hadrons have all low momenta we need to integrate over all momenta of the internal γ or W^+. This means that all regimes come into play and that they need to be connected properly to each other. The last is known as matching.

The main part is in Sect. I where I show how we can explain the mass difference, $m_{\pi^+}^2 - m_{\pi^0}^2$ using this class of methods. Here we can also see how the model approach and the correct answer agree. Sect. II then covers the extra problems involved in non-leptonic weak decays and how the X-boson method of [2] can be used to solve those. Finally I present numerical results for this case and conclusions.

I A SIMPLE EXAMPLE: THE π^+-π^0 MASS DIFFERENCE.

This non-leptonic matrix element has several features that make it simpler.

1. We can neglect m_u and m_d to a rather good approximation. This then allows current algebra to relate the electromagnetic mass difference to a vacuum to vacuum matrix element only [5]. This can then be related to the measured hadronic cross-sections in electron-positron annihilation so in this case we know the correct answer.

2. There are no large masses involved so there are no large logarithms that need resummation.

3. The photon itself provides for an easy identification of correct scales.

Basically the procedure is now to evaluate

$$m_{em}^2 = -\langle M | e^2 \int \frac{d^4q}{(2\pi)^4} \frac{J_\mu(q) J_\nu(-q)}{q^2} \left(g_{\mu\nu} - \xi \frac{q_\mu q_\nu}{q^2} \right) | M \rangle . \tag{1}$$

where M stands for the meson under consideration and J_μ for the electromagnetic current. ξ is a gauge parameter. The procedure is now as follows:
1: We rotate the integral over photon momenta in Eq. (1) to Euclidean space. This has two advantages, in Euclidean space thresholds and poles are smoothed out making treatment of these easier and Euclidean space momenta have all components small if q_E^2 is small. The latter allows for a simpler identification of long and short-distance than in Minkowski space.
2: The final step is now to set

$$\int d^4 q_E = \int_0^\infty q_E^3 dq_E \int d\Omega = \underbrace{\int_0^\mu q_E^3 dq_E \int d\Omega}_{\text{long-distance}} + \underbrace{\int_\mu^\infty q_E^3 dq_E \int d\Omega}_{\text{short-distance}} \quad (2)$$

and perform both integrals separately. Notice that the scale μ is just a splitting scale in the integral and is not directly related to any subtraction scale in the calculation itself. Therefore, if both the long-distance (from 0 to μ) and the short-distance are calculated with high enough precision the final result should be independent of the value of μ. We check this by varying μ in all our calculations, i.e. we check the matching.

A Short-Distance

The short-distance contribution was first calculated in [6] using the sum rule by Das et al. [5]. It was later rederived using the Operator Product expansion in [7]. The diagrams in Fig. 1 depict the main contributions. Performing the photon integral leads to a set of four-quark operators that can be evaluated in leading $1/N_c$ since we can then apply factorization. The result is [6,7]

$$\left(m_{\pi^+}^2 - m_{\pi^0}^2\right)_{SD} = \frac{3\alpha_S \alpha}{\mu^2} F^2 B_0^2 = \frac{3\alpha_S \alpha}{\mu^2} \frac{\langle \bar{q}q \rangle^2}{F^2}, \quad (3)$$

with F the pion decay constant in the chiral limit and B_0 the parameter in lowest order CHPT describing the quark condensate.

B Long-Distance

In the previous subsection we could use perturbative QCD but that is not possible in the long distance domain. So here we have to put in the things we know and try various models.

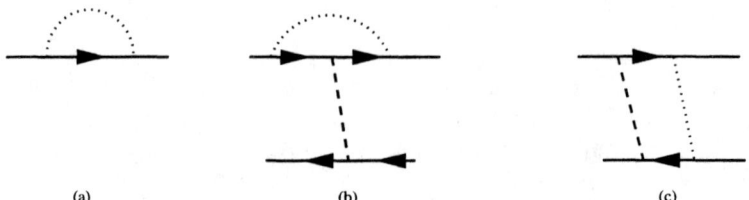

FIGURE 1. The three short-distance contributions, (a) electromagnetic quark-mass corrections. (b) Penguin-like diagrams (c) Box Diagrams. The dashed line is a gluon, the dotted line a photon and the full lines are quarks.

FIGURE 2. The long-distance contributions to $(m_{\pi^+}^2 - m_{\pi^0}^2)$. The dotted line is a photon and the full lines are pions.

CHPT: This can be done for μ rather small and leads to

$$\left(m_{\pi^+}^2 - m_{\pi^0}^2\right)_{LD} = \frac{3\alpha}{4\pi}\left(\mu^2 + \frac{2L_{10}}{F^2}\mu^4\right). \qquad (4)$$

The first term was first done in [6] and the chiral correction in [8,9]. The two contributing diagrams are depicted in Fig. 2. Something that is important is that the gauge dependence only cancels between the two diagrams in Fig. 2.

Chiral Quark Model: This was done in [10] and gives only a marginal improvement. Note that we cannot use the usual dimensional regularization here but must use the cut-off in the photon propagator. There is the additional problem that at first sight only the equivalent of the diagram of Fig 2(a) appears, which is a two-loop diagram, and the result is not gauge invariant. Only after the equivalent of (b) is added, which is a three-loop diagram, does the gauge dependence cancel as required [10].

With Vector-axial-vector Mesons: We have to include here Weinberg's constraint on the couplings to obtain a unique result otherwise the result will be very dependent on the specific model used. E.g. a hidden gauge model with only vector mesons is still quadratic in μ^2 but with a negative coefficient. Using Weinberg's constraints leads to

$$\left(m_{\pi^+}^2 - m_{\pi^0}^2\right)_{LD} = \frac{3\alpha}{2\pi}M_V^2 \log\left[\frac{M_V^2 + \mu^2}{M_A^2 + \mu^2}\frac{M_A^2}{M_V^2}\right]. \qquad (5)$$

But beware of partial results. Using a linear vector representation only even gave a quartic dependence on μ [8]. The result in (5) for $\mu \to \infty$ is basically the result of [5] and was also obtained in [11]. It has also several nice features. Expanding in μ for small μ reproduces the CHPT result with the meson dominated value for L_{10}. For large μ it goes as $1/\mu^2$ so it can match on very well to the earlier short-distance result.

ENJL This basically coincides with the previous result and was first obtained in [12].

Extensions of the above exists for nonzero-quark masses [7,13] and references therein and also with more large N_c arguments to underpin the matching [9].

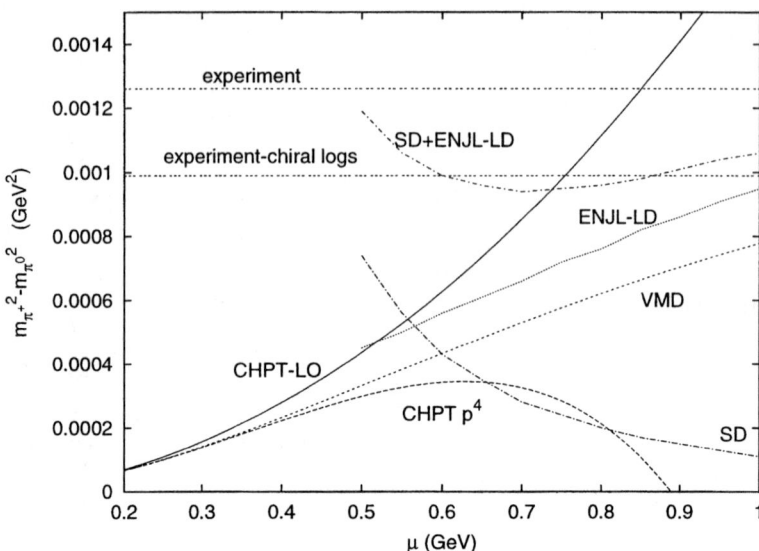

FIGURE 3. The long-distance contributions and short-distance contribution compared with the experimental values.

C Discussion

Numerical results are shown in Fig 3 for all cases. The experimental value and the one with the sub-leading in $1/N_c$ subtracted are shown as the horizontal lines. The subtracted part is the chiral logarithm contribution as estimated in [13]. Notice that CHPT starts deviating quickly from the VMD and ENJL results. The CHPT result is only reliable up to about 500 MeV. The VMD result and the ENJL result basically coincide here, the difference is due to the precise input values. Both these curves also follow essentially the correct answer as obtained from electron-positron annihilation and the sum-rule of [5].

Notice that the ENJL model has the correct matching on to the low μ CHPT result and is a considerable improvement over it at higher μ. Notice also the almost perfect agreement with the estimated part leading N_c part of the mass-difference.

From this section we can conclude:

1. Different Low energy models give quite different results and we have to use short-distance constraints and phenomenological inputs to improve the long-distance contribution to above the regime where CHPT is applicable.

2. CHPT alone for the long-distance regime is as a first guestimate acceptable but start differing from the correct answer at a scale of about 500 MeV.

3. Even for this low-momentum dominated observable the short-distance contributions are sizable at scales around 800 MeV.

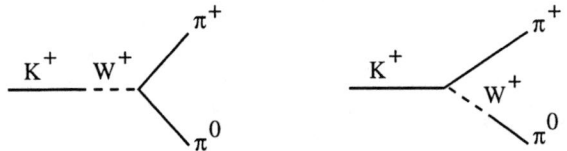

FIGURE 4. The naive W^+ exchange contribution to $K^+ \to \pi^+\pi^0$.

II KAON NON-LEPTONIC DECAYS

One of the difficult unresolved problems is to understand the origin of the $\Delta I = 1/2$ rule. The underlying process is W^+-exchange leading to an operator of the quark-structure $(\bar{s}u)(\bar{u}d)$ which has both isospin 1/2 and isospin 3/2 pieces. If we assume the W^+ couples directly to hadrons the process $K^+ \to \pi^+\pi^0$ goes simply via the diagrams in Fig. 4, but there are no such diagrams for $K^0 \to \pi^0\pi^0$ because of charge conservation. So we would expect that $\Gamma(K^+ \to \pi^+\pi^0) \gg \Gamma(K^0 \to \pi^0\pi^0)$. The experimental numbers are $\Gamma(K^+ \to \pi^+\pi^0) = 1.1\ 10^{-14}$ MeV and $\Gamma(K^0 \to \pi^0\pi^0) = \frac{1}{2}\Gamma(K_S \to \pi^0\pi^0) = 2.3\ 10^{-12}$ MeV, precisely the opposite. Translated into isospin amplitudes for the decays, see e.g. [14] for the precise definitions, we obtain $|A_0/A_2|_{\exp} = 22.1$. The problem is not due to chiral corrections since using the estimate of [14,15] we can extract them and get

$$|A_0/A_2|_{\text{chiral}} = 16.4 \underbrace{= \sqrt{2}}_{\text{naive}}. \tag{6}$$

where the last number is the one using naive W^+-exchange as depicted in Fig. 4. In the notation used in [2,14] we have

$$A_0 = C(9G_8 + G_{27})\sqrt{6}/9\ F_0(m_K^2 - m_\pi^2) \quad A_2 = C10G_{27}\sqrt{6}/9\ F_0(m_K^2 - m_\pi^2) \tag{7}$$

which after subtracting the estimated chiral corrections from experiment yields

$$G_8 = 6.2 \pm 0.7 \quad G_{27} = 0.48 \pm 0.06\ . \tag{8}$$

Both G_8 and G_{27} are equal to one in the W^+-exchange limit, the constant C was chosen to have this. We thus have to explain the large deviation from 1 using the corrections suppressed by $1/N_c$.

This is not a hopeless task as the sub-leading corrections coming from the diagrams in Fig. 1 with the photon replaced by the gluon are of order

$$\frac{\alpha_S}{N_c} \log \frac{M_W^2}{\mu^2} \tag{9}$$

compared to the leading contribution and this is in fact larger than one.

Luckily we know how to resum this type of logarithms [16]. At a high scale we can replace the effect of W^+-exchange by a sum of local operators by virtue of

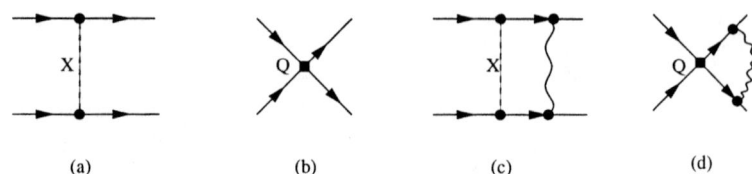

FIGURE 5. The diagrams needed for the identification of the local operator Q with X-boson exchange in the case of only one operator and no Penguin diagrams. The wiggly line denotes gluons, the square the operator Q and the dashed line the X-exchange. External lines are quarks.

the operator product expansion. We can then use the whole renormalization group machinery to run this sum over local four-quark operators down to a low scale μ_R. This is explained in great detail in [16]. The end result is

$$H_W = \sum_{i=1,10} C_i(\mu_R) Q_i \tag{10}$$

with a series of known coefficients $C_i(\mu_R)$, the Wilson coefficients. The final answer is then the matrix element of this sum over four-quark operators, $\langle \pi\pi | H_W | K \rangle$.
But here we have two problems:

1. In the previous section the short and long-distance contributions were separated via the photon momentum. Here we have to link this somehow to the scale μ_R appearing in the weak Hamiltonian H_W.

2. To next-to-leading order in the renormalization group the coefficients $C_i(\mu_R)$ also depend rather strongly on the precise definition of the local four quark operators Q_i in QCD perturbation theory.

In [18] we showed how the method of [1] supplemented with the correct momentum routing [6,17] solved problem 1. In [2] and in the various already published talks [3] we showed how a careful identification across the long-short-distance boundary is also possible in this case. The basic idea is to go at the scale μ_R back from the local four-quark operators to the exchange of a series of X-bosons. These X-bosons can then be treated in exactly the same way as we did the photon in the previous section, thus allowing a correct calculation at all length scales.

So we replace, using a single operator as an example

$$C_1 Q_1 = (\bar{s}_L \gamma_\mu d_L)(\bar{u}_L \gamma_\mu u_L) \iff X_\mu \left[g_1 (\bar{s}_L \gamma^\mu d_L) + g_2 (\bar{u}_L \gamma^\mu u_L) \right]. \tag{11}$$

Using the tree level diagrams of Fig. 5 this gives $C_1 = g_1 g_2 / M_X^2$. If we now include the one loop diagrams we obtain instead:

$$C_1 \left(1 + \alpha_S(\mu_R) r_1\right) = \frac{g_1 g_2}{M_X^2} \left(1 + \alpha_S(\mu) a_1 + \alpha_S(\mu_R) b_1 \log \frac{M_X^2}{\mu_R^2}\right). \tag{12}$$

On the l.h.s. the scheme dependence disappears but there is a dependence in r_1 on the choice of external states. The exact same dependence in a_1 cancels this.

We now split the integral over the X-boson momentum as in the previous section

$$\int_0^\infty dp_X \Longrightarrow \int_0^\mu dp_X + \int_\mu^\infty dp_X \tag{13}$$

In the final answer all M_X dependence drops out, the logarithm proportional to b_1 shows up in precisely the same way in the evaluation of the short distance part of (13) which is proportional to $g_1 g_2/M_X^2 \{\alpha_S(\mu) a_2 + \alpha_S(\mu) b_1 \log(M_X^2/\mu^2)\}$ The coefficients r_1, a_1 and a_2 give the corrections to the naive $1/N_c$-method.

We now use the X-boson method described above and put $\mu = \mu_R$. The low energy part can be calculated using CHPT, this is the approach used originally by [1] and presently pursued by [19] without including the corrections due to the change in scheme when going to the long-distance part. Their results coincide with ourss when we restrict our results to their approximations. We obtain [2]

$$B_K = 0.6\text{---}0.8 \quad B_K^\chi = 0.25\text{---}0.40 \quad G_8 = 4.3\text{---}7.5$$
$$G_{27} = 0.25\text{---}0.40 \quad G_8' = 0.8\text{---}1.1 \quad B_6(\mu = 0.8 \text{ GeV}) \approx 2.2 \tag{14}$$

B_K is the bag-parameter relevant for $K^0 - \overline{K^0}$ mixing at the physical quark masses and B_K^χ the same in the chiral limit. The quark mass corrections are quite sizable. The results for G_8 and G_{27} are obtained without any free input and agree within the uncertainties of the method with the experimental values. We conclude that we basically now have a first principle understanding of the $\Delta I = 1/2$ rule. We discuss the various contributions below. G_8' is the coefficient of the weak mass term, it contributes at leading order to processes like $K_L \to \gamma\gamma$ [14] and is often forgotten in those analyses. Finally B_6 is much larger than used in all the analysises of the recent experimental results for ϵ'/ϵ [20] which is very encouraging.

The final result for G_8 is depicted in Fig. 6. We have shown the one-loop result (1-loop), the two-loop result with NDR Wilson coefficients (2-loop) and the two-loop results with correction for the long-distance scheme added (SI) using our results for the long distance part. We also showed what naive factorization with the SI Wilson coefficients would give and what the method of [19] would give in the chiral limit with the same Wilson coefficients (SI quad).

If we look at the various contributions to G_8 we see in Fig. 7 that the contribution of Q_1 and Q_2 are both large and fairly constant while Q_6 contributes 20% or less. If we look inside the calculation we see that the difference with the G_{27} evolution is mainly given by the long-distance Penguin-like contributions to Q_2. The behaviour of B_6 is more difficult, it is ill-defined in the chiral limit in the factorizable approximation [2] and we can thus only define it with respect to the full large N_c limit. Calculating it in CHPT only then gives fairly low values as is visible in the second line of Table 1. Adding higher order corrections we immediately obtain a strongly enhanced value as is obvious from the second line in Table 1.

TABLE 1. B_6 as a function of μ using the results of [2].

μ (GeV)	0.6	0.7	0.8	0.9	1.0
CHPT	1.19	0.93	0.70	0.50	0.36
ENJL	2.27	2.16	2.11	2.11	2.14

III CONCLUSIONS

a The X-boson method in combination with large N_c arguments allows to correctly identify quantities across theory boundaries assuming we can identify currents across the boundary.

b The mass difference $m_{\pi^+}^2 - m_{\pi^0}^2$ is well described by these methods with a surprisingly large short-distance contribution.

c The $\Delta I = 1/2$ rule is now quantitatively understood to about 30% with NO free input. This calculation passes *all* requirements usually asked of in this context but there are many technical subtleties.

d $B_6 \approx 2.2$ is good news for those trying to explain the observed values of ϵ'/ϵ within the standard model.

e This program has been quite successful but we need new ideas to calculate more complex processes.

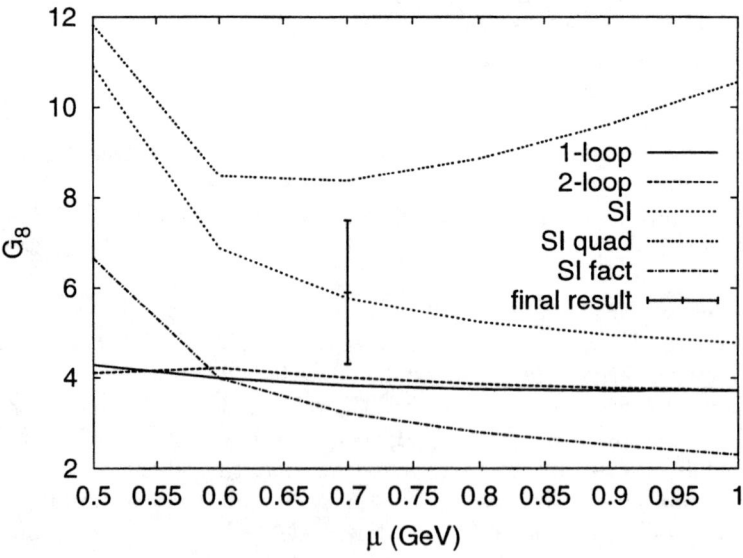

FIGURE 6. The octet coefficient G_8 as a function of μ using the ENJL model and the one-loop Wilson coefficients, the 2-loop ones and those including the r_1 (SI). In the latter case also the factorization (SI fact) and the approach of [19] (SI quad) are shown.

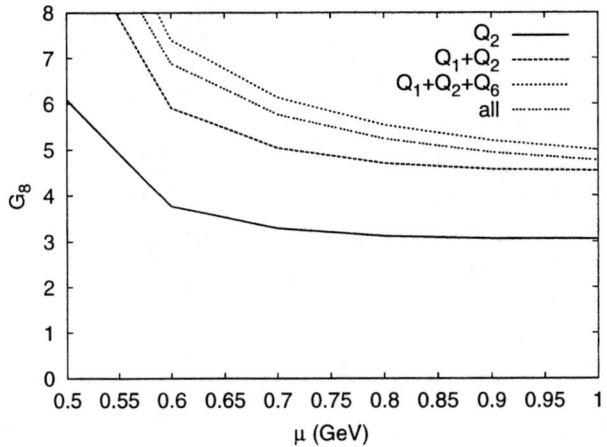

FIGURE 7. The composition of G_8 as a function of μ. Shown are Q_2, $Q_1 + Q_2$, $Q_1 + Q_2 + Q_6$ and all 6 Q_i. The coefficients r_1 are included in the Wilson coefficients.

REFERENCES

1. Bardeen W.A., et al., Nucl. Phys. B293 (1987) 787; Phys. Lett. B192 (1987) 138.
2. Bijnens J., and Prades J., JHEP 01 (1999) 023 [hep-ph/9811472].
3. Bijnens J.,hep-ph/9907514,hep-ph/9907307,hep-ph/9910263;Prades J.,hep-ph/9909245.
4. See e.g. Ecker G., Prog.Part.Nucl.Phys. 35 (1995)1 [hep-ph/9501357].
5. Das T.,et al., Phys.Rev.Lett. 18 (1967) 759:
6. Bardeen W.A., et al., Phys. Rev. Lett. 62 (1989) 1343
7. Bijnens J., Phys.Lett. B306(1993)343 [hep-ph/9302217].
8. Bijnens J., unpublished.
9. Peris S, et al, JHEP 05(1998)011 [hep-ph/9805442]; Knecht M., et al., Phys. Lett. B443 (1998) 255 [hep-ph/9809594]; Phys. Lett. B457 (1999) 227 [hep-ph/9812471]; hep-ph/9910396; de Rafael E., hep-ph/9909210 .
10. Bijnens J., and de Rafael E., Phys.Lett. B273 (1991) 483.
11. Ecker G., et al., Nucl.Phys. B321 (1989) 311.
12. Bijnens J., et al., Z.Phys. C62 (1994) 437 [hep-ph/9306323].
13. Bijnens J., and Prades J., Nucl. Phys. B490 (1997) 239. [hep-ph/9610360].
14. Bijnens J., et al., Nucl. Phys. B521 (1998) 305 [hep-ph/9801326].
15. Kambor J., et al., Phys. Lett. B261(1991)496.
16. Buras A., Les Houches Lectures[hep-ph/9806471] and references therein.
17. Bijnens J., et al., Phys. Lett. B257 (1991) 191.
18. Bijnens J., and Prades J., Nucl. Phys. B444 (1995) 523. [hep-ph/9502363]; Phys. Lett. B342 (1995) 331 [hep-ph/9409255].
19. Hambye T.,et al., Phys. Rev. D58 (1998) 014017 [hep-ph/9802300]; hep-ph/9902334, hep-ph/9906434, hep-ph/9908232.
20. KTEV, Phys.Rev.Lett. 83 (1999) 22 [hep-ex/9905060]; NA48, hep-ex/9909022.

Anomalies for Nonlocal Dirac Operators

E. Ruiz Arriola[1] and L. L. Salcedo

Departamento de Física Moderna. Universidad de Granada
E-18071 Granada, Spain

Abstract. The anomalies of a very general class of non local Dirac operators are computed using the ζ-function definition of the fermionic determinant and an asymmetric version of the Wigner transformation. For the axial anomaly all new terms introduced by the non locality can be brought to the standard minimal Bardeen's form. Some extensions of the present techniques are also commented.

INTRODUCTION

Local field theories provide the commonly accepted setup where the implementation of space-time symmetries becomes rather simple. On the other hand, effective theories are not necessarily local, although an appropriate choice of degrees of freedom can make them almost local [1]. In low energy QCD, light quarks and gluons are dressed by the interaction in a way that the effective theory looks highly nonlocal [2], and a kind of dynamical perturbation theory would be needed [3]. In terms of pions and (heavy) nucleons the theory becomes weakly nonlocal and a chiral perturbation theory becomes of practical interest [4]. In a Dyson-Schwinger setting [5] most information about such a non local theory comes from the constraints imposed by the relevant Ward and Slavnov-Taylor identities [5], perturbation theory to some finite order and hadronic phenomenology. These approaches are necessary if one wants to know, for instance, about the momentum distribution of a quark in a hadron; neither perturbative QCD nor chiral perturbation theory can properly handle this problem. In this regard, anomalies are particulary interesting because their existence is linked to a violation of classical symmetries by high energy regulators although their physical effect is formulated as a low energy theorem. At the one loop level, anomalies for nonlocal models have been previously discussed [6–8] for some specific processes like e.g. $\pi^0 \to 2\gamma$, $\gamma \to 3\pi$ and $2K \to 3\pi$. We refer to those works and [9] for further motivation. Rather than computing all specific processes one by one we prove that the new terms generated by the non locality can be subtracted by adding suitable counterterms.

---

[1] Talk given at the "Workshop on Hadron Physics", Coimbra (Portugal), 10-15 September 1999.

I THE ONE LOOP EFFECTIVE ACTION

Our starting point is the effective action of Dirac fermions in the flat Euclidean space-time endowed with internal degrees of freedom collectively referred to as "flavor",

$$W(\mathbf{D}) = -\log \int \mathcal{D}\bar{\psi}\mathcal{D}\psi \exp\left\{-\int d^D x\, \bar{\psi}(x)\mathbf{D}\psi(x)\right\} = -\operatorname{Tr}\log \mathbf{D}. \tag{1}$$

Here, Tr stands for trace over all degrees of freedom and $\mathbf{D}$ is the Dirac operator to be specified below. The definition of the fermion determinant requires some renormalization of the ultraviolet divergences. The (consistent) chiral anomaly is defined as the variation of the effective action under infinitesimal chiral transformations, given by

$$\psi(x) \to e^{i\beta - i\alpha\gamma_5}\psi(x), \qquad \bar{\psi}(x) \to \bar{\psi}(x)e^{-i\beta - i\alpha\gamma_5} \tag{2}$$

where $\alpha(x)$ and $\beta(x)$ are Hermitian matrices in flavor space only, regarded as multiplicative operators on the fermionic wave functions. The particular cases $\alpha = 0$ and $\beta = 0$ correspond to vector and axial transformations, respectively. This induces transformations of the Dirac operator

$$\mathbf{D} \to e^{i\beta - i\alpha\gamma_5}\mathbf{D} e^{-i\beta - i\alpha\gamma_5}, \tag{3}$$

which infinitesimally become

$$\delta \mathbf{D} = \delta_V \mathbf{D} + \delta_A \mathbf{D} - [i\beta, \mathbf{D}] \quad \{i\alpha\gamma_5, \mathbf{D}\}. \tag{4}$$

Since we will be considering a ζ-function renormalization of W (see below), there will be no vector anomaly,

$$\delta_V W = 0, \quad \delta_A W = \mathcal{A}_A. \tag{5}$$

Correspondingly, the same current conservation formulas valid for the local case can be written here,

$$0 = \int d^D x\, \langle \bar{\psi}(x)[i\beta, \mathbf{D}]\psi(x)\rangle_Q \tag{6}$$

$$-\mathcal{A}_A = \int d^D x\, \langle \bar{\psi}(x)\{i\alpha\gamma_5, \mathbf{D}\}\psi(x)\rangle_Q. \tag{7}$$

(The symbol $\langle\ \rangle_Q$ stands for quantum vacuum expectation value.) In particular the term $\gamma_\mu P_\mu$ in $\mathbf{D}$ in the right-hand side yields, after integration by parts, the divergence of the fermionic vector and axial currents whereas the other terms in $\mathbf{D}$, local and non local, represent the explicit chiral symmetry breaking due to the external fields. On the other hand, the left-hand side shows the anomalous breaking of the axial current conservation.

The class of Dirac operators to be considered here is

$$D = D_L + M. \tag{8}$$

The term D_L, the local component of D, is a standard Dirac operator

$$D_L = \gamma_\mu P_\mu + Y \tag{9}$$

We will follow the conventions of [10,11]), $P_\mu = i\partial_\mu$ and Y is an arbitrary matrix-valued function in flavor and Dirac spaces. Y is a function of the position operators X_μ, defined by $X_\mu \psi(x) = x_\mu \psi(x)$, so that Y is a multiplicative operator in the Hilbert space of fermions. The term M is a purely non local [2], more precisely bilocal, operator also with arbitrary structure in flavor and Dirac spaces,

$$(Y\psi)(x) = Y(x)\psi(x), \qquad (M\psi)(x) = \int d^D y M(x,y) \psi(y). \tag{10}$$

More restrictive assumptions on M are spelled out in ref. [11]. For the purpose of doing detailed calculations we will assume that the non local operator M admits an expansion in inverse powers of P_μ for large P_μ of the form

$$M = M_\mu \frac{P_\mu}{P^2} + M_{\mu\nu} \frac{P_\mu P_\nu}{P^4} + M_{\mu\nu\rho} \frac{P_\mu P_\nu P_\rho}{P^6} + \cdots \tag{11}$$

The coefficients $M_{\mu_1 \ldots \mu_n}$ are multiplicative operators and they are completely symmetric under permutation of indices. For convenience the P_μ has been put at the right [3]. The better way to obtain the transformation properties of M is by introducing a family of operators associated to M as

$$\tilde{M}(p) = e^{ipX} M e^{-ipX} \tag{12}$$

where the momentum p_μ is just a constant c-number. Effectively, $\tilde{M}(p)$ corresponds to make the replacement $P_\mu \to P_\mu + p_\mu$ in M. The function $\tilde{M}(p)$ admits an expansion in inverse powers of p_μ similar to that in eq. (11), namely

$$\tilde{M}(p) = \tilde{M}_\mu \frac{p_\mu}{p^2} + \tilde{M}_{\mu\nu} \frac{p_\mu p_\nu}{p^4} + \cdots. \tag{13}$$

We will adopt the ζ-function renormalization prescription combined with an asymmetric Wigner transformation. This method, as well as several of its applications, is presented in great detail in [10]. The ζ-function effective action is given by [13,14]

---

[2] M is softer in the ultraviolet sector than any multiplicative operator, that is, the distribution $M(x,y)$ is less singular than the Dirac delta $\delta(x-y)$.

[3] This choice does not exhaust all possible non local operators, but it is realistic since it accommodates the operator product expansion estimate of the quark self-energy $\Sigma(p^2) \sim_{p^2 \to \infty} (\log p^2)^{d-1}/p^2$ with d the anomalous dimension of the quark condensate $\bar\psi\psi$ (see ref. [12])

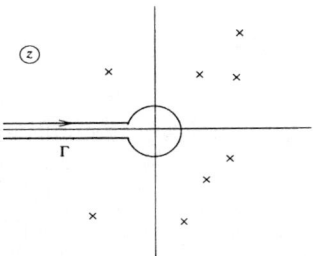

FIGURE 1. The Γ contour in the complex z-plane. Crosses represent isolated eigenvalues of $\boldsymbol{D}$.

$$W(\boldsymbol{D}) = -\text{Tr}\log \boldsymbol{D} = -\frac{d}{ds}\text{Tr}\,(\boldsymbol{D}^s)_{s=0}\,, \tag{14}$$

where $s = 0$ is to be understood as an analytical extension on s from the ultraviolet convergent region $\text{Re}(s) < -D$. The operator $\boldsymbol{D}^s$ can be obtained from

$$\boldsymbol{D}^s = -\int_\Gamma \frac{dz}{2\pi i}\frac{z^s}{\boldsymbol{D}-z} \tag{15}$$

where the integration path Γ starts at $-\infty$, follows the real negative axis, encircles the origin $z=0$ clockwise and goes back to $-\infty$ (see figure 1). The key point is that for sufficiently negative s there are no ultraviolet divergences and formal operations become justified. By construction, the ζ-function renormalized effective action is invariant under all symmetry transformations associated to similarity transformations of $\boldsymbol{D}$, thus in particular it is vector gauge invariant. On the other hand, the variable z plays the role of a mass and hence breaks explicitly both chiral and scale invariance.

The operator $(\boldsymbol{D}-z)^{-1}$ can be conveniently expressed by means of an asymmetric version of the Wigner representation [15]. For any operator A, let

$$A(x,p) = \int d^D y\, e^{iyp}\langle x|A|x-y\rangle = \frac{\langle x|A|p\rangle}{\langle x|p\rangle} \tag{16}$$

be its (asymmetric) Wigner representation. $|p\rangle$ is the momentum eigenstate with $\langle x|p\rangle = e^{-ixp}$. From this definition

$$\langle x|A|x\rangle = \int \frac{d^D p}{(2\pi)^D} A(x,p)\,,\quad \text{Tr}\,A = \int \frac{d^D x\, d^D p}{(2\pi)^D}\text{tr}\,A(x,p)\,, \tag{17}$$

where tr acts on internal and Dirac spinor degrees of freedom only, and the product of two operators satisfy the following formula

$$(AB)(x,p) = \exp(i\partial_p^A \cdot \partial_x^B) A(x,p)B(x,p)\,, \tag{18}$$

where ∂_p^A acts only on the p-dependence in $A(x,p)$ and ∂_x^B on the x-dependence in $B(x,p)$. Let the propagator or resolvent of $\boldsymbol{D}$, be $G(z) = (\boldsymbol{D}-z)^{-1}$ and $G(x,p;z)$ its Wigner representation. Applying Eq.18 to $(\boldsymbol{D}-z)G = 1$ one obtains

$$G(x,p;z) = \langle x|(\slashed{p}+\boldsymbol{D}-z)^{-1}|0\rangle \tag{19}$$

where $|0\rangle$ is the state of zero momentum, $\langle x|0\rangle = 1$. In practice this implies that $i\partial_\mu$ derivates every x dependence at is right, until it annihilates $|0\rangle$, $i\partial_\mu|0\rangle = 0$. This method is very efficient for a derivative expansion for it computes directly, that is non recursively, each of the terms [10].

The definition given for $A(x,p)$ is not gauge covariant because $|p\rangle$ is not. We will consider only local objects of the form $\langle x|f(\boldsymbol{D})|x\rangle$ as given by the formula

$$\langle x|f(\boldsymbol{D})|x\rangle = -\int \frac{d^D p}{(2\pi)^D} \int_\Gamma \frac{dz}{2\pi i} f(z) G(x,p;z) \tag{3.7}$$

We will assume that the function $f(z)$ is sufficiently convergent at infinity or else that it can be obtained as a suitable analytical extrapolation from a parametric family $f(z,s)$ in the variable s. In either case the integration over z should be performed in the first place, to yield the Wigner representation of the operator $f(\boldsymbol{D})$. Afterwards, the p integration is carried out, corresponding to take the diagonal matrix elements of $\langle y|f(\boldsymbol{D})|x\rangle$, hence restoring gauge covariance. This obviously means that the gauge breaking piece of $G(x,p;z)$ is a total derivative in the momentum variable. Recently, a method has been developed [16] where this total derivative is, by construction, gauge invariant.

II ANOMALIES

Because the chiral transformations are local, both $\boldsymbol{D}_L$ and $\boldsymbol{M}$ transform covariantly separately, that is,

$$\delta \boldsymbol{D}_L = [i\beta, \boldsymbol{D}_L] - \{i\alpha\gamma_5, \boldsymbol{D}_L\}, \quad \delta \boldsymbol{M} = [i\beta, \boldsymbol{M}] - \{i\alpha\gamma_5, \boldsymbol{M}\}. \tag{20}$$

Note that the bilocal structure of M implies that local factors at each side of the operator are taken at different points, i.e. $\boldsymbol{M}(x,x') \to e^{i\beta(x)-i\alpha(x)\gamma_5}\boldsymbol{M}(x,x')e^{-i\beta(x')-i\alpha(x')\gamma_5}$.

The two lowest coefficients are given by

$$\tilde{\boldsymbol{M}}_\mu = \boldsymbol{M}_\mu, \quad \tilde{\boldsymbol{M}}_{\mu\nu} = \boldsymbol{M}_{\mu\nu} + t_{\mu\nu\rho\sigma}\boldsymbol{M}_\rho P_\sigma, \tag{21}$$

where we have introduced $t_{\mu\nu\rho\sigma} = \delta_{\mu\nu}\delta_{\rho\sigma} - \delta_{\mu\rho}\delta_{\nu\sigma} - \delta_{\mu\sigma}\delta_{\nu\rho}$. It should be noted that the coefficients $\tilde{\boldsymbol{M}}_{\mu_1\ldots\mu_n}$ are not multiplicative operators. One useful property of $\tilde{\boldsymbol{M}}(p)$ is that it transforms covariantly under chiral transformations. Indeed, if $\boldsymbol{M}_\Omega = \Omega_1 \boldsymbol{M} \Omega_2$ for two multiplicative operators $\Omega_{1,2}$,

$$\tilde{\boldsymbol{M}}_\Omega(p) = e^{ipX}\boldsymbol{M}_\Omega e^{-ipX} = \Omega_1 \tilde{\boldsymbol{M}}(p)\Omega_2. \tag{22}$$

As a consequence, the coefficients are also chiral covariant

$$\delta \tilde{M}_{\mu_1...\mu_n} = [i\beta, \tilde{M}_{\mu_1...\mu_n}] - \{i\alpha\gamma_5, \tilde{M}_{\mu_1...\mu_n}\}. \qquad (23)$$

From here it is immediate to derive the transformation of the original coefficients $M_{\mu_1...\mu_n}$. For the two lowest order coefficients one finds

$$\begin{aligned}\delta M_\mu &= [i\beta, M_\mu] - \{i\alpha\gamma_5, M_\mu\}, \\ \delta M_{\mu\nu} &= [i\beta, M_{\mu\nu}] - \{i\alpha\gamma_5, M_{\mu\nu}\} + t_{\mu\nu\rho\sigma} M_\rho (\partial_\sigma \beta + \partial_\sigma \alpha\gamma_5).\end{aligned} \qquad (24)$$

In general, the variation of each coefficient involves those of lower order.

The scale transformation $\psi(x) \to e^{-\alpha_S(D-1)/2}\psi(e^{-\alpha_S}x)$ induces the corresponding transformation in D, namely,

$$Y(x) \to e^{-\alpha_S} Y(e^{-\alpha_S}x), \quad M_{\mu_1...\mu_n}(x) \to e^{-\alpha_S(n+1)} M_{\mu_1...\mu_n}(e^{-\alpha_S}x). \qquad (25)$$

Infinitesimally, it implies

$$\delta_S D = -\alpha_S(D - i[X_\mu P_\mu, D]). \qquad (26)$$

A Chiral anomaly

Due to the ζ regularization the chiral anomaly becomes an axial anomaly,

$$\mathcal{A}_A = \text{Tr}\,(2i\alpha\gamma_5 D^s)_{s=0}. \qquad (27)$$

Using the Wigner transformation technique [10], the anomaly can be written as (a similar expression holds for the effective action)

$$\mathcal{A}_A = -\int \frac{d^D p}{(2\pi)^D} \int_\Gamma \frac{dz}{2\pi i} z^s \text{tr}\langle 0|2i\alpha\gamma_5 \frac{1}{\tilde{D}(p) - z}|0\rangle\Big|_{s=0}. \qquad (28)$$

Here tr stands for trace over Dirac and flavor degrees of freedom, $|0\rangle$ is the zero momentum state normalized as $\langle x|0\rangle = 1$, thus $P_\mu|0\rangle = \langle 0|P_\mu = 0$. Further

$$\tilde{D}(p) = e^{ipX} D e^{-ipX} = \slashed{p} + D_L + \tilde{M}(p). \qquad (29)$$

The integration over z should be performed first, since it defines the operator D^s, then the integral over p which corresponds to take the trace over space-time degrees of freedom and finally s is to be analytically extended to $s = 0$. The simplest way to proceed is to introduce a mass term, i.e., to apply the formula to the Dirac operator $D + m$ and then make an expansion in powers of $D_L + \tilde{M}(p)$, letting $m \to 0$ at the end. In this way the following expression is derived

$$\mathcal{A}_A = \sum_{N\geq 0} \int \frac{d^D p}{(2\pi)^D} \int_\Gamma \frac{dz}{2\pi i} z^s \text{tr}\langle 0|2i\alpha\gamma_5 \frac{(D_L + \tilde{M}(p))\left((\slashed{p}+z-m)(D_L + \tilde{M}(p))\right)^N}{(p^2 + (z-m)^2)^{N+1}}|0\rangle\Big|_{s=0, m=0}. \qquad (30)$$

Simplification has been achieved by using the cyclic property for the trace in Dirac space.

After an angular average over p_μ, the indicated integrals on p_μ and z can be carried out directly with the integral I_1 given in [10]. The result for the four dimensional chiral anomaly is

$$\mathcal{A}_A = \Big\langle 2i\alpha\gamma_5 \Big[\frac{1}{2}\boldsymbol{D}_L^4 + \frac{1}{12}\boldsymbol{D}_L\{\gamma_\mu, \boldsymbol{D}_L\}^2 \boldsymbol{D}_L + \frac{1}{4}\tilde{\boldsymbol{M}}_\mu^2 + \frac{1}{4}\{\boldsymbol{D}_L, \tilde{\boldsymbol{M}}_{\mu\mu}\}$$
$$+\frac{1}{8}\Big(\tilde{\boldsymbol{M}}_\mu\{\gamma_\mu, \boldsymbol{D}_L\}\boldsymbol{D}_L + \boldsymbol{D}_L\{\gamma_\mu, \tilde{\boldsymbol{M}}_\mu\}\boldsymbol{D}_L + \boldsymbol{D}_L\{\gamma_\mu, \boldsymbol{D}_L\}\tilde{\boldsymbol{M}}_\mu\Big)\Big]\Big\rangle \quad (31)$$

The notation $\langle f \rangle$ stands for $\langle f \rangle = \frac{1}{(4\pi)^{D/2}}\text{tr}\langle 0|f(X)|0\rangle$. Note that, even for non local Dirac operators, the anomaly is a local polynomial of dimension D constructed with P_μ and the external fields $\boldsymbol{Y}$ and $\boldsymbol{M}_{\mu_1\ldots\mu_n}$. This is a general property of all anomalies since only ultraviolet divergent terms can contribute to them. The expressions found for the anomaly can be put in a more usual form, in terms of vector and axial fields, scalar fields, etc [4], but it is preferable to use compact notation. Since the regularization preserves vector gauge invariance, the axial anomaly is also invariant. In our expression for the anomaly, this is a direct consequence of the operators there being multiplicative. Indeed, any operator f constructed with the gauge covariant blocks $\boldsymbol{D}_L$ and $\tilde{\boldsymbol{M}}_{\mu_1\ldots\mu_n}$ is also covariant, i.e., $f \to \Omega f \Omega^{-1}$. If in addition f is multiplicative $\langle f \rangle$ is invariant. Note that $\langle\ \rangle$ is not a trace and so the cyclic property does not hold for arbitrary non multiplicative operators.

B Trace anomaly

The corresponding trace anomaly, within the ζ-function method is [10]

$$\mathcal{A}_S = \delta_S W = \alpha_S \text{Tr}(\boldsymbol{D}^s)_{s=0}. \quad (32)$$

The calculation is entirely similar to that of the axial anomaly, yielding

$$\mathcal{A}_S = \alpha_S \Big\langle \frac{1}{2}\boldsymbol{D}_L^4 + \frac{1}{12}(\boldsymbol{D}_L^2\{\gamma_\mu,\boldsymbol{D}_L\}^2 + \{\gamma_\mu,\boldsymbol{D}_L\}\boldsymbol{D}_L^2\{\gamma_\mu,\boldsymbol{D}_L\} + \{\gamma_\mu,\boldsymbol{D}_L\}^2\boldsymbol{D}_L^2)$$
$$+\frac{1}{96}(\{\gamma_\mu,\boldsymbol{D}_L\}^2\{\gamma_\nu,\boldsymbol{D}_L\}^2 + (\{\gamma_\mu,\boldsymbol{D}_L\}\{\gamma_\nu,\boldsymbol{D}_L\})^2 + \{\gamma_\mu,\boldsymbol{D}_L\}\{\gamma_\nu,\boldsymbol{D}_L\}^2\{\gamma_\mu,\boldsymbol{D}_L\})$$
$$+\frac{1}{12}(\gamma_\mu\tilde{\boldsymbol{M}}_\mu\boldsymbol{D}_L^2 + \gamma_\mu\boldsymbol{D}_L^2\tilde{\boldsymbol{M}}_\mu + \gamma_\mu\boldsymbol{D}_L\tilde{\boldsymbol{M}}_\mu\boldsymbol{D}_L)$$
$$-\frac{1}{24}(\tilde{\boldsymbol{M}}_\mu\gamma_\mu\boldsymbol{D}_L^2 + \boldsymbol{D}_L^2\gamma_\mu\tilde{\boldsymbol{M}}_\mu + \boldsymbol{D}_L\{\gamma_\mu,\tilde{\boldsymbol{M}}_\mu\}\boldsymbol{D}_L + \boldsymbol{D}_L\gamma_\mu\boldsymbol{D}_L\tilde{\boldsymbol{M}}_\mu + \tilde{\boldsymbol{M}}_\mu\boldsymbol{D}_L\gamma_\mu\boldsymbol{D}_L)$$
$$+\delta_{\mu\nu\alpha\beta}\Big(\frac{1}{36}(\gamma_\mu\tilde{\boldsymbol{M}}_\nu\gamma_\alpha\boldsymbol{D}_L\gamma_\beta\boldsymbol{D}_L + \gamma_\mu\boldsymbol{D}_L\gamma_\alpha\tilde{\boldsymbol{M}}_\nu\gamma_\beta\boldsymbol{D}_L + \gamma_\mu\boldsymbol{D}_L\gamma_\alpha\boldsymbol{D}_L\gamma_\beta\tilde{\boldsymbol{M}}_\nu)$$
$$+\frac{1}{24}\gamma_\mu\tilde{\boldsymbol{M}}_\nu\gamma_\alpha\tilde{\boldsymbol{M}}_\beta + \frac{1}{24}\{\gamma_\alpha\tilde{\boldsymbol{M}}_{\mu\nu},\gamma_\beta\boldsymbol{D}_L\} + \frac{1}{12}\gamma_\mu\tilde{\boldsymbol{M}}_{\nu\alpha\beta}\Big)\Big\rangle. \quad (33)$$

---

[4] See ref [10] for an explicit expression in the local case and the remarks of ref [11] for the non local case.

Where $\delta_{\mu\nu\alpha\beta} = \delta_{\mu\nu}\delta_{\alpha\beta} + \delta_{\mu\alpha}\delta_{\nu\beta} + \delta_{\mu\beta}\delta_{\alpha\nu}$. The result is again a local polynomial of dimension D in the external fields and their derivatives. Unlike the axial case, the coefficients $M_{\mu\nu\alpha}$ in four dimensions do contribute to the scale anomaly.

Because scale and chiral transformations commute (in a properly defined sense), the crossed variations $\delta_S \mathcal{A}_{V,A}$ and $\delta_{V,A} \mathcal{A}_S$ coincide and they vanish since the axial anomaly is scale invariant. Thus the scale anomaly must be chiral invariant. The vector gauge invariance of the previous expressions is easy to check noting that the operators inside $\langle \rangle$ are multiplicative. Axial invariance is much more involved in general. In four dimensions it is relatively easy to check that the trace anomaly is axially invariant in the particular case of $M_\mu = 0$, which defines a class of operators invariant under chiral and scale transformations.

C Counterterms and minimal form of the anomaly

Presumably due to its topological connection [17], the axial anomaly is a very robust quantity. It is not affected by higher order radiative corrections [18], and remains unchanged at finite temperature and density [19]. It gets no contributions from scalar and pseudo scalar fields [20], tensor fields [21,22,10] or internal gauge fields, i.e, transforming homogeneously under gauge transformations [23,24]. In all known cases, the anomaly only affects the imaginary part of the effective action in Euclidean space and only involves vector and axial fields. The counter terms can always be chosen so that the axial anomaly adopts the minimal or Bardeen's form [20]. Not surprisingly, the new terms introduced in the anomaly by the non local component of the Dirac operator are also unessential, that is, they can be removed by adding a suitable local and polynomial counter term to the effective action. In other words, all new terms can be derived as the axial variation of an action which is a polynomial constructed with the external fields Y and $M_{\mu_1...\mu_n}$ and their derivatives. The dimension of the polynomial can be at most D.

The general proof that the anomaly can always be brought to Bardeen's form has already been presented [11]. The actual construction of the counter terms can be done using the method in ref. [10] (see some further details in ref. [27]). One interesting insight in the local case [10] is that the needed counterterms require not only the Dirac operator D but also its adjoint $D^\dagger$ not related to the original theory.

The scale anomaly is already minimal. It can be modified by adding polynomial counter terms of dimension smaller than D but this would add terms of the same type to the scale anomaly.

III EXTENSION TO FINITE TEMPERATURE

The Wigner transformation method combined with the ζ-function regularization has been further extended to the finite temperature case in [28,29]. As is well known, in the imaginary time formulation of finite temperature field theory, the field configurations are periodic or antiperiodic functions of the Euclidean time for bosons and fermions respectively and thus the frequency running in the fermion loop takes discrete values only, $\omega_n = \pi(2n+1)T$ (where T stands for the temperature and n is any integer) which are known as Matsubara frequencies. At finite temperature, the trace of an operator $f(x_0, \boldsymbol{x}; i\partial_0, i\boldsymbol{\nabla})$, acting on the Hilbert space of $d+1$-dimensional fermions with possible internal degrees of freedom, becomes

$$\text{Tr}(f) = T \sum_n \int \frac{d^d p}{(2\pi)^d} \text{tr}\langle 0|f(x_0, \boldsymbol{x}; \omega_n + i\partial_0, \boldsymbol{p} + i\boldsymbol{\nabla})|0\rangle. \tag{34}$$

This formula generalizes that for zero temperature. Note that f is a periodic function of x_0, and $|0\rangle$ is the state with zero momentum and energy normalized to $\langle x_0, \boldsymbol{x}|0\rangle = 1$, and thus it is periodic too.

In Ref. [28] this method has been applied to compute the anomalous component of the effective action of two- and four-dimensional fermions at finite temperature in the presence of arbitrary vector and axial gauge fields and scalar and pseudoscalar fields on the chiral circle. The computation is carried out to leading order in a suitable commutator which preserves chiral symmetry. As is well known, at zero temperature the gauge Wess-Zumino-Witten (WZW) action, which saturates the chiral anomaly, is the only leading contribution in the anomalous sector; further terms must be Lorentz and chiral invariant and they vanish identically unless they have more gradients thereby being sub-leading terms. At finite temperature the situation is different since Lorentz invariance is partially broken and this allows to have new chiral invariant contributions[5] of the same order as the WZW action. In particular these terms modify the $\pi \to \gamma\gamma$ amplitude which is no longer determined by the chiral anomaly [30]. The calculation in [28] confirms previous results [31,32] that this amplitude vanishes in a chiral symmetric phase (see also [33,34]).

In Ref. [29] the same technique is applied to the study of $2+1$-dimensional fermions at finite temperature in the presence of arbitrary background gauge fields. The use of the ζ-function regularization guarantees the gauge invariance of the result under topologically small and large transformations. This has allowed to solve a long standing puzzle, namely, the apparent renormalization of the Chern-Simons coefficient at finite temperature, which has been shown to be a perturbation theory artifact [35]. In Ref. [29] all ultraviolet divergent terms of the effective action, within a strict gradient expansion, have been computed. The result preserves gauge and parity symmetries (up to the standard temperature independent parity anomaly) and display the correct $2\pi i$ multivaluation introduced by the Chern-Simons term. The known exact result for massless fermions [36] is also reproduced.

ACKNOWLEDGMENTS

One of us (E.R.A.) acknowledges the organizers for the stimulating atmosphere during the workshop. This work is supported in part by funds provided by the Span-

---

[5] It is well established that the chiral anomaly is temperature independent (see e.g. [19]).

ish DGICYT grant no. PB95-1204 and Junta de Andalucía grant no. FQM0225.

REFERENCES

1. S. Weinberg, Physica **96 A** (1979) 327.
2. See e.g. W. Marciano and H. Pagels, Phys. Rep. **3** (1978) 137 and references therein.
3. H. Pagels and S. Stokar, Phys. Rev. **D 20** (1979) 2947.
4. J. Gasser and H. Leutwyler, Ann. Phys. NY **158** (1984) 142.
5. For a review see e.g. C. D. Roberts and A. G. Williams, Prog. Part. Nucl. Phys. **33** (1994) and references therein.
6. C. D. Roberts, R. T. Cahill and J. Prashifka, Ann. Phys. (N.Y.) **188** (1988) 20.
7. B. Holdom, J. Terning and K. Verbeek, Phys. Lett. **B 232** (1989) 351.
8. R. D. Ball and G. Ripka, in *Many Body Physics*, World Scientific, 1994. C. Fiolhais, M. Fiolhais, C. Sousa and J. N. Urbano (Eds).
9. B. Golli, W. Broniowski and G. Ripka, Phys. Lett. **437** (1998) 24.
10. L.L. Salcedo and E. Ruiz Arriola, Ann. Phys. (N.Y.) **250** (1996) 1.
11. E. Ruiz Arriola and L. L. Salcedo, Phys. Lett. **B 450** (1999) 225.
12. H. D. Politzer, Nucl. Phys. **B 117** (1976) 397.
13. R.T. Seeley, Amer. Math. Soc. Proc. Symp. Pure Math. **10** (1967) 288.
14. S.W. Hawking, Comm. Math. Phys. **55** (1977) 133.
15. E. P. Wigner, Phys. Rev. **40** (1932) 749.
16. N. G. Pletnev and A. T. Banin, hep-ph/9811031.
17. L. Álvarez-Gaumé and P. Ginsparg, Ann. Phys. (N.Y.) **161** (1985) 423.
18. S.L. Adler and W.A. Bardeen, Phys. Rev. **182** (1969) 1517.
19. A. Gómez Nicola and R.F. Álvarez-Estrada, Int. J. Mod. Phys. **A 9** (1994) 1423.
20. W.A. Bardeen, Phys. Rev. **184** (1969) 1848.
21. T.E. Clark and S.T. Love, Nucl. Phys. **B 223** (1983) 135.
22. J. Minn, J. Kim and C. Lee , Phys. Rev. **D 35** (1987) 1872.
23. J. Bijnens and J. Prades, Phys. Lett. **B 320** (1993) 130.
24. E. Ruiz Arriola and L.L. Salcedo, Nucl. Phys. **A 590** (1995) 703.
25. W.A. Bardeen and B. Zumino, Nucl. Phys. **B 244** (1984) 421.
26. H. Leutwyler, Phys. Lett. **B 152** (1985) 78.
27. E. Ruiz Arriola and L. L. Salcedo, hep-th/9811073.
28. L.L. Salcedo, Phys. Rev. **D58** (1998) 125007.
29. L.L. Salcedo, Nucl. Phys. **B549** (1999) 98.
30. R.D. Pisarski and M. Tytgat, Phys. Rev. **D56** (1997) 7077.
31. R.D. Pisarski, Phys. Rev. Lett. **76** (1996) 3084.
32. R. Baier, M. Dirks and O. Kober, Phys. Rev. **D54** (1996) 2222.
33. S. Gupta and S.N. Nayak, hep-ph/9702205.
34. F. Gelis, Phys. Rev. **D59** (1999) 076004.
35. G. Dunne, K. Lee and C. Lu, Phys. Rev. Lett. **78** (1997) 3434.
36. L. Álvarez-Gaumé, S. Della Pietra and G. Moore, Ann. Phys. **163** (1985) 288.

Meson Properties in a renormalizable version of the NJL model.

André L. Mota[*,1], M. Carolina Nemes[*],
Brigitte Hiller[†] and Hans Walliser[†]

[*]*Departamento de Física, Instituto de Ciências Exatas, Universidade Federal de Minas Gerais, Belo Horizonte, CEP 30.161-970, C.P. 702, MG, Brazil*
[†]*Departamento de Física, Universidade de Coimbra, P-3000 Coimbra, Portugal*

Abstract. We construct a non-trivial and renormalized extension of the Nambu and Jona-Lasinio (NJL) model. This model is analogous in form to the linear σ-model. However, given the Nc dependence of the coupling constant, diagrams involving mesonic internal lines are absent in leading order, as opposed to the linear σ-model. Meson properties are systematically improved in the present calculations as compared to usual Pauli-Villars regularized versions.

I INTRODUCTION

The Nambu and Jona-Lasinio model (NJL) [1] has undoubtedly played an important role in our understanding of low and medium energy hadronic physics [2,3]. Although it is non-renormalizable in perturbation theory, it becomes renormalizable in the mean field expansion for $d > 2$ [4–6]. As will be shown in what follows, this renormalized model becomes trivial in the continuous limit. We next implement a non-trivial renormalizable version of the model, showing that all meson properties which depend on finite integrals are systematically improved.

II MEAN FIELD EXPANSION OF THE NJL MODEL AND TRIVIALITY

In this section we briefly recapitulate the main features of the renormalization procedure for the NJL model using the mean–field expansion. Following Eguchi [4] we isolate the UV singularities, but we consider also finite contributions which enter the renormalization scheme in order to demonstrate the equivalence to the

---

[1)] also at DCNat - Departamento de Ciências Naturais. FUNREI - Fundação de Ensino Superior de São João del Rei. Praça Dom Helvecio, CEP : 36300-000, São João del Rei - M.G. Brazil; e-mail : motaal@funrei.br

procedure presented by Guralnik and Tamvakis [5]. In addition we allow the symmetry to be broken explicitly by a current quark mass. Finally we discuss the issue of triviality and the introduction of a cut-off Λ preventing the model from the collapse. The following derivation applies to SU(2) with N_C colors, an extension to SU(3) is straightforward.

Starting point is the NJL Lagrangian with a local four-quark interaction

$$L = \bar{q}(i\gamma^\mu \partial_\mu - \hat{m}_0)q + \frac{G_0}{2}\left[(\bar{q}q)^2 + (\bar{q}i\gamma_5\boldsymbol{\tau}q)^2\right],\tag{1}$$

where we consider the isospin symmetric limit $\hat{m}_u = \hat{m}_d = \hat{m}$. To allow for wave function renormalizations later on and also in order to trace the N_C orders, it is convenient to replace $G_0 = g_0^2/\mu_0^2$ with $g_0^2 \sim 1/N_C$. The subscript zero denotes bare (infinite) quantities everywhere. With boson fields introduced in the standard way the Lagrangian becomes

$$L = \bar{q}\left[i\gamma^\mu \partial_\mu - \hat{m}_0 - g_0(\sigma_0 + i\gamma_5\boldsymbol{\tau}\boldsymbol{\pi}_0)\right]q - \frac{\mu_0^2}{2}(\sigma_0^2 + \boldsymbol{\pi}_0^2)$$

$$L = \bar{q}\left[i\gamma^\mu \partial_\mu - g_0(\sigma_0 + i\gamma_5\boldsymbol{\tau}\boldsymbol{\pi}_0)\right]q - \frac{\mu_0^2}{2}(\sigma_0^2 + \boldsymbol{\pi}_0^2) + \frac{\mu_0^2}{g_0}\hat{m}_0\sigma_0,\tag{2}$$

where the latter representation is obtained by shifting the scalar field $\sigma_0 \to \sigma_0 - \hat{m}_0/g_0$ and a term independent of the dynamical fields is omitted. Integrations over the fermion fields q and $\bar{q}$ may now be performed in the path integral such that the resulting effective Lagrangian collects the corresponding trace log contribution

$$L = -iTr\ell n\left[i\gamma^\mu\partial_\mu - g_0(\sigma_0 + i\gamma_5\boldsymbol{\tau}\boldsymbol{\pi}_0)\right] - \frac{\mu_0^2}{2}(\sigma_0^2 + \boldsymbol{\pi}_0^2) + \frac{\mu_0^2}{g_0}\hat{m}_0\sigma_0.\tag{3}$$

Expecting the scalar field to possess a nonvanishing vacuum expectation value we expand $\sigma_0 = m/g_0 + \sigma_0'$, where m represents the (finite) constituent mass

$$L = i\sum_{n=1}^{\infty}\frac{1}{n}Tr\left[(i\gamma^\mu\partial_\mu - m)^{-1}g_0(\sigma_0' + i\gamma_5\boldsymbol{\tau}\boldsymbol{\pi}_0)\right]^n - \frac{\mu_0^2}{2}(\sigma_0'^2 + \boldsymbol{\pi}_0^2) - \frac{\mu_0^2}{g_0}(m - \hat{m}_0)\sigma_0'.\tag{4}$$

To evaluate this sum is now quite straightforward. The terms for $n = 1,\ldots,4$ contain UV divergencies showing up as

$$I_{quad} = i\int\frac{d^4q}{(2\pi)^4}\frac{1}{q^2 - m^2},\quad I_{log} = i\int\frac{d^4q}{(2\pi)^4}\frac{1}{(q^2 - m^2)^2}\tag{5}$$

quadratically and logarithmically divergent integrals respectively. The effective Lagrangian is then given by

$$L = \frac{1}{2}(-4N_C g_0^2 I_{log})(\partial_\mu \sigma_0' \partial^\mu \sigma_0' + \partial_\mu \pi_0 \partial^\mu \pi_0) - \frac{N_C g_0^2}{12\pi^2} \partial_\mu \sigma_0' \partial^\mu \sigma_0'$$
$$- \frac{1}{2}(\mu_0^2 - 8N_C g_0^2 I_{quad})(\sigma_0'^2 + \pi_0^2) - \frac{1}{2} 4m^2(-4N_C g_0^2 I_{log})\sigma_0'^2 + \ldots$$
$$+ 8N_C g_0^3 I_{log} m \sigma_0'(\sigma_0'^2 + \pi_0^2) + 2N_C g_0^4 I_{log}(\sigma_0'^2 + \pi_0^2)^2 + \ldots$$
$$- \left[\frac{\mu_0^2}{g_0}(m - \hat{m}_0) - 8N_C g_0 m I_{quad}\right] \sigma_0'. \tag{6}$$

The dots after the first two lines denote finite higher derivative terms of $g_0^2 \sigma_0^2$ and $g_0^2 \pi_0^2$ proportional to N_C^0 not explicitly shown, but a finite kinetic term for the scalars is kept and finally leads to different wave–function renormalizations $\sigma' = Z_\sigma^{-1/2} \sigma_0'$ and $\pi = Z_\pi^{-1/2} \pi_0$. The dots in the third line indicate finite higher order terms also not shown. The renormalized parameters may now be introduced as follows

$$Z_\sigma^{-1} = \frac{g_0^2}{g_\sigma^2} = -4N_C g_0^2 I_{log} - \frac{N_C g_0^2}{6\pi^2} \tag{7}$$

$$Z_\pi^{-1} = \frac{g_0^2}{g_\pi^2} = -4N_C g_0^2 I_{log} \tag{8}$$

$$\frac{\mu_\sigma^2}{Z_\sigma} = \mu_0^2 - 8N_C g_0^2(I_{quad} + 2m^2 I_{log}) \tag{9}$$

$$\frac{\mu_\pi^2}{Z_\pi} = \mu_0^2 - 8N_C g_0^2 I_{quad}. \tag{10}$$

Together with the gap–equation

$$\frac{\mu_0^2}{g_0^2}(m - \hat{m}_0) = 8N_C m I_{quad}, \tag{11}$$

which makes the linear term in σ_0' vanish, we obtain the renormalized Lagrangian in its final form

$$L = \frac{1}{2}(\partial_\mu \sigma' \partial^\mu \sigma' + \partial_\mu \pi \partial^\mu \pi) - \frac{1}{2}(\mu_\sigma^2 \sigma'^2 + \mu_\pi^2 \pi^2)$$
$$- \frac{2m g_\sigma}{g_\pi^2} \sigma'(g_\sigma^2 \sigma'^2 + g_\pi^2 \pi^2) - \frac{1}{2g_\pi^2}(g_\sigma^2 \sigma'^2 + g_\pi^2 \pi^2)^2 + \ldots \tag{12}$$

The current quark mass has disappeared and it is noticed that the remaining parameters are related according to eqs. (7-10)

$$\frac{1}{g_\sigma^2} = \frac{1}{g_\pi^2} - \frac{N_C}{6\pi^2}, \qquad \frac{\mu_\sigma^2}{g_\sigma^2} = \frac{4m^2 + \mu_\pi^2}{g_\pi^2} \tag{13}$$

such that the renormalized model is characterized by three parameters (m, g_π, μ_π) as the regularized one $(G, m, \hat{m})$. From these equations using the gap–equation (11) we may also define a renormalized four-fermion coupling G

$$\hat{m}\frac{\mu_\pi^2}{g_\pi^2} = \hat{m}_0 \frac{\mu_0^2}{g_0^2} = \frac{\hat{m}_0}{G_0} = \frac{\hat{m}}{G} \tag{14}$$

by reintroduction of the renormalized (physical) current quark mass. Although this relation is beyond the scope of the renormalized model it will nevertheless prove useful for the evaluation of the quark condensate.

The N_C orders are now carried by the couplings $g_\pi^2 \sim g_\sigma^2 \sim 1/N_C$. For practical calculations we consider only the leading order N_C for each process. From the quadratic terms in the Lagrangian we read off the renormalized meson propagators of order N_C^0

$$\Delta_\sigma^{-1}(p^2) = p^2 - \mu_\sigma^2 - 4N_C g_\sigma^2 \left[(p^2 - 4m^2)Z_0(p^2) - \frac{p^2}{24\pi^2}\right]$$
$$\Delta_\pi^{-1}(p^2) = p^2 - \mu_\pi^2 - 4N_C g_\pi^2 p^2 Z_0(p^2), \tag{15}$$

where the momentum dependent terms due to the finite higher derivative terms not explicitely shown in eq.(12) are contained in the finite function

$$Z_0(p^2) = \frac{1}{16\pi^2} \int_0^1 dz \ell n\left[1 - \frac{p^2}{m^2}z(1-z)\right]$$
$$= \frac{1}{8\pi^2} \begin{cases} \sqrt{1 + \frac{4m^2}{|p^2|}} \operatorname{arsinh}\sqrt{\frac{|p^2|}{4m^2}} - 1 & p^2 \leq 0 \\ \sqrt{\frac{4m^2}{p^2} - 1} \arcsin\sqrt{\frac{p^2}{4m^2}} - 1 & 0 < p^2 \leq 4m^2 \\ \sqrt{1 - \frac{4m^2}{p^2}} \left(\operatorname{arcosh}\sqrt{\frac{p^2}{4m^2}} - i\frac{\pi}{2}\right) - 1 & 4m^2 < p^2 \end{cases}$$
(16)

which possesses different branches. The physical meson masses are defined as usual via the poles of the propagators

$$m_\sigma^2 = 4m^2 + \mu_\sigma^2 + 4N_C g_\pi^2(m_\sigma^2 - 4m^2)Z_0(m_\sigma^2)$$
$$m_\pi^2 = \mu_\pi^2 + 4N_C g_\pi^2 m_\pi^2 Z_0(m_\pi^2). \tag{17}$$

In the chiral limit ($\mu_\pi^2 = 0$) we obtain $m_\pi = 0$ and $m_\sigma = 2m$.

Similarly, also in the chiral limit the leading N_C 3- and 4-boson vertex functions at zero momenta are obtained from the cubic and quartic self-couplings in accordance with Guralnik and Tamvakis [5], who derive these results from the corresponding Ward identities.

In the following we want to discuss the triviality of the NJL model which is suggested by lattice calculations [9]. In the mean-field expansion it follows immediately from (8)

$$g_\pi^2 = -(4N_C I_{log})^{-1}, \tag{18}$$

namely in the continuum limit the couplings g_π and hence also g_σ are driven to zero rendering the Lagrangian (12) a theory of non–interacting mesons. This is caused by the mesonic kinetic terms and quartic self couplings in (12) being created purely by radiative corrections: they were not present in the original Lagrangian. To avoid the collapse of the model, g_π must be kept fixed at some finite value. For that purpose a cut-off Λ has to be introduced in order to keep the logarithmically divergent integral in (18) finite (note that the quadratic divergence has already disappeared in the renormalized parameters). In the continuum limit $g_\pi^2 \sim (4\pi/N_C)/\ell n(\Lambda/m)$ tends to zero logarithmically, and a finite coupling g_π requires also a finite Λ, in fact, in order to reproduce a reasonable coupling strength a rather low cut-off of the order of 1GeV is needed in contrast to the situation in QED where the collaps is prevented by a cut-off located way above all physical energies of interest. Nevertheless, if it were possible to choose Λ large enough such that all finite integrals may be evaluated in the continuum limit then the cut-off would disappear from the theory being only implicitly contained in the coupling g_π which is kept finite [14]. Exactly this is achieved by adding mesonic kinetic terms and quartic self-couplings to the model which are capable to absorb the troublesome radiative terms in (12) leading to a non-trivial renormalizable extension of the NJL discussed in the following section. Mesonic properties calculated in the two versions of the model are then presented in section IV.

III NON-TRIVIAL EXTENSION OF THE NJL MODEL

We have seen in the previous section that the triviality of the NJL model is connected with the fact that the mesonic kinetic and interaction terms are created purely by radiative corrections. In fact triviality may be avoided by adding these contributions to the Lagrangian (3) from the beginning [15]

$$L = -iTr\ell n\left[i\gamma^\mu\partial_\mu - g_0(\sigma_0 + i\gamma_5\boldsymbol{\tau}\boldsymbol{\pi}_0)\right] + \frac{f_0^2}{2}(\partial_\mu\sigma_0\partial^\mu\sigma_0 + \partial_\mu\boldsymbol{\pi}_0\partial^\mu\boldsymbol{\pi}_0)$$
$$-\frac{\mu_0^2}{2}(\sigma_0^2 + \boldsymbol{\pi}_0^2) - \frac{\lambda_0}{2}(\sigma_0^2 + \boldsymbol{\pi}_0^2)^2 + \frac{\mu_0^2}{g_0}\hat{m}_0\sigma_0 + \ldots \quad (19)$$

Of course this may lead beyond the NJL model, we will comment on this later. Repeating the steps which lead to eq. (6) we find the renormalized parameters as

$$Z_\sigma^{-1} = \frac{g_0^2}{g_\sigma^2} = f_0^2 - 4N_C g_0^2 I_{log} - \frac{N_C g_0^2}{6\pi^2} \quad (20)$$

$$Z_\pi^{-1} = \frac{g_0^2}{g_\pi^2} = f_0^2 - 4N_C g_0^2 I_{log} \quad (21)$$

$$\frac{\mu_\sigma^2}{Z_\sigma} = \mu_0^2 - 8N_C g_0^2(I_{quad} + 2m^2 I_{log}) + \frac{6\lambda_0}{g_0^2}m^2 \quad (22)$$

$$\frac{\mu_\pi^2}{Z_\pi} = \mu_0^2 - 8N_C g_0^2 I_{quad} + \frac{2\lambda_0}{g_0^2}m^2 \quad (23)$$

$$\frac{\lambda}{g_\pi^2} = \frac{\lambda_0}{g_0^4} - 4N_C I_{log} \quad (24)$$

together with the gap–equation

$$\frac{\mu_0^2}{g_0^2}(m - \hat{m}_0) - 8N_C m I_{quad} + \frac{2\lambda_0}{g_0^4} m^3 = 0. \qquad (25)$$

The renormalized Lagrangian in the shifted scalar fields becomes

$$L = \frac{1}{2}(\partial_\mu \sigma' \partial^\mu \sigma' + \partial_\mu \boldsymbol{\pi} \partial^\mu \boldsymbol{\pi}) - \frac{1}{2}(\mu_\sigma^2 \sigma'^2 + \mu_\pi^2 \boldsymbol{\pi}^2)$$
$$-\frac{2m\lambda}{g_\pi^2} g_\sigma \sigma'(g_\sigma^2 \sigma'^2 + g_\pi^2 \boldsymbol{\pi}^2) - \frac{\lambda}{2g_\pi^2}(g_\sigma^2 \sigma'^2 + g_\pi^2 \boldsymbol{\pi}^2)^2 + \ldots \qquad (26)$$

formally identical to the bosonized NJL Lagrangian (12) if we choose $\lambda = 1$. In general the model parameters are now related

$$\frac{1}{g_\sigma^2} = \frac{1}{g_\pi^2} - \frac{N_C}{6\pi^2}, \qquad \frac{\mu_\sigma^2}{g_\sigma^2} = \frac{4\lambda m^2 + \mu_\pi^2}{g_\pi^2} \qquad (27)$$

quite similar to (13) and the relation (14) is unchanged. In analogy to (17) we obtain for the physical meson masses

$$m_\sigma^2 = 4\lambda m^2 + \mu_\pi^2 + 4N_C g_\pi^2 (m_\sigma^2 - 4m^2) Z_0(m_\sigma^2)$$
$$m_\pi^2 = \mu_\pi^2 + 4N_C g_\pi^2 m_\pi^2 Z_0(m_\pi^2). \qquad (28)$$

While the pion mass remains unchanged, the Nambu relation in the chiral limit $m_\sigma = 2m$ does no longer hold when $\lambda \neq 1$.

We want to emphasize here, that the Lagrangian (26) does not represent the trivial NJL model and constitutes instead a different *non–trivial* theory. As a consequence the couplings g_π and g_σ do no longer vanish in the continuum limit. The numerical results of this model with $\lambda = 1$ are of course identical to those obtained from the NJL model (12) with g_π and g_σ kept fixed at some finite values as discussed in the preceeding section.

Concluding, there seems to be two options to treat the problem of triviality which appears in the NJL model:

(i) A cut–off Λ is retained to prevent the model from the collapse. Because numerically the cut–off is of the order of 1 GeV only, it has to be kept also in all finite integrals.

(ii) The NJL model is augmented by kinetic terms and mesonic self–interactions. This results in the linear sigma model coupled to quarks and constitutes a perfect non–trivial field theory, renormalizable to all loop orders [10].

The latter theory contains one additional parameter λ. For $\lambda = 1$ (ii) gives the same results as the conventional NJL in case of a large cut–off. In principle the assumption $\lambda = 1$ may be tested in $\pi\pi$ scattering, of course not in the leading chiral order which is fixed by a famous low energy theorem [16], but in the next to leading orders (subsection IV.B). Many other meson properties are quite independent of this parameter. In the following section we compare some mesonic observables calculated in the two versions of the NJL model.

IV MESON PROPERTIES

In this section the parameters of both versions of the model are fixed. We show that in the renormalized version the chiral expansion becomes quite simple and we discuss the issue of the additional parameter λ appearing in the linear sigma model with quarks. Finally we are going to calculate several mesonic observables.

A Determination of the model parameters

For both versions of the model we used two sets of model parameters : one with a low constituent quark mass fixed at $m = 210$ MeV, and the other with the constituent quark mass fixed at $m = 350$ MeV, values which are used widely in the literature. The remaining model parameters are adjusted to reproduce the pion decay constant $f_\pi = 93.3$ MeV and the pion mass $m_\pi = 139$ MeV. As a result, one obtains for the Pauli–Villars regularized version a large dimensionless ratio $\Lambda/m \simeq 6$ in the small constituent quark mass case, and a small one for the large mass case, $\Lambda/m \simeq 2$. Numerical values, including those of the corresponding model parameters G and $\hat{m}$, are given in Table I. It should be mentioned here, that the NJL extended by inclusion of spin–one mesons allows for larger values of the cut-off.

As will be discussed in section IV.B, chiral expansions of the renormalized and regularized versions coincide in the $\Lambda/m \to \infty$ limit, showing that this ratio is a measure for the deviations in the two models. On the other hand, in the renormalized version the pion decay constant is

$$f_\pi = g_{\pi qq} \frac{m}{g_\pi^2} \frac{\mu_\pi^2}{m_\pi^2} , \qquad (29)$$

and the pion mass is given by (17). Here and in the following we use the pion quark and sigma quark couplings

$$g_{\pi qq}^{-2} = g_\pi^{-2} - 4N_C \left[Z_0(m_\pi^2) + m_\pi^2 Z_0'(m_\pi^2) \right]$$
$$g_{\sigma qq}^{-2} = g_\pi^{-2} - 4N_C \left[Z_0(m_\sigma^2) + (m_\sigma^2 - 4m^2) Z_0'(m_\sigma^2) \right] , \qquad (30)$$

for abbreviation. Thus, f_π and m_π fix the parameters g_π and μ_π listed also in Table I. Furthermore, for the evaluation of the quark condensate according to (14)

$$\hat{m} <\bar{q}q> = -(m - \hat{m})\frac{\hat{m}}{G} = -m(m - \hat{m})\frac{\mu_\pi^2}{g_\pi^2} \qquad (31)$$

a value for the current quark mass has to be adopted which strictly speaking is not a parameter of the renormalized version of the model. The standard value of $\hat{m} = 7.5$ MeV leads to the result quoted in Table II.

TABLE 1. Parameters of the regularized and renormalized versions of the NJL model.

	regularized model			renormalized model		
model parameters	$m = 350$ $G = 17.6$ $\hat{m} = 8.5$	(210) (5.11) (4.1)	MeV GeV^{-2} MeV	$m = 350$ $g_\pi = 3.752$ $\mu_\pi = 141$	(210) (2.250) (141)	MeV MeV
related parameters	$\Lambda = 769$	(1190)	MeV	$g_\sigma = 7.006$ $\mu_\sigma = 1333$ $\hat{m} = 7.5$	(2.610) (513.8)	MeV MeV

B Sigma meson properties

The sigma meson mass can be evaluated using eq.(28). In order to obtain this mass, we made use of the real part of the propagator only, in both, the regularized and the renormalized version. The imaginary part related to the decays into $\bar{q}q$ pairs is very small and can be neglected [19]. The decay width of $\sigma \to \pi\pi$ is :

$$\Gamma_{\sigma\pi\pi} = \frac{3\sqrt{m_\sigma^2 - 4m_\pi^2}}{8\pi m_\sigma^2} f_{\sigma\pi\pi}^2 . \tag{32}$$

All the expressions for the Pauli–Villars regularized NJL model used here and in the following sections may be found in [20] and are not repeated here. The amplitude $f_{\sigma\pi\pi}$ reads in the renormalized version

$$f_{\sigma\pi\pi} = 16mN_C g_{\sigma qq} g_{\pi qq}^2 \left[\frac{\lambda}{4N_C g_\pi^2} - Z_0(m_\sigma^2) + \frac{2m_\pi^2 - m_\sigma^2}{2} I_3(m_\sigma^2, m_\pi^2, m_\pi^2) \right],$$

$$I_3(p^2, p_1^2, p_2^2) = i \int \frac{d^4q}{(2\pi)^4} \frac{1}{(q^2 - m^2)[(q-p_1)^2 - m^2][(q-p_2)^2 - m^2]} \tag{33}$$

with $p^2 = (p_1 - p_2)^2$. Both, the sigma mass and its decay width into two pions depend on the additional parameter λ. In particular the decay width increases rapidly if λ is allowed to become larger than 1. The values quoted in Table II are calculated for $\lambda = 1$ because the experimental evidence for the sigma meson is too weak to serve for a determination of this parameter. From this Table it is also noticed that the scalar decay width depends most sensitively on the two versions of the model, specially of course in the large mass (low cut–off) case. This difference persists also in the chiral limit.

C Pion charge formfactor

The pion properties are independent of the additional parameter λ. In the renormalized version, the pion electromagnetic formfactor is given by

$$F_\pi(p^2) = 2N_C g_{\pi qq}^2 \left[\frac{1}{4N_C g_\pi^2} + m_\pi^2 \left(Z_0'(m_\pi^2) - I_3(p^2, m_\pi^2, m_\pi^2) \right) \right], \quad (34)$$

and the corresponding pion electromagnetic radius is defined as usual. The charge radius turns out too small compared to its experimental value, see Table II. We also see that the renormalizable version yields a value close to the chiral limit result

$$<r^2>_\pi = \frac{3N_C}{4\pi^2 f_\pi^2} \simeq (0.59 fm)^2, \quad (35)$$

whereas the regularized version tends to decrease this value further with decreasing ratio Λ/m. The formfactors in the two versions of the model are very similar for the lower mass case, but for the larger mass case the renormalized version improves on the results, whereas the regularized version does not yield a satisfactory fit as noticed already in [11].

D Anomalous $\pi^0 \to \gamma\gamma$ decay

The formfactor associated with the anomalous process $\pi_0 \to \gamma^*\gamma$, with one of the photons being off shell is

$$F_{\pi\gamma^*\gamma}(p^2) = -\frac{8N_C e^2}{3} g_{\pi qq} m I_3(0, p^2, m_\pi^2). \quad (36)$$

With both photons on–shell one obtains the anomalous pion decay $\pi_0 \to \gamma\gamma$ analytically in the renormalized version, given by the following expression:

$$\Gamma_{\pi^0 \to \gamma\gamma} = \frac{m_{\pi^0}^3}{64\pi} F_{\pi\gamma\gamma}^2, \quad F_{\pi\gamma\gamma} = -\frac{N_C e^2 g_{\pi qq}}{3\pi^2 m_\pi} \frac{m}{\sqrt{4m^2 - m_\pi^2}} \arctan \frac{m_\pi}{\sqrt{4m^2 - m_\pi^2}}. \quad (37)$$

In the chiral limit, this reduces to the well–known result [22]

$$\Gamma_{\pi^0 \to \gamma\gamma} = \frac{m_{\pi^0}^3}{64\pi} \left(\frac{N_C e^2}{12\pi^2 f_\pi} \right)^2. \quad (38)$$

The model independent amplitude for the anomalous $\pi^0 \to \gamma\gamma$ decay in the chiral limit is obtained exactly only in the renormalized version. The regularized version renders the decay width strongly cut–off dependent (see Table II). This formal result is one of the obvious advantages of the renormalized model. In this context we should mention that in the spirit of regularized models a consistent treatment of

TABLE 2. Some meson properties calculated in the regularized and renormalized versions of the NJL model are compared to experimental data [26,27]. The asterisks indicate quantities which served as input to determine the model parameters. Results are calculated for a constituent quark mass of $m = 350$ MeV and $m = 210$ MeV (in brackets).

		regularized model		renormalized model		experiment
f_π	[MeV]	93.3*		93.3*		93.3
m_π	[MeV]	139*		139*		139
m_σ	[MeV]	705	(434)	705	(433)	~ 700
$\Gamma_{\sigma \to \pi\pi}$	[MeV]	647	(336)	457	(307)	~ 700
$\Gamma_{\pi^0 \to \gamma\gamma}$	[eV]	3.7	(7.2)	7.6	(8.3)	7.7 ± 0.6
$<r^2>_\pi^{1/2}$	[fm]	0.49	(0.59)	0.58	(0.59)	0.678 ± 0.012
$<\bar{q}q>$	[MeV^3]	-265^3	(-340^3)	-283^3	(-281^3)	$-(296 \pm 25)^3$

anomalous processes has been forwarded in [23], where the quark–loop is regulated dynamically by the intrinsic non–locality of the quark–meson interaction. In an extended version of the regularized NJL model to include spin–one mesons this is achieved by a subtraction imposed by the anomalous Ward identity to obtain the correct QCD flavor anomaly [24].

For the larger Λ/m ratio the results in Table II are quite similar in the two versions as expected. For the larger constituent mass case, however, the results for the anomalous pion decay and the pion electromagnetic radius are substantially improved in the renormalized version.

V CONCLUSIONS

We have considered a non–trivial and renormalizable extension of the NJL model obtained by adding the necessary counter terms, namely the mesonic kinetic energies and quartic mesonic interactions to the originial Lagrangian. This amounts to a linear sigma model coupled to quarks with fixed strength of the quartic mesonic self–interactions. It was shown that this version of the model coincides with its familiar regularized versions provided the cut–off is large enough.

We presented a comparative and quantitative analysis of the most relevant mesonic observables with an effective Pauli–Villars regularized version in leading $1/N_C$ order. We found that in the renormalized version of the model the electromagnetic pion decay is always (independently of other parameters) in agreement with experiment in contrast to results obtained in regularized versions (usually $\geq 30\%$ off). A reasonable description of the corresponding transition formfactor is obtained. In the regularized versions the quality of the results depends crucially on the ratio Λ/m where Λ is the four-dimensional cut-off and m is the constituent quark mass. The larger the ratio, the better the agreement with phenomenology. This is in a way reflected in the renormalized version, where the agreement is systematically better when compared with the regularized version for the same m value. We noticed that the quantitative results become quite similar in the two versions for ratios $\Lambda/m \gtrsim 6$.

Further applications of the extended non-trivial and renormalizable NJL model are the study of the SU(3) flavor case and the inclusion of vectors and axialvectors mesons. One of the most interesting issues related to the SU(3) sector is connected with the kaon decay constant which persistently turns out to be too small ($f_K \simeq f_\pi$) unless one is willing to accept a small nonstrange constituent quark mass of the order $m \simeq 210$MeV which brings along its own difficulties (small scalar meson masses, low $\bar{q}q$ threshold, etc.). Unfortunately the renormalizable version presented here is not capable to resolve this problem, the solution has to be sought elsewhere.

H.W. would like to thank G. Holzwarth for clarifying comments on the renormalization of the model. We thank O.A. Battistel for critical comments on regularization techniques [28]. The present research has been supported partly by FCT, Portugal (Contract PRAXIS/4/4.1/BCC/2753 and PCERN/S/FIS/1162/97) and by CNPq-Brazil.

REFERENCES

1. Y. Nambu and G. Jona-Lasinio, Phys. Rev. **122** (1961) 345.
2. T. Hatsuda and T. Kunihiro, Phys. Rep. **247** (1994) 221.
3. J. Bijnens, Phys. Rep. **265** (1996) 369.
4. T. Eguchi, Phys. Rev. **D14** (1976) 2755.
5. G.S.Guralnik and K. Tamvakis, Nucl. Phys. **B148** (1979) 283.
6. B. Rosenstein, B.J. Warr and S.H. Park, Phys. Rev. Lett. **62** (1989) 1433.
7. K.G. Wilson, Phys. Rev. **D7** (1973) 2911.
8. T. Eguchi, Phys. Rev. **D17** (1978) 611.
9. S. Kim, A. Kocić and J. Kogut, Nucl. Phys. **B429** (1994) 407.
10. J.-L. Gervais and B.W. Lee, Nucl. Phys. **B12** (1969) 627.
11. A.H. Blin, B. Hiller and M. Schaden, Z. Phys. **A331** (1988) 75.
12. R.J. Perry, Phys. Lett. **B199** (1987) 489; R.J. Furnstal and C. Horowitz, Nucl. Phys. **A485** (1988) 632.
13. K. Langfeld, C. Kettner and H. Reinhardt, Nucl. Phys. **A608** (1996) 331.
14. S. Kawati and H. Miyata, Phys. Rev. **D23** (1981) 3010.
15. S. Weinberg, Phys. Rev. **D56** (1997) 2303.
16. S. Weinberg, Phys. Rev. Lett. **17** (1966) 616.
17. V. Bernard, A.A. Osipov and U.-G. Meissner, Phys. Lett. **B285** (1992) 119.
18. J. Gasser and H. Leutwyler, Phys. Lett. **125B** (1983) 325.
19. V. Bernard, A.H. Blin, B. Hiller, Y.P. Ivanov, A. A. Osipov and U. -G. Meissner, Phys. Lett. **B409** (1997) 483.

20. V. Bernard, A.H. Blin, B. Hiller, Y.P. Ivanov, A.A. Osipov and U.-G. Meissner, Ann. Phys. **249** (1996) 499.
21. S. R. Amendolia et al., Nucl. Phys. **B277** (1986) 168.
22. S. Adler, Phys. Rev. **177** (1969) 2426; J.S. Bell and R. Jackiv, Nuovo Cimento LXA (1969) 47; W. Bardeen, Phys. Rev. **184** (1969) 1848.
23. R.D. Ball and G. Ripka, hep-ph/9312260 and "Proceedings of the Int. Conf. in many-body Physics", 1993, Coimbra, Portugal, World Scientific.
24. J. Bijnens and J. Prades, Z. Phys. **C64** (1994) 475.
25. H.-J. Behrend et al., Z. Phys. **C49** (1991) 401.
26. J. Bijnens, J. Prades and E. de Rafael, Phys. Lett. **B348** (1995) 226.
27. Particle data group, Phys.Rev. **D54** (1996) 1.
28. O.A. Battistel, M.C. Nemes, hep-th/9811154, to appear on Phys. Rev. D, february 1999; see also O. A. Battistel, *Ph.D. Thesis*, UFMG (1999);
397.

Mesons in non-local chiral quark models[1]

Wojciech Broniowski

H. Niewodniczański Institute of Nuclear Physics, PL-31-342 Kraków, Poland
broniows@solaris.ifj.edu.pl

Abstract. After briefly reviewing chiral quark models with non-local regulators and listing their advantages over the conventional Nambu-Jona-Lasionio-like models, we study vector meson correlators in both types of approaches. Since effective chiral quark models are valid in the Eulidean domain only, with $-0.5\text{GeV}^2 \leq q^2 \leq 0$, in our study the meson correlators are not described directly, but with help of a method based on dispersion relations for the physical correlators. A set of sum rules is derived, which allows for a comparison of model predictions to data. We find that both the local and non-local models fail to satisfy the sum rules, unless very low values of the constituent quark mass parameter are used. We also show that the two Weinberg sum rules hold in the non-local model.

This research has been done in collaboration with Maxim Polyakov from Bochum.

I WHY NON-LOCAL REGULATORS?

First, let me recall the reasons why we wish to consider chiral quark models with non-local regulators (see also the contribution by Bojan Golli and Georges Ripka to these proceedings):

1. Non-locality arises naturally in several approaches to low-energy quark dynamics, such as the instanton-liquid model [1–3] or Schwinger-Dyson resummations [4]. For the discussions of various "derivations" and applications of non-local quark models see, *e.g.*, [5–14]. Hence, we should cope with non-local regulators from the outset.

2. Non-local interactions regularize the theory in such a way that the anomalies are preserved [10,15] and charges are properly quantized. With other methods, such as the proper-time regularization or the quark-loop momentum cut-off [13,14,16,17] the preservation of the anomalies can only be achieved if the (finite) anomalous part of the action is left unregularized, and only the

---

[1] Research supported by the Polish State Committee for Scientific Research, grant 2P03B-080-12, and by DSF and BMBF.

non-anomalous (infinite) part is regularized. If both parts are regularized, anomalies are violated badly [18,19]. We consider such a division artificial and find it quite appealing that with non-local regulators both parts of the action can be treated on equal footing.

3. With non-local interactions the effective action is finite to all orders in the loop expansion ($1/N_c$ expansion). In particular, meson loops are finite and there is no more need to introduce extra cut-offs, as was necessary in the case of local models [20-22]. As the result, non-local models have more predictive power.

4. As Bojan Golli, Georges Ripka and WB have shown [23], stable solitons exist in a chiral quark model with non-local interactions without the extra constraint that forces the σ and π fields to lie on the chiral circle. Such a constraint is external to the known derivations of effective chiral quark models.

5. The empirical values of the low-energy constants g_8 and g_{27} of the effective weak chiral Lagrangian are better reproduced within the non-local model [24] compared to the conventional NJL model.

In view of these improvements it becomes important to look at all other applications of effective quark models and compare the predictions of non-local and local versions.

II DISPERSION-RELATION SUM RULES FOR MESON CORRELATORS

In this talk we will present results for the vector-meson correlators. The basic object of our study is the meson correlation function, defined as

$$\Pi^{AB}(q^2) = \langle 0|i \int d^4x \, e^{iq\cdot x} \mathrm{T}\left\{j^A(x), j^B(0)\right\}|0\rangle, \tag{1}$$

where T denotes the time-ordered product, and $j^A(x)$ describes a color-singlet quark bilinear with appropriate Lorentz and flavor matrix, e.g. in the ρ-meson channel we have $j_\mu^a(x) = \frac{1}{2}\bar\psi(x)\gamma_\mu \tau^a \psi(x)$. The tensor structure can be taken away, with $\Pi^{AB}(q^2) = t^{AB}\Pi(q^2)$, where t^{AB} is a tensor in the Lorentz and isospin indices, and $\Pi(q^2)$ is a scalar function. The powerful feature of $\Pi(q^2)$ is its analyticity in the q^2 variable, which will be used shortly. In Figure 1 we display various regions in the complex q^2 plane: the positive real axis is the physical region, with poles and cuts corresponding to physical states in the particular channel. In that physical region we have (for certain channels) direct experimental information. Far to the left, at the end of the negative real axis, is the deep-Euclidean region, where perturbative QCD and its operator product expansion can be applied. There is yet another region, close to 0 on the negative real axis: the *shallow-Euclidean* region. This is the playground of the effective chiral models. Indeed, the most successful attempt to derive such a model from QCD, namely, the instanton-liquid model, has a natural limitation to that region of momenta [2,3]. More to the point, we believe that all effective chiral quark models should be used *in and only in* the shallow Euclidean domain. There have been many attempts, however, to apply such models directly to the physical region. In our opinion these are doomed to fail because of the lack of confinement, which is a key player in the physical region. With unconfined quarks, the unphysical $q\bar{q}$ continuum obstructs any model calculation for $q^2 > (2M)^2$, where $M \sim 300 - 400$MeV is the "constituent" quark mass. For these reasons of principles we always remain with the model at low negative q^2 [25,26], and compare to the data via the dispersion relation which holds for the *physical* correlator. In other words, we bring the physical data to the shallow Euclidean region via the dispersion relation. This is similar in spirit to the QCD sum rule approach, where one compares the physical spectrum to the deep Euclidean region via (Borelized) dispersion relation (see Figure 1).

The correlators considered here satisfy the twice-subtracted dispersion relation

$$\Pi(Q^2) = c_0 + c_1 Q^2 + \frac{Q^4}{\pi} \int_0^\infty ds \frac{\text{Im}\Pi(s)}{s^2(s+Q^2)}, \qquad (2)$$

where $Q^2 = -q^2$, and c_i are subtraction constants. Relation (2) holds for the *physical* correlators, and does not in general hold for the correlators evaluated in models [27]. For some channels (vector channels) $\text{Im}\Pi(s)$ is obtained directly from experiment, in other channels we have indirect information only, *e.g.* from QCD sum rules. Let us take the ρ-meson channel, where $j_a^\mu(x) = \frac{1}{2}\bar{\psi}(x)\gamma^\mu \tau_a \psi(x)$ and

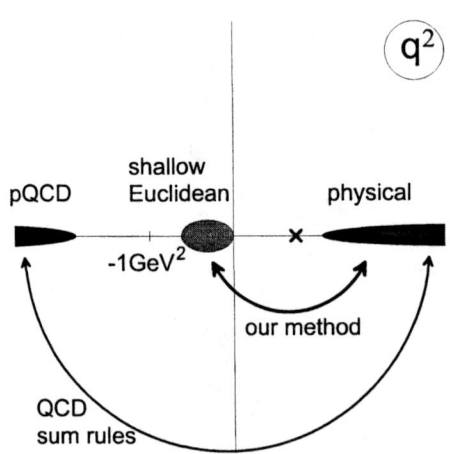

FIGURE 1.

$\Pi_{ab}^{\mu\nu}(Q^2) = \delta_{ab}(Q^\mu Q^\nu/Q^2 - g^{\mu\nu})\Pi_\rho(Q^2)$. The spectral strength in related to the ratio

$$\text{Im}\Pi_\rho^{\text{phen}}(s) = \frac{s}{6\pi} \frac{\sigma(e^+e^- \to n\pi)}{\sigma(e^+e^- \to \mu^+\mu^-)}, \quad n = 2, 4, 6, \ldots \tag{3}$$

known very accurately from experiment. The spectral strength peaks at the position of the ρ-meson pole, and at large s assumes the perturbative-QCD value. For our task it is more convenient to use the simple pole+continuum parameterization of $\text{Im}\Pi_\rho(s)$, such as used e.g. in QCD sum rules. In a given channel the fit has the form

$$\text{Im}\Pi^{\text{phen}}(s) = \frac{\pi s^2}{g^2} \delta(s - m^2) + a\, s\, \theta(s - s_0), \tag{4}$$

where a is known from perturbative QCD, and g, m and s_0 are chosen such that the experimental data are reproduced. In the ρ-channel we have $a = \frac{1}{8\pi}(1 + \alpha_s/\pi + \ldots)$, $m = 0.77\text{GeV}$, $g^2/(4\pi) = 2.36$, and $s_0 = 1.5\text{GeV}^2$ [28]. Since all our calculations will be done in the leading-N_c level, we drop the α_s correction in a.

With the parametrization (4) we readily obtain from (2)

$$\Pi^{\text{phen}}(Q^2) = c_0 + c_1 Q^2 + \frac{Q^4}{g^2(m^2 + Q^2)} + \frac{a}{\pi} Q^2 \log\left(1 + \frac{Q^2}{s_0}\right). \tag{5}$$

On the other hand, $\Pi(Q^2)$ can be calculated directly in chiral quark models in the shallow Euclidean space. We denote this model correlation $\Pi^{\text{mod}}(Q^2)$, as want to compare it somehow to $\Pi^{\text{phen}}(Q^2)$. One possibility is to Fourier-transform to coordinate space [29–32]. Here we apply a simpler method, which relies on just Taylor-expanding $\Pi^{\text{phen}}(Q^2)$ and $\Pi^{\text{mod}}(Q^2)$ in the Q^2 variable. For the phenomenological correlator we get from Eq. (5)

$$\Pi^{\text{phen}}(Q^2) = \sum_{k=1}^{\infty}(-)^{k+1} b_k^{\text{phen}} = c_0 + c_1 Q^2 + Q^2 \sum_{k=1}^{\infty}(-)^{k+1}\left[\left(\frac{Q^2}{m^2}\right)^k + \frac{a}{k\pi}\left(\frac{Q^2}{s_0}\right)^k\right], \tag{6}$$

whereas for the model correlator we can write

$$\Pi^{\text{mod}}(Q^2) = \sum_{k=1}^{\infty}(-)^{k+1} b_k^{\text{mod}}, \tag{7}$$

with the expansion coefficients b_k^{mod} yet to be determined. We can now compare the coefficients b_k^{phen} and b_k^{mod}, and form a set of "sum rules". With two subtractions in (2) we can start at $k = 2$: $b_k^{\text{phen}} = b_k^{\text{mod}}$, $k \geq 2$. With the explicit form (6) this gives

$$b_2^{\text{mod}} = \frac{1}{g^2 m^2} + \frac{a}{\pi s_0}, \quad b_3^{\text{mod}} = \frac{1}{g^2 m^4} + \frac{a}{2\pi s_0^2}, \quad \ldots \quad (8)$$

Sum rules for higher values of k are sensitive to the details of the phenomenological spectrum, hence are not going to be of much help.

In some channels the coupling constant g is not well known. We can then eliminate it from Eqs. (8) to obtain

$$m^2 = \frac{b_2^{\text{mod}} - \frac{a}{\pi s_0}}{b_3^{\text{mod}} - \frac{a}{2\pi s_0^2}}, \quad m^2 = \frac{b_3^{\text{mod}} - \frac{a}{2\pi s_0^2}}{b_4^{\text{mod}} - \frac{a}{3\pi s_0^3}}, \quad \ldots \quad (9)$$

Sum rules (8) or (9) can be used to verify model predictions for meson correlators.

III RESULTS FOR THE LOCAL MODEL

The model evaluation of meson correlators is well known. At the leading-N_c level one has

The diagrams with the dashed line occur only if the coupling constant G is non-zero in a given channel. This is the case of the of the σ and π channels. In the vector channels they may or may not be present, depending on whether we allow for explicit vector interactions between the quarks.

We first consider the ρ-channel in the local NJL model with the proper-time regularization [13,14,16,17]. After some simple algebra we get

$$b_2^{\text{mod}} = \frac{1}{8\pi^2} e^{-M^2/\Lambda^2} \frac{1}{5M^2}, \quad b_3^{\text{mod}} = \frac{1}{8\pi^2} e^{-M^2/\Lambda^2} \frac{3(M^2 + \Lambda^2)}{140 M^4 \Lambda^2}, \quad \ldots \quad (10)$$

where M is the constituent quark mass generated by the spontaneous breaking of the chiral symmetry, and Λ is the proper-time cut-off, adjusted such that the pion decay constant has its experimental value, $F_\pi = 93$MeV. Here we work in the strict chiral limit, with the current quark mass set to zero. The results for sum rules (9) are shown in Table 1. We can see that the ratios of phenomenological to model coefficients b_2 and b_3 are larger than 1, and increase rather rapidly with increasing M. Thus the sum rules (8) favor lower values of M. However, even for such low values as $M = 250$MeV the ratios $b_2^{\text{phen}}/b_2^{\text{mod}}$ and $b_3^{\text{phen}}/b_3^{\text{mod}}$ are still significantly above 1. We conclude that the model is far from satisfying the sum rules (8).

TABLE 1. ρ channel sum rules in the NJL model with the proper-time regulator ($m = 0$, $F_\pi = 93$MeV).

M [GeV]	Λ [GeV]	$b_2^{\rm phen}/b_2^{\rm mod}$	$b_3^{\rm phen}/b_3^{\rm mod}$
0.25	0.79	1.8	1.4
0.3	0.69	2.8	3.0
0.35	0.65	4.2	5.7
0.4	0.64	6.1	9.9

Next, we repeat our calculation for the variant of the model where vector interactions are included [14,33,34] in the Lagrangian: $-\frac{1}{2}G_\rho\left((\bar\psi\gamma_\mu\tau^a\psi)^2 + (\bar\psi\gamma_\mu\gamma_5\tau^a\psi)^2\right)$. In that model the formulas for the axial coupling constant of the quark, g_A^Q, and for F_π read

$$g_A^Q = \left(1 + G_\rho \frac{N_c M^2}{\pi^2}\Gamma(0, M^2/\Lambda^2)\right)^{-1}, \quad F_\pi^2 = g_A^Q \frac{N_c M^2}{4\pi^2}\Gamma(0, M^2/\Lambda^2), \quad (11)$$

with $\Gamma(0, x) = \int_x^\infty dt\, e^{-t}/t$. The results are shown in Table 2. We can see that the ratio $b_2^{\rm phen}/b_2^{\rm mod}$ decreases as G_ρ increases. However, uncomfortably large values of G_ρ are needed in order to satisfy the sum rule, i.e. to make $b_2^{\rm phen}/b_2^{\rm mod} \sim 1$. The conclusion is that at moderate values of M the model needs very large values of G_ρ to describe properly the vector channel.

TABLE 2. Same as Table 1 with vector interactions included.

G_ρ [GeV^{-2}]	$M = 0.3$GeV				$M = 0.35$GeV			
	Λ [GeV]	$\frac{b_2^{\rm phen}}{b_2^{\rm mod}}$	$\frac{b_3^{\rm phen}}{b_3^{\rm mod}}$	g_A^Q	Λ [GeV]	$\frac{b_2^{\rm phen}}{b_2^{\rm mod}}$	$\frac{b_3^{\rm phen}}{b_3^{\rm mod}}$	g_A^Q
0	0.69	1.8	3.0	1	0.65	4.2	5.7	1
4	0.78	2.2	2.1	0.86	0.72	3.4	4.0	0.86
8	0.91	1.6	1.4	0.72	0.81	2.5	2.8	0.72
12	1.14	1.0	1.0	0.58	0.97	1.7	1.9	0.58

IV RESULTS FOR THE NON-LOCAL MODEL

The *non-local chiral quark model* differs from the local versions in the fact that the interaction vertex carries momentum-dependent factors $r(p_i)$:

$$= G\delta(p_1 + p_2 - p_3 - p_4)r(p_1^2)r(p_2^2)r(p_3^2)r(p_4^2)$$

For some more details see the contribution of Bojan Golli and Georges Ripka. We use here the following form of the regulator, $r(p^2) = 1/(1 + p^2/\Lambda^2)^2$, which is simpler that the instanton-model expression but is known to reproduce well its basic predictions. We use the notation $M_k = Mr(k^2)^2$, $D_k = k^2 + M_k^2$. In the vacuum sector one finds that

$$\frac{1}{G} = 4N_c N_f \int \frac{d_4 k}{(2\pi)^4} \frac{r(k^2)^4}{D_k}, \quad \langle \bar{u}u \rangle = \langle \bar{d}d \rangle = -4N_c \int \frac{d_4 k}{(2\pi)^4} \frac{M_k}{D_k}, \quad (12)$$

$$\langle \frac{\alpha_s}{8\pi} G^a_{\mu\nu} G^{\mu\nu}_a \rangle = 4N_c \int \frac{d_4 k}{(2\pi)^4} \frac{M_k^2}{D_k}, \quad F_\pi^2 = 4N_c \int \frac{d_4 k}{(2\pi)^4} \frac{M_k^2 - k^2 M_k M_k' + k^4 M_k'^2}{D_k^2},$$

where the first equation is the stationary point condition, expressing the quark coupling constant via the parameter M, the second equation is the quark condensate in the chiral limit, the third is the gluon condensate in the chiral limit, [35], and the last one gives the pion decay constant [11,12] in the chiral limit. We use the notation $M_k' = dM_k/dk^2$.

There is a complication associated with non-local interactions, namely the Noether currents pick up extra contributions. Furthermore, the transverse parts of these currents are not uniquely determined and their choice constitutes a part of the model. Here we use the so-called straight-line P-exponent prescription for gauging the model. For a discussion of this and related issues see Refs. [11,12,36]. We then find that

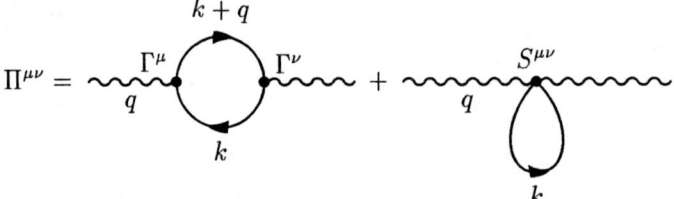

where

$$\Gamma^\mu = \gamma^\mu - \int_0^1 d\alpha \frac{dM(k+\alpha q)}{dk_\mu}, \quad S^{\mu\nu} = -\int_0^1 d\alpha \int_0^1 d\beta \frac{d^2 M(k+\alpha q - \beta q)}{dk_\mu dk_\nu}. \quad (13)$$

It is simple to check that the current is conserved, $q_\mu \Pi^{\mu\nu} = 0$. Coming back to

TABLE 3. Same as Table 1 for the non-local model.

M [GeV]	Λ [GeV]	b_2^{phen}/b_2^{mod}	b_3^{phen}/b_3^{mod}
0.25	1.35	1.5	1.4
0.3	1.0	1.9	2.9
0.35	0.83	2.2	3.9
0.4	0.72	2.3	3.8

the sum rules (8), we note by comparing Tables 1 and 3 that the results are a bit better in the non-local model. Especially at larger values of M, around 350MeV, we gain about a factor of 2. The discrepancy leaves room for such effects as the vector-channel interactions and $1/N_c$ corrections, which should be the object of a further study.

The effects of non-localities in currents, as depicted in the figure for $\Pi^{\mu\nu}$, bring about 15 % to the results. More precisely, the calculation with γ^μ instead of Γ^μ in the vertices (and without the sea-gull term $S^{\mu\nu}$) yields b_2^{mod} and b_3^{mod} roughly 15% lower.

V WEINBERG SUM RULES

Now we turn to a more formal aspect of our study. An appealing feature of non-local regulators is that now both Weinberg sum rules hold. The famous sum rules (I) and (II) are:

$$\frac{1}{\pi}\int_0^\infty \frac{ds}{s}[\text{Im}\Pi_\rho(s) - \text{Im}\Pi_{A_1}(s)] = F_\pi^2, \tag{I}$$

$$\frac{1}{\pi}\int_0^\infty ds\,[\text{Im}\Pi_\rho(s) - \text{Im}\Pi_{A_1}(s)] = -m\langle \bar{u}u + \bar{d}d\rangle. \tag{II}$$

Whereas (I) holds in all variants of the NJL model, in local models (II) picks up M instead of m on the right-hand side, thus is violated badly. To prove the sum rules in the non-local model we consider the dispersion relation

$$\Pi_\rho(Q^2) - \Pi_{A_1}(Q^2) = \frac{1}{\pi}\int_0^\infty \frac{ds}{s+Q^2}[\text{Im}\Pi_\rho(s) - \text{Im}\Pi_{A_1}(s)]. \tag{14}$$

No subtractions are necessary. We set $Q^2 = 0$. An explicit evaluation gives

$$\Pi_\rho(0) - \Pi_{A_1}(0) = 4N_c\int \frac{d_4k}{(2\pi)^4}\frac{M_k^2 - k^2 M_k M_k' + k^4 M_k'^2}{D_k^2}, \tag{15}$$

in which we recognize our formula for F_π^2 (12), thus verifying (I). To prove WSR II we multiply both sides of (14) by Q^2 and take the limit of $Q^2 \to \infty$. We find, to the first order in the current quark mass m,

$$\lim_{Q^2\to\infty} Q^2\left(\Pi_\rho(Q^2) - \Pi_{A_1}(Q^2)\right) =$$
$$\lim_{Q^2\to\infty} Q^2 \int \frac{d_4k}{(2\pi)^4}\frac{4N_c[M_k M_{k+Q} + m(M_k + M_{k+Q})]}{D_k D_{k+Q}} =$$
$$4N_c m \lim_{Q^2\to\infty} Q^2 \int \frac{d_4k}{(2\pi)^4}\left(\frac{M_k}{D_k D_{k+Q}} + \frac{M_k}{D_k D_{k-Q}}\right) =$$
$$8mN_c \int \frac{d_4k}{(2\pi)^4}\frac{M_k}{D_k} = -2m\langle \bar{u}u + \bar{d}d\rangle. \tag{16}$$

In passing from the second to the third line in the above derivation we have used the fact that M_p is strongly concentrated around $p = 0$, thus we could drop the term with $M_k M_{k+Q}$ at $Q^2 \gg \Lambda^2$. By a similar argument in the third line we have replaced $Q^2 / \left((k \pm Q)^2 + M_{k+Q}^2 \right)$ by 1. Finally, in the last line we have recognized our expression for the quark condensate (12). We note that non-local contributions to the Noether currents are suppressed and do not contribute to (16). Similarly, rescattering diagrams as displayed in the equation for Π^{AB} can be dropped. This is because the vertex Γ contains the regulator, hence the diagram is strongly suppressed at large Q^2.

Clearly, the reason for the compliance with the second Weinberg sum rule is the fact that the momentum-dependent mass of the quark becomes asymptotically, in the deep-Euclidean region, just the current quark mass. This is not the case of local models [14,37], where the mass is constant, and this is why these models violate Eq. (II).

VI FINAL REMARKS

To end this talk we describe shortly the results for other channels. In the ω-meson channel the model results are, at the leading-N_c level, exactly the same as for the ρ-channel. In the pseudoscalar and scalar channels we do not know the the corresponding parameter g from experiment, hence we consider sum rules (9). In the pion channel the pion pole entirely dominates the sum rules, *i.e.* the continuum contribution is negligeable and we get $m_\pi^2 \simeq b_2^{\text{mod}}/b_3^{\text{mod}} \simeq b_3^{\text{mod}}/b_4^{\text{mod}}...$ in both the local and non-local variants of the model. In the σ-meson channel things are more interesting. Whereas in the local model the sum rules (9) simply give the pole at twice the quark mass, $m_\sigma = 2M$, in the non-local model the predicted value of m_σ ranges from 400MeV at $M = 300$MeV to 470MeV at $M = 450$MeV and is insensitive to the value of the threshold parameter s_0.

One of us (WB) is grateful to Bojan Golli, Georges Ripka, Enrique Ruiz Arriola and Mike Birse for many useful conversations.

REFERENCES

1. D. I. Diakonov and V. Y. Petrov, Nucl. Phys. **B 272**, 457 (1986).
2. T. Schäfer and E. V. Shuryak, Rev. Mod. Phys. **70**, 323 (1998).
3. D. Diakonov, in *Selected topics in nonperturbative QCD*, p.397, talk given at International School of Physics, Enrico Fermi, Varenna, Italy, 27 June - 7 July 1995, hep-ph/9602375.
4. C. D. Roberts, in *QCD Vacuum Structure* (World Scientific, Singapore, 1992), p.114.
5. G. Ripka, *Quarks Bound by Chiral Fields* (Oxford University Press, Oxford, 1997).
6. J. Praschifka, C. D. Roberts, and R. T. Cahill, Phys. Rev. **D36**, 209 (1987).
7. B. Holdom, J. Terning, and K.Verbeek, Phys. Lett. **B232**, 351 (1989).
8. R. D. Ball, Int. Journ. Mod. Phys. **A5**, 4391 (1990).

9. M. Buballa and S. Krewald, Phys. Lett. **B294**, 19 (1992).
10. R. D. Ball and G. Ripka, in *Many Body Physics (Coimbra 1993)*, edited by C. Fiolhais, M. Fiolhais, C. Sousa, and J. N. Urbano (World Scientific, Singapore, 1993).
11. R. D. Bowler and M. C. Birse, Nucl. Phys. **A582**, 655 (1995).
12. R. S. Plant and M. C. Birse, Nucl. Phys. **A628**, 607 (1998).
13. D. Diakonov, Prog. Part. Nucl. Phys. **36**, 1 (1996).
14. J. Bijnens, C. Bruno, and E. de Rafael, Nucl. Phys. **B390**, 501 (1993).
15. E. R. Arriola and L. L. Salcedo, Phys. Lett. **B450**, 225 (1999).
16. C. Christov, A. Blotz, H. Kim, P. Pobylitsa, T. Watabe, Th. Meissner, E. Ruiz Arriola, and K. Goeke, Prog. Part. Nucl. Phys. **37**, 1 (1996).
17. R. Alkofer, H. Reinhardt, and H. Weigel, Phys. Rep. **265**, 139 (1996).
18. Ö. Kaymakcalan, S. Rajeev, and J. Schechter, Phys. Rev.**D31**, 1109 (1985).
19. A. H. Blin, B. Hiller, and M. Schaden, Z. Phys. **A331**, 75 (1988).
20. E. N. Nikolov, W. Broniowski, C. Christov, G. Ripka, and K. Goeke, Nucl. Phys. **A608**, 411 (1996).
21. V. Dmitrašinović, H.-J. Schulze, R. Tegen, and R. H. Lemmer, Ann. of Phys. (NY) **238**, 332 (1995).
22. W. Florkowski and W. Broniowski, Phys. Lett. **B386**, 62 (1996).
23. B. Golli, W. Broniowski, and G. Ripka, Phys. Lett. **B437**, 24 (1998).
24. M. Franz, H.-C. Kim, and K. Goeke, preprint RUB-TPII-09/99, Ruhr-University Bochum, **hep-ph/9908400**.
25. C. V. Christov, K. Goeke, and M. Polyakov, Ruhr-Universität-Bochum preprint, **hep-ph/9501383**.
26. W. Broniowski, M. Polyakov, H.-C. Kim, and K. Goeke, Phys. Lett. **B438**, 242 (1998).
27. W. Broniowski, G. Ripka, E. N. Nikolov, and K. Goeke, Z. Phys. **A354**, 421 (1996).
28. M. A. Shifman, A. I. Vainshtein, and V. I. Zakharov, Nucl. Phys. **B147**, 385, 448, 519 (1979).
29. E. V. Shuryak, Rev. Mod. Phys. **65** 1 (1993).
30. E. V. Shuryak and J. J. M. Verbaarschot, Nucl. Phys. **B410** 37 and 55 (1993).
31. R. M. Davidson and E. Arriola, Phys. Lett. **B359** 273 (1995).
32. M. Jaminon and E. R. Arriola, Phys. Lett. **B443**, 33 (1998).
33. S. Klimt, M. Lutz, U. Vogl, and W. Weise, Nucl. Phys. **A516**, 429 (1990).
34. U. Vogl, M. Lutz, S. Klimt, and W. Weise, Nucl. Phys. **A516**, 469 (1990).
35. D. I. Diakonov, M. V. Polyakov, and C. Weiss, Nucl. Phys. **B461**, 539 (1996).
36. W. Broniowski, INP Cracow preprint 1828/PH, talk presented at the Mini-Workshop on *Hadrons as Solitons*, Bled, Slovenia, 6-17 July 1999, **hep-ph/9909438**.
37. V. Dmitrašinović, S. P. Klevansky, and R. H. Lemmer, Phys. Lett. **B386**, 45 (1996).

Workshop Program

Friday, September 10

08:30 – Registration
09:30 – Opening Session

BARYONS

Chair: D.V. Shirkov

10:00 - 10:40 – G. Ripka
Quantum fluctuations and center of mass corrections in the soliton sector

10:40 - 11:10 – coffee break

11:10 - 11:50 – H. Walliser
Baryons as solitons

11:50 - 12:30 – K. Goeke
Deeply virtual lepton scattering and the chiral quark soliton model

12:30 – lunch

Chair: W. Broniowski

14:00 - 14:40 – F. Stancu
Stability of multiquark systems

14:40 - 15:05 – A. Arriaga
Quantum Monte Carlo studies of relativistic effects in light nuclei

15:05 - 15:30 – H. Sprenger
Computation of the nucleon properties in the chiral quark model

15:30 - coffee break

Chair: J.A. McGovern

15:50 - 16:15 – N.N. Scoccola
Multibaryons in the Skyrme model

16:15 - 16:40 – M. Oettel
Electromagnetic and strong form factors of octet baryons and strangeness production in a covariant and confining diquark-quark model

16:40 – coffee break

Chair: E. Ruiz Arriola

17:00 - 17:25 – D. Bartz
Nucleon-nucleon interaction in a chiral constituent quark model

17:25 - 17:50 – B. Krippa
Chiral NN interactions in nuclear matter

17:50 - 18:15 H.W. Griesshammer
Deuteron properties from perturbative nuclear physics

18:30 – Welcome Cocktail

Saturday, September 11

VARIOUS APPROACHES TO QCD

Chair: H. Walliser

09.00 - 09.40 – D.V. Shirkov
Invariant analytic approach to perturbative QCD

09:40 - 10:20 – M.C. Birse
A renormalization-group treatment of two-body scattering

10:20 – coffee break

Chair: A.A. Osipov

10:50 - 11:30 – A.V. Smilga
QCD at $\theta \sim \pi$

11:30 - 12:10 – D. Ebert
Confining properties and string representation of dual superconductor type of gauge model

12:10 – lunch

Chair: U.-G. Meissner

14:00 - 14:40 – H. Reinhardt
Magnetic monopoles, vortices and confinement

14:40 - 15:20 – A.V. Efremov
How could quark polarization be measured

15:20 - coffee break

Chair: K. Goeke

15:50 - 16:30 – P.J. Mulders
Structure of hadrons measured in hard processes

16:30 - 17:10 – W.L. van Neerven
Heavy flavor contributions to QCD sum rules and the running coupling constant

17:10 - 17:35 – Y. Shnir
The color-flavor transformation and low-energy QCD effective action

18.30 - Visit to the Physics Museum

Sunday, September 12

09:00 - Excursion

Monday, September 13

CHIRAL PERTURBATION THEORY

Chair: M.C. Birse

09:00 - 9:40 U.-G. Meissner
Chiral QCD dynamics: recent results

09:40 - 10:20 J.A. McGovern
Fifth-order contribution to the nucleon mass in heavy-baryon chiral perturbation theory

10:20 - 10:45 A. Gómez Nicola
Chiral perturbation theory out of thermal equilibrium

10:45 - coffee break

EFFECTS OF HOT AND DENSE MATTER

Chair: H. Reinhardt

11:15 - 11:55 T. Kunihiro
The sigma meson and chiral restoration in the hot and/or dense matter

11:55 - 12:35 W. Florkowski
Matter induced hadronic processes

12:35 - lunch

Chair: W.L. van Neerven

14:30 - 15:10 D. Blaschke
Hadron properties and deconfinement in the quark matter phase diagram

15:10 - 15:50 S. Raha
The deconfinement phase transition, hadronization and the NJL model

15:50 - coffee break

Chair: M.C. Nemes

16:20 - 17:00 M.C. Ruivo
Interplay between kaons and kaon like excitations of the medium

17:00 - 17:25 A. Bhattacharyya
ρ-mass modification in dense medium - some recent studies in finite nucleus

17:25 - 17:50 K.G. Klimenko
Magnetic oscillations in the Nambu – Jona-Lasinio models with nonzero particle density

19:30 - Departure to Palácio de São Marcos
20:00 - Dinner at Palácio de São Marcos

Tuesday, September 14

MESONS

Chair: G. Ripka

09:00 - 09:40 S.B. Gerasimov
On valence gluons in vector quarkonia: implications for spectra and decays

09:40 - 10:20 M. Jaminon
Brief review of modified versions of the Nambu–Jona- Lasinio model

10:20 - coffee break

Chair: W. Florkowski

10:50 - 11:30 A.A. Osipov
The invariant regularization of one-loop determinants in non-renormalizable theories

11:30 - 12:10 J. Schechter
Possible nonet of low lying scalar mesons

12:10 - lunch

Chair: M. Jaminon

14:00 - 14:40 M.K. Volkov
Radial excitations of scalar and η, η' mesons in a quark model

14:40 - 15:05 G. Rupp
Scalar mesons as "simple" $Q\bar{Q}$ states

15:05 - coffee break

Chair: D. Blaschke

15:35 - 16:00 S. Nishiyama
The resonating mean field theoretical approach to the Nambu - $\sigma - \pi$ sector -

16:00 - 16:25 V.A. Andrianov
Extended non-chiral quark models confronting QCD

16:25 - 16:50 M. Mathot
Strong decays of the scalar glueball in a confining effective lagrangian (NJL model)

Wednesday, September 15

Chair: A.V. Efremov

09:00 - 09:40 J. Bijnens
Models of low energy effective theory applied to kaon nonleptonic decays and other matrix elements

09:40 - 10:20 E. Ruiz Arriola
Chiral and scale anomalies of non local Dirac operators

10:20 - coffee break

Chair: D. Ebert

10:50 - 11:30 M.C. Nemes
Meson properties in a renormalizable version of the NJL model

11:30 - 12:10 W. Broniowski
Mesons in non-local chiral quark models

12:10 - 12:20 Closing Session

LIST OF PARTICIPANTS

Pedro Alberto
Departamento de Física
Universidade de Coimbra
P-3004-516 Coimbra
Portugal
E-mail: pedro@teor.fis.uc.pt

Vladimir A. Andrianov
Department of Theoretical Physics
University of St.-Petersburg, NIIF
Uljanovskaja 1
198904 St. Petersburg
Russia
E-mail: andrian@hen.gc.spb.ru

Ana Arriaga
Centro de Física Nuclear
Universidade de Lisboa
Av. Gama Pinto 2
P-1649-003 Lisboa
Portugal
E-mail: arriaga@alf1.cii.fc.ul.pt

Daniel Bartz
University of Liège
Institut de Physique B. 5
Sart Tilman
B - 4000 Liège 1
Belgium
E-mail: d.bartz@ulg.ac.be

Eef van Beveren
Departamento de Física
Universidade de Coimbra
P-3004-516 Coimbra
Portugal
E-mail: eef@teor.fis.uc.pt

Abhijit Bhattacharyya
Variable Energy Cyclotron Centre
1/AF, Bidhannagar
Calcutta - 700 064
India
E-mail: abhijit@veccal.ernet.in

Johan Bijnens
Lund University
Department of Theoretical Physics 2
Sölvegatan 14A
S 22362 Lund
Sweden
E-mail: bijnens@thep.lu.se

Michael C. Birse
Department of Physics and Astronomy
University of Manchester
Manchester, M13 9PL
UK
E-mail: mike.birse@man.ac.uk

David Blaschke
Fachbereich Physik
Universität Rostock
D-18051 Rostock
Germany
E-mail: david@darss.mpg.uni-rostock.de

Alex H. Blin
Departamento de Física
Universidade de Coimbra
P-3004-516 Coimbra
Portugal
E-mail: alex@teor.fis.uc.pt

Lucília Brito
Departamento de Física
Universidade de Coimbra
P-3004-516 Coimbra
Portugal
E-mail: lucilia@teor.fis.uc.pt

Wojciech Broniowski
Institute of Nuclear Physics
ul. Radzikowskiego 152
PL-31-342 Cracow
Poland
E-mail: broniows@solaris.ifj.edu.pl

Matthias Buechler
University of Zürich
Physikinstitut
Winterthurerstr.190
CH-8057 Zürich
Switzerland
E-mail: buechler@physik.unizh.ch

Filipe Carvalheiro
Departamento de Física
Universidade de Coimbra
P-3004-516 Coimbra
Portugal

Rita Coimbra
Departamento de Física
Universidade de Coimbra
P-3004-516 Coimbra
Portugal

Pedro Costa
Departamento de Física
Universidade de Coimbra
P-3004-516 Coimbra
Portugal

Dietmar Ebert
CERN-Theory Division
CH-1211 Genève 23
Switzerland
E-mail: dietmar.ebert@cern.ch

Anatoli Efremov
Joint Institute for Nuclear Research
Dubna 141980
Russia
E-mail: efremov@thsun1.jinr.ru

Manuel Fiolhais
Departamento de Física
Universidade de Coimbra
P-3004-516 Coimbra
Portugal
E-mail: tmanuel@teor.fis.uc.pt

Wojciech Florkowski
ul. Radzikowskiego 152
PL-31-342 Kraków
Poland
E-mail: florkows@solaris.ifj.edu.pl

Sergo B. Gerasimov
Joint Institute for Nuclear Research
141980 Dubna, Moscow Region
Russia
E-mail: gerasb@thsun1.jinr.ru

Klaus Goeke
Institut für Theoretische Physik II
Ruhr Universität Bochum
D-44780 Bochum
Germany
E-mail: klaus.goeke@ruhr-uni-bochum.de

Angel Gomez Nicola
Departamento de Fisica Teorica
Facultad de Ciencias Fisicas
Universidad Complutense
E-28040 Madrid
Spain
E-mail: gomez@eucmax.sim.ucm.es

Harald W. Griesshammer
Institut für Theoretische Physik
Physik-Department der Technischen Universität München
D-85748 Garching
Germany
E-mail: hgrie@physik.tu-muenchen.de

Brigitte Hiller
Departamento de Física
Universidade de Coimbra
P-3004-516 Coimbra
Portugal
E-mail: brigitte@teor.fis.uc.pt

Yuri P. Ivanov
Institut für Theoretische Physik
Universität Heidelberg
D-69120 Heidelberg
Germany
E-mail: yupi@tphys.uni-heidelberg.de

Martine Jaminon
Université de Liège
Institut de Physique B5
Sart Tilman
B-4000 Liège 1
Belgium
E-mail: martine.jaminon@ulg.ac.be

Konstantin G. Klimenko
Institute for High Energy Physics
142284, Protvino, Moscow Region
Russia
E-mail: kklim@mx.ihep.su

Boris Krippa
Dep. of Physics and Astronomy
Free University of Amsterdam
De Boelelaan 1081
NL-1081 HV Amsterdam
The Netherlands
E-mail: krippa@nat.vu.nl

Teiji Kunihiro
Faculty of Science and Technology
Ryukoku University
Seta, Ohtsu, 520-2194
Japan
E-mail: kunihiro@rins.ryukoku.ac.jp

Pedro Martins
Departamento de Física
Universidade de Coimbra
P-3004-516 Coimbra
Portugal

Michel Mathot
Université de Liège
Institut de Physique B5
Sart Tilman
B-4000 Liège 1
Belgium
E-mail: mmathot@ulg.ac.be

Judith A. McGovern
Department of Physics and Astronomy
University of Manchester
Manchester M13 9PL
UK
E-mail: judith.mcgovern@man.ac.uk

Ulf-G. Meissner
IKP (Th)
FZ Jülich
D-52425 Jülich
Germany
E-mail: u.meissner@fz-juelich.de

Piet J. Mulders
Div. Physics and Astronomy
Faculty of Sciences, VU
De Boelelaan 1081
NL-1081 HV Amsterdam
The Netherlands
E-mail: mulders@nat.vu.nl

Willie L. van Neerven
Instituut-Lorentz, University of Leiden
Niels Bohrweg 2
NL-2333 CA Leiden
The Netherlands
E-mail: neerven@lorentz.leidenuniv.nl

M. Carolina Nemes
Departamento de Física
Universidade Federal de Minas Gerais
Belo Horizonte, CEP 30.161-970, CP 702
Brazil
E-mail: carolina@fisica.ufmg.br

Seiya Nishiyama
2-5-1 Akebono-cho
Kochi 780-8520
Japan
E-mail: nisiyama@cc.kochi-u.ac.jp

Martin Oettel
Institute for Theoretical Physics
Tübingen University
Auf der Morgenstelle 14
D-72076 Tübingen
Germany
E-mail: oettel@pthp3.tphys.physik.uni-tuebingen.de

Frederik Orellana
University of Zürich
Institute of Theoretical Physics
Winterthurerstr.190
CH-8057 Zürich
Switzerland
E-mail: fjob@physik.unizh.ch

Alexander A. Osipov
Departamento de Física
Universidade de Coimbra
P-3004-516 Coimbra
Portugal
E-mail: alexguest@teor.fis.uc.pt

Humberto Pascoal
Departamento de Física
Universidade de Coimbra
P-3004-516 Coimbra
Portugal
E-mail: pascoal@teor.fis.uc.pt

Constança Providência
Departamento de Física
Universidade de Coimbra
P-3004-516 Coimbra
Portugal
E-mail: constanca@teor.fis.uc.pt

João da Providência
Departamento de Física
Universidade de Coimbra P-3004-516 Coimbra
Portugal
E-mail: providencia@teor.fis.uc.pt

João Providência Jr.
Departamento de Física
Universidade de Coimbra
P-3004-516 Coimbra
Portugal
E-mail: joao@teor.fis.uc.pt

Sibaji Raha
Physics Department, Bose Institute
93/1 A.P.C. Road
Calcutta 700 009
India
E-mail: sibaji@bosemain.boseinst.ernet.in

Gilberto Ramalho
Complexo Interdisciplinar da Univ. de Lisboa
Centro de Física Nuclear da Universidade de Lisboa
Av. Prof. Gama Pinto, 2
P-1649-003 Lisboa
Portugal
E-mail: gilberto@default.cii.fc.ul.pt

Hugo Reinhardt
Institute for Theoretical Physics
Tübingen University
Auf der Morgenstelle 14
D-72076 Tübingen
Germany
E-mail: hugo.reinhardt@uni-tuebingen.de

Georges Ripka
Service de Physique Théorique
Centre d'Etudes de Saclay
F-91191 Gif-sur-Yvette Cedex
France
E-mail: ripka@spht.saclay.cea.fr

Maria C. Ruivo
Departamento de Física
Universidade de Coimbra
P-3004-516 Coimbra
Portugal
E-mail: maria@teor.fis.uc.pt

Enrique Ruiz Arriola
Departamento de Física Moderna
Universidad de Granada
E-18071 Granada
Spain
E-mail: earriola@goliat.ugr.es

George Rupp
Centro de Física das Interacções Fundamentais
Instituto Superior Técnico, Edifício Ciência
P-1049-001 Lisboa
Portugal
E-mail: george@ajax.ist.utl.pt

Joseph Schechter
Department of Physics
Syracuse University
Syracuse, NY 13244-1130
USA
E-mail: schechte@suhep.phy.syr.edu

Norberto N. Scoccola
Physics Department
Comissión Nacional de Energía Atómica
Av. Libertador 8250
(1429) Buenos Aires
Argentina
E-mail: scoccola@tandar.cnea.gov.ar

Dmitri V. Shirkov
Bogoliubov Lab at JINR
Dubna, 141980
Russia
E-mail: shirkovd@thsun1.jinr.ru

Ya. Shnir
Institute for Theoretical Physics
University of Cologne
Zuelpicher Str. 77
D-50937 Köln
Germany
E-mail: shnir@thp.uni-koeln.de

Gil Silva
Departamento de Física
Universidade de Coimbra
P-3004-516 Coimbra
Portugal

Paulo Silva
Departamento de Física
Universidade de Coimbra
P-3004-516 Coimbra
Portugal

Ângela Silveirinha
Departamento de Física
Universidade de Coimbra
P-3004-516 Coimbra
Portugal

Andrei Smilga
Universit de Nantes
2, rue de la Houssinière, BP 92208
F-44322 Nantes
France
E-mail: andrei.smilga@math.univ-nantes.fr

Célia A. de Sousa
Departamento de Física
Universidade de Coimbra
P-3004-516 Coimbra
Portugal
E-mail: celia@teor.fis.uc.pt

Hendrik Sprenger
Institut für Physik
Universität Dortmund
Otto-Hahn-Str. 4
D-44221 Dortmund
Germany
E-mail: sprenger@hall.physik.uni-dortmund.de

Floarea Stancu
Institute of Physics, B.5
University of Liège
Sart Tilman
B-4000 Liège 1
Belgium
E-mail: stancu@baryon.theo.phys.ulg.al.be

José N. Urbano
Departamento de Física
Universidade de Coimbra
P-3004-516 Coimbra
Portugal
E-mail: urbano@teor.fis.uc.pt

Mikhail K. Volkov
Bogoliubov Laboratory of Theoretical Physics
Joint Institute for Nuclear Research
Dubna, 141980
Russia
E-mail: volkov@thsun1.jinr.ru

Hans Walliser
Fachbereich Physik
Universität Siegen
D-57068 Siegen
Germany
E-mail: walliser@physik.uni-siegen.de

AUTHOR INDEX

A

Ahlig, S., 73
Alkofer, R., 73
Andrianov, A. A., 328
Andrianov, V. A., 328
Antonov, D., 132
Arriaga, A., 43

B

Baacke, J., 53
Bartz, D., 83
Bhattacharyya, A., 246
Bijnens, J., 348
Birse, M. C., 113, 195
Black, D., 290
Blin, A.H., 283
Broniowski, W., 218, 380
Budzcies, J., 172

D

da Providência, J., 318
Dressler, B., 23

E

Ebert, D., 132, 253
Efremov, A. V., 152
Engelhardt, M., 142

F

Fariborz, A. H., 290
Fischer, C., 73
Florkowski, W., 218

G

Gerasimov, S. B., 265
Goeke, K., 23

Golli, B., 3
Gómez Nicola, A., 204
Grießhammer, H. W., 100

H

Hiller, B., 218, 283, 368

J

Jaminon, M., 273, 338

K

Klimenko, K. G., 253
Krippa, B., 94
Kumar, K. B. V., 195
Kunihiro, T., 217

L

Langfeld, K., 142

M

Mathot, M., 338
McGovern, J. A., 113, 195
Meißner, U.-G., 185
Mota, A. L., 368

N

Nemes, M.C., 368
Nishiyama, S., 318

O

Oettel, M., 73
Ohno, O., 318
Osipov, A. A., 283

P

Polyakov, M. V., 23

Q

Quandt, M., 142

R

Raha, S., 226
Reinhardt, H., 142
Richardson, K. G., 113
Ripka, G., 3
Ruivo, M. C., 237
Ruiz Arriola, E., 358
Rupp, G., 310

S

Salcedo, L. L., 358
Schäfke, A., 142
Schechter, J., 290

Schweitzer, P., 23
Scoccola, N. N., 63
Shnir, Y., 172
Smilga, A.V., 122
Sprenger, H., 53
Stancu, F., 34, 83
Strikman, M., 23

V

van Beveren, E., 310
Van den Bossche, B., 338
van Neerven, W. L., 162
Volkov, M. K., 300

W

Walliser, H., 13, 368
Weiss, C., 23

Y

Yudichev, V. L., 300